Dynamics of Flexible Aircraft

Coupled Flight Mechanics, Aeroelasticity, and Control

Explore the connections among aeroelasticity, flight dynamics, and control with an up-to-date multidisciplinary approach. New insights into the interaction between these fields, which is a distinctive feature of many modern aircraft designed for very high aerodynamic efficiency, are fully illustrated in this one-of-a-kind book.

Presenting basic concepts in a systematic and rigorous, yet accessible way, this book builds up to state-of-the-art models through an intuitive step-by-step approach. Both linear and nonlinear attributes are covered and, by revisiting classical solutions using modern analysis methods, this book provides a unique perspective to bridge the gap between disciplines.

Numerous original numerical examples, including online source codes, help to build intuition through hands-on activities. This book will empower the reader to design better and more environmentally friendly aircraft, and is an ideal resource for graduate students, researchers, and aerospace engineers.

Rafael Palacios is Professor of Computational Aeroelasticity and Director of the Brahmal Vasudevan Institute for Sustainable Aviation at Imperial College, London. He is a fellow of the Royal Aeronautical Society and an associate fellow of the American Institute of Aeronautics and Astronautics.

Carlos E. S. Cesnik is the Clarence L. (Kelly) Johnson Collegiate Professor and the Richard A. Auhll Department Chair of Aerospace Engineering at the University of Michigan, Ann Arbor. He is a fellow of the American Institute of Aeronautics and Astronautics, the Royal Aeronautical Society, and the Vertical Flight Society.

Cambridge Aerospace Series

Editors: Wei Shyy and Vigor Yang

Dynamics of Flexible Aircraft

Coupled Flight Mechanics, Aeroelasticity, and Control

RAFAEL PALACIOS

Imperial College of Science, Technology and Medicine

CARLOS E. S. CESNIK

University of Michigan

CAMBRIDGE
UNIVERSITY PRESS

CAMBRIDGE
UNIVERSITY PRESS

Shaftesbury Road, Cambridge CB2 8EA, United Kingdom

One Liberty Plaza, 20th Floor, New York, NY 10006, USA

477 Williamstown Road, Port Melbourne, VIC 3207, Australia

314–321, 3rd Floor, Plot 3, Splendor Forum, Jasola District Centre,
New Delhi – 110025, India

103 Penang Road, #05–06/07, Visioncrest Commercial, Singapore 238467

Cambridge University Press is part of Cambridge University Press & Assessment,
a department of the University of Cambridge.

We share the University's mission to contribute to society through the pursuit of
education, learning and research at the highest international levels of excellence.

www.cambridge.org
Information on this title: www.cambridge.org/9781108420600

DOI: 10.1017/9781108354868

First published 2023

A catalogue record for this publication is available from the British Library.

Library of Congress Cataloging-in-Publication Data
Names: Palacios, Rafael, 1974– author. | Cesnik, Carlos E. S., 1963– author.
Title: Dynamics of flexible aircraft : coupled flight mechanics, aeroelasticity,
and control / Rafael Palacios, Imperial College of Science, Technology and Medicine,
London, Carlos E. S. Cesnik, University of Michigan, Ann Arbor.
Description: Cambridge, United Kingdom ; New York, NY, USA :
Cambridge University Press, 2023. | Series: Cambridge aerospace series ;
volume 52 | Includes bibliographical references and index.
Identifiers: LCCN 2022049327 | ISBN 9781108420600 (hardback) |
ISBN 9781108354868 (ebook)
Subjects: LCSH: Airplanes – Control surfaces. | Flying machines – Design and
construction. | Equilibrium of flexible surfaces | Surfaces, Deformation of. |
Flutter (Aerodynamics)
Classification: LCC TL574.C6 P35 2023 | DDC 629.134/33–dc23/eng/20230118
LC record available at https://lccn.loc.gov/2022049327

ISBN 978-1-108-42060-0 Hardback

For
Nelya and Sasha
and
Cibele and Hannah

Somewhat as a voice in the wilderness,
I would like to pass on a few thoughts.

Letter to Robert Seamans, NASA
associate administrator
John Houbolt, 1961.

Contents

Preface

No other issue is more pressing for mankind to maintain, not to mention improve, its standard of living in the second half of this century than climate change. At the time of writing, aviation contributes only a relatively small fraction to global carbon (or more generally, carbon-equivalent) emissions, yet a clear pathway for decarbonization of air transportation does not yet exist. This is a collective challenge for the aerospace community, and it is our responsibility to identify solutions that will enable today's children to continue enjoying into their adulthood the joy and opportunities provided by air travel, and affordably so.

In response to those renewed, and urgent, environmental concerns, aircraft manufacturers are currently in a race to identify emission-free energy carriers. Hydrogen, electrochemical storage, and various synthetic fuels are all being considered at the time of writing. Success on this quest will likely need radical changes on how aircraft both look like and operate, not only to seek additional efficiencies that make the cost and performance of any new solutions comparable to the current ones, but also to accommodate the design and operational requirements of those new propulsion systems. A direct implication of this is that traditional methods for design and analysis, which so successfully steered us into the current golden age of air travel, will be insufficient to propel aviation into a sustainable future. This book is written to support the search for those additional efficiencies in the synergistic interactions among structural sizing, aerodynamic design, and flight control systems. Consequently, it takes a multidisciplinary view of the aircraft, which blends traditional concepts of flight dynamics, control, and aeroelasticity to describe the rich and possibly nonlinear dynamics of more flexible vehicles and their interactions with the atmosphere.

Anthropogenic climate change also manifests itself in a different manner in the pages of this book. The warming of the atmosphere is already a sad reality, and this brings changes to the environment in which aircraft operate. Not only do we need to design more efficient vehicles, but we also need them to fly in more adverse conditions as high turbulence events become more frequent. The implication of this is a need for a better understanding of the non-stationary atmospheric conditions, which would enable aerospace engineers to challenge the current design paradigms, should they be necessary. We take in this book a broad view in our exploration of the atmospheric conditions, which is not restricted by current certification processes, to try to equip our readers with a basic understanding of the atmospheric physics and its impact on aviation.

The book draws from over 50 years of combined research efforts in these matters by the authors. It also builds on our teaching and research experiences bringing those concepts into the graduate curriculum. The result is a synthesis of the current disciplinary knowledge in flight dynamics, aeroelasticity, atmospheric turbulence, and controls, from which we will systematically build increasingly complex models to explain and explore the stronger interactions that appear among those individual constituents. The range of topics is vast, and we have concentrated on the tools and methods that we consider to be more suitable for multidisciplinary analysis of aircraft dynamics, while giving pointers to other important solutions in the literature, including some promising ones that still have not achieved a similar level of maturity. We have also made an effort to ease in the transition from linear to nonlinear thinking in aircraft design, which is one of the major undercurrents of twenty-first-century aerospace engineering. We need to acknowledge, however, a bias toward the problems that we are more familiar with, but we can only write about what we know.

We cannot obviate the conditions in which the manuscript took the final form. The roots of this book should be found in the early work of the second author with Prof. Dewey Hodges. Sadly, Dewey has recently left us, but his outstanding intellectual contributions stay behind and many of his ideas will be found in the pages in this book. The final push has occurred during the coronavirus pandemic that has overwhelmed us since early 2020. These terrible events have highlighted once again the precariousness of our relationship with the natural environment but also the enormous power of science for good. It has made us optimists about our ability to overcome the climate crisis. Eventually.

The material in this book should be accessible to final-year and postgraduate students in aerospace engineering, as well as practicing engineers. It builds on the core knowledge in an aerospace engineering degree and expects good command of classical mechanics, aerodynamics, and linear systems. We have included mathematical derivations whenever we believe they are key to grasp concepts and point to references in the literature otherwise. Often, we present the mathematical apparatus together with its application to tackle a problem, so that we can directly draw from relevant examples while establishing a strong link between a general theory and the particular problem of interest.

As a final point, we would like to thank many colleagues for their help and support while preparing the manuscript. The first draft of Chapter 12 was prepared by Dr. Hector Climent, from Airbus Defense and Space. Professor Eli Livne, from the University of Washington, has been a constant source of wisdom in all things aeroelastic. The sections on nonlinear control build upon a long and fruitful collaboration of the first author with Dr. Andrew Wynn, from Imperial College. Many of the figures have been meticulously prepared by Kelvin Cheng. Dr. Alessandra Vizzacaro, from Imperial College, kindly produced the backbone curves in Chapter 8. We would like to thank, for reading parts of the manuscript: Stefanie Duessler, Drs. Marc Artola, Alvaro Cea, and Norberto Goizueta, while at the LoCALab, Prof. Cristina Riso, Dr. Mateus Pereira, Bilal Sharqi, Prof. Leandro Lustosa, while at A^{2}SRL. Dr. Arturo Munoz for his help in setting up the aerodynamic simulations examples. Dr. Jessica Jones generated the

X-56A results of Chapter 6, Dr. Salvatore Maraniello computed the T-Tail results of Chapter 7, Drs. Alfonso del Carre and Patricia Texeira built the models and computed the results on the X-HALE UAV (Chapter 8), Prof. Weihua Su generated the blended wing body results (Chapter 9), Dr. Alvaro Cea produced the sample results on a commercial transport aircraft, Dr. Norberto Goizueta computed the Pazy wing results, and Drs. Marc Artola and Robert Simpson computed the nonlinear control results. Our students have greatly inspired us and they make our job exciting and, in many ways, unique.

Online material

We have developed numerous numerical examples for this book. The captions to many of the figures include, in square brackets, the name of the file with the software used to generate them. All the scripts have been made available as open source with a permissive license and can be found at

https://github.com/ImperialCollegeLondon/flexibleaircraftbook

Symbols

$\mathcal{A}$	Aerodynamic influence coefficient matrix, generalized aerodynamic forces
$\mathcal{I}$	Unit or identity matrix
$\mathcal{C}$	Beam cross-sectional compliance matrix
$\mathcal{M}$	Beam cross-sectional mass matrix
$\mathbf{A}$	State matrix
$\mathbf{B}$	Input matrix
$\mathbf{C}$	Output matrix
$\mathbf{D}$	Damping matrix; feedthrough matrix
$\mathbf{e}_i$	Unit vector along axis i, for $i = 1, 2, 3$
$\mathbf{f}_B$	Components of the applied force vector in frame B
$\mathbf{f}_q$	Generalized modal forces; state weighting matrix
$\bar{\mathbf{f}}_S$	Components of the applied distributed force vector in frame S
$\mathfrak{f}_S$	Components of the beam internal force vector in frame S
$\mathbf{h}_B$	Components of the angular momentum vector in frame B
$\mathbf{I}_B$	Inertia tensor in body-fixed axes
$\mathbf{K}$	Stiffness matrix; feedback gain matrix
$\mathbf{k}_S$	Components of the local curvature vector of a beam in frame S
$\mathbf{M}$	Mass matrix
$\mathbf{m}_B$	Components of the applied moment vector in frame B
$\bar{\mathbf{m}}_S$	Components of the applied distributed moment vector in frame S
$\mathfrak{m}_S$	Components of the beam internal moment vector in frame S
$\mathbf{n}$	Uncorrelated white noise signals
$\mathbf{p}_B$	Components of the linear momentum vector in frame B
$\mathfrak{p}_E$	Components of the absolute position vector in Earth frame
$\mathbf{q}$	Modal degrees of freedom
$\mathbf{R}$	Control weighting matrix
$\mathbf{r}_B$	Components of the position vector in frame B
$\mathbf{R}_{BA}$	Coordinate transformation matrix from frame A to B
$\mathbf{T}$	Tangential operator
$\mathbf{u}$	Input vector
$\mathbf{u}_B$	Components of the displacement vector in frame B
$\mathbf{v}_B$	Components of the linear velocity vector in frame B

$\mathbf{v}_{gB}$	Components of the gust velocity vector in frame B
$\bar{\mathbf{v}}_S$	Components of the beam linear velocity vector in frame S
$\mathbf{w}$	Disturbance vector, wind velocity
$\mathbf{W}_c$	Controllability Gramian
$\mathbf{W}_o$	Observability Gramian
$\mathbf{x}$	State vector
$\mathbf{y}$	Output vector
$\bar{\boldsymbol{\omega}}_S$	Components of the beam angular velocity vector in frame S
χ	Generalized forces associated with the nodal elastic degrees of freedom
χ_e	Internal loads
$\boldsymbol{\delta}_c$	Control surface inputs
γ	Beam force strains
κ	Beam moment strains
$\boldsymbol{\Lambda}$	Matrix of eigenvalues
$\boldsymbol{\Omega}$	Diagonal matrix of natural frequencies in vibration analysis
ω_B	Components of the angular velocity vector in frame B
ω_{gB}	Components of the angular gust velocity (or rotary gust) vector in frame B
$\boldsymbol{\Phi}$	Matrix of eigenmodes; modal matrix in vibration analysis
$\boldsymbol{\Phi}_w(\omega)$	Spectral density of the random signal $\mathbf{w}$
$\phi_w(\tau)$	Correlation function of the random signal $\mathbf{w}$
ψ	Cartesian rotation vector
$\boldsymbol{\Sigma}_w$	Autocorrelation matrix of the random signal $\mathbf{w}$
ξ	(Nodal) elastic degrees of freedom
ζ	Amplitude of the elastic modes
α	Angle of attack
η	Curvilinear coordinate along beam axis
β	Sideslip angle
δ_a	Aileron input
δ_e	Elevator input
δ_r	Rudder input
δ_T	Throttle input
ℓ	Turbulence length scale
γ	Climb angle
$\mathcal{J}$	Jacobian operator originating from linearization
κ	Wavenumber or spatial frequency
$\mathcal{C}(ik)$	Theodorsen's lift deficiency function
$\mathcal{H}$	Heaviside or unit step function
$\mathcal{S}_x$	Sears' function at location x/c from airfoil leading edge
$\mathcal{T}$	Kinetic energy
$\mathcal{U}$	Internal or strain energy
$\mathcal{K}$	Kernel function or matrix in potential flow aerodynamics

η	Beam cross-sectional coordinates
μ	Mass parameter
ω	Angular frequency
ω_k	Natural frequency of the k th vibration mode
ϕ	Bank (roll) Euler angle
ψ	Heading (azimuth) Euler angle
ρ	Air density
σ_w^2	Mean square value of signal $\mathbf{w}$
θ	Pitch (elevation) Euler angle; twist angle of a flexible beam
A	Beam cross-sectional area
b	Wing semispan
c	Wing chord
C_D	Drag Coefficient
c_d	Sectional drag coefficient
C_L	Lift coefficient
c_l	Sectional lift coefficient
C_M	Aerodynamic moment coefficient
c_m	Sectional moment coefficient about the aerodynamic center
g	Magnitude of the Gravitational acceleration
i	imaginary unit
I_{xx}	Moment of inertia about axis x
I_{xy}	Product of inertia about axes x and y
k	Reduced frequency
K_g	Critical gust alleviation factor
m	Mass
n	Load factor
N_m	Number of normal modes
N_u	Number of inputs
N_x	Number of states
N_y	Number of outputs
q	Dynamic pressure ($q = \frac{1}{2}\rho V^2$)
S	Wing reference area
s	Dimensionless time (reduced time)
s	Laplace variable
T	Magnitude of the total thrust force
V	Airspeed
w_b	Relative normal flow velocity or upwash (of airfoil or wing)
w_g	Vertical gust velocity
CM	Center of mass
Ma	Mach number
Re	Reynolds number
$\bullet^*$	Transpose conjugate of a complex matrix
$\bullet^\top$	Transpose of a matrix

$\bullet_B$	Components in body-attached frame of reference, stability axes
$\bullet_E$	Components in Earth (inertial) frame of reference
$\bullet_S$	Components in local material frame of reference
$E[\bullet]$	Expectation operator
$\mathrm{Im}(\bullet)$	Imaginary part of a complex number
$\mathrm{Re}(\bullet)$	Real part of a complex number
$\mathrm{tr}(\bullet)$	Trace of a matrix
$\tilde{\bullet}$	Cross-product (or skew-symmetric) operator

1 Introduction

1.1 Flexible Aircraft Dynamics: Expanding Aircraft Designers' Toolkit

Flexible aircraft dynamics is the study of air vehicles in atmospheric flight that *simultaneously* display both rigid-body and elastic behavior. Since all aircraft deform and vibrate to some extent under aerodynamic forces, we can safely state that all aircraft are flexible. However, for most aircraft currently in operation, airframe flexibility has a negligible effect in the vehicle's flight dynamics, and most of the time, both effects can be independently analyzed. As we argue below, this is unlikely to stay in the next-generation aircraft, and this book presents an analysis framework that enables the rigid and elastic contributions to the aircraft dynamics to be considered at once. While this is of high relevance for both aircraft design and operation, including pilot training, our focus will be on the design phase, and therefore, it is convenient to outline first the stages of the design cycle for a new aircraft. *Appropriate fidelity* models and tools are employed based on the knowledge and information available at that particular stage of the design.

The initial, exploratory phase is known as *conceptual design*. At this stage, target performance requirements for a given mission are selected (for a transport aircraft, they would be metrics like range, payload, and operational cost), and the basic features of an aircraft that would meet them, as well as generic road maps for its commercial and technological viability, are established. In the conceptual design, multiple competing configurations are typically investigated, and the main design trade-offs (e.g., whether the engines should be wing or tail mounted) are analyzed for their sizing. Experimental tests, such as wind-tunnel tests, may be spun out to investigate critical enabling technologies in some detail, but a key feature of this phase is the use of *low-order* models. This phase is followed by the *preliminary design* stage, in which disciplinary analysis is carried out to establish confidence in the viability in the design. Specialist teams from structures, aerodynamics, flight dynamics and control, manufacturing, systems, etc. assess the initial design to identify potential roadblocks, for which they attempt to propose potential solutions. Those disciplinary models produce a first full description of the main components of the aircraft, including its external geometry and key manufacturing and assembly choices. In the last two decades, cross-disciplinary teams have started to appear at this stage to systematically investigate some key trade-offs using multidisciplinary design optimization (MDO) strategies. A successful preliminary design with a plausible business case may result in the full

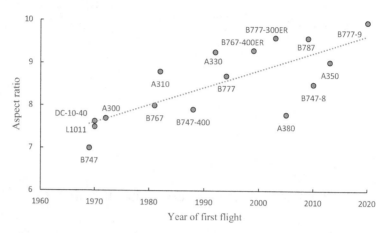

Figure 1.1 Aspect ratio by year of first flight of long-range transport aircraft.

development of the vehicle, and this is done in the final *detail design* phase. Detail design performs all the necessary analysis to manufacture and certify the vehicle, and it is supported by extensive ground-based and in-flight testing of components and full-vehicle prototypes. It is much more complex and costly than the previous two stages, and it stretches over several years.

Design is often described as the art of compromise, and aircraft designers are constantly faced, particularly in conceptual studies, with some fundamental trade-offs. The oldest and arguably most important aircraft design trade-off of all is that between aerodynamic and structural efficiency. Most aircraft are designed using the *tube-and-wing configuration*, which decouples the generation of lift in the wings from the hosting of the payload in the fuselage. This also means that lift is generated in a different part of the aircraft to where it is actually needed (to counteract the weight of passengers and payload, of course), from which the basic layout of their structural design starts to emerge. The historical trend is to improve aerodynamic efficiency while reducing airframe weight, as both directly contribute to an improvement in fuel efficiency. Aerodynamic efficiency is greatly improved by increasing the wing aspect ratio, defined as the ratio of the square of the wingspan over the wing area. However, longer and lighter wings are also necessarily more flexible, and structural considerations put an upper limit to the final wingspan in the design (the second limit comes from ground infrastructure). The maximum aspect ratio considered in aircraft design has steadily increased through the last decades as materials and manufacturing methods have improved. This can be seen in Figure 1.1, which shows the aspect ratio of most modern wide-body airliners against the year of their first flight. Note, however, that very large aircraft, such as the Airbus A380, are constrained in the wingspan by the ground infrastructure, which produces some scattering in the graph (and worse fuel efficiency).

It is interesting to present some numbers here to understand certain underlying design choices. The maximum takeoff weight (MTOW) of the largest Airbus aircraft at the time of writing (the A380-800) is 575 ton, which is 28% higher than that of

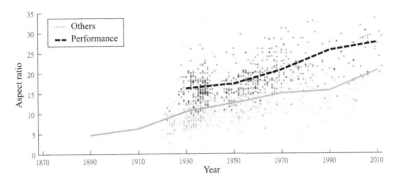

Figure 1.2 Aspect ratio by year in gliders. Each dot represents a glider design that was successfully flown.

the largest Boeing aircraft (the B747-8). Therefore, even though the A380-800 has a much longer wingspan (79.6 m for 68.4 m for the B747-8), the need for a very large area to sustain flight and the hard constraint on wingspan coming from runway width more than counteract that effect and result in a relatively low aspect ratio design. An alternative solution has been sought in the Boeing 777-9 (shown in Figure 1.3), which achieves the largest aspect ratio of any airliner to date by folding the wingtips when on the ground. By doing so, the aircraft can park in the same gates as the original B777 aircraft. Most striking still is a direct comparison between the older and newer aircraft in that chart. The wingspan of the Boeing 787 is roughly the same as that of the first-generation B747 from the 1970s, yet its MTOW is approximately a third smaller than that of the B747-100. The result is that the wings of the B787 have a much smaller area than the original Jumbo Jet for the same wingspan, and that gives the aircraft a huge boost in aerodynamic efficiency.

A similar trend can be observed in Figure 1.2, which includes almost all the gliders that have been built since the dawn of aviation.[1] A subset of gliders, designed to achieve maximum performance (one such typical vehicle is shown in Figure 1.3), is also identified in the figure. Moreover, the average aspect ratio of all vehicles flown in each decade is also included to show the underlying trend. Contrary to commercial vehicles, glider construction has little restrictions from the airport constraints, and it is of course unaffected by advances in engine technology. Therefore, gliders push the limits for aerodynamic-driven design. Comparing Figures 1.1 and 1.2 gives us a glimpse of how air vehicle configurations have evolved under different design trade-offs.

The wing aspect ratio of tube-and-wing configurations has historically been a conceptual design choice. It has been driven by the state of the art on wing manufacturing technology, and therefore follows the slow but steady increase that can be observed in Figures 1.1 and 1.2. However, the current demand for emission-free flight travel and the subsequent development of new propulsion systems, possibly based on hydrogen or electrochemical storage, may upend this trajectory. The integration of new

---

[1] The raw data in the graph have been retrieved from www.j2mcl-planeurs.net.

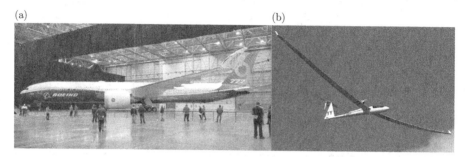

Figure 1.3 (a) Rollout of the Boeing 777-9, with folded wingtips (photo by Dan Neville). (b) A high-performance glider with the aspect ratio of 25 (photo by Juergen Lehle (albspotter.eu)).

propulsion systems brings a huge challenge to an aircraft design office, as even lighter airframes with much higher aerodynamic efficiency are needed for the viability of most new designs. A basic question arises here: How can we still ensure safe flight operation of a design built under such constraints? A large part of the answer lies in a deeper integration of the underlying disciplines, to enable the necessary trade-offs by opening the design space. More flexible aircraft break the boundary between flight dynamics and the low-frequency vibrations of the aircraft; they break the constraints to small displacements used in aerodynamic predictions, and put additional strain on the flight control systems, which need to deal with both airframe flexibility and vehicle dynamics. Navigating that larger design space is a major engineering challenge, but also a huge opportunity, the objective of this book is to give aerospace engineers the basic tools for that task.

1.2   Rigid Aircraft, Flexible Aircraft

The design offices for major aircraft manufacturers sprawl over multiple buildings in multiple sites, sometimes in multiple countries. Specialist teams have honed their skills through the cumulative effort of generations of bright engineers. The result is a very intricate but well-oiled machine with established process to ensure the safe operation of new aircraft. It is also a working environment that tends to reinforce the traditional separation between disciplines.

While there is only one end product, different representations of the aircraft are used within each discipline. On one end, we have the flight dynamics perspective. Flight dynamicists are concerned with the in-flight performance of the aircraft, and the design of the flight control systems. The aircraft is then often represented as a rigid body moving in the atmosphere and its aerodynamics given in the form of look-up tables. On the other extreme, we encounter the stress analysts, which model minute details of the airframe to understand and prevent any potential structural failure. The aircraft is now a large elastic object, typically considered in static equilibrium and under fixed forces. There are many shades of gray between both extremes and it is therefore convenient first to define the traditional scope of those key disciplines, which

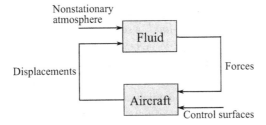

Figure 1.4 Feedback interactions between an aircraft and its surrounding fluid.

will be the starting point for the exploration in this book. Knowing where the current boundaries are will enable us to build bridges.

As we have just mentioned, *flight dynamics* is concerned with the time evolution of an aircraft in atmospheric flight, that is, establishing its trajectory and attitude under aerodynamic forces, as well as modifying those forces to steer the vehicle. It assesses the ability of the vehicle to meet its intended mission (performance) and the level of effort needed from the pilot to maneuver the vehicle (handling qualities), while providing rules for acceptable weight distribution (static stability) and configuration layout (to ensure dynamic stability). The basic interaction between the aircraft and the surrounding fluid is represented in Figure 1.4. In flight dynamics, the aircraft is typically modeled as a rigid body and the description of its interactions with the fluid is assumed to occur only through its six degrees of freedom. Moreover, in most flight regimes, the atmosphere can be considered stationary and the force resultants on the aircraft (lift, drag, moments) are uniquely determined from the instantaneous relative position and orientation (and the corresponding velocities and accelerations) of the aircraft. As a result, the fluid forces can be easily tabulated as a function of relatively few coefficients and be retrieved for flight simulation and analysis.

The evaluation of those fluid forces on the aircraft in flight is the remit of *aerodynamics*. The key objective of an aerodynamicist is to ensure that flight can be sustained across a number of design conditions at a minimum cost (drag), and also that those forces can be easily modified during maneuvers. Yet aerodynamics produces not only the force resultants that enable its flight dynamics but also the distributed loading that deforms its structure. If those deformations are sufficiently large to generate noticeable changes on the external structure, then a second feedback loop appears between structural deformations and aerodynamic forces that is studied by *aeroelasticity*. It can equally be described by Figure 1.4, although now the aircraft is also changing its geometry. We further distinguish between static and dynamic aeroelasticity, depending on whether the vehicle is in static equilibrium or not.

Regarding *static aeroelasticity*, the main interest is on the deformations that appear on the wing as it produces lift, and their effect both on (1) the final lift and drag on the wing and (2) the internal stress appearing on its airframe. Note that drag forces are affected by the shape changes due to the deformation, but they rarely deform the wing on their own. In other words, drag comes from a one-way interaction, which facilitates its study. (A corollary of this is that drag prediction is rarely within the remit of aeroelastic analyses.) Regarding the stresses on the airframe, they are investigated by the computation of *load cases* that size the structure. In particular, a number of steady maneuvers need to be considered for which the static aeroelastic equilibrium is sought under suitable control surface deflections and from which the distribution of applied forces (the loads) on the airframe is investigated. The worst-case conditions on each structural element define its *limit loads*, which needs to be supported without "detrimental permanent deformation" (FAA, 2011). The airframe also needs to support the *ultimate loads*, defined as the limit loads times a safety factor, without structural failure (Niu, 1999, Ch. 3). For commercial aviation, the typical safety factor is 1.5, that is, the aircraft is designed to support loads (the *ultimate loads*) that are 50% higher than the highest loads that are predicted to appear in service (its limit loads).

Often the aircraft is subjected to nonstationary atmospheric conditions, which may have a significant impact on its aerodynamic forces. When the timescales of those interactions are very large, the change of forces may mainly affect the flight dynamics, but most commonly it will also deform the airframe. In that case, we are within the remit of *dynamic aeroelasticity*. The feedback process in Figure 1.4 now needs to consider both elastic and inertial effects on the aircraft response, as well as steady and unsteady effects on the fluid. The systematic study of the response to nonstationary atmospheric conditions results in *gust loads* on the structural elements, which may also become the limit loads on certain component. Finally, as in flight dynamics, one needs to assess the dynamic aeroelasticity stability of the system, known as *flutter*.

1.3　　Brief Historical Overview

Since the early days of heavier than air flights, the challenges of flexible aircraft were clearly present. Then, there was only a rudimentary understanding of aero-structural interaction and stability and controls, and most of it was empirical in nature. Here we would like to review some of the key developments along more than a century that led us to the dynamics of flexible aircraft as discussed in this book. There are several review papers and some books covering the area of aeroelasticity and/or stability and control of aircraft. Among those, we recommend the reader to explore Garrick and Reed (1981) (this one in particular provides a great list of references), Collar (1978), Perkins (1970), Abzug and Larrabee (2002), and von Kármán (1954) that served the basis for our attempt to merge these two major development paths so to better understand how we ended up where we are.

1.3.1　　The Early Years

In the early 1900s, Samuel P. Langley, then secretary of the Smithsonian Institute and one of the most preeminent scientists in the United States, had been invested in the development of an airplane. The *Aerodrome* was conceived to be flown from a platform on top of a houseboat on the Potomac River. After successful subscale and unmanned flights, the second attempt to fly the manned Aerodrome on December 8,

1903, that is, nine days prior to the famous *Wright Flyer* flight at Kitty Hawk, suffered a structural failure. As the craft accelerated during launch at an untrimmed setting, the aerodynamic loads on the front wing deformed it beyond its strength limits and resulted in its collapse. This is the first example of the importance of aero-structural considerations in trimmed flight, something that is done today for any performance assessment. It is worth noticing that in a lecture delivered to the Western Society of Engineers in Chicago in 1901 (Wright, 2001, p. 99), Wilbur Wright delineated three main issues related to fly:

The difficulties which obstruct the pathway to success in flying machine construction are of three general classes: (1) Those which relate to the construction of the sustaining wings. (2) Those which relate to the generation and application of the power required to drive the machine through the air. (3) Those related to the balancing and steering of the machine after it is actually in flight.

He believed that the first two had been solved, and the third one was key. Indeed, the difficulties in maneuvering an aircraft were already identified during Otto Lilienthal's glider experiments several years earlier. Lilienthal's glider was made stable by positioning the center of gravity forward and controlled by moving it through body motion. The control authority for such arrangement was limited. The Wright Brothers had a very different solution to the problem: focus on the controllability and not its stability. Maybe due to their extensive bicycle experience, where the person is responsible to provide stability to the vehicle, the same idea was applied to their Wright Flyer. Particularly challenging was the lateral control, something they achieved by allowing differential elastic twist of the wings (known as *wing warping*) to change the spanwise lift distribution. This was realized by making the wing stiff in bending but soft in torsion. By taking advantage of the twist flexibility of the wing in a controlled manner, something that the Aerodrome did not have a chance to show, they were able to maneuver their aircraft. The stability was provided by the pilot employing constant corrections through the various control inputs. This highlights the two main approaches to stability: inherent stability requirement vs. stability created by the pilot (or autopilot). The wing warping concept was later employed in some successful European aircraft of the time, and this concept was also at the center of patent fights promoted by the Wright Brothers in the early years of aviation.

One of the first European concepts to employ wing warping was the Bleriot XI, shown in Figure 1.5, which was the first airplane to successfully cross the English Channel in 1909. The monoplane configuration employed wing warping for roll control and initially cruised at approximately 40 mph. The success of the Channel crossing propelled the Bleriot XI as a popular airplane, and re-engine efforts were undertaken to increase its speed. As the flight speed increased to approximately 80 mph, the wing collapsed. It was thought that it was the strength of the construction that needed improvement. So they reinforced the guy wires that supported the wings in an attempt to eliminate the problem. Unfortunately, it did not solve it and questioned the feasibility of monoplanes. In fact, the issue was found to be so critical that the British government banned monoplanes in 1912, which lasted until the approach of the Great

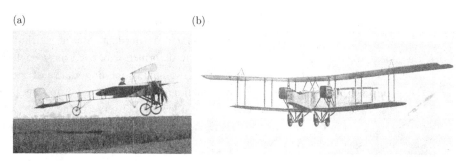

Figure 1.5 (a) Bleriot XI (photo from the Library of Congress catalogue). (b) Handley-Page Type 0 (photo from the National Archives and Records Administration).

War. This is one of the earliest records of wing divergence, a static aeroelastic instability, and, therefore, an issue of *stiffness and not strength*. A remarkable event also related to static aeroelastic instability happened closer to the end of the Great War. By 1918, the D-VIII Fokker monoplane had been designed for aerial superiority. The prototype plane flew very well with the wing structure made of two spars: the front one with a larger cross section when compared with the rear one. The concept was moved to production, but with a small requirement modification by the German Army. In order to strengthen the wing, the rear spar was to be of the same, larger cross section, as the stronger front spar. The brand new high-performance planes were given to the top pilots, who started facing wing collapse during flight. Inadvertently, by strengthening the rear-wing spar, its elastic axis was also moved aft and reduced its divergence dynamic pressure. It was not until later that this phenomenon was completely understood. Reissner (1926) presented a detailed analysis of wing torsional divergence, describing the importance of the relative position of the aerodynamic center[2] and the wing elastic axis.

Still at the time of the Great War (1914–1918), aeroelastic dynamic issues started appearing as well. A classic example is the fuselage-tail instability of the Handley Page 0/400 bomber in the 1914–17 timeframe, shown in Figure 1.5. At a certain critical speed, large couple oscillations appeared between the aft fuselage in torsion and the antisymmetric elevator motion of the biplane tail. At the time, the right and left elevators were independent from each other and connected to cables that were actuated in a way to deflect them symmetrically. Lanchester (1916) evaluated the situation conceptually and in a short technical report to the Advisory Committee for Aeronautics stated that, "[w]ithout making calculations," he concluded that "it is clear to my mind that if the elevator flaps were fixed ... an oscillation such as that described should be damped with great rapidity; in other words, to maintain such an oscillation would require some enormous direct application of energy...." And he went on to conclude that "... when the oscillation is taking place the damping action above mentioned must be absent, and one is led to look for the cause of the oscillation in some reversal of this

---

[2] Joukowski's theory had already showed that the pitching moment about the point one-quarter chord behind the leading edge of monoplane wings would not change with changes in the angle of attack.

damping effect." The flexibility of the cables that made the control system allowed for asymmetric deflections of the elevators, creating rolling moment at the tail end of the fuselage. This seems to be the first formally identified flutter phenomenon in aircraft. A similar issue was encountered a year later in the DH-9 aircraft. Lanchester recommended that a single stiff torque tube should be introduced to connect the upper right and upper left elevators,[3] along with few reinforcements with the lower part of the tail. Moreover, Lanchester also recommended "for the dynamics of the tail to be studied at the National Physical Laboratory, and the matter more fully investigated there." This was undertaken by Bairstow and Fage (1916) and they confirmed Lanchester's view.

The mathematical developments related to flight lagged the initial flight experiments, that is, the experimentalists were building, testing, and flying airplanes before the mathematical models were available to help understand the various pieces of the puzzle. In 1904, British mathematicians G. H. Bryan and W. E. Williams (Bryan and Williams, 1904) were the first to apply the mathematical theory of dynamic stability previously formulated by their compatriot, Edward J. Routh (Routh, 1877). The problem of longitudinal stability and the identification of the "phugoid"[4] mode was published by Lanchester (1908). While it is acknowledged that Lanchester arrived to this conclusion on his own, Nikolai Joukowski had already identified this motion back in 1891 (Joukowski, 1891). Lanchester made observations about trajectories of a glider when "launched at the exact or natural speed, it would glide steady as if on rails." He was able to identify oscillations and their damping when perturbed from those conditions, bringing into the argument the interchange of potential and kinetic energies that determine the transient dynamics. The first complete airplane dynamics model in a rigorous mathematical way was presented by Bryan (1911), who also introduced the concept of *stability derivatives*, which separated the airplane motion into two parts, "symmetrical" and "rotative" components (now known as longitudinal and lateral, respectively, see Chapter 4), and uncovered its natural frequencies. There was no consideration of aircraft flexibility on those formulations, something that would not show up for another five decades or so. His equations of motion for a rigid body with six degrees of freedom were brought into a dimensional form in Bairstow et al. (1913), and these are still the familiar equations we use today and will see them in this book. Bairstow and his colleagues at the National Physics Laboratory also separated the equations into two independent solutions, leading to today's longitudinal and lateral ones along with the basic natural modes and frequencies. Further "wind channel" tests were conducted by them using a model scale of the Bleriot monoplane to try to correlate the stability characteristics being analyzed mathematically with the experimental observations in an effort to understand how the early airplanes were able to fly with such poor stability characteristics. Perkins (1970) presented a very nice discussion on the various consequences of those studies in the understanding of airplane stability

---

[3] This has become a common practice, and it can still be found in general aviation aircraft today.
[4] von Kármán (1954) argued that Lanchester may have misinterpreted the Greek word "phygoid" that "means 'to fly' in the sense of fleeing before a menace and not flying as a bird." But that has since being widely accepted to represent this longitudinal flight dynamic mode.

and controls. As he stated there, "the perplexing fact, ... when airplanes encountered the instability directly forecast by the mathematicians, the airplanes go along quite well and flew anyway." It was not until the mid of the twentieth century when stability and controls, and the role of the human pilot in the loop, were brought together in the studies at the National Advisory Committee for Aeronautics (N.A.C.A.) "at Langley and Ames laboratories [in the U.S.A.] and by researchers of the Royal Aircraft Establishment (RAE) in Great Britain."

It was during the 1920s that a series of flutter problems came to plague aviation, mainly due to wing-aileron flutter, which was first identified in 1922. This new dynamic instability involved the inertial forces on the wing deflecting the aileron in a particular phase in order to increase the motion in the direction of the initial motion. As was in the case of the Handley Page 0/400, the flexibility of the cables allows aileron deformation around the setting point when subject to inertial accelerations. von Baumhauer and Koning (1922) pointed out the importance of the inertia coupling for the flutter onset and proposed to solve it by *mass balancing*: by bringing the center of mass of the control surface to be at its hinge location or forward of it. During this decade, significant aerodynamic developments were achieved. Since the characteristic timescales of the structural motions may be comparable to the convective timescales in the flow, these led to the need of theories of nonstationary flows around moving airfoils. Studies were initially developed by Ludwig Prandtl and his students in Göttingen. The first person to successfully complete his dissertation on airfoil theory was W. Birnbaum, who published the classical vortex theory of two-dimensional (2-D) steady flow of thin airfoils (Birnbaum, 1923),[5] followed by the harmonically oscillating airfoil in uniform motion (Birnbaum, 1924). In his work, he expressed the wake vorticity as the function of the airfoil-bound vorticity in order to obtain an integral equation relating pressure over the airfoil with its normal velocity–a key step employed by Theodorsen over 10 years later for harmonically oscillating airfoils (Theodorsen, 1935). Furthermore, the concept of *reduced frequency* was already employed in his work. The approach, however, did not quite work for reduced frequencies (as defined in Section 3.3.3) greater than 0.2. Also from Prandtl's group, Wagner (1925) presented an alternative solution to the problem of harmonic nonstationary flow by introducing the solution to a step change in the angle of attack, which is known today as the indicial solution. The resulting function satisfying Kutta-Joukowski[6] trailing-edge condition and giving the growth of lift with distance traveled is known as the Wagner's function (see Section 3.4.3). Glauert (1929), following Wagner's methods, addressed flat plate airfoils undergoing steady pitch oscillations. Glauert was then able to numerically solve the problem without the convergence difficulties experienced by Birnbaum. In that same year, Hans-Georg Küssner (1929) presented a seminal paper on flutter prediction utilizing improvements on Birnbaum's method. He showed improved numerical convergence to reduced frequency of order

---

[5] Max Munk (1922) had published his thin airfoil theory shortly before Birnbaum, while H. Glauert (1923) provided an alternative formulation the year after.
[6] We follow here von Kármán (1954, Ch. 2), even though many references simply refer to it as the Kutta condition.

1 in beam-type problems with bending and torsion and including aileron. At the end of this decade, known as the "flutter decade," Frazer and Duncan (1928) summarized their studies in flutter of wings that, according to Collar (1959), became known to aeroelasticians as the "Flutter Bible."

The advent of stressed skin construction finally brought about the end of the biplane dominance and the dominance of the monoplane around 1930s. By then, the flying speeds were such that the flexibility could not be ignored anymore. Flutter assessment in every design became a necessity. That was also the decade of intensive theoretical developments. Duncan and Collar (1932) extended the theory of Glauert to include wing translation and rotations. Theodorsen (1935) at the National Advisory Committee for Aeronautics (N.A.C.A.) Langley Research Center worked on the flutter problem and presented the theory of the 2-D oscillating thin airfoil in potential flow undergoing simple harmonic motion in pitch and plunge, and aileron-type motions. He separated the noncirculatory terms from the circulatory ones and, like Birnbaum, by imposing the Kutta-Joukowski condition at the trailing edge he connected the wave vorticity with the bound vorticity through an integral relation. The solution led to a combination of Bessel functions that is now known as the *Theodorsen function* (see Section 3.3.3). These extended quasi-steady constants used before have now become frequency dependent. This exact solution (in contrast to the numerical approximations made by Glauert and Küssner that limited the reduced frequency applicability of their approaches) for the simple harmonic motion of thin flat airfoil in potential flow has become the cornerstone for the *strip theory* widely used in industry for simple flutter estimation, particularly for high-aspect-ratio wings. While there are some clear similarities between Theodorsen's approach and the earlier ones, particularly Birnbaum's, Wagner's, and Glauert's, here there is no use of the Routh's discriminants as widely used before since the flutter determinant is complex and both real and imaginary parts must be satisfied simultaneously. This led to approximate ways of determining the flutter solution, including the so-called k-method introduced by Smilg and Wasserman (1942). It is worth mentioning that about the same time, Cicala (1935) in Italy independently developed the oscillating flat plate solution. Theodorsen and Garrick (1938) conducted numerous trend studies for the various parameters of a typical section: center of mass, elastic axis, mass ratio, bending/torsion frequency ratio, etc. At some point, it became apparent from experiments that flutter usually involved more degrees of freedom than the usual two or three used in the typical section calculations. Duncan (1939) suggested then to use normal modes to represent the complete participation of the aircraft elasticity into the problem and that would save on computation. Indeed, normal modes have been used up to this day for describing the response of the linear deformed aircraft.

An initial theoretical study of aircraft encountering discrete gust was published in the first N.A.C.A. report to the U.S. Congress (Wilson, 1916). It contains the aircraft equations of motion with its stability derivatives and a very simple model of the aircraft response to discrete gust encounter without unsteady aerodynamic effects. The analytical solution for the interaction of a gust with an airfoil was derived by Bill Sears in his doctoral research under von Kármán's supervision at Caltech's Guggenheim Aeronautical Laboratory (GALCIT) (Sears and von Kármán, 1938). In his memoirs, Sears

(1994) recounts how von Kármán proposed unsteady aerodynamics as a research topic to him after a visit to Göttingen in 1936. Küssner (1936) had showed to him his solution to the "sharp-edge" gust problem that seemed to have shown that the problem of a body moving on a fluid and its reciprocal (the fluid moving with respect to the body) had different solutions. Very soon after this, Sears identified a sign error in Küssner theory! In other words, the career of one prominent aerodynamicist (and one of the key solutions in this book) is the result of an algorithmic mistake by another now-famous aerodynamicist. The frequency-domain solutions derived during this time have been a cornerstone of dynamic aeroelastic studies ever since.

It was also during the 1930s and 1940s that elastic distortions began to affect rigid-body stability and control of the aircraft. One of the first documented work discussing stability with aeroelasticity was presented by Pugsley (1933), who was part of the Royal Aircraft Establishment (RAE). He investigated the impact of the elastic changes in wing twist and its consequence in the longitudinal stability of the aircraft. Following this, Bryant and Pugsley (1936) investigated the impact of the antisymmetric wing deformation on the lateral stability of the aircraft. The elastic distortion on the longitudinal stability has proven to be much more serious (Collar, 1946), and that included the effects of tail and fuselage distortion. At the time, design criteria had been developed for wing stiffness, but not for other parts of the aircraft, including the tail. From a coupling with controls, Cox and Pugsley (1932) and Duncan and McMillan (1932) investigated the then newly discovered aeroelastic-induced control problem: *aileron reversal*. In this phenomenon, with the increase in dynamic pressure, the change in lift generated by the deflection of the trailing-edge control surface is counteracted by an opposite twisting of the wing. While at the beginning this presents itself as a reduction in control effectiveness, at a particular value of the dynamic pressure, one effect can cancel the other, and there is no rolling moment resulting from the aileron deflection; this dynamic pressure is known as the aileron reversal dynamic pressure. Although not a stability issue per se, this was the first important coupled aeroelastic–flight mechanics problem which became a critical design consideration for World War II (WWII) fighters.

During the war period, the trend to higher speeds and all-metal aircraft persisted. Many flutter problems occurred during this time, particularly related to stores being carried under the wing (e.g., a tip-tank flutter problem occurred with the P-80 aircraft) and various battle damages and field fixes that altered the mass and stiffness of the airframe. Another issue that started concerning the engineers of the time was the compressibility effects in aileron reversal, further augmented by the tapered wings. At this point in time, the structural dynamics of the aircraft was still in its infancy and mostly unconnected with the overall aeroelastic analysis of the aircraft. In fact, the airplanes of the World War II that took flutter into consideration in their design had the analyses done for the wing and/or tail (separately) assuming a representative section. That section, converted into a "typical section" had no more than three degrees of freedom: plunge (for bending), pitch (for twisting), and control surface (for the elastic deformation of aileron or elevators). The idea of bringing the rigid-body degrees of freedom to the mix still seemed unnecessary. A very nice account of the efforts at the

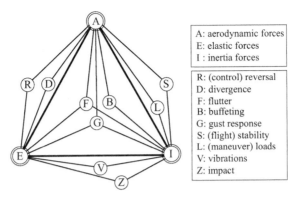

Figure 1.6 *Collar's triangle* of aeroelastic interactions.

time can be found in a review paper by Collar (1946). There he proposed a triangular graphical representation of the various disciplines that constituted *aeroelasticity*[7] and became known as the *Collar's triangle*, as shown in Figure 1.6. The figure describes the three main forces that make the field of aeroelasticity: elastic, aerodynamic, and inertial forces, and from there, various problems were derived, either inside or outside the triangle. Most of them are discussed in later chapters in this book. In a later account (Collar, 1978), Garrick, a research scientist and manager first at N.A.C.A. and later at National Aeronautics and Space Administration (NASA), and a coworker of Theodorsen, qualified this graphical representation as "it is both a chart and a compass for aeroelasticians." It seems Collar was the first to formally present the idea of combining elasticity and stability/controls together, thus resulting in the *dynamics of deformable aeroplanes*. According to Collar:

... dynamics of a deformable aeroplane – suggests the formal solution at once. We select as generalised co-ordinates quantities defining the translational and angular freedoms of the aeroplane as a rigid body; we add co-ordinates representing the control surface angles, tab angles, stick and pedal movements, and with these we include co-ordinates to represent the freedoms due to automatic controls; finally we introduce a number of co-ordinated sufficient to describe the modes–normal or otherwise–of elastic deformation of the aircraft. We then derive, in terms of the inertia and elastic properties of the aircraft and the forces, the complete Lagrangian equations of motion ... (1946, p. 630).

While visionary, Collar also faced the limitations at the time: "... we obtain a set of simultaneous differential equations, probably in twenty to thirty degrees of freedom, ... From the practical viewpoint, of course the labour of dealing with such a set of equations is prohibited at present." Just for reference, today we deal with thousands of equations routinely, empowered by modern computers and numerical methods.

---

[7] According to Collar (1978), the term *aeroelasticity* was coined by Roxbee Cox and Alfred Pugsley in the early 1930s, and it may have been inspired by the term *photoelasticity* in vogue among experimentalists at the time.

1.3.2   The Jet Age

Up to the mid-1950s the aeroelastic and flight dynamics developments continued their courses separately, except for the control effectiveness and stability considerations that bridged the two. As von Kármán highlighted in his book (von Kármán, 1954), "The science of aeroelasticity, including flutter theory, is now in a period of active development." Small corrections to the effect of flexibility on the airplane stability were made assuming small aeroelastic contributions to the rigid airplane stability derivatives obtained from wind-tunnel testings. However, the increase in aircraft size and flying speed led to combined studies. One of such studies is presented in the report by the J. B. Rea Company (Rea, 1957), where a comprehensive study was conducted to bridge the aeroelasticity developments with the flight dynamics of the time. More like a handbook, this report was created by several authors, including Y. C. Fung from Caltech, for the practicing engineer to "give methods for incorporating aeroelastic effects in equations of motion as well as techniques for obtaining the solutions," mostly through matrix methods. As also emphasized there, "many of the techniques of analysis are still in a state of development."

The early airplanes were far from rigid, but their low speeds masked the elastic effects, particularly in view of all the other challenges they were facing then. With the increase in flying speed along the years, two relevant effects came to play: the relative increased flexibility effects in the structure and the compressibility effects in the aerodynamics. The relative flexibility of a structure is directly related to the dynamic pressure and is dependent on the air density and the square of the airspeed. The compressibility effects come in the form of change in aerodynamic coefficients, center of pressure, and aerodynamic center, ultimately impacting the stability of the aircraft. Aircraft aeroelastic problems were markedly accentuated with the higher speeds provided by jet propulsion. And with those higher speeds, the wings were swept, bringing additional challenges as well. The airplane that brought the aeroelastic problem to everyone's attention during that phase was the Boeing XB-47 bomber, as shown in Figure 1.7. It was known that the airplane would have flexible wings, but at the end,

(a)   (b)

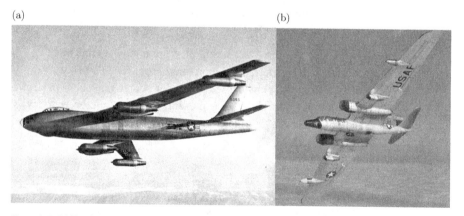

Figure 1.7 (a) Boeing XB-47. (b) Martin RB-57F Canberra (photos from the U. S. Air Force).

for a 116-foot span wing, it showed deflections of 35 ft. between the maximum positive and negative loads (Perkins, 1970), representing a tip-deflection amplitude of the order of 30% of its semispan! Along with that, came additional challenges with stability and control. For example, during pull-up maneuvers, the wings would deflect up, inducing an effective down twist, which in turn would alter the wing aerodynamic center. This would have a direct impact on the tail downwash and its effectiveness. The elastic wing deflection would also move the center of gravity by 15% forward during a pull-up maneuver, altering the aircraft stability (Cook, 1991).

A major milestone in bridging aeroelasticity and flight dynamics came with the work of R. D. Milne (1962). In his work, Milne took the equations of motion of Bairstow et al. (1913) and augmented them with the structural equations in the form of beam or plate theories solved with the Rayleigh–Ritz method, or the use of equivalent influence functions for more general structures. He also introduced the powerful concept of *mean axes*, which will be further discussed in Chapter 6. During this time, a new aeroelastic instability was found: *body-freedom flutter*. This is a flutter mechanism that involves the participation of at least one rigid-body motion (typically pitch) of the aircraft along with elastic modes. This coupling between the rigid-body motion and elastic motion is facilitated either by a reduction in pitch inertia of flying-wing configurations, where the resulting increase in the short-period frequency gets closer to the first bending mode of the wings, or by increasing the wing flexibility and bringing its frequency down and in the vicinity of the short-period mode. While this was already faced by the Horten high-aspect-ratio flying-wing sailplane in the pre-WWII Germany, it was not until the General Dynamics RB-57F that the problem came screaming to the aeroelasticians. This aircraft, as shown in Figure 1.7, was a new high-altitude reconnaissance aircraft based on a conversion of the existing Martin B-57 but with larger wings, bringing its first bending frequency to 1.5 Hz (Love et al., 2005). The instability was discovered during flight tests and just beyond the limit speed of the aircraft – with two more additional flutter incidents as a result of unintentional overspeed. In fact, a similar problem was later found in the Northrop B-2 bomber but, once again, outside its operational envelope (Jacobson et al., 1998).

By the 1960s, closer connections between aeroelastic and flight dynamics considerations were clearly needed due to: "1. The considerable increase in speed. 2. The thinner shapes required for transonic and supersonic flight. 3. The new airframe designs such as variable geometry. 4. The new generation of commercial aircraft: stretching of present long range liners, Jumbojets, Supersonic transports. 5. The growing interest in low level flight for military purposes" as identified and addressed by an Advisory Group for Aerospace Research and Development meeting in 1969 (Anonymous, 1969). This was a coordinated effort among the North Atlantic Treaty Organization nations to bring theoretical and experimental aeroelastic methods into flight mechanics. An example is the work of Dusto (1969) from the Boeing Company, who employed Milne's mean-axes approximation along with structural influence coefficients to obtain steady and unsteady stability derivatives through residualization. This approach was applied to evaluate the stability of two aircraft of the time: B707 and the SST. Chevalier et al. (1969) found the elasticity effects on the static longitudinal stability to be "large and unfavorable" while not greatly affecting their dynamic stability characteristics. However, the authors make the point that "results indicate that as the structural frequencies approach the frequency of motion of the airplane (ratios less than 4:1), major effects might occur."

There were finally two main developments that took place in the late 1960s that have impacted the way we evaluate the flutter today. One was the development of the doublet-lattice method (DLM) by Albano and Rodden (1969), and the other was the initial development of a common structural dynamics analysis software for NASA's various centers, named NASA STRucture ANalysis or NASTRAN. This eventually went commercial (MacNeal and McCormick, 1972), evolved along the years, and through its aeroelastic solver incorporated the DLM solution. This has become the most used aeroelastic solver in the aerospace industry.

With the development of high-bandwidth actuators and control systems theory and hardware, the possibilities of altering the dynamics and response of an aircraft through active controlled closed-loop action of its control effectors became a reality.[8] The increased importance of the active control system (ACS) and its interplay with the flexible airframe has led to an explicit inclusion of their combined effects in the analysis, design, and certification (when applicable) of new aircraft. Known as aeroservoelasticity (ASE), this area of study encompasses the alleviation of loads, both maneuver and gust, fatigue life extension, control of elastic modes and overall deformation, stability and flutter margin augmentation, and improvements in ride quality (Hönlinger et al., 1994).

Although some theoretical studies were conducted in the previous decade, one early example of experimental tests on the impact of the ACS on the airframe loads can be found in the work of Payne (1953). Although the autopilot was not designed for load alleviation, the series of flight tests with a small transport plane[9] resulted in gust loads being attenuated by approximately 7% when the autopilot was engaged. A more comprehensive gust-load alleviation (GLA) study was presented by Phillips (1957). In his report, a review of the various effects related to controlling the load is presented, including sensors and actuators. Two dedicated flight tests had just finished: one targeting improvements in ride quality while measuring the impact on the structural loads due to gust, and the other measuring the impact of a yaw damper on reducing vertical tail loads. For the former, the vane-controlled gust-alleviation system flight test was studied by Chris Kraft (Kraft, Jr., 1956), later to become the mastermind of space flight management for all future NASA space manned missions. The gust-load investigation showed mixed results: there was actually an increase in structural loads while improving passenger comfort during light turbulence, and for severe turbulence, the

---

[8] It is interesting to notice that the idea of active feedback control had been known for a while (e.g., Millikan, 1947). In terms of flutter control, for example, Bisplinghoff et al. (1955) indicate that "... exists an excellent possibility of improving flutter performance by ... rapidly responding automatic control system, actuated in closed-loop fashion by the motion to be stabilized. However, we cannot count on the human pilot to compensate for flutter ... because the frequencies are too high ...."

[9] The report does not specify the airplane model, but it contains drawing of it. It is our understanding that the aircraft used for these tests was a version of the Convair CV-240.

system was expected to reduce the wing loads but increase the loads in the tail and other structural components–a well-known phenomenon understood today.

With larger aircraft being developed, the U.S. Air Force established the Load Alleviation and Mode Stabilization (LAMS) program in the second half of the 1960s to study the impact of ACS to alleviate gust loads and enhance fatigue life in more flexible aircraft. The final report from Burris and Bender (1969) describes the extent of the study involving U.S. industry and government to demonstrate the technology in a modified B-52E test vehicle. That included the addition of hydraulic powered controls and fly-by-wire pilot station, besides the instrumentation required for flight tests, specifically designed to alleviate structural loads during atmospheric turbulence encounters. The program successfully showed that the LAMS combined with the ACS provided significant reduction in the fatigue damage rate while satisfying the design and performance criteria established at the time. A parallel analysis-only study was also conducted in the C-5A Galaxy to show that the technology could be used in other large aircraft.

There were significant ASE activities in the 1970s, and some key examples are presented in the review report of Regan and Jutte (2012) as well as in the comprehensive review of Livne (2018). The B-52 LAMS studies continued with the B-52 Control Configuration Vehicle (CCV) program, a pioneering effort on active aeroservoelastic control. The testing aircraft was modified to include canards and ventral vanes, and external stores were mass balanced to bring flutter into the operational envelope of the aircraft. In this way, flight tests for active flutter control could be safely conducted since in the case of instability, the stores could be quickly ejected from the aircraft, restoring its initial stable characteristics. The CCV program was able to successfully show simultaneous improvements in flutter mode control, maneuver load alleviation (MLA), ride control, fatigue reduction, and stability augmentation (Roger and Hodges, 1975). Another early example of an aircraft incorporating active control is the Lockheed C-5A, also a direct extension of the initial analytical studies as part of the LAMS program (Burris and Bender, 1969). The aircraft was suffering from fatigue life problems related to wing bending loads. The Active Lift Distribution Control System (ALDCS) was developed to reduce wing stress during turbulence encounters and maneuvers (Disney, 1975). By sensing forward and aft vertical wingtip accelerations along with inertial reference system signals, the ALDCS actuated ailerons and inboard elevators to reduce the wing root bending moment by more than 30%, although it also observed increases in torsional loads. The system was eventually superseded by a structural modification that added 5.5% of the initial empty weight of the aircraft. An interesting example of increasing wing span (5.8%) to reduce total drag (3%) and avoid a mass increase (1.25% of gross weight) is found in the Lockheed L-1011-500. The ACS was put in place to provide maneuver load alleviation (MLA) and gust load alleviation (GLA) by using the horizontal stabilizers and ailerons without significant structural modifications. Another example of GLA is found in the Northrop B-2 Spirit aircraft, where gust loads were sizing the vehicle (Britt et al., 2000). The GLA system employs the inboard elevons and a dedicated GLA surface at the aircraft centerline. Due to undesirable excitation of the aircraft's first symmetric bending mode, the outer elevons are used out of phase to damp the vibration. In terms of controlling the ride quality, an early example is the structural mode control system present in the Boeing's B-1 Lancer aircraft. By using dedicated active canard-control surfaces and colocated accelerometers, the system suppressed vibrations at the pilot station, particularly during gust encounters. Parallel studies were also being carried out in Europe. In particular, the German Aerospace Center (DLR, formerly known as DFVLR) conducted flight tests targeting ride quality and the control of flight dynamics characteristics, particularly the short-period mode, as exemplified by the Open Loop Gust Alleviation in the late 1970s and early 1980s, followed by the Load Alleviation and Ride Smoothing system in the 1980s and early 1990s. In this latter program, full fly-by-wire capability was deployed with additional direct-lift-control flaps in DLR's twin-jet test aircraft, which showed reduction in the longitudinal and vertical accelerations with a combination of lift and drag control devices but marginal reductions in the wing root bending moment (Hahn and König, 1992). Many other examples can be found from a more fundamental research on exploring ACS for aeroelastic and flight dynamics improvements. One example is the successful series of wind-tunnel studies of two SensorCraft concepts (Lucia, 2005) for load alleviation and stability enhancements. The blended wing body concept (Vartio et al., 2005) and the joined-wing configuration (Scott et al., 2011) were tested both with a rigid support and with attachment to a pitch-and-plunge apparatus in the Transonic Dynamics Tunnel at NASA Langley. The aeroelastically scaled models showed the importance of including rigid-body degrees of freedom when assessing and controlling the stability (flutter) and gust response of flying wings and more flexible concepts.

While these are examples of solutions developed to mitigate a design problem, recent commercial aircraft have been developed with these technologies in mind. An early example is the Airbus A320 (entered service in 1987) that incorporated a load alleviation function (LAF) that actuated a combination of ailerons, spoilers, and elevators while sensing the inertial load factor. The increase in actuator utilization might have led to its de-activation in its derivatives and later reintroduction in the Airbus A330 (entered service in 1994) and A340 (entered service in 1993) for MLA and improved flying qualities (Regan and Jutte, 2012). Similarly, the Boeing B787 (entered service in 2011) includes MLA and ride quality enhancements through the use of ailerons, spoilers, and flaperons. According to Norris and Wagner (2009, p. 87), several thousand pounds of weight were taken out of the design by actively reducing the maneuver loads.

1.3.3   Towards Sustainable Flight

After some niche human-powered aircraft developments primarily led by Paul Mac-Cready in the 1970s and 1980s, where the extremely light structural constructions and flexible airframe brought challenges in the coupled aeroelastic–flight dynamics behavior, it was really the high-altitude, long-endurance (HALE) planes from the late 1980s and until now that have really challenged the then state of the art in this field.

During the 1990s and early 2000s, a joint NASA and industry initiative established a long-duration Earth science and environmental missions at high altitudes known as Environmental Research Aircraft and Sensor Technology (ERAST) program.[10] Among the industry participants, Aerovironment (founded by MacCready) had developed Helios, an ultra-lightweight flying-wing solar-powered aircraft with a constant chord wing of 247 ft. (75.3 m) span and an aspect ratio of 30.9. The airplane had many successful flights, including an altitude record of 96,863 ft. in 2001. In 2003, equipped with a fuel cell for longer duration flight, Helios encountered a shear layer at approximately 3,000 ft. over Hawaiian waters, which excited an instability that led to the loss of the aircraft. As part of the mishap investigation, Noll et al. (2004) concluded "[that] more advanced, multidisciplinary (structures, aeroelastic, aerodynamics, atmospheric, materials, propulsion, controls, etc.) time-domain analysis methods appropriate to highly flexible, morphing vehicles [be developed]." They also stated that "[t]he Helios accident highlighted our limited understanding and limited analytical tools necessary for designing very flexible aircraft and to potentially exploit aircraft flexibility." This motivated many in the research community to look into the problem, and for the next decade, many studies were conducted coupling nonlinear aeroelasticity and flight dynamics, primarily targeted to the HALE class of aircraft. Most of the work has been numerical in nature, and few experimental studies in wind-tunnel (e.g., Tang and Dowell, 2001) and flight tests (e.g., Cesnik et al., 2012; Ryan et al., 2014) attempted to support those developments. During that time period, there were also several studies in ASE targeting control of loads (e.g., Boeing/AFRL/NASA Active Aeroelastic Wing Program) and stability (e.g., AFRL/Boeing/NASA B-52E flutter, loads and ride control program) (Friedmann, 1999; Livne, 2018; Livne and Weisshaar, 2003), and we encourage the reader to explore the rich literature available in the topic.

While climate change had been identified as a threat to life on Earth for quite some time, it was not until the early 2010s that the aviation industry committed itself to reduce emissions. NASA conducted studies in what they referred to as the N+1, N+2, and N+3 aircraft of the future. Part of these studies was the identification of various technologies that could reduce fuel burn (and consequently emissions) and noise. Among the investigated technologies, higher-aspect-ratio wings for lower induced drag was one of the key airframe concepts with the largest impact.[11]

An example of how the aspect ratio can impact fuel burn for a long-range commercial transport aircraft is included in Figure 1.8. It shows curves for optimum designs (i.e., Pareto fronts) obtained from a blend between minimizing fuel burn

---

[10] Early concepts of this class of aircraft, including Theseus, Pathfinder, and Helios, were in part the motivation for this book's second author to spearhead his research interests in this area. He first involved his Ph.D. advisor, Dewey Hodges, while a postdoctoral fellow at Georgia Tech (Cesnik, 2023). As he moved to the faculty at MIT and later to the University of Michigan, he extended his research activities in the field of coupled nonlinear aeroelasticity and flight dynamics. This book's first author's doctoral and postdoctoral studies at the University of Michigan were directly connected to those activities.

[11] More recent designs such as the Boeing 787 and the 777X are examples of higher aspect-ratio configurations that resulted in more flexible wings, as shown in Figure 1.1.

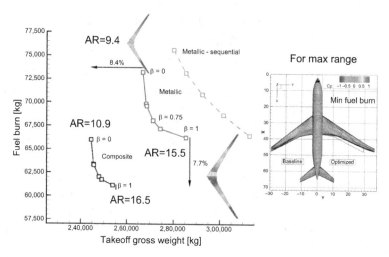

Figure 1.8 Highly efficient flight performance leads to high-aspect-ratio wings (adapted from Kennedy et al. (2014)).

and takeoff gross weight, where β is the weight between both cost functions, that is, cost metric $= \beta$(fuel burn) $+ (1 - \beta)$(takeoff gross weight) (Kennedy et al., 2014). For $\beta = 1$, that is, when all the emphasis is on minimizing fuel burn, the optimized wing aspect ratio for a simultaneous aero-structural optimization is 65% higher than the one when takeoff gross weight is minimized. While the high-fidelity MDO behind these calculations did not include dynamic stability considerations (i.e., flutter), it clearly shows the trends. Moreover, it also shows the importance of a simultaneous optimization instead of a sequential one. To achieve this higher aerodynamic efficiency while keeping the structural weight low, the resulting designs are naturally more flexible and more prone to aeroelastic effects. Moreover, the fact that the airframe elastic frequencies go down implies that they now cannot be clearly separated from the flight dynamics frequencies. Therefore, the problem has become intertwined and the coupled nonlinear aeroelastic/flight dynamics needs to be considered when sizing and certifying these new designs. These challenges have brought a new wave of research and development needs, and while these are still ongoing, we expect the new methods and capabilities to directly impact our next generation commercial transport aircraft development and support the zero-emission efforts our industry is committed to. Moreover, the coupled nature of the problem put even more demand on a holistic and integrative approach to the problem. The originally separated objectives of load alleviation, flutter enhancement, and ride and handling qualities, all need to be treated and addressed simultaneously by the appropriate manipulation of the rigid and elastic modes. The deeply coupled nature of this multidisciplinary problem creates new demands on the way the engineers are educated and the way aircraft companies must organize their engineering departments. We are seeing a renascence of aeroelasticity through the future environmentally designed "deformable aerial bodies." These are very exciting times indeed.

1.4   Some Basic Concepts

This book is mostly concerned with aircraft dynamics, which will be described within the framework of classical mechanics. Some relevant conventions and formalisms are therefore introduced first in Section 1.4.1. Moreover, for the analysis of the vehicle performance and/or design of control systems, it is often also convenient to consider its linearized response. Here, the mathematical framework of linear time-invariant (LTI) dynamical systems is particularly useful, and some basic definitions are introduced in Section 1.4.2.

1.4.1   Dynamics

The most general description of the dynamics of a deforming aircraft is obtained within the framework of continuum mechanics; yet, in practice, we always use a finite-dimensional representation of the vehicle obtained from a spatial discretization of the physical domain. The resulting set of parameters is known as *generalized coordinates*, and the vector space in which they lie is called the *configuration space*. Let N_q be the number of generalized coordinates. We aggregate them into a column vector, $\mathbf{q} = \left\{ q_1 \quad q_2 \quad \cdots \quad q_{N_q} \right\}^\top$, and the problem of describing the flexible aircraft dynamics becomes the evaluation of the time history $\mathbf{q}(t)$ during the events of interest. For example, in a finite-element discretization of a wing, the generalized coordinates are the time-dependent displacements at all the grid nodes, and we may be interested in tracking their values as the aircraft performs a certain maneuver. The generalized coordinates, however, do not uniquely define the instantaneous state of a moving or vibrating wing, which also depends on the instantaneous velocities at the nodes. We define then the *phase space* (or state space) as a representation of the mechanical system that uniquely determines its instantaneous state, given by a state variable $\mathbf{x}^\top = \left\{ \mathbf{q}^\top \quad \dot{\mathbf{q}}^\top \right\}$. Consequently, we often need to distinguish between *degrees of freedom*, $\mathbf{q}(t)$, which are the generalized coordinates that need to be solved in the equations of motion, and *states*, $\mathbf{x}(t)$, which are needed for phase-space descriptions.

Hamilton's principle provides the framework to determine the dynamics of a flexible body within a finite interval $0 \leq t \leq t_f$ for given initial conditions and external forces. It states that the trajectory of a mechanical system in the configuration space is a stationary point of the system Lagrangian, $\mathcal{L} = \mathcal{T} - \mathcal{U}$, where $\mathcal{T}$ and $\mathcal{U}$ are the kinetic and internal (or strain) energy of the system, respectively. For a flexible body, this results in enforcing that perturbations of the system Lagrangian cancel the virtual work of the external forces, $\delta \mathcal{W}$, when obtained under the same virtual displacement field, that is,

$$\int_0^{t_f} (\delta \mathcal{T} - \delta \mathcal{U} + \delta \mathcal{W}) \, \mathrm{d}t = 0. \tag{1.1}$$

This expression is valid independently of the selection of reference frame or any other parameterization of the flexible body kinematics. If a phase-space description (with the generalized velocities replaced by generalized momentum using a Legendre transformation) is now introduced, it results in the formalism of Hamiltonian mechanics, which will be used later in this book in Section 8.6. The starting point in most of this book is, however, the framework of Lagrangian mechanics, in which the solution is sought in terms of a finite number of generalized coordinates. In that case, Hamilton's principle results in *Lagrange's equations of motion* (see Chapter 1 of Géradin and Rixen (1997) for the derivation), which are written as

$$\frac{\mathrm{d}}{\mathrm{d}t}\frac{\partial \mathcal{T}}{\partial \dot{q}_i} - \frac{\partial \mathcal{T}}{\partial q_i} + \frac{\partial \mathcal{U}}{\partial q_i} = f_{qi}, \ \text{ for } i = 1,\ldots,N_q, \tag{1.2}$$

with f_{qi} being the generalized force associated with the generalized coordinate q_i. They satisfy $\delta\mathcal{W} = \delta\mathbf{q}^\top\mathbf{f}_q$. If $\|\mathbf{q}\| = \sqrt{\mathbf{q}^\top\mathbf{q}}$ is sufficiently small, linear dynamics can be assumed. In that case, the kinetic and strain energies are written, respectively, as

$$\mathcal{T} = \frac{1}{2}\dot{\mathbf{q}}^\top\mathbf{M}\dot{\mathbf{q}} \ \text{ and } \ \mathcal{U} = \frac{1}{2}\mathbf{q}^\top\mathbf{K}\mathbf{q}, \tag{1.3}$$

where $\mathbf{M}$ and $\mathbf{K}$ are the mass and stiffness matrices associated with the generalized coordinates. The linear dynamics of a flexible body in an inertial frame can then be approximated as

$$\mathbf{M}\ddot{\mathbf{q}} + \mathbf{K}\mathbf{q} = \mathbf{f}_q, \tag{1.4}$$

which is solved together with the initial conditions $\mathbf{q}(0) = \mathbf{q}_0$ and $\dot{\mathbf{q}}(0) = \dot{\mathbf{q}}_0$. This basic description is often augmented with a model for viscous damping, as will be done later in this book.

1.4.2   Linear Dynamical Systems

Under sufficiently small perturbations, and if there are no explicit functional dependencies with time in the underlying physical processes (e.g., mass assumed to remain constant), we can treat many problems in this book as finite-dimensional LTI systems with multiple inputs and outputs. The definition of inputs and outputs depends on the system and the particular problem of interest. For a wing described by Equation (1.4), the inputs may be deflections of a trailing-edge flap that modify the force vector $\mathbf{f}_q$, and the outputs may be the resulting elastic displacements at certain points of the wing, which can be obtained from the generalized coordinates $\mathbf{q}$. Let N_u be the number of independent inputs and N_y the number of outputs. We define the input and output vectors as $\mathbf{u}(t)$ and $\mathbf{y}(t)$, respectively, and describe their relation using block diagrams that encapsulate the internal dynamical processes, such as the basic one shown in Figure 1.9.

In this book, we consider three different, but also equivalent, representations of the input–output dynamics of an LTI system, namely by means of a state-space

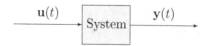

Figure 1.9 Input and output on a dynamical system.

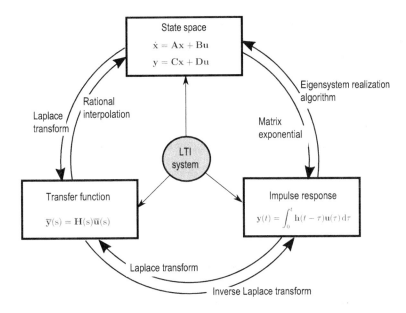

Figure 1.10 Three representations of linear time-invariant systems (after Brunton and Kutz (2019)).

representation, its transfer function, or its impulse response. They are schematically shown in Figure 1.10. As will be made apparent later, each one of them offers advantages over the other two for specific analyses. The rest of this section will discuss them, introducing the key definitions in each formulation, as well as the transformations to switch between any two of those three representations for a given system.

State-Space Representation

As we have seen in Section 1.4.1, finite-dimensional dynamical systems result in ordinary differential equations in time. After linearization about an equilibrium point, they can be written in a state-space form as

$$\dot{\mathbf{x}} = \mathbf{A}\mathbf{x} + \mathbf{B}\mathbf{u},$$
$$\mathbf{y} = \mathbf{C}\mathbf{x} + \mathbf{D}\mathbf{u}, \tag{1.5}$$

where $\mathbf{x}(t)$ is the state vector, of dimension N_x, and $\mathbf{A}$ is the $N_x \times N_x$ state matrix, $\mathbf{B}$ is the $N_x \times N_u$ input matrix, $\mathbf{C}$ is the $N_y \times N_x$ output matrix, and $\mathbf{D}$ is the $N_y \times N_u$ feedthrough matrix. For a given physical system, and input and output magnitudes of interest, the definition of the state vector is not unique, that is, there are multiple, but equivalent, realizations of Equation (1.5) for a given problem. In particular, a *minimal realization* is obtained when the system dynamics are described with the smallest possible number of states.[12]

---

[12] The formal definition is that the system realization has no uncontrollable and unobservable states (see Section 7.4 for further details).

Once a suitable state vector is chosen, Equation (1.5) can be integrated in time for the given initial conditions $\mathbf{x}(0) = \mathbf{x}_0$ and the input signal $\mathbf{u}(t)$. Time-marching schemes, such as those of the Runge–Kutta family of algorithms, are readily available for this task.

Equation (1.5) is the continuous-time state-space representation of an LTI system. It is sometimes more suitable to write it as a *discrete-time system* in which all signals are only evaluated at sampling points with, typically, a constant sampling rate Δt. Let $\mathbf{x}_n = \mathbf{x}(t_n)$; with $t_n = n\Delta t$, the integration of Equation (1.5) between t_n and t_{n+1} results in the difference equations

$$\mathbf{x}_{n+1} = \hat{\mathbf{A}}\mathbf{x}_n + \hat{\mathbf{B}}\mathbf{u}_n, \tag{1.6}$$

with $\hat{\mathbf{A}} = e^{\mathbf{A}\Delta t}$ and $\hat{\mathbf{B}} = \int_0^{\Delta t} e^{\mathbf{A}\tau}\mathbf{B}\,d\tau$, while the output equations remain unchanged. This is the *discrete-time state-space* representation of the system.

The state-space representation of an LTI system also allows establishing its linear (or asymptotic) stability from the eigenvalues of $\mathbf{A}$ or $\hat{\mathbf{A}}$, and its controllability and observability from the Gramians of the system, which are discussed in Section 7.4. To define linear stability, we study the response under small perturbations. They are introduced as the nonzero initial conditions $\mathbf{x}(0) = \mathbf{x}_0$ in Equation (1.5), and without forcing terms, that is, with $\mathbf{u} = \mathbf{0}$. A linear system is stable when it settles down to zero in its long-term response for any choice of $\mathbf{x}_0$. The solution of this problem can be directly obtained as

$$\mathbf{x}(t) = e^{\mathbf{A}t}\mathbf{x}_0, \tag{1.7}$$

where the *matrix exponential* of a matrix $\mathbf{A}$ is defined as

$$e^{\mathbf{A}} = \exp\mathbf{A} = \mathcal{I} + \sum_{n=1}^{\infty} \frac{\mathbf{A}^n}{n!}. \tag{1.8}$$

The evaluation of this series is in general something to be avoided, and for small systems that can be achieved by the use of the Cayley–Hamilton theorem (see Section 5.4.2). Alternatively, we can seek the solution by means of the modal projection of the system. For that, consider first the eigendecomposition

$$\mathbf{A}\mathbf{\Phi} = \mathbf{\Phi}\mathbf{\Lambda}, \tag{1.9}$$

where $\mathbf{\Lambda}$ is the diagonal matrix of, in general, the complex eigenvalues of $\mathbf{A}$ and $\mathbf{\Phi}$ is a matrix whose columns are its associated eigenmodes.[13] We introduce then the change of variable $\mathbf{x} = \mathbf{\Phi}\mathbf{z}$, which, for the unforced problem, results in the system dynamics written as $\dot{\mathbf{z}} = \mathbf{\Lambda}\mathbf{z}$. This equation can be integrated independently for each mode since $\mathbf{\Lambda}$ is diagonal, and this finally enables explicit solutions in the original variables to be obtained, thus resulting in

$$e^{\mathbf{A}t} = \mathbf{\Phi}e^{\mathbf{\Lambda}t}\mathbf{\Phi}^{-1}. \tag{1.10}$$

---

[13] This assumes that there are no repeated eigenvalues in the system; otherwise, we need the Jordan form of the eigendecomposition. This technical difference, however, effectively changes little in our discussion.

(a)   (b)

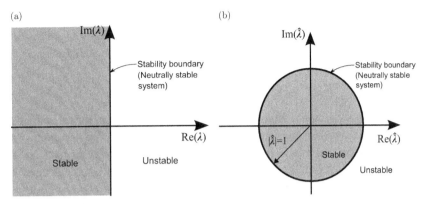

Figure 1.11 Asymptotic stability of linear time-invariant systems. (a) Continuous system, $\lambda = \mathrm{eig}(\mathbf{A})$ and (b) discrete system, $\hat{\lambda} = \mathrm{eig}(\hat{\mathbf{A}})$.

Since $\boldsymbol{\Phi}$ is a constant matrix, the solution to Equation (1.10) returns to zero as time grows if and only if all the eigenvalues of $\mathbf{A}$ are all in the left-half complex plane, as shown in Figure 1.11a. Finally, note that if λ is an eigenvalue of $\mathbf{A}$ in Equation (1.5), then $e^{\lambda \Delta t}$ is an eigenvalue of $\hat{\mathbf{A}}$ in Equation (1.6). Therefore, a discrete-time system is stable if and only if all the eigenvalues of $\hat{\mathbf{A}}$ are within the unit circle in the complex plane. This is graphically illustrated in Figure 1.11b.

Impulse Response

Given initial conditions $\mathbf{x}(0) = \mathbf{x}_0$, the solution to the state equation (1.5) can be obtained in terms of

$$\mathbf{x}(t) = e^{\mathbf{A}t}\mathbf{x}_0 + \int_0^t e^{\mathbf{A}(t-\tau)}\mathbf{B}\mathbf{u}(\tau)\,\mathrm{d}\tau. \tag{1.11}$$

Equation (1.11) is easily verified by substituting it into the differential equation (1.5). We say then that the exponential matrix $e^{\mathbf{A}(t-\tau)}$ is the *state-transition matrix* of the continuous-time LTI system. The output variable is finally obtained as

$$\mathbf{y}(t) = \mathbf{C}e^{\mathbf{A}t}\mathbf{x}_0 + \int_0^t \mathbf{C}e^{\mathbf{A}(t-\tau)}\mathbf{B}\mathbf{u}(\tau)\,\mathrm{d}\tau + \mathbf{D}\mathbf{u}(t). \tag{1.12}$$

We can now define the *impulse response* as the matrix function with the time history of the outputs to a unit impulse in the inputs, $\mathbf{u}(0) = \delta(t)$, with $\delta(t)$ being the Dirac delta function over an N_u-dimensional space. Thus, the impulse response, $\mathbf{h}(t)$, can be written as

$$\mathbf{h}(t) = \mathbf{C}e^{\mathbf{A}t}\mathbf{B} + \mathbf{D}\delta(t). \tag{1.13}$$

The impulse response of a system is therefore a unique matrix function with time as the independent variable, which has N_y rows and N_u columns. Using this definition, the output signal for a system initially at rest ($\mathbf{x}_0 = \mathbf{0}$) can be written as a *convolution integral*,

$$\mathbf{y}(t) = \int_0^t \mathbf{h}(t-\tau)\mathbf{u}(\tau)\,\mathrm{d}\tau. \tag{1.14}$$

For a discrete-time system, the impulse response results in the series $\mathbf{h}_0 = \mathbf{D}$ and $\mathbf{h}_n = \mathbf{C}e^{\mathbf{A}(n-1)\Delta t}\mathbf{B}$ for $n > 0$, which is known as the *Markov parameters* of Equation (1.6). The convolution operation can then be written as

$$\mathbf{y}_n = \sum_{m=1}^{n} \mathbf{h}_m \mathbf{u}_{n-m}, \quad \text{for} \quad n > 0, \tag{1.15}$$

and $\mathbf{y}_0 = \mathbf{h}_0 \mathbf{u}_0$. Note that, as they are instantiations of the impulse response, the Markov parameters are matrices whose number of rows and columns are the number of outputs and inputs, respectively. As has just been seen, the evaluation of the impulse response from a state-space model (the differential equations describing the system dynamics) is a straightforward process. The converse, that is, effortlessly obtaining a state-space description from either an impulse response or Markov parameters is generally not true, and it becomes a form of the realization problem that we discuss in Section 1.4.3.

Transfer Function

Finally, it is often convenient to study the dynamics of a system in the frequency domain. The Laplace transform gives a generic tool to enable that transformation. For a time-domain function $g(t)$, its Laplace transform, $\overline{g}(s)$, is defined as

$$\overline{g}(s) = \int_0^{\infty} g(t)e^{-st}dt, \tag{1.16}$$

where s is a complex variable known as the Laplace variable. A major advantage of Equation (1.16) is that it transforms differential equations into algebraic equations, since it is easy to see that the Laplace transform of $\frac{d}{dt}g(t)$ is $g(0) + s\overline{g}(s)$. Considering now the LTI system given by Eq. (2.19) with zero initial conditions, its Laplace transform results in

$$\begin{aligned} s\overline{\mathbf{x}}(s) &= \mathbf{A}\overline{\mathbf{x}}(s) + \mathbf{B}\overline{\mathbf{u}}(s), \\ \overline{\mathbf{y}}(s) &= \mathbf{C}\overline{\mathbf{x}}(s) + \mathbf{D}\overline{\mathbf{u}}(s), \end{aligned} \tag{1.17}$$

where $\overline{\mathbf{u}}(s)$ and $\overline{\mathbf{y}}(s)$ are the Laplace transforms of the input and output vectors, respectively. It is possible to solve $\overline{\mathbf{x}}(s)$ in the first equation and substitute in the second, which results in a closed-form relation between inputs and outputs in the Laplace domain $\overline{\mathbf{y}}(s) = \mathbf{H}(s)\overline{\mathbf{u}}(s)$. Here we have introduced the $N_y \times N_u$ matrix $\mathbf{H}(s)$ as the system *transfer function*

$$\mathbf{H}(s) = \mathbf{C}(s\mathcal{I} - \mathbf{A})^{-1}\mathbf{B} + \mathbf{D}. \tag{1.18}$$

Stable linear systems, which have been defined above, can additionally be studied in the frequency domain. This is the restriction of the Laplace domain to the imaginary axis, that is, $s = i\omega$, which results in the *frequency-response function* (FRF) of the system

$$\mathbf{H}(i\omega) = \mathbf{C}(i\omega\mathcal{I} - \mathbf{A})^{-1}\mathbf{B} + \mathbf{D}. \tag{1.19}$$

The FRF can be determined, both experimentally and numerically, by exciting the system with harmonic inputs at the frequencies of interest and considering the steady-state response in the outputs after the initial transient dynamics have dissipated (thus the condition of asymptotic stability). In mechanical systems, the FRF is also referred to as *admittance*.

Replacing the complex-valued Laplace variable in the Laplace transform by $i\omega$ defines the Fourier transform for continuous-time systems, and the Z-transform for discrete-time ones. Both transformations are discussed in some detail in Appendix A. In particular, Equation (A.8) derives the Fourier transform of the convolution integral, which has been introduced in Equation (1.14). From that, it can be seen that the FRF of a linear system is the Fourier transform of its impulse response (scaled by a factor of 2π), that is,

$$\mathbf{H}(i\omega) = \int_{-\infty}^{\infty} \mathbf{h}(t)e^{-i\omega t}\,dt. \tag{1.20}$$

We have by now identified the transformations that define the FRF of a system from either its state-space representation, Equation (1.19), or its impulse response description, Equation (1.20). The impulse response is equally obtained from the inverse Fourier transform of the FRF; yet, the general problem of obtaining the state-space realization of a system given by its frequency-response or its transfer function can only be solved approximately (see Section 1.4.3). An important exception to this occurs when $\mathbf{H}(s)$ is given by proper rational entries, that is, when the transfer function can be written as

$$\mathbf{H}(s) = \mathbf{D} + \frac{1}{\mathbf{r}(s)}\mathbf{P}(s), \tag{1.21}$$

where $\mathbf{r}(s) = s^n + \sum_{k=0}^{n-1} s^k r_k$ is a polynomial expression of order n with real coefficients r_k, and $\mathbf{P}(s) = \sum_{k=0}^{n-1} s^k \mathbf{P}_k$ is a matrix of polynomials of, at most, order $n-1$. The roots of $\mathbf{r}(s)$ are the poles of the system, while the roots of $\mathbf{P}(s)$ are the zeros of each input/output pair. The linear stability of the system corresponds to all poles being in the left-hand complex plane. When the zeros are also in the left-hand plane, the system is known as *minimum-phase system*. In such a case, the inverse system is also stable and unique, and therefore it is possible to determine uniquely its inputs from its outputs. Minimum-phase systems are causal, that is, their current outputs only depend on past and current inputs. In fact, an LTI system is minimum-phase if and only if the system and its inverse both are stable and causal.

A state-space realization for a system such as that of Equation (1.21) can be obtained with

$$\mathbf{A} = \begin{bmatrix} \mathbf{0} & \mathcal{I} & \mathbf{0} & \cdots & \mathbf{0} \\ \mathbf{0} & \mathbf{0} & \mathcal{I} & \cdots & \mathbf{0} \\ \vdots & \vdots & \vdots & \ddots & \vdots \\ \mathbf{0} & \mathbf{0} & \mathbf{0} & \cdots & \mathcal{I} \\ -r_0\mathcal{I} & -r_1\mathcal{I} & -r_2\mathcal{I} & \cdots & -r_{n-1}\mathcal{I} \end{bmatrix}, \quad \mathbf{B} = \begin{bmatrix} \mathbf{0} \\ \mathbf{0} \\ \cdots \\ \mathbf{0} \\ \mathcal{I} \end{bmatrix}, \tag{1.22}$$

$$\mathbf{C} = \begin{bmatrix} \mathbf{P}_0 & \mathbf{P}_1 & \mathbf{P}_2 & \cdots & \mathbf{P}_{n-1} \end{bmatrix},$$

as well as the feedthrough matrix $\mathbf{D}$, that was already introduced in Equation (1.21). All the block matrices in Equation (1.22) have the same dimensions as $\mathbf{H}$. This realization however may not be minimal, thus benefiting from additional model-order reduction strategies to remove unnecessary internal dynamics. Reduced-order model (ROM) generation is discussed in this book in Section 7.4 in the context of unsteady aerodynamics.

Example 1.1 Flapped Airfoil with Vertical Displacements in a Horizontal Flow.
Consider the moving airfoil of a chord c and a unit span as shown in Figure 1.12. The airfoil has mass m, and it is suspended by a spring k_h; therefore, when in vacuum, it is a spring-mass mechanical system with natural frequency $\omega_h = \sqrt{\frac{k_h}{m}}$. It is then positioned in a horizontal airstream with velocity V and density ρ and actuated by a flap. In the first approximation, the flap generates lift proportional to its deflection, $\beta(t)$, while the motions of the airfoil generate lift proportional to the effective angle of attack $\dot{h}/V$. As a result, the equation of motion for $V > 0$ of the resulting aeroelastic system can be written as

$$m\ddot{h} + k_h h = -\frac{1}{2}\rho V^2 c \left(c_{l_\alpha} \frac{\dot{h}}{V} + c_{l_\beta}\beta \right). \tag{1.23}$$

The right-hand side in this equation is the instantaneous lift per unit span on the airfoil, which appears with a negative sign due to the sign convention in the definition of the vertical coordinate, $h(t)$ (see Figure 1.12). We can rewrite it as

$$\ddot{h} + (\mu_\alpha/V)\,\dot{h} + \omega_h^2 h = -\mu_\beta\beta, \tag{1.24}$$

where we have defined $\mu_\bullet = \frac{\rho V^2 c c_{l_\bullet}}{2m} > 0$. This can be written in a state-space form by defining the state vector $\mathbf{x}^\top = \{h \quad \dot{h}\}$, input $u = \beta$, and choosing an output of interest, for example, $y = h$. This results in the dynamics written as Equation (1.5), with

$$\mathbf{A} = \begin{bmatrix} 0 & 1 \\ -\omega_h^2 & -\mu_\alpha/V \end{bmatrix}, \qquad \mathbf{B} = \begin{Bmatrix} 0 \\ -\mu_\beta \end{Bmatrix}, \qquad \mathbf{C} = \begin{Bmatrix} 1 & 0 \end{Bmatrix}. \tag{1.25}$$

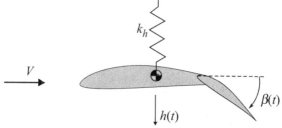

Figure 1.12 Airfoil with a trailing-edge flap suspended by a vertical spring.

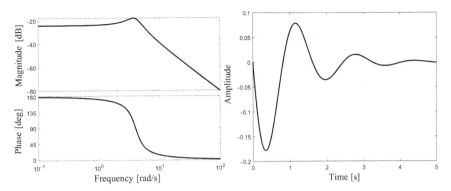

Figure 1.13 Frequency response function, $H(i\omega)$, and impulse response, $h(t)$, for $\mu_\alpha/V = 2$, $\mu_\beta = 1$, and $\omega_h = 4$. (a) Bode plot and (b) impulse response.

For zero initial conditions, applying the Laplace transform to Equation (1.24) results in

$$\left(s^2 + (\mu_\alpha/V)\,s + \omega_h^2\right)\overline{h}(s) = -\mu_\beta\overline{\beta}(s), \tag{1.26}$$

which defines the following (scalar) transfer function for the single-input–single-output system

$$H(s) = \frac{\overline{h}(s)}{\overline{\beta}(s)} = -\frac{\mu_\beta}{s^2 + (\mu_\alpha/V)\,s + \omega_h^2}. \tag{1.27}$$

The FRF is then directly obtained from the restriction of Equation (1.27) to the imaginary axis, that is, $H(i\omega)$. Finally, the impulse response function can be obtained from either the inverse Laplace transform of Equation (1.27) or from an analytical solution of Equation (1.24), with $u = \delta(t)$ being the Dirac delta function. In either case, it results in

$$h(t) = -\frac{\mu_\beta}{\omega_h\sqrt{1-\zeta^2}}e^{-\zeta\omega_h t}\sin\left(\sqrt{1-\zeta^2}\omega_h t\right), \tag{1.28}$$

with $\zeta = \frac{\mu_\alpha}{2V\omega_h}$. Figure 1.13a shows the amplitude and phase of the FRF in log–log scale (known as the *Bode plot*) for $\mu_\alpha/V = 2$, $\mu_\beta = 1$, and $\omega_h = 4$, while Figure 1.13b shows the corresponding impulse response function. This simple aeroelastic system is a damped oscillator, with the damping generated from the aerodynamic forces.

1.4.3   Dynamical System Identification

In the previous section, we have discussed three equivalent representations of LTI systems. The first of them, the state-space model, uses inputs, states, and outputs, while the last two, namely the impulse response and the transfer function descriptions, only use inputs and outputs. Consequently, the state-space representation is often referred to as the *internal* description, while the other two (which are a Laplace/Fourier transform pair) are known as *external* descriptions (Antoulas, 2005, Ch. 4). We have seen that the

transformation from internal to external descriptions is obtained from direct manipu-
lation of the system matrices $(\mathbf{A},\mathbf{B},\mathbf{C},\mathbf{D})$. The inverse problem, that is, obtaining an
internal description from an external one, is in general significantly more involved,
and it is known as the *realization problem*. Moreover, in most practical applications,
input/output information is only available either at discrete times or for a finite range of
frequencies and, therefore, the realization problem can only be approximately solved.
In fact, it becomes a data-driven system identification problem, with the data being a
finite number (at sampled frequencies or times) of input/output pairs of the dynamical
system. The literature on system identification methods is vast, and here we focus on
two methods of particular importance to unsteady aerodynamics, namely the eigensystem realization algorithm (ERA) and rational interpolation methods, which will be
used later in this book.

Eigensystem Realization Algorithm

The ERA builds a state-space representation of a given dynamical system from its
impulse response, which may have been obtained either experimentally or numerically. As the data series are necessarily finite, the approach works in discrete time
and generates the internal model from a finite series of Markov parameters, Equation (1.15). It has been extensively used to analyze experimental vibration, for which it
was originally developed (Juang and Pappa, 1985), and to construct low-order models
from (expensive) computational fluid-dynamics simulations (Silva and Bartels, 2004),
which are discussed in Section 6.5.4.

The ERA seeks to identify the dominant underlying features in the data series. For
that purpose, it starts by building a rectangular matrix with all the output measurements
to the impulse response, which are shifted by one time step on each column, as

$$\mathcal{H}_k = \begin{bmatrix} \mathbf{h}_{k+1} & \mathbf{h}_{k+2} & \cdots & \mathbf{h}_{k+N} \\ \mathbf{h}_{k+2} & \mathbf{h}_{k+3} & \cdots & \mathbf{h}_{k+N+1} \\ \vdots & \vdots & \ddots & \vdots \\ \mathbf{h}_{k+M} & \mathbf{h}_{k+M+1} & \cdots & \mathbf{h}_{k+N+M-1} \end{bmatrix}. \tag{1.29}$$

This is known as the truncated *Hankel matrix* for the system at the reference time
t_k. Here, M defines the time window where the dynamics of interest occur, which
are normally determined in terms of frequency bandwidth. The number of columns,
N, determines the number of measurements used to extract the dominant temporal
patterns. As a result, $N+M-1$ time samples are needed to populate the Hankel matrix.
Their dominant features are then extracted from a singular value decomposition (SVD)
with reference to the origin of times, $k=0$, that is,

$$\mathcal{H}_0 = \mathbf{U}\mathbf{\Sigma}\mathbf{V}^\top, \tag{1.30}$$

where $\mathbf{\Sigma}$ is the diagonal matrix of the (real and nonnegative) singular values, and
$\mathbf{U}$ and $\mathbf{V}$ are real orthogonal matrices whose columns are the left and right singular
vectors, respectively. The SVD is the essential algorithm in modern data-driven methods (Brunton and Kutz, 2019, Ch. 1), as it enables the reduction of high-dimensional
datasets into low-order approximations. To achieve this, the SVD representation is

partitioned such that only the largest singular values are retained. Let $\hat{\boldsymbol{\Sigma}}$ be the resulting truncated set of singular values. Since it is a diagonal matrix, it is easy to see that only the corresponding singular vectors need to be retained in a truncated approximation. This results in very narrow matrices $\hat{\mathbf{U}}$ and $\hat{\mathbf{V}}$ that approximate the Hankel matrix as $\mathcal{H}_0 \approx \hat{\mathbf{U}}\hat{\boldsymbol{\Sigma}}\hat{\mathbf{V}}^\top$. Finally, it can be shown (Juang and Pappa, 1985) that a low-order discrete-time state-space model of the system can be constructed from the SVD as

$$\hat{\mathbf{A}} = \hat{\boldsymbol{\Sigma}}^{-1/2}\hat{\mathbf{U}}^\top \mathcal{H}_1 \hat{\mathbf{V}}\hat{\boldsymbol{\Sigma}}^{-1/2},$$

$$\hat{\mathbf{B}} = \hat{\boldsymbol{\Sigma}}^{1/2}\hat{\mathbf{V}}^\top \begin{bmatrix} \mathcal{I}_u \\ \mathbf{0} \end{bmatrix},$$

$$\hat{\mathbf{C}} = \begin{bmatrix} \mathcal{I}_y & \mathbf{0} \end{bmatrix} \hat{\mathbf{U}}\hat{\boldsymbol{\Sigma}}^{1/2},$$

$$\hat{\mathbf{D}} = \mathbf{0},$$

(1.31)

where $\mathcal{I}_u$ and $\mathcal{I}_y$ are unit matrices whose dimension is the number of inputs and outputs, respectively.

Rational Interpolation

Rational interpolation methods build a state-space model from a sampled transfer function. They are therefore the frequency domain equivalent to the ERA considered above. In a similar manner to the Hankel matrix for the time-domain problem, a matrix of sampled frequency-domain data, known as the Löwner matrix, can be assembled (Mayo and Antoulas, 2007) and used in conjunction with a truncated SVD to derive a low-order internal model. This approach has been recently explored by Yue and Zhao (2020) and Quero et al. (2021) to obtain an elegant process of system identification from frequency-domain unsteady aerodynamics datasets.

In this book, however, we will restrict ourselves to the well-established approach used in aeroelasticity, known as the *rational-function approximation*. Here, an Ansatz is chosen for the FRF of the system (e.g., a structure such as Equation (1.19) with a diagonal state matrix), whose coefficients are then obtained by least-squares fitting on the sampled data. The resulting description is finally expanded to the full complex plane using analytical continuation. Further details of this method are discussed in Section 6.5.3.

1.5 Outline of This Book

This chapter has laid the groundwork for this book. We have seen that it comes in response to both some long-term trends in aircraft design to build vehicles with increasingly dominant aeroelastic effects and the more short-term disruptions associated with the search for zero-emission solutions to mitigate climate change. Our interest throughout is on the dynamics of air vehicles, and consequently, the book makes the extensive use of the standard analysis methods for dynamical systems throughout. The readers are expected to be familiar with state-space equations, transfer functions, and Fourier transforms. A basic introduction has been included in this chapter, and some additional material can be found in the appendices. A more detailed description of some other, more specific, mathematical tools of relevance to the analysis of the flexible aircraft dynamics is inserted throughout the book.

The rest of the material in this book has been structured into 10 additional chapters that collectively build a rather generic framework for modeling, simulation, and control of flexible aircraft. A consistent notation is used throughout; yet most chapters are also self-contained to enable readers to focus on specific topics. As we are often concerned here with flight in a nonstationary atmosphere, an overview of the most relevant atmospheric conditions for aircraft design is first introduced in Chapter 2. The main results of that chapter are later used in subsequent chapters to investigate the dynamic response of both the rigid and flexible air vehicles.

Chapter 3 is mostly a self-contained introduction to the problems in unsteady aerodynamics and aeroelasticity that are relevant to this book, including dynamic stability and response to atmospheric gusts. It introduces those concepts using the analytical solutions for thin airfoils of Theodorsen and Sears, which serve to build time-domain formulations of the simple aeroelastic systems and their response to atmospheric disturbances. It can be used as the basis on a short introductory course to aeroelasticity.

Chapter 4 reviews the equations of motion that describe the flight dynamics of rigid aircraft. As a characteristic feature of this book, this is done in the context of flight in a nonsteady atmosphere, with still-air conditions presented as a particular case. The general Newton–Euler equations of motions for the rigid body are derived for an aircraft with arbitrary kinematics, as well as the linearized solutions for the longitudinal and lateral problems, which are written in a state-space form. Linear quasi-steady aerodynamics is used, with the stability and control derivatives assembled into matrices of aerodynamic influence coefficients. The material is self-contained and can serve as a brief course in flight mechanics. It also introduces both inertial and body-attached reference frames that are needed again in later chapters to describe the unsteady aerodynamics and the flexible aircraft dynamics, respectively.

The effect of flexibility on the aircraft dynamics is first introduced in Chapter 5 under certain restrictive assumptions that will be systematically removed in the following chapters. In particular, Chapter 5 considers flight dynamics for the case in which there is a sufficient frequency separation between the rigid-body and the vibration characteristics of the aircraft. This situation occurs in many aircraft over a large portion of its flight and maneuvering envelope and results in the so-called elastified equations of motion. In that solution, the degrees of freedom are still those describing the rigid-body kinematics of the aircraft but with modified coefficients that account for the aeroelastic effects. This formulation is also generally valid for most static problems and is therefore used to introduce aeroelastic trim and control reversal.

A more general description is needed when the elastic deformations have a nonnegligible effect of on the inertia of the vehicle, and this will be the focus of Chapter 6. The assumption in this chapter is of small-amplitude vibrations, in which case the vehicle dynamics can be written in a compact form in modal coordinates. Consequently, we will define linear normal modes in structural dynamics and use them to project the general equations of motion of a flexible unsupported body. As we consider a larger frequency space, unsteady aerodynamics need to be considered. The focus is on the nonstationary aerodynamic forces resulting from the small-amplitude wall motions, which are obtained from applying the data-driven system identification techniques of Section 1.4.3 to suitable fluid-dynamics models. This defines the generalized aerodynamic forces associated with the modal description from results obtained from either computational fluid dynamics (using ERA) or potential flow aerodynamics (using rational interpolation). Many of the aeroelastic and flight dynamics concepts first introduced in Chapter 3 and Chapter 4, respectively, are revisited for flexible aircraft displaying dynamic couplings between its rigid and flexible modes.

While there is a restriction on the amplitude of elastic deformations in Chapter 6, the aerodynamic models are generally valid for subsonic and transonic flight. Subsequent chapters deal with arbitrary large wing deformations, although still with attached flows, as the interest is enabling deformation to achieve structural efficiency. They will, however, assume an incompressible flow regime, as most vehicles currently displaying large deformations fly at low speed. Chapter 7 reviews time-domain unsteady aerodynamic models for such conditions, with a particular focus on the unsteady vortex-lattice method. As in the previous chapter, the interest in this chapter is on the aerodynamic forces resulting from the motions of the structure, and the high computational cost of evaluating them on the general approach is lowered by considering reduced-order modeling techniques, in particular, balanced realizations of the aerodynamic equations.

The structural models for aircraft with large elastic deformations are considered in Chapter 8. As before, we restrict ourselves to a commonly found strategy that is suitable for most aircraft displaying geometric nonlinearity. In particular, the chapter presents several flexible multibody dynamic modeling approaches based on geometrically nonlinear composite beams. The mean assumption of a beam model is that the structural dynamics can be well approximated by tracking a reference line with suitable stiffness and inertial properties. In the case of composite beams, that description also allows elastic couplings between the different degrees of freedom (e.g., coupling between the bending and torsion of the wing). Three alternative solutions to the beam equations are presented, namely a conventional finite-element model based on nodal displacement and rotations and two alternative methods that use spatial and time derivatives of the displacement field as primary variables. They enable different analysis techniques for very flexible aircraft in the final chapters and are therefore jointly presented.

Chapter 9 combines the unsteady aerodynamic models of Chapter 7 and the composite beams models of Chapter 8 to describe the dynamics of very flexible aircraft. This approach naturally captures the nonlinear couplings between flight dynamics and the aeroelastic response of the vehicle, including the expense of a higher computational time. To address that, we include an approximated solution method on the basis of the modal projection of the nonlinear equations and linearized aerodynamics, which hugely reduces the computational burden on the simulation. Next, typical situations are exemplified using numerical results on several recent prototypes that were purposely built to explore geometric nonlinearities in highly efficient wings and aircraft. Finally, the chapter reviews some analysis methods in aircraft design that are strongly affected by airframe deformations (static and dynamic stability, loads, flight control, etc.) and suggests methodological improvements to accommodate the more complex physics associated with increased flexibility.

Feedback control is then considered in Chapter 10. Even though we make extensive use of methods and language of control theory throughout this book, all the previous chapters to Chapter 10 consider the open-loop dynamics of the aircraft. Here, we introduce first the common elements to a control architecture for flexible aircraft, and the focus is then on optimal control methods to determine a suitable logic. Both linear and nonlinear methods are outlined, with examples and applications in aeroelasticity and the flexible aircraft dynamics. The methodological sections of the chapter start with a summary of the linear quadratic regulator, for which some details of the solution process using Lagrange multipliers are included. That is then expanded to the nonlinear problem, which is formulated within the paradigm of model predictive control, with an efficient internal model derived in Chapter 9. Equally, the estimation problem, that is, the approximate reconstruction of the state from limited measurements is first presented for linear problems, with a discussion on Kalman filters. This is finally expanded to a nonlinear problem with an outline of the moving-horizon estimator. The strong connections between the optimal control and estimation problems for both linear and nonlinear problems are highlighted.

Finally, Chapter 11 outlines the current industrial methods for experimental modal analysis of air vehicles. Both ground and flight vibration tests are discussed, with a focus on large transport aircraft with moderately flexible wings.

2 The Nonstationary Atmosphere

2.1 Introduction

Aircraft encounter nonstill air conditions during most flight segments: side winds in takeoff and landing, atmospheric turbulence when cruising, and the wake of a second nearby vehicle in congested airspace are some typical examples. Any of those situations creates a nonstationary disturbance on the vehicle dynamics that may affect passenger comfort, pilot handling, or even aircraft safe operation. For example, the relatively large time separation between successive landings in airports is mainly defined by the time that the wakes generated during landings take to dissipate. The strength of the wake vortices is proportional to the aircraft weight, and larger aircraft consequently need longer airport slots, impacting their operation costs. Other situations, such as encounters with clear-air turbulence (CAT), which is atmospheric turbulence occurring without clouds and thunderstorms, are very difficult to detect and avoid and can have a significant impact on the safety and cost of operation. In the three decades to 2008, the Federal Aviation Authority (FAA) logged on average about 10 accidents per year[1] involving US airlines. This will unfortunately get much worse due to anthropogenic climate change. Climate models are currently predicting a 10%–40% increase in strength and a 40%–170% increase in frequency of occurrence for CAT events in North Atlantic flights by 2050 (Williams and Joshi, 2013). This book is often concerned with the dynamic response of aircraft in those conditions, and therefore we will first introduce the environment in which the various analysis models are developed in subsequent chapters.

The effect of nonstationary atmospheric conditions on the aircraft has been investigated since the dawn of aviation (Hunsaker and Wilson, 1917). The key features of the interactions between the aircraft and the atmosphere depend on both the spatial and temporal scale of the air velocity fluctuations with respect to those of the aircraft. At the lower-frequency end of the spectrum of atmospheric disturbances, changes to the incoming velocity are a quasi-steady aerodynamic effect that can be compensated by pilot (or autopilot) commands. In the low-to-mid-frequency range, the disturbance may excite the aircraft in the form of elastic vibrations, which gives rise to dynamic

---

[1] An *accident* is defined by the U.S. National Transportation Safety Board as an occurrence associated with the operation of an aircraft in which any person suffers death or serious injury, or the vehicle receives substantial damage (see www.faa.gov/travelers/fly_safe/turbulence for more information).

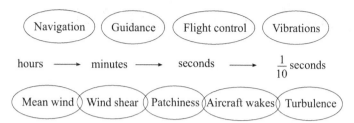

Figure 2.1 Spatial scales in atmospheric conditions and typical corresponding timescales in flight physics.

loading. This is the situation experienced by passengers of airliners in bad weather conditions (particularly those sitting in the front and rear parts of the fuselage), and it is the most challenging from the point of view of analysis and design. At high enough frequencies, the inertia of the aircraft acts like a low-pass filter and air disturbances mostly go unnoticed.

While frequency of encounter becomes the key feature in the forward flight direction, a similar analysis can be done regarding the spatial scales in the flow disturbance in the spanwise and vertical directions: If they are large compared to the aircraft length scales, then the aircraft can be considered to be subjected to a uniform instantaneous flow velocity; however, if they are comparable with the aircraft dimensions, then the spatial distribution of flow velocities across the aircraft can have an important effect on its response and needs to be analyzed. For wavelengths of excitations that are very small compared to the aircraft, the disturbance is averaged out to zero and has no direct effect on the aircraft dynamics, although it may still affect aerodynamics features, such as the transition from laminar to turbulent flow over the wings (Reeh and Tropea, 2015).

In general, we can identify multiple relevant nonstationary wind situations that affect aircraft in flight. They are shown diagrammatically in Figure 2.1, which includes their typical frequency length scales for subsonic aircraft, and will be discussed at length in this chapter. They are:

1. **Mean wind**. Aircraft often fly in a steady wind due to very large scale atmospheric phenomena. This could be head or tail wind, crosswind, or any combination of the three, and it is very important in aircraft guidance and navigation. Its mean values, in conjunction with weather forecasts, are regularly used in route planning. They have, however, no direct effect on the aircraft flight dynamics if the inertial reference frame is chosen to move with the mean wind speed. Crosswind, resulting from a dominant direction in the atmospheric boundary layer (ABL), needs to be considered during takeoff and landing.
2. **Wind shear**. This includes changes in wind speed (with changes either in vertical or horizontal components, meaning that the spatial gradient of the wind velocity is nonzero, and therefore the wind is "sheared") due to local weather conditions or the presence of ground obstacles. A very important wind shear condition in aircraft design and operations is the local "patchiness" (MIL-HDBK-1797, 1997),

which is typically investigated using *discrete gusts*, or "bumps in the air" (see Section 2.4.2), whose severity varies across the flight envelope.

3. **Aircraft wakes**. This is a man-made but no less important source of disturbance in aircraft flight. Wake encounters occur when an aircraft crosses the wake of another one (or its own one during a long or very tight maneuver). They are particularly important in congested airspace, particularly near airports, where expected wake dissipation times are used to define the spacing between aircraft in takeoff and landing.

4. **Atmospheric turbulence**. This is due to natural variability (or randomness) of the velocity field in all fluids. Its intensity may vary depending on the atmospheric conditions and they need to be investigated throughout the flight envelope. It is often referred to as *clear-air turbulence* in weather forecasting (Sharman et al., 2012) and as *continuous turbulence* in aircraft loads analysis.

Local changes in air speed result in changes of the instantaneous velocity vector between the aircraft and the surrounding air, and therefore of the aerodynamic forces on the aircraft. A common assumption when studying all the previous scenarios is that this *background flow* is not affected by the presence of the aircraft. This will be true for the flow features of larger scales, but it is increasingly less valid at smaller wavelengths. Moreover, the aircraft itself will dynamically respond to the change of atmospheric conditions (e.g., by deforming its wings or modifying its trajectory), and this will also modify its instantaneous aerodynamic loads. Therefore, to investigate the dynamic loads on a vehicle in a nonstationary atmosphere, one needs to consider not only the external disturbance but also the resulting vehicle dynamics. Many of these interactions will be discussed in subsequent chapters in some detail, but, in general, it can be said that unsteady atmospheric conditions may result in one or several of the following situations:

1. Mission replanning or *trajectory rerouting* to avoid heavy turbulence areas. This is imperative with severe weather phenomena and relies both on forecasting and real-time updates from ground control, pilot reports (so-called PIREP), and/or on-board radar.

2. *Poor control response*, thus requiring bigger effort from the pilot with associated increase in the pilot's workload.

3. *Degradation of ride quality* and overall passenger comfort, either from the excitation of the airframe structural modes, thus making tasks like reading/sleeping difficult, or the generation of rigid-body motions that result in motion sickness.

4. Flight control system saturation and potential stability risks (e.g., phase and/or gain margins may be exceeded), which may imply a temporary *loss of vehicle control*.

5. *Structural integrity risks*, due to either reduced fatigue life or resulting loads above the strength limit of the airframe. Gust and continuous turbulence loading is critical in the structural sizing of many parts of the airframe of modern aircraft.

The rest of this chapter describes the various situations in which the incoming flow on an air vehicle may be nonstationary, as well as the mathematical models typically used to incorporate them in engineering analysis. Later in this book, we use them to describe the disturbance inputs that go into the aircraft equations of motion in order to, first, investigate their effect on the vehicle dynamics, and second, develop correcting measures via either vehicle redesign or *aerodynamic load control*. The starting point, however, is the reference still-air conditions in the atmosphere, which are discussed first.

2.2   The Standard Atmosphere

The aerodynamic and propulsive performance of aircraft in flight depends on the local properties of the atmosphere. However, even in calm atmosphere those conditions vary with the position on the Earth, time of the day, seasons, and weather patterns. To establish a common reference, the International Civil Aviation Organization adopted a *standard atmosphere* in 1952, known as the International Standard Atmosphere (ISA). It defines a unique (average) air temperature profile with altitude (ICAO, 1993). In the ISA, the troposphere is the lower layer of the atmosphere up to 11 km from an averaged sea level mark. In this model of the troposphere, temperature drops linearly from 15 °C at sea level to −56.5 °C at 11 km, that is, it assumes a drop of 6.5 °C for every 1,000 m. The next layer is known as the stratosphere and extends in the ISA model from 11 to 50 km. There, the atmospheric temperature is considered to be constant (at −56.5 °C) until 20 km and then raise again to −44.5 °C at 32 km, that is, an increase of 1 °C for every 1000 m. Powered flight above this altitude (roughly 100,000 ft.) has been very rarely attempted. For most practical purposes, the Earth's gravity can be considered constant within the 0–32 km altitude range.

Flight performance is particularly affected by the changes in air density with altitude, which in the ISA drops from 1.225 kg/m^3 at sea level to just 0.0167 kg/m^3 at 100,000 ft. A less dramatic but equally important drop occurs on the speed of sound, which goes from 340 to 302 m/s between those two altitudes, and therefore reduces the maximum flight speed before shock waves start appearing in the fluid. Expressions for the evolution of those magnitudes with altitude can be found in most introductory books to aeronautics (e.g., Brandt et al., 2006, Ch. 2) and are readily available in online calculators.

From Bernoulli's principle, it is known that aerodynamic forces on an aircraft flying at constant speed scale with the local *dynamic pressure*, that is, $\frac{1}{2}\rho V^2$, where V is the airspeed. It is then customary to correct for changes of air density when reporting aircraft speed. The true airspeed (V_{TAS}) of an aircraft is defined as its actual speed with respect to the airmass in which the vehicle flies. The corresponding equivalent airspeed (V_{EAS}) is the flight speed that would generate the same dynamic pressure as V_{TAS} at sea level, that is,

$$\frac{1}{2}\rho_0 V_{EAS}^2 = \frac{1}{2}\rho V_{TAS}^2 \tag{2.1}$$

and

$$V_{EAS} = \sqrt{\frac{\rho}{\rho_o}} V_{TAS}. \tag{2.2}$$

Therefore, the equivalent airspeed is effectively a normalized dynamic pressure, and it is often used to compare flight performance parameters corresponding to different altitudes.

Two final remarks are relevant here. First, it is important to note that most weather events, which are behind many of the nonstationary atmospheric conditions discussed in this chapter, occur in the troposphere. The stratosphere, on the contrary, is characterized by high-speed yet steady winds that are driven by the Earth's rotation (a noninertial reference frame). Those high-altitude wind currents result in the stark flight time differences of flights in opposite directions in long transcontinental routes. Second, it is somewhat fortunate that density drops with altitude. This means that aircraft can sustain flight at relatively low velocities in the ground, while the low density at high altitude implies that the aircraft must fly at a much higher speed to achieve the same dynamic pressure. The optimum flight altitude for cruise is sought from a trade-off between maximum speed and minimum drag (in practice, minimum fuel cost), while being sufficiently high to avoid nonstationary atmospheric conditions as much as possible.

2.3   Continuous Turbulence

In the study of the dynamics of air vehicles, *continuous turbulence response* refers to the effect on the aircraft of an encounter with atmospheric turbulence. The exact instantaneous values of the turbulent velocity field impinging on the aircraft are fundamentally unpredictable, and consequently the study of continuous turbulence response is based on statistical analysis. The assumption is that the profile of incoming velocities in atmospheric turbulence is a random (or stochastic) process, which implies that individual measurements under the same conditions would not, in general, give the same results – contrary to a deterministic process. Furthermore, it can often be assumed that it is a *stationary random process*, that is, a process whose statistical properties do not change with time (when sampled for sufficiently long times). Regions of high turbulence intensity typically appear in patches during flight, but those are assumed to be sufficiently long such that the statistical properties of the flow do not change with the spatial location either.[2] We say in that case that the random process is *statistically homogeneous*.

The mathematical description of atmospheric turbulence and its impact on the aircraft dynamics requires a basic understanding of some concepts in probability theory on stochastic system dynamics. They are summarized in the next two sections, before we introduce the one- and two-dimensional (1-D and 2-D) models used for analysis of continuous turbulence response.

---

[2] Although this will not be discussed here, the nominal time that an aircraft is expected to be in regions with high turbulence is a critical aspect to consider in its design, as it defines potential damage by *structural fatigue* (Hoblit, 1988, Ch. 4). Moreover, in-flight monitoring of high turbulence events in the latest generation aircraft allows for adjusting processes for maintenance, repair, and overhaul (MRO) according to the expected cumulative loads on each individual aircraft.

2.3.1   Random Processes

Second-Moment Analysis

Let $\big(w_1(t), w_2(t), \ldots, w_N(t)\big)$ be N time histories obtained from the *realization* of N independent stationary random processes. They may represent the time evolution of the local velocities at different points in the atmosphere or the components along different spatial directions of the velocity vector at a single location. We can define the *continuous random variable* $\mathbf{w}(t)$ as the column vector function of dimension N obtained by the assembly of those time functions. Every realization of the random variable would define different time histories; however, they will always conform to a certain probability function and we are therefore interested in their statistical characterization. This will be done by its first- and second-order moments, which correspond to averages of individual and pairs of random variables, respectively.

We define first the *expected value*, or mean, as the average of all possible values of the random variable, that is,

$$\bar{\mathbf{w}} = \mathrm{E}[\mathbf{w}(t)] = \lim_{T \to \infty} \frac{1}{2T} \int_{-T}^{T} \mathbf{w}(t)\, \mathrm{d}t. \tag{2.3}$$

Here, we have also introduced the expectation operator $\mathrm{E}[\bullet]$. Measured signals are given as discrete series of finite size, n, typically with a constant sampling rate, Δt, and therefore at times $(\Delta t, 2\Delta t, \ldots, n\Delta t)$. The expected value is then estimated, for each component w_i of $\mathbf{w}$, as

$$\bar{w}_i = \mathrm{E}[w_i(t)] \approx \frac{1}{n} \sum_{k=1}^{n} w_i(k\Delta t), \quad \text{for } i = 1, \ldots, N. \tag{2.4}$$

The second moment is the so-called *correlation* function, which for a stationary random process is defined as the expected value of the outer product between realizations of the random variable $\mathbf{w}(t)$ and its value shifted in time, that is,

$$\phi_w(\tau) = \mathrm{E}[\mathbf{w}(t+\tau)\mathbf{w}^\top(t)]. \tag{2.5}$$

The correlation is therefore a matrix function of dimension $N \times N$ that provides information about the relation between all pairs of instances of the random variable, both with respect to the different components in the vector (via the outer product) and with respect to different points in time (via the time *lag*, τ). Noting that for a stationary random process one can shift the origin of time without affecting its statistical properties, the correlation function with a negative lag satisfies

$$\phi_w(-\tau) = \mathrm{E}[\mathbf{w}(t-\tau)\mathbf{w}^\top(t)] = \mathrm{E}[\mathbf{w}(t-\tau+\tau)\mathbf{w}^\top(t+\tau)]$$
$$= \mathrm{E}[\mathbf{w}(t)\mathbf{w}^\top(t+\tau)] = \phi_w^\top(\tau). \tag{2.6}$$

It is worth remarking that the convention followed in this book is not universally accepted. Etkin (1972) and Burl (1999), for example, define the correlation function as the transpose of Equation (2.5), resulting in the opposite convention for a positive lag.

For two signals w_i and w_j given in the discrete form, and for which n samples are known with a constant time step Δt, the (unbiased) correlation function is defined at the sampling times as

$$\phi_{ij}(l\Delta t) \approx \frac{1}{n-l}\sum_{k=1}^{n-l} w_i(k\Delta t + l\Delta t)w_j(k\Delta t), \tag{2.7}$$

for nonnegative integer values of l and n, with $0 < l < n$ (and typically $l \ll n$). Note that we have dropped the subindex $\bullet_w$ to indicate the components i and j of the matrix function. Equation (2.6) can then be used to obtain estimates of the correlation for negative lags.

The diagonal terms of the matrix defined by Equation (2.5), that is, functions of the form $\phi_{ii}(\tau)$, are referred to as the *autocorrelation* functions. They give information on each individual component of the random variable.[3] The off-diagonal terms, that is, the correlation between any two components i and j, with $i \neq j$, of the random variable, $\phi_{ij}(\tau)$, are usually referred to as *cross-correlation* functions. Two random signals $w_1(t)$ and $w_2(t)$ are said to be uncorrelated if $E[w_1(t)w_2(t)] = E[w_1(t)]E[w_2(t)]$ or equivalently if $\phi_{12}(0) = \bar{w}_1\bar{w}_2$. Furthermore, if any of the uncorrelated signals has a zero mean, their cross-correlation is zero.

The values taken by a random variable at times that are sufficiently far apart are uncorrelated, that is, $\lim_{\tau \to \infty} \phi_w(\tau) = E[\mathbf{w}(t)]E[\mathbf{w}(t)]^\top$. Note, however, that in practical cases, τ does not need to be very large at all for this asymptotic behavior to be found. Therefore, there is a characteristic timescale associated with the signal correlation found within a random variable. It is referred to as the *integral timescale*, and it is associated with each individual component of the random variable from its autocorrelation function:

$$T_i = \frac{1}{\phi_{ii}(0)}\int_0^\infty \phi_{ii}(\tau)\,\mathrm{d}\tau. \tag{2.8}$$

A graphical representation of the integral timescale is shown in Figure 2.2. The integral timescale T_i for a signal $w_i(t)$ is obtained from a rectangle of side $\phi_{ii}(0)$ that defines the same area as that under the corresponding autocorrelation function. Finally, we define the time delay between each pair of component signals in $\mathbf{w}(t)$ as the argument of the maximum of their cross-correlation function:

$$\tau_{ij} = \arg\max_\tau \phi_{ij}(\tau), \quad \text{for } i \neq j. \tag{2.9}$$

Both integral timescale and time delay play important roles in the characterization of random processes, in general, and of atmospheric turbulence, in particular. The first one gives a typical timescale in which there is some structure inside the signal and, therefore, the relevant timescale of the physical phenomena that are described by that signal. The second is often linked to having different measurements of the same (or related) physical phenomena with sensors that are not collocated (Beck, 1981). It is indeed easy to see that $\tau_{ji} = -\tau_{ij}$, that is, if $w_i(t)$ is "behind" $w_j(t)$, then $w_j(t)$ is "ahead" of $w_i(t)$.

---

[3] Note that we use the standard definition used in linear systems theory. In fluid dynamics (see, e.g., Pope, 2000), the autocorrelation is normally defined on the fluctuating velocities and normalized such that its maximum value is 1.

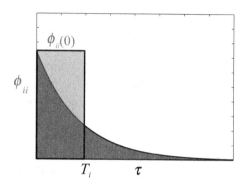

Figure 2.2 Graphical interpretation of the integral timescale T_i: the area under the curve defined by $\phi_{ii}(\tau)$ has the same area as that of the rectangle defined by T_i and $\phi_{ii}(0)$.

The correlation function with zero lag, $\tau = 0$, gives the usual definitions of second-order moments used in statistics, which we will often use to characterize random signals. First, we can obtain the *mean square value* of the signals as the expected value of the inner product of the signal $\mathbf{w}(t)$, that is,

$$\sigma_w^2 = \mathrm{E}[\mathbf{w}^\top(t)\mathbf{w}(t)] = \mathrm{tr}(\boldsymbol{\phi}_w(0)), \tag{2.10}$$

where the last equality comes directly from the definition of the correlation function. The mean square value is, therefore, also the sum of the terms of the diagonal (the trace) of the correlation function for $\tau = 0$. This defines a norm, that is, a non-negative number that gives a measure of the amplitude of the random signal. Note that for a univariate signal, $w(t)$, the mean square value is the length of the vertical side of the rectangle in Figure 2.2.

We finally define the *autocorrelation matrix* from the values of the correlation function with zero lag:

$$\boldsymbol{\Sigma}_w = \mathrm{E}[\mathbf{w}(t)\mathbf{w}^\top(t)] = \boldsymbol{\phi}_w(0), \tag{2.11}$$

which is a symmetric matrix defined on each pair of components of the random variable. If written in terms of fluctuations with respect to the mean, it becomes the *covariance matrix* of the random process, $\mathrm{E}[(\mathbf{w}(t) - \bar{\mathbf{w}})(\mathbf{w}(t) - \bar{\mathbf{w}})^\top]$, sometimes also referred to as the covariance intensity matrix. Note finally that covariance and autocorrelation matrices coincide when the mean is zero, $\bar{\mathbf{w}} = \mathbf{0}$.

Example 2.1 Second-Moment Analysis of Two Random Signals. Consider a situation such as that described in Figure 2.3, where two measurements are taken in a turbulent flow at two different locations. They correspond to two random signals with zero mean, which are shown in Figure 2.4. Both signals need to be captured over a long enough period of time for any result to be statistically significant, and only the first 10 seconds are shown in the figure.

Using Equation (2.5), we can now assess their temporal correlation. This is shown in Figure 2.5, which shows the correlation function between both signals

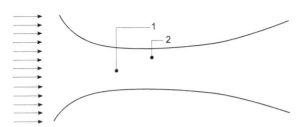

Figure 2.3 Two measurement points in a flow.

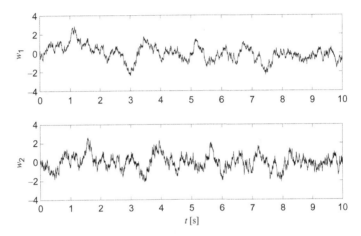

Figure 2.4 Two random signals. [02NonSteadyAtmosphere/randomsignals.m]

(a 2 × 2 matrix). Note how all functions go to zero in a rather short period of time, which is a characteristic feature of random processes. The functions in the diagonal show the autocorrelation function for each individual signal. Their peak value is at $\tau = 0$ and gives the mean square value of each signal. It can be seen how $\phi_{11}(\tau)$ has a larger area integral than $\phi_{22}(\tau)$, which indicates that the characteristic timescale in the internal flow structures w_1 is larger than that in w_2, which may be due, for example, to one probe being within or near the boundary layer. Equation (2.8) for this problem yields $T_1 = 0.43$ s and $T_2 = 0.10$ s.

The off-diagonal functions in Figure 2.5 are the cross-correlations between both signals. They satisfy $\phi_{12}(\tau) = \phi_{21}(-\tau)$, and in this particular case, there is a rather strong correlation between both random signals, with a time delay, determined by the time for the maximum of the curve (Equation (2.9)), as $\tau_{21} = 0.5$ s (or $\tau_{12} = -0.5$ s). This indicates that, in this example, w_2 likely corresponds to instantaneous measurements on the same physical process of w_1, since they are highly correlated signals. However, at sensor 1, the fluid arrives 0.5 s earlier than at sensor 2 ($\tau_{12} < 0$). This corresponds to a time history of w_2 that is shifted to the right (toward positive t) with respect to the time history in the first probe, w_1.

Finally, the autocorrelation matrix is defined by the values at $t = 0$ of the autocorrelation function. Here it is $\Sigma_w = \begin{pmatrix} 0.81 & 0.01 \\ 0.01 & 0.56 \end{pmatrix}$. The mean square value is $\sigma_w = 1.37$. Because of the delay between both functions, the autocorrelation matrix does not

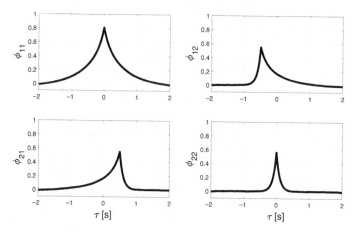

Figure 2.5 Correlation function for the signals of Figure 2.4.

capture the strong correlation of both signals. Indeed, the usual practice would be first to remove the delay by shifting the signal in sensor 2 to be synchronized with that of sensor 1, and then to compute the autocorrelation matrix of the resulting signal.

Spectral Description

It is often more interesting to investigate the frequency–domain characteristics of stationary random processes. However, since a stationary random signal is not absolutely integrable (see Appendix A), its Fourier transform does not exist. Instead, the Fourier transform of its second moment, proved to exist by the Wiener–Khinchin theorem (see, e.g., Wirsching et al., 1995, Ch. 5), is used.

The mean value, Equation (2.3), is a constant and does not bring any useful spectral information. This is obtained from the Fourier transform of the correlation function defined in Equation (2.5). The resulting metric is the *spectral density*, which is defined as

$$\Phi_w(\omega) = \frac{1}{\pi} \int_{-\infty}^{\infty} \phi_w(\tau) e^{-i\omega\tau} \, d\tau. \tag{2.12}$$

This definition includes a factor of 2 in the Fourier transform, Equation (A.1). However, this is the definition commonly used in both fluid mechanics (Pope, 2000) and mechanical vibrations (Hoblit, 1988; Wirsching et al., 1995), and it is, therefore, also used here. As with the correlation, the spectral density is a matrix function. From Equation (2.6) and the definition above, it is easy to see that it is a Hermitian matrix, that is, it is a square matrix equal to its conjugate transpose, $\Phi_w^*(\omega) = \Phi_w(\omega)$. It can also be seen that Equation (2.6) implies $\Phi_w(-\omega) = \Phi_w^\top(\omega)$.

The elements in its diagonal are the Fourier transform of the autocorrelation functions, and they are referred to as the *power spectral density* (PSD) of the components of **w**. Since $\phi_{ii}(-\tau) = \phi_{ii}(\tau)$, they are real functions that can be written as

$$\Phi_{ii}(\omega) = \frac{2}{\pi} \int_{0}^{\infty} \phi_{ii}(\tau) \cos(\omega\tau) \, d\tau. \tag{2.13}$$

The PSD of a univariate time series is an even function of angular frequency that is normally given only for nonnegative angular frequencies, $\omega \geq 0$. It characterizes the frequency content for each measurement in a stochastic system and is used in Section 2.3.3 to describe atmospheric turbulence models.

The off-diagonal terms in Equation (2.12), corresponding to the Fourier transform of the cross-correlation functions of the random process $\mathbf{w}(t)$, are, in general, complex numbers and are seldom used directly. Instead, we define the *coherence*, $\gamma_{ij}(\omega)$ (often also written as coh_{ij}), between two signals $w_i(t)$ and $w_j(t)$ as the squared modulus of the cross-spectral density function, $\left|\Phi_{ij}(\omega)\right|^2 = \Phi_{ij}(\omega)\Phi_{ij}(-\omega)$, normalized by the corresponding PSDs, that is,

$$\gamma_{ij}(\omega) = \frac{\left|\Phi_{ij}(\omega)\right|^2}{\Phi_{ii}(\omega)\Phi_{jj}(\omega)}. \tag{2.14}$$

The coherence is a real function that satisfies $0 \leq \gamma_{ij}(\omega) \leq 1$, with the limit values corresponding to uncorrelated ($\gamma_{ij} = 0$) or perfectly correlated ($\gamma_{ij} = 1$) signals in all or in part of the frequency spectrum. The coherence is used in Section 2.3.4 to relate the instantaneous velocities at two different points in the atmosphere.

To complete the discussion in this section, we note that the inverse Fourier transform of the spectral density, defined as in Equation (A.2) in the appendices, recovers the correlation function

$$\phi_w(\tau) = \frac{1}{2}\int_{-\infty}^{\infty} \Phi_w(\omega) e^{i\omega\tau}\, d\omega. \tag{2.15}$$

The symmetry properties of the spectral density lead to two particular solutions of interest. First, the autocorrelation matrix (or the covariance if the mean is zero) can be directly obtained from the spectral density as

$$\Sigma_w = \phi_w(0) = \frac{1}{2}\int_{-\infty}^{\infty} \Phi_w(\omega)\, d\omega = \frac{1}{2}\int_0^{\infty}\left(\Phi_w(\omega) + \Phi_w^{\top}(\omega)\right) d\omega, \tag{2.16}$$

which is always real valued. Second, the autocorrelation function satisfies

$$\phi_{ii}(\tau) = \int_0^{\infty} \Phi_{ii}(\omega)\cos(\omega\tau)\, d\omega. \tag{2.17}$$

Finally, setting $\tau = 0$ and summing in i, we have that the integral of the PSD is linked to the mean square value, Equation (2.10), as

$$\sigma_w^2 = \int_0^{\infty} \mathrm{tr}(\Phi_w(\omega))\, d\omega. \tag{2.18}$$

The physical interpretation of this expression is that the net power (total energy per unit time) of a given stationary process is the integral of its power density across all angular frequencies. It can be finally noted that the scaling of the Fourier transform in Equation (2.12) results in a definition of the PSD that gives the power per unit frequency using only positive frequencies.

Example 2.2 Spectral Analysis of Two Random Signals.   Consider again the two random signals introduced in Figure 2.4. The Fourier transform of their correlation

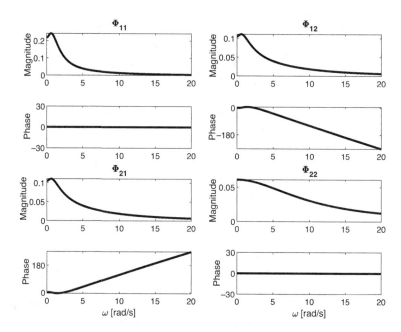

Figure 2.6 Spectral density for the signals given in Figure 2.4.

function, Figure 2.5, gives the spectral density for this random variable, as shown in Figure 2.6.

The terms in the diagonal are the PSD of $w_1(t)$ and $w_2(t)$. The area integral for those functions is equal to $\sigma_1^2 = 0.81$ and $\sigma_2^2 = 0.56$, respectively, according to Equation (2.18). As indicated in the analysis of their autocorrelation functions, $w_2(t)$ has a larger frequency bandwidth (but also smaller amplitudes of the PSD, since the area is constant) than $w_1(t)$. The off-diagonal functions in Figure 2.6 are complex functions. In this example, the coherence is exactly one for all frequencies.

2.3.2   Response of Linear Systems to Random Excitations

In this book, we are often concerned with dynamics under random excitations, in general, and the aircraft dynamics under atmospheric disturbances, in particular. We have already introduced continuous-time LTI systems in Section 1.4.2, for which we need now statistical properties when the input is a random variable. We restrict ourselves here to LTI systems with no feedthrough, initially at rest, $\mathbf{x}(0) = \mathbf{0}$, whose dynamics are given in a state-space form as

$$
\begin{aligned}
\dot{\mathbf{x}} &= \mathbf{A}\mathbf{x} + \mathbf{B}\mathbf{u}, \\
\mathbf{y} &= \mathbf{C}\mathbf{x}.
\end{aligned}
\tag{2.19}
$$

It will be further assumed that the system is asymptotically stable, that is, that all the eigenvalues of the system matrix $\mathbf{A}$ have a negative real part.

Figure 2.7 Linear system with a white-noise input.

White Noise

Consider first the case in which the only input is a particular random process known as *white noise*. A white noise is a signal with zero mean such that its realizations are uncorrelated even for a very small lag τ. We will define more generally a vector variable of uncorrelated white noise inputs $\mathbf{n}(t)$, with $\mathrm{E}[\mathbf{n}(t)] = 0$, and $\phi_n(\tau) = \Sigma_n \delta(\tau)$, with Σ_n being a diagonal matrix of positive numbers, which is normally referred to as the *intensity* of the white noise, and $\delta(t)$ is the Dirac delta function.

From Equation (2.12), the spectral density of a white noise process is constant with the angular frequency, that is, $\Phi_n(\omega) = \Sigma_n/\pi$. This implies that a white noise has equal power density at all frequencies (and indeed its name comes from an analogy with white light).[4] However, this also means that the integral in Equation (2.18) results in an infinite value or, equivalently, infinite net power. White noise signals have very convenient mathematical properties, but they do not exist in the physical world. In practice, white noise is defined over a large enough bandwidth to cover all frequencies of interest, and it is zero at higher frequencies.

We consider now the response to the linear system, Equation (2.19), to an input $\mathbf{u}(t) = \mathbf{n}(t)$, as illustrated in Figure 2.7.

Note first that, as the expected value of the white noise is zero, this will also be the expected value of the state from Equation (1.11) and, consequently, of the output, $\mathrm{E}[\mathbf{y}(t)] = \mathbf{0}$. The relevant metrics will again come from the second-order moments.

Output Correlation Matrix and the Lyapunov Equation

We obtain first the *output correlation matrix*, using the definition of Equation (2.11). Note that since this is a problem with zero mean, the output correlation matrix is also the output covariance matrix. With reference to Figure 1.10, it is convenient here to use the impulse response of the LTI system given in Equation (2.19). From Equation (1.13), the impulse response is $\mathbf{h}(t) = \mathbf{C}e^{\mathbf{A}t}\mathbf{B}$, and the output correlation matrix is obtained from convolution as

$$
\begin{aligned}
\Sigma_y &= \mathrm{E}[\mathbf{y}(t)\mathbf{y}^\top(t)] \\
&= \mathrm{E}\left[\left\{\int_0^t \mathbf{h}(t-\tau_1)\mathbf{n}(\tau_1)\,\mathrm{d}\tau_1\right\}\left\{\int_0^t \mathbf{h}(t-\tau_2)\mathbf{n}(\tau_2)\,\mathrm{d}\tau_2\right\}^\top\right] \\
&= \mathrm{E}\left[\int_0^t\int_0^t \mathbf{h}(t-\tau_1)\mathbf{n}(\tau_1)\mathbf{n}^\top(\tau_2)\mathbf{h}^\top(t-\tau_2)\,\mathrm{d}\tau_1\,\mathrm{d}\tau_2\right] \\
&= \lim_{t\to\infty}\int_0^t\int_0^t \mathbf{h}(t-\tau_1)\mathrm{E}\left[\mathbf{n}(\tau_1)\mathbf{n}^\top(\tau_2)\right]\mathbf{h}^\top(t-\tau_2)\,\mathrm{d}\tau_1\,\mathrm{d}\tau_2,
\end{aligned}
\tag{2.20}
$$

---

[4] This more intuitive property is often used to define a white noise signal in the literature, instead of via the correlation function as was done here. Both are completely equivalent.

where we have finally interchanged the order of the integral and the expectation operator. From the properties of the white noise, it becomes

$$E[\mathbf{n}(\tau_1)\mathbf{n}^\top(\tau_2)] = E[\mathbf{n}(\tau_1)\mathbf{n}^\top(\tau_1 + (\tau_2 - \tau_1))] = \Sigma_n \delta(\tau_1 - \tau_2), \qquad (2.21)$$

and therefore the previous expression becomes

$$\Sigma_y = \lim_{t \to \infty} \int_0^t \mathbf{h}(t - \tau) \Sigma_n \mathbf{h}^\top(t - \tau)\, d\tau. \qquad (2.22)$$

This limit only exists if the system is asymptotically stable, which was an initial assumption in this section. Equation (2.22) can be directly integrated, but this is normally very costly. Instead, one can obtain a more useful expression by some algebraic manipulations. First, we note that the output and state correlation system matrices can be related by

$$\Sigma_y = E[\mathbf{y}(t)\mathbf{y}^\top(t)] = \mathbf{C}E[\mathbf{x}(t)\mathbf{x}^\top(t)]\mathbf{C}^\top = \mathbf{C}\Sigma_x\mathbf{C}^\top. \qquad (2.23)$$

Next, we focus on the state-correlation matrix that from Equations (1.13) and (2.22) is given by

$$\Sigma_x = \lim_{t \to \infty} \int_0^t e^{\mathbf{A}(t-\tau)} \mathbf{B}\Sigma_n\mathbf{B}^\top e^{\mathbf{A}^\top(t-\tau)}\, d\tau, \qquad (2.24)$$

which, after differentiation with time, results in

$$\mathbf{0} = \frac{d}{dt} \lim_{t \to \infty} \int_0^t e^{\mathbf{A}(t-\tau)} \mathbf{B}\Sigma_n\mathbf{B}^\top e^{\mathbf{A}^\top(t-\tau)}\, d\tau. \qquad (2.25)$$

The order of the differentiation and the limit can now be exchanged, and using Leibniz's rule for differentiation under the integral sign, one has

$$\begin{aligned}
\mathbf{0} = &\lim_{t \to \infty} e^{\mathbf{A}(t-t)} \mathbf{B}\Sigma_n\mathbf{B}^\top e^{\mathbf{A}^\top(t-t)} \\
&+ \lim_{t \to \infty} \int_0^t \mathbf{A}e^{\mathbf{A}(t-\tau)} \mathbf{B}\Sigma_n\mathbf{B}^\top e^{\mathbf{A}^\top(t-\tau)}\, d\tau \\
&+ \lim_{t \to \infty} \int_0^t e^{\mathbf{A}(t-\tau)} \mathbf{B}\Sigma_n\mathbf{B}^\top e^{\mathbf{A}^\top(t-\tau)} \mathbf{A}^\top\, d\tau.
\end{aligned} \qquad (2.26)$$

Noting that the integrals can be written in terms of Equation (2.24) and reordering the terms, this equation finally results in

$$\mathbf{A}\Sigma_x + \Sigma_x\mathbf{A}^\top + \mathbf{B}\Sigma_n\mathbf{B}^\top = \mathbf{0}, \qquad (2.27)$$

which is a linear algebraic expression that determines the state-correlation matrix in response to a white noise. An equation of this form is known as a *Lyapunov equation*.[5] It is a matrix equation whose unknowns are the coefficients in the symmetric matrix Σ_x. Several algorithms are readily available to solve this equation (Penzl, 1998), but they will not be discussed here.

In summary, given a linear time-invariant system with no feedthrough, subject to a (in general, multivariate) white noise signal of known intensity (determined by

---

[5] This is the continuous-time Lyapunov equation. See Section 7.4.1 for discrete-time systems.

its covariance matrix, $\boldsymbol{\Sigma}_n$), solving Equations (2.23) and (2.27) uniquely determines the covariance of the system outputs. The importance of this result is that we can directly determine the statistical characteristics in the system dynamics without having to march it forward in time. This is critical in airframe design, as it dramatically accelerates the numerical evaluation of the vehicle response to atmospheric turbulence (Hoblit, 1988).

Output Spectral Density

In order to obtain the spectral response to a random excitation, it is convenient to consider first the system output to a white noise. This requires the evaluation of the output correlation function, which is obtained in the same manner as the output correlation matrix obtained from Equation (2.24). However, to consider the whole range of values of the time lag τ, we consider now an integration time spanning from $-\infty$ to $+\infty$, that is,

$$\phi_y(\tau) = \mathrm{E}[\mathbf{y}(t+\tau)\mathbf{y}^\top(t)] = \int_{-\infty}^{\infty} \mathbf{h}(t+\tau-\tau)\boldsymbol{\Sigma}_n\mathbf{h}^\top(t-\tau)\,\mathrm{d}t, \qquad (2.28)$$

and the corresponding spectral density is given by Equation (2.12),

$$\begin{aligned}\boldsymbol{\Phi}_y(\omega) &= \frac{1}{\pi} \int_{-\infty}^{\infty}\int_{-\infty}^{\infty} \mathbf{h}(t)\boldsymbol{\Sigma}_n\mathbf{h}^\top(t-\tau)e^{-\mathrm{i}\omega\tau}\,\mathrm{d}t\,\mathrm{d}\tau \\ &= \frac{1}{\pi} \int_{-\infty}^{\infty}\int_{-\infty}^{\infty} \mathbf{h}(t)e^{-\mathrm{i}\omega t}\boldsymbol{\Sigma}_n\mathbf{h}^\top(t-\tau)e^{\mathrm{i}\omega(t-\tau)}\,\mathrm{d}t\,\mathrm{d}\tau \qquad (2.29) \\ &= \frac{1}{\pi} \int_{-\infty}^{\infty} \mathbf{h}(t)e^{-\mathrm{i}\omega t}\boldsymbol{\Sigma}_n\left\{\int_{-\infty}^{\infty} \mathbf{h}^\top(t-\tau)e^{\mathrm{i}\omega(t-\tau)}\,\mathrm{d}\tau\right\}\mathrm{d}t.\end{aligned}$$

In the last expression, we have exchanged the order of integration. Using now Equation (1.20) and knowing that $\boldsymbol{\Phi}_n = \boldsymbol{\Sigma}_n/\pi$, we can recursively solve both integrals as

$$\boldsymbol{\Phi}_y(\omega) = \int_{-\infty}^{\infty} \mathbf{h}(t)e^{-\mathrm{i}\omega t}\boldsymbol{\Sigma}_n\mathbf{H}^\top(-\mathrm{i}\omega)\,\mathrm{d}t = \mathbf{H}(\mathrm{i}\omega)\boldsymbol{\Phi}_n\mathbf{H}^\top(-\mathrm{i}\omega). \qquad (2.30)$$

The last term is the conjugate transpose of the admittance, and so we finally have

$$\boldsymbol{\Phi}_y(\omega) = \mathbf{H}(\mathrm{i}\omega)\boldsymbol{\Phi}_n\mathbf{H}^*(\mathrm{i}\omega). \qquad (2.31)$$

This solution can be generalized into the *spectral factorization theorem* (Anderson and Moore, 1979). It states that the spectral density of any random process, $\mathbf{w}(t)$, can be written in the following form:

$$\boldsymbol{\Phi}_w(\omega) = \mathbf{G}_{wn}(\mathrm{i}\omega)\boldsymbol{\Omega}\mathbf{G}_{wn}^*(\mathrm{i}\omega), \qquad (2.32)$$

with $\mathbf{G}_{wn}$ being a square transfer function with stable poles and zeros (i.e., minimum phase) and $\boldsymbol{\Omega}$ a symmetric positive definite matrix. Consequently, any stationary random process, $\mathbf{w}(t)$, can be obtained from the output of a filter $\mathbf{G}_{wn}(\mathrm{i}\omega)$ whose input is a white noise with (constant) spectral density $\boldsymbol{\Omega}$. The filter $\mathbf{G}_{wn}(\mathrm{i}\omega)$ is often referred to as either a *coloring filter* or a spectrum shaping filter, and the output is a *colored noise*. This property will allow building synthetic turbulence models in Section 2.3.3.

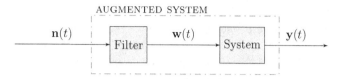

Figure 2.8 Response of a linear system to a random process via a coloring filter.

The response of a linear system to an arbitrary random process, $\mathbf{w}(t)$, can now be obtained by considering the augmented system of Figure 2.8 in its response to a white noise. If $\Phi_w(\omega)$ is the spectral density of the random signal, given by Equation (2.32), the response of the system Equation (2.19) to $\mathbf{w}(t)$ becomes

$$\Phi_y(\omega) = \mathbf{H}(\mathrm{i}\omega)\Phi_w(\omega)\mathbf{H}^*(\mathrm{i}\omega). \tag{2.33}$$

Equation (2.33) is a fundamental relation in the study of *random vibrations* of mechanical systems, which is a major cause of fatigue damage in engineering applications. See, for example, Wirsching et al. (1995) for an extended discussion on the topic and the related methods in the design of structures. A typical problem is included in Exercise 2.8, page 71.

Example 2.3 Spectral Factorization.   Consider a scalar random signal with PSD of the form

$$\Phi_w(\omega) = \frac{\omega^2 + \alpha^2}{\omega^2 + \beta^2}, \tag{2.34}$$

with $\alpha > 0$ and $\beta > 0$. It can be factorized as

$$\Phi_w(\omega) = \frac{(\alpha + \mathrm{i}\omega)(\alpha - \mathrm{i}\omega)}{(\beta + \mathrm{i}\omega)(\beta - \mathrm{i}\omega)}.$$

Choosing $\Omega = 1$ in Equation (2.32), possible filter choices for spectral factorization can be any of $H = (\mathrm{i}\omega \pm \alpha)/(\mathrm{i}\omega \pm \beta)$. However, only those with a positive sign in the numerator and denominator are of the minimum phase, that is, have zeros and poles on the left-hand complex plane. Consequently, the FRF (the coloring filter) that factorizes the given PSD is

$$H(\mathrm{i}\omega) = \frac{\mathrm{i}\omega + \alpha}{\mathrm{i}\omega + \beta}.$$

A linear system with this FRF outputs a random signal with spectral density given by Equation (2.34) when its input is a white noise with unit spectral density.

Example 2.4 Response of a 1-DoF Oscillator to Exponentially Correlated Noise.   An *exponentially correlated noise* is a scalar signal with autocorrelation given as

$$\phi_w(\tau) = \sigma_w^2 \exp\left(-\frac{|\tau|}{\tau_0}\right), \tag{2.35}$$

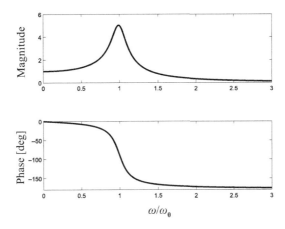

Figure 2.9 Frequency-response function of the mass–spring–damper system ($\xi = 0.1$).

with $\tau_0 > 0$ being a time constant that defines the integral timescale of the process. This type of noise is commonly found in signal processing. Its PSD is given by Equation (2.12), which becomes

$$
\begin{aligned}
\Phi_w(\omega) &= \frac{1}{\pi} \int_{-\infty}^{0} \sigma_w^2 e^{-\left(i\omega - \frac{1}{\tau_0}\right)\tau} \, d\tau + \frac{1}{\pi} \int_{0}^{\infty} \sigma_w^2 e^{-\left(i\omega + \frac{1}{\tau_0}\right)\tau} \, d\tau \\
&= \frac{\sigma_w^2}{\pi} \left(\frac{\tau_0}{1 - i\omega\tau_0} + \frac{\tau_0}{1 + i\omega\tau_0} \right) = \frac{2\sigma_w^2}{\pi} \frac{\tau_0}{1 + \omega^2\tau_0^2}.
\end{aligned}
\tag{2.36}
$$

Consider now a single degree-of-freedom (1-DoF) mass–spring–damper system with angular frequency ω_0, damping coefficient ξ, and stochastic excitation $w(t)$. Its equation of motion can be written as

$$
\ddot{x}(t) + 2\xi\omega_0\dot{x} + \omega_0^2 x(t) = \omega_0^2 w(t),
\tag{2.37}
$$

where $w(t)$ is further assumed to be an exponentially correlated noise signal. The PSD of the output can be computed using the augmented system description of Figure 2.8. The FRF between the excitation $w(t)$ and the response $x(t)$ is

$$
H(i\omega) = \frac{\omega_0^2}{\omega_0^2 + 2i\xi\omega_0\omega + (i\omega)^2},
\tag{2.38}
$$

and from Equation (2.33), one has

$$
\Phi_y(\omega) = \frac{2\sigma_w^2}{\pi} \frac{\tau_0}{1 + \omega^2\tau_0^2} \frac{1}{\left(1 - (\omega/\omega_0)^2\right)^2 + 4\xi^2 (\omega/\omega_0)^2}.
\tag{2.39}
$$

Figures 2.9 and 2.10 show the results in all the three steps in the solution process. Figure 2.9 shows the magnitude and phase of the FRF for $\xi = 0.1$, with frequencies normalized by the undamped natural frequency ω_0. A peak of the magnitude of the response is found just below this frequency, and no response is obtained at high frequencies. Figure 2.10a shows the spectral density of the exponentially correlated

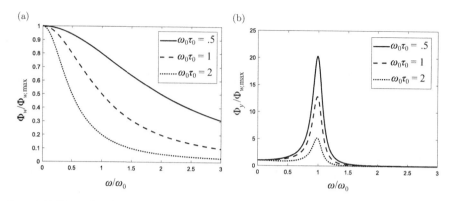

Figure 2.10 Normalized PSD of the mass–spring–damper system ($\xi = 0.1$) for a random excitation. (a) PSD of the excitation and (b) PSD of the response. [02NonSteadyAtmosphere/expcorr.m]

noise input, for different values of $\omega_0 \tau_0$, which compares the timescales in the excitation with the natural period of the response, which goes as $1/\omega_0$. Finally, Figure 2.10b includes the PSD of the response for each of those excitations. As can be seen, larger timescales imply smaller frequencies and so the effect of the excitation around the natural frequency becomes increasingly smaller.

2.3.3   One-Dimensional Models for Atmospheric Turbulence

As discussed in the introduction to this section, the flow velocities due to natural atmospheric turbulence can be often analyzed as a stationary random process. We consider first the case in which the aircraft is very small compared with the spatial scales in the atmospheric turbulence. Under that assumption, the aircraft can be seen as a point mass moving with constant velocity V, along the x axis, through a nonstationary atmosphere. The velocity of the aircraft is typically much higher than gust velocities, and the gust profile is then assumed to be a space-varying but time-independent velocity field with the aircraft flying through it. This is known as the *frozen turbulence* hypothesis (Taylor, 1938), and we refer to this field as a *background velocity field*, as it is often also assumed that it remains unperturbed by the presence of the aircraft. The gust profile is then defined in terms of its wavenumber, κ, which has units of radians per unit distance.[6] The effective frequency of the disturbances on the aircraft will depend on both the spatial frequency of the gust and the velocity of the aircraft. The angular frequency of excitations on the vehicle the also known as *frequency of encounter* (with units of [rad/s]), is then given as $\omega = V\kappa$.

In this 1-D model, the atmospheric turbulence is described by the three components of the instantaneous velocity at the location of the (point) vehicle, $\mathbf{w}(t) = (u, v, w)$ as shown in Figure 2.11. As the gust is effectively known as a spatial distribution

---

[6] It is also known as the *spatial frequency*. If the corresponding wavelength for spatial flow disturbances is λ, then $\kappa = \frac{2\pi}{\lambda}$.

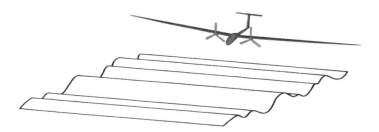

Figure 2.11 Aircraft encountering 1-D continuous turbulence (only vertical component of the velocities is included in the sketch).

of velocities (along the x axis), we define the wavenumber-based spectral density as $\hat{\mathbf{\Phi}}_w(\kappa) = V\mathbf{\Phi}_w(\omega)$, which is a matrix of dimension 3. From Equation (2.18), the mean square of the gust velocity is then given by

$$\sigma_w^2 = \int_0^\infty \mathrm{tr}\left(\hat{\mathbf{\Phi}}_w(\kappa)\right) d\kappa = \int_0^\infty \mathrm{tr}\left(\mathbf{\Phi}_w(\omega)\right) d\omega. \tag{2.40}$$

The positive square root of this expression, that is, the root mean square (RMS) of the gust velocity, σ_w, is known as the *turbulence intensity*. It has units of speed. Continuous turbulence models are defined by a PSD in terms of κ and are obtained from a mix of measurements and theoretical considerations on the rate of dissipation of turbulence at high frequencies. At altitudes above approximately 2,500 ft., there are no relevant boundary effects or stratification, and atmospheric turbulence is approximated as homogeneous and isotropic, that is, independent of the direction. Kolmogorov's theory can be used to model the statistical properties of the local kinetic energy (Pope, 2000, Ch. 6), from which those of the velocity field can be estimated. Furthermore, the spectral density matrix for 1-D isotropic turbulence is diagonal, that is, there are no cross-correlations between the components of the local velocity vector. We define the elements of the matrix as

$$\hat{\mathbf{\Phi}}_w = \mathrm{diag}(\hat{\Phi}_u, \hat{\Phi}_v, \hat{\Phi}_w). \tag{2.41}$$

A key result for isotropic turbulence is that the velocity spectra are uniquely determined by that of the longitudinal component, $\hat{\Phi}_u(\kappa)$, with the relation between the lateral and longitudinal spectra being (see Appendix B)

$$\hat{\Phi}_v = \tfrac{1}{2}\hat{\Phi}_u - \tfrac{1}{2}\kappa\tfrac{d}{d\kappa}\hat{\Phi}_u, \tag{2.42}$$

which also applies to the z component, $\hat{\Phi}_w(\kappa) = \hat{\Phi}_v(\kappa)$. The most common model defined in this way is the *von Kármán turbulence model*. For the 1-D model, it assumes the following shape for the PSD of the longitudinal velocity:

$$\hat{\Phi}_u(\kappa) = \sigma_w^2 \frac{2\ell}{\pi} \frac{1}{\left[1 + (a\kappa\ell)^2\right]^{5/6}}, \tag{2.43}$$

where a is a constant (see below) and ℓ is the *turbulence length scale* that is defined from the normalized integral of the spatial autocorrelation function, Equation (2.8), and gives the average eddy size in the turbulent flow. At the low-frequency end, the PSD is roughly a constant function; however, at higher wavenumbers, it asymptotes to a function with a –5/3 exponent. This is the result predicted by Kolmogorov's theory on the rate of dissipation of the turbulent kinetic energy in the inertial subrange, that is, for moderately large wavenumbers.

The transverse spectra are then obtained from Equation (2.42), that is,

$$
\hat{\Phi}_v(\kappa) = \sigma_w^2 \frac{\ell}{\pi} \frac{1 + \frac{8}{3}(a\kappa\ell)^2}{\left[1 + (a\kappa\ell)^2\right]^{11/6}},
$$
$$
\hat{\Phi}_w(\kappa) = \sigma_w^2 \frac{\ell}{\pi} \frac{1 + \frac{8}{3}(a\kappa\ell)^2}{\left[1 + (a\kappa\ell)^2\right]^{11/6}}.
\tag{2.44}
$$

Now the constant a can be determined by enforcing Equation (2.40) that yields $a = 1.339$. See Appendix B for details. The convention of aircraft load analysis (Hoblit, 1988) is to assign different gust intensities in each spatial direction, so that they can be studied independently. However, this is not done here to keep a physically consistent description.[7] In particular, note that this would imply that the turbulence intensity no longer satisfies Equation (2.40), that is, σ_w cannot be recovered as the RMS of the time- or frequency-domain signals. Note finally that, from Equation (2.40), the units of $\hat{\Phi}_w$ are [(m^2/s^2)(m/rad)].

A second, less accurate model for continuous turbulence is the *Dryden turbulence model*. It defines the spectrum for the longitudinal, lateral, and vertical gusts, respectively, as (Press et al., 1955)

$$
\hat{\Phi}_u(\kappa) = \sigma_w^2 \frac{2\ell}{\pi} \frac{1}{\left[1 + \kappa^2\ell^2\right]^2},
$$
$$
\hat{\Phi}_v(\kappa) = \sigma_w^2 \frac{\ell}{\pi} \frac{1 + 3\kappa^2\ell^2}{\left[1 + \kappa^2\ell^2\right]^2},
$$
$$
\hat{\Phi}_w(\kappa) = \sigma_w^2 \frac{\ell}{\pi} \frac{1 + 3\kappa^2\ell^2}{\left[1 + \kappa^2\ell^2\right]^2}.
\tag{2.45}
$$

Note that the Dryden spectrum also satisfies Equations (2.40) and (2.42). It is often used to test the effect of turbulence in flight dynamic applications (MIL-HDBK-1797, 1997; Schmidt, 2012).

The normalized spectra for vertical and longitudinal continuous turbulence obtained from both von Kármán and Dryden models are shown in Figure 2.12. Note that the lateral component has the same PSD as the vertical one, and it is not included. While both models diverge considerably in the longitudinal component, they are reasonably close for the vertical (and lateral) components, which are the most critical in aircraft design. Both atmospheric turbulence models are uniquely defined by two parameters:

---

[7] As an example, MIL-HDBK-1797 (1997) defines three different length scales, $\ell_u, \ell_v,$ and ℓ_w, with the lateral ones rescaled by a factor of 1/2, which implies that, to retrieve Equations (2.43) and (2.44), one needs to have $\ell = \ell_u = 2\ell_v = 2\ell_w$.

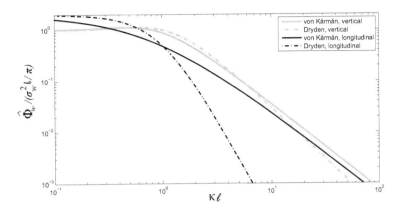

Figure 2.12 Normalized PSD of von Kármán and Dryden continuous turbulence models (ℓ is the turbulence length scale). [02NonSteadyAtmosphere/karmandryden.m]

the turbulence length scale, ℓ, and the turbulence intensity, σ_w. For aircraft design, both values are specified in the airworthiness regulations, as we discuss in the following section.

Continuous Turbulence in Aircraft Design

Airworthiness regulations specify a length scale $\ell = 2,500$ ft. for the von Kármán model (EASA, 2017; FAA, 2011) and $\ell = 1,750$ ft. for the Dryden model (MIL-HDBK-1797, 1997). For altitudes below ℓ, atmospheric turbulence is no longer isotropic and increasingly depends on the local terrain characteristics as one approaches the ground. In that region, the previous models for clean-air turbulence are thus no longer valid, and one needs to consider the detailed flow features appearing in the ABL, which will be discussed in Section 2.4.1. Nevertheless, 1-D continuous turbulence models are sometimes used, with reduced length scales, as a first approximation. For example, for the Dryden model and altitude h below 1,750 ft., an orthotropic distribution is often considered (even though it is not justified by the theory), with $\ell = h$ in the vertical velocity and $\ell = 145h^{1/3}$ ft. in the horizontal plane (Schmidt, 2012).

The turbulence intensity, σ_w, used in aircraft design (known in this context as the *limit turbulence intensity*) is determined to be a function of the altitude and the probability of encounter. At higher altitudes, atmospheric turbulence levels drop, and regulation accounts for that effect by considering $\sigma_w = 90$ ft./sec at sea level and decreasing linearly up to 79 ft./sec at 24,000 ft. Those values are also slightly modified based on the aircraft characteristics (EASA, 2017; FAA, 2011). The probability of encounter is indirectly considered by decreasing the limit turbulence intensity with aircraft speed, from cruise (where the aircraft operates the vast majority of its time) to dive speed (which is effectively never reached).

For military aircraft, recommended values are given for three levels of turbulence that put different constraints on vehicle handling and maneuverability, among others. Further discussion of the acceptable performance levels under adverse atmospheric

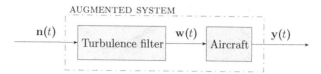

Figure 2.13 Turbulence filter connected to aircraft dynamic model ($\mathbf{n}(t)$ represents the white noise).

conditions, from a flying qualities point of view, can be found in the military standard handbook and, for instance, in Cook (2013).

The turbulence models presented in this section are the most commonly found models in the literature and the ones used in aircraft certification. In particular, the von Kármán model given in Equation (2.43) is used in conjunction with Equation (2.33) to determine the statistics of aircraft response to continuous turbulence modeling. Many other 1-D models for atmospheric turbulence have been proposed in the literature, particularly for flight dynamics applications, but they will not be discussed here. A summary of the most common ones can be found in Wang and Frost (1980).

Synthetic Turbulence Models

The semiempirical models for atmospheric turbulence have defined a spectrum density for the expected velocity profile. As we have just seen, this then can be used to obtain the corresponding statistical characterization of the dynamics of aircraft flying through it. Equation (2.33), however, is only valid under linear assumptions for the aircraft dynamics. For situations when the vehicle response requires nonlinear equations or for time-domain simulation (e.g., to reproduce atmospheric conditions in a flight simulator), one can use spectral factorization, as defined in Equation (2.32). It results in frequency-domain filters (known as *turbulence filters*) whose output to a white noise is a random signal with the given turbulence spectrum and that can be used as input to an aircraft dynamics model, as shown schematically in Figure 2.13.

As our interest is in the effect of continuous turbulence on aircraft, we will first need to evaluate the excitation in the time domain. If a vehicle flies through continuous turbulence at constant speed V, the turbulence models can be written in terms of the angular frequency of encounter as seen by the aircraft, that is, $\omega = V\kappa$. The 1-D von Kármán frequency spectra then becomes

$$\Phi_u(\omega) = \sigma_w^2 \frac{2T}{a\pi} \frac{1}{\left[1+(\omega T)^2\right]^{5/6}},$$

$$\Phi_v(\omega) = \Phi_w(\omega) = \sigma_w^2 \frac{T}{a\pi} \frac{1+\frac{8}{3}(\omega T)^2}{\left[1+(\omega T)^2\right]^{11/6}}.$$

(2.46)

We have defined a modified turbulence integral timescale as $T = a\ell/V$, which corresponds to a characteristic period of encounter. Let now $G_{un}(i\omega)$ be the FRF of the turbulence filter in the longitudinal direction. Choosing unit PSD for the input white noise, $\Phi_n = 1$, from Equation (2.33), the output PSD becomes

$$\Phi_u(\omega) = G_{un}(i\omega)G_{un}(-i\omega) = \left| G_{un}(i\omega) \right|^2. \tag{2.47}$$

Similar definitions can be introduced to define FRFs of turbulence filters in the transverse directions y and z. The FRFs can be chosen to give the PSD of the von Kármán turbulence model, Equations (2.43) and (2.44). They are usually picked as

$$
\begin{aligned}
G_{un}(i\omega) &= \sigma_w \sqrt{\frac{T}{\pi a}} \frac{\sqrt{2}}{(1+i\omega T)^{5/6}}, \\
G_{vn}(i\omega) = G_{wn}(i\omega) &= \sigma_w \sqrt{\frac{T}{\pi a}} \frac{1 + \frac{2\sqrt{2}}{\sqrt{3}}i\omega T}{(1+i\omega T)^{11/6}}.
\end{aligned}
\tag{2.48}
$$

Turbulence filters can now be used to generate representative time histories for time-domain simulation, which are referred to as *synthetic turbulence* signals. However, for this, one needs additional assumptions of the statistics of turbulence, as higher order moments in isotropic turbulence are nonzero. The usual one is that synthetic turbulence is Gaussian, which means that the probability of the random variable to take any given value follows a normal distribution. A Gaussian random process is fully specified, in a statistical sense, by its first two moments (mean and correlation), and this allows the generation of synthetic turbulence signals from the Gaussian white noise and a turbulent filter, as shown in Figure 2.13.[8]

As we saw in Section 1.4.2, when utilizing frequency-domain filters for time-domain applications, we first need to identify a state-space model of the system, which can then be integrated in time for simulation. Otherwise, we would need to constantly perform Fourier transforms on the inputs and outputs of the filter to generate time-domain information. We then seek to approximate the transfer function of the turbulence filters by means of rational functions. In particular, the semiempirical filters obtained in Equation (2.48) have the exponent 5/6 in the denominator, and a rational function approximation would be needed on that term.

Several approximations have been proposed (Beal, 1993; Gage, 2003). A popular and easy to implement approximation is that of Campbell (1986), which uses a generic expression to approximate functions of the form $f(s) = (1+s)^\nu$, where s is the Laplace variable, by continued fraction expansions up to any desired order. In particular, a third-order expansion has been shown to give a good approximation,

$$(1+s)^{-5/6} \approx \frac{\frac{91}{12}s^2 + 52s + 60}{\frac{935}{216}s^3 + \frac{561}{12}s^2 + 102s + 60}. \tag{2.49}$$

---

[8] Jones (2007) advocates that this does not result in a conservative procedure, since Gaussian processes do not predict the so-called *black swan* events, which may actually lead to catastrophic performance. Gaussian turbulence is still assumed for certification and analysis, augmented with safety factors and discrete gust conditions. All evidence points to the current design approach providing sufficiently safe designs.

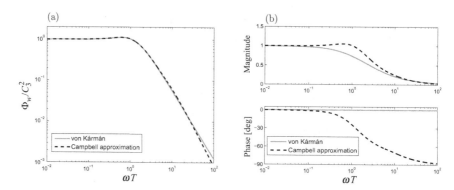

Figure 2.14 Power spectral density and frequency-response function of the von Kármán turbulence model (vertical gust) and a rational function approximation using Equation (2.49). (a) Normalized PSD and (b) FRF, $G_{wn}(i\omega T)/C_3$. [02NonSteadyAtmosphere/ karmanrfa.m]

This function has all poles and zeros on the left half-plane and therefore is minimum phase (i.e., stable and causal), as required by the spectral factorization theorem, Section 2.3.2. Substituting Equation (2.49) into Equation (2.48) gives a rational function approximation to the von Kármán turbulence filter. The corresponding PSD is shown in Figure 2.14a, where it is compared to the exact solution with $C_3 = \sigma_w \sqrt{\frac{T}{\pi a}}$. Because the PSD in the von Kármán's model is not a rational function, the FRF for the approximating filters is not unique. A comparison between the FRF of Equation (2.48) and Campbell's approximation is given in Figure 2.14b. Note that there is an amplitude change around $\omega T \approx 1$ and a phase delay at larger frequencies. As the interest is in the statistics of the response, there is no conflict in this mismatch: The statistics of the response, over a sufficiently large sample, still have the expected PSD.

We can finally generate time histories of continuous turbulence from a synthetic turbulence filter. An example is shown, in a nondimensional form, in Figure 2.15 for an aircraft flying with constant horizontal speed, V, through vertical turbulence. Figure 2.15a shows the time history of a Gaussian white noise that goes into the filter. Figure 2.15b shows the corresponding vertical turbulent velocity obtained by Campbell's approximation to the von Kármán turbulent spectrum. Note the fluctuations in the Gaussian noise have no apparent relation to the timescale $T = a\ell/V$, but this is much clearer in the gust profile. As the turbulence scale ℓ is constant, the effective frequency of the excitations on the aircraft will grow with its forward flight velocity.

Synthetic turbulence generated in this way is extensively used in flight dynamic simulations and is particularly relevant when the vehicle dynamics are described by nonlinear equations of motion, in which case, Equation (2.33) is no longer valid, and direct simulation in the time domain is typically needed to assess vehicle performance.

2.3.4 Two-Dimensional Models for Isotropic Atmospheric Turbulence

For aircraft with wings of a sufficiently high aspect ratio, the point-mass assumption used in the 1-D continuous turbulence models may no longer be valid. In such cases,

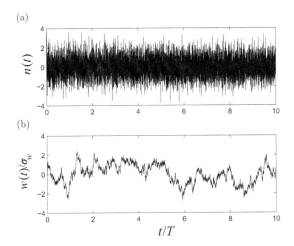

Figure 2.15 (a) Time history of a Gaussian white noise, $n(t)$, and (b) its corresponding normalized synthetic (vertical) turbulence using the filter of Equation (2.49). [02NonSteadyAtmosphere/karmantime.m]

one one would need, in general, a full 3-D description of the isotropic turbulence. However, as the aircraft spanwise dimensions are typically much larger than their vertical ones, it is often sufficient to consider only variations of the incoming wind velocity along the spanwise direction.

This results in a velocity distribution that shows both temporal (in t) and spatial (along y) correlations, which therefore defines 2-D (in both t and y) spectral density functions between all three components of the velocity. Full results, for both the von Kármán and Dryden models, were obtained by Eichenbaum (1971), and the main steps are summarized in Appendix B. As an example, we present only the spatial cross-spectrum of the vertical velocity between two points at a distance $\Delta y \geq 0$ for the von Kármán model, which, for airspeed V, becomes (Sleeper, 1990)

$$\Phi_{ww}(\omega, \Delta y) = \sigma_w^2 \frac{2^{1/6} T}{\Gamma(\frac{1}{3})\sqrt{\pi}} \left[\frac{8}{3} r_2^{5/6} K_{5/6}(r_1) - r_2^{11/6} K_{11/6}(r_1) \right]. \tag{2.50}$$

This is a real-valued function where $K_\nu(x)$ is a modified Bessel function of the second kind (of rational order ν), $\Gamma(x)$ is the Gamma function, $r_1 = \frac{\Delta y}{a\ell} \left(1 + (\omega T)^2\right)^{1/2}$, and $r_2 = \frac{\Delta y}{a\ell} \left(1 + (\omega T)^2\right)^{-1/2}$. Using the approximation $K_\nu(\epsilon) = \frac{\Gamma(\nu)}{2} \left(\frac{2}{\epsilon}\right)^\nu$, for $0 < \epsilon \ll 1$, it is easy to prove that in the limit $\Delta y \to 0$, the spectrum given in Equation (2.50) coincides with Equation (2.46), that is, $\Phi_{ww}(\omega, 0) = \Phi_w(\omega)$, the 1-D von Kármán turbulence spectrum.

Equation (2.14) has defined the coherence between two random processes. That definition is typically generalized for 2-D turbulence using a coherence function $\gamma_{ww}(\omega, \Delta y) = \frac{|\Phi_{ww}(\omega, \Delta y)|^2}{|\Phi_{ww}(\omega, 0)|^2}$. Figure 2.16 shows the frequency content of the square root of the coherence function for discrete values of Δy. For $\Delta y = 0$, there is a perfect correlation, while $\sqrt{\gamma_{ww}}$ drops to around 0.1 for $\Delta y = a\ell$. There is also a local maximum

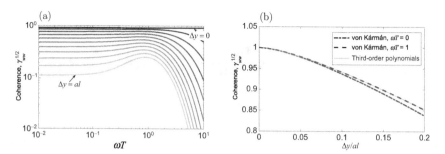

Figure 2.16 Coherence in 2-D von Kármán turbulence for pairs of points at distances Δy. (a) Frequency dependency for $0 \leq \Delta y \leq al$ in increments of $al/10$ and (b) spatial dependency for ωT and 1. [02NonSteadyAtmosphere/karman2d.m]

of correlation for all distances near $\omega T = 1$, beyond which the coherence drops rather abruptly. Importantly, for sufficiently small distances, the coherence function is nearly constant for frequencies up to $\omega T = 1$. This is illustrated in Figure 2.16b that shows the coherence for $\Delta y < 0.2a\ell$ for two frequencies. Note that for the turbulence length scale in aircraft regulations, we get $a\ell \approx 1$ km. Consequently, for aircraft flying above the ABL, the coherence in von Kármán's model can be assumed to be constant at low enough frequencies (see Figure 2.16).

As can also be seen in Figure 2.16, the coherence is a nonlinear function of the distance. A cubic polynomial fit of Equation (2.50), also included in the figure, gives a good approximation. As an example, consider an aircraft with a wingspan of 35 m, which is typical of single-aisle airliners. The coherence between wingtips of the vertical component of the atmospheric turbulence in that case is $\gamma_{ww} \approx 0.97$. However, for the largest air vehicles currently in operation, whose wingspans are above 100 m, this value is closer to 0.88. In those cases, one may need to consider (2-D) spanwise changes of the incoming wind velocity.

In practice, the 2-D cross-spectrum obtained in Equation (2.50) is discretized at a finite number of points on an aircraft. Time histories of flow velocities define a finite-dimensional random variable $\mathbf{w}(t)$, for which the spectral density is a full matrix with the 1-D von Kármán spectrum, Equation (2.43), in the diagonal and Equation (2.50) elsewhere, with Δy being the positive distance between each pair of points. Different sampling criteria have been proposed for the spanwise velocity fields: Houbolt (1973) proposed the discretization of the wing in multiple strips, each of them with constant gust velocities, while Eichenbaum (1971) assumed also a chordwise discretization. Etkin (1981) highlighted that assuming a linear variation of velocities could often be sufficient as a first approximation and consequently proposed a four-point method in which velocities are only evaluated at the front and rear of the fuselage and both wingtips.

The statistical characterization of the response to 2-D continuous turbulence is then obtained from a straightforward application of Equation (2.33). In this case, the admittance is a matrix with the flow velocity at the points used for sampling the continuous turbulence field (i.e., Etkin's four points) as its inputs, and the outputs are the relevant *interesting quantities* in the aircraft dynamic response (e.g., wing root

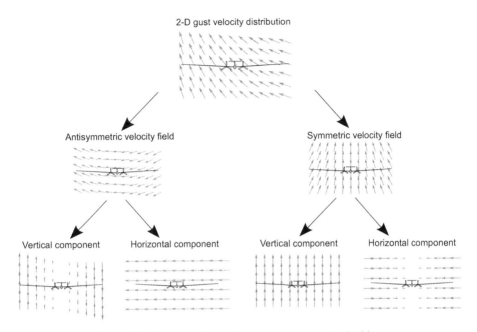

Figure 2.17 Decomposition of linearly varying gust velocities.

bending moments). This approach holds for any given linear time-invariant model of the dynamics of an aircraft, from simplified 2-D flight mechanics models to very large finite-element-based aeroelastic descriptions (Teufel et al., 2000).

However, as the spectral density is a 2-D model and no longer diagonal, the generation of synthetic turbulence filters becomes substantially more involved. Mouyon and Imbert (2002) have successfully generated finite-state approximations that converge to the 2-D nonrational wind spectra using techniques of signal processing. A commonly used approximation for wind turbine applications was obtained by Veers (1988). Noting that the cross-spectral densities given by Equation (2.50) are real, he proposed a lower triangular matrix $\mathbf{H}(i\omega)$ in the spectral factorization of Equation (2.32). This defines a simple recursive process to evaluate the individual FRFs in the resulting multidimensional turbulence filter.

When only a linear variation of the background velocity field is considered, it is then convenient to split this velocity field into its symmetric and antisymmetric contributions, and then into its respective lateral and vertical components (Eichenbaum, 1971). This is illustrated in Figure 2.17. While the symmetric horizontal component is typically negligible, the vertical antisymmetric component is not. In this linear approximation, we define the *gust angular velocity components* as the spatial derivatives of the velocity field at a given location (i.e., the center of mass of the aircraft) as

$$p_g = \frac{\partial w_g}{\partial y}, \quad q_g = \frac{\partial w_g}{\partial x}, \quad \text{and} \quad r_g = -\frac{\partial v_g}{\partial x}. \tag{2.51}$$

From the 2-D cross-spectrum, additional 1-D PSD functions are then obtained for the gust angular velocity. For the von Kármán model, they are (MIL-HDBK-1797, 1997)

$$\hat{\Phi}_p(\kappa) = \frac{\sigma_w^2}{\ell} \frac{0.8\left(\frac{\pi\ell}{4b}\right)^{1/3}}{1+\left(\frac{4}{\pi}b\kappa\right)^2},$$

$$\hat{\Phi}_q(\kappa) = \frac{\pm\kappa^2}{1+\left(\frac{4}{\pi}b\kappa\right)^2}\hat{\Phi}_w(\kappa), \qquad (2.52)$$

$$\hat{\Phi}_r(\kappa) = \frac{\mp\kappa^2}{1+\left(\frac{3}{\pi}b\kappa\right)^2}\hat{\Phi}_v(\kappa),$$

with b being the wingspan. Similar expressions can be obtained for Dryden's model.

2.4   Wind Shear

Wind shear, sometimes referred to as a *wind gradient*, is defined as a spatial variation of the mean wind velocity. Atmospheric wind shear can be vertical, when the wind speed changes with the altitude, or horizontal, when changes in the wind velocity vector are seen by an aircraft flying at constant altitude. Wind shear is also an important mechanism in turbulence generation, and the changes to both mean and fluctuating components of the wind velocity often need to be considered simultaneously (Houbolt, 1973).

Horizontal wind shear appears in weather fronts, where two masses of air with different properties coalesce, and in coastal areas, due to the daily changes between sea and land temperatures (as well as the different surface roughness between the two). Vertical shear mainly occurs due to the presence of thermals in the lower atmosphere (e.g., in the vicinity of thunderstorm cloud formation), from a convective ABL (see Section 2.4.1) or from low-altitude winds on the lee side of mountains, among others.

The key magnitude in the description of vertical wind shear is the rate of change of wind speed with respect to altitude (a similar metric is used for horizontal shear). There is, however, no established engineering model for the wind gradient due to weather or geographical features. Instead, the analysis of its effect on aircraft dynamics is based on tabulated datasets from experimental campaigns. Overall, it is generally accepted that a severe vertical wind shear corresponds to a drop of 0.2 m/s per m (Nelson, 1998). In aircraft design, this has informed the definition of standardized discrete gusts that are discussed in Section 2.4.2. The situation changes for wind shear due to the ABL, which has well known features, and it can be estimated based on the local surface roughness. This will be discussed in Section 2.4.1.

2.4.1   The Atmospheric Boundary Layer

Wind conditions near the ground (often taken to be below 2,500 ft.) are very different from those at higher altitudes. We have seen in Section 2.3.3 that the constants in atmospheric turbulence models are often adjusted for low altitudes, but this is just a rather rough approximation to the more complex physical phenomena that appear in the ABL. Within the ABL, the flow physics are determined by the local topography and the heating and cooling of the Earth's surface. Assuming a flat Earth, the main factors that define the structure of the ABL are the surface roughness of the ground and the rate of change of temperature with height (known as the *lapse rate*).

Mean wind speeds as a function of altitude in the ABL are often approximated by either a logarithmic or a power law. The log law is derived from Prandtl's boundary layer theory (Manwell et al., 2009, Ch. 2) and defines an explicit relation with the surface roughness parameter (typical length scale of the ground that may vary from millimeter scale in a calm open sea to several meters near forests or a built environment). The power law, on the other hand, is a simpler model that is often used for engineering purposes. It relates wind speed, U_w, with altitude, h, to those measured or known at certain reference altitude, that is,

$$\frac{U_w}{U_{ref}} = \left(\frac{h}{h_{ref}}\right)^\alpha. \tag{2.53}$$

The power-law index α is obtained empirically and depends mostly on surface roughness but also on temperature and weather conditions. Typical values range between 0.1 and 0.4 (Etkin, 1981).

Temperature effects define three distinctive states of the ABL. If thermal gradients are negligible, one has a *neutral boundary layer* where turbulence is generated purely due to mechanical processes. In a neutral boundary layer, air particles move adiabatically due to the vertical distribution of wind shear, Coriolis forcing, and surface friction. Strictly neutral boundary layers are very rarely found in the Earth's atmosphere, which typically displays a nonnegligible vertical gradient of temperatures. This nonzero lapse rate determines a gradient of atmospheric pressure with altitude, which then may be sufficient to stabilize air parcels perturbed along the vertical axis. A *stable boundary layer* occurs when the air particles tend to return to their original position when perturbed along the vertical direction. Their dynamics can be estimated assuming an adiabatic process, and the resulting condition displays strong temperature stratification and wind shear, but relatively small eddies. In other conditions, we have a *convective boundary layer* that is dominated by large vertical air masses (thermals) that transport air away from the Earth's surface. Convective boundary layers, which are typical at sunrise, generate much stronger turbulence than stable ones (Wyngaard, 2010). This results in a higher dissipation rate for aircraft wakes (Ahmad and Proctor, 2013) but also adverse conditions for very slow aircraft (Deskos et al., 2020). Estimation of the instantaneous lapse rate is, therefore, key to identify the local ABL characteristics, and this is particularly critical near airports. There, ground-based meteorological towers and Light Detection and Ranging (LIDAR) scanners, supported by weather balloons, are used to measure the lapse rate, and other local conditions. Observation campaigns with remotely piloted aircraft have increasingly provided a cost-effective alternative to obtain high-quality datasets (Bell et al., 2021).

Field data are fed into fast (semiempirical) models for wake transport and decay used at airports (Sarpkaya et al., 2001), possibly supported by mesoscale atmospheric models that give spatial details (Ahmad and Proctor, 2013). Estimation of the loads on the aircraft requires more detailed information at higher wavenumbers. Here, two approaches for ABL turbulence simulation are generally considered. The most common solution is to use stochastic analysis to generate wind fields based on second-order statistics, in a similar manner to the 2-D isotropic models of Section 2.3.4. Stochastic generation of ABL turbulence has been an active area of research over several decades, particularly for the development of large wind turbines, where a substantial amount of effort has been undertaken to characterize the wind environment (Kelley and Jonkman, 2007; Mann, 1994, 1998; Muñoz-Esparza et al., 2015; Veers, 1988). In particular, the model of Kaimal et al. (1972) is typically used for wind engineering applications.

The second approach for ABL turbulence simulation directly solves the fluid dynamics equations for a given temporal and spatial resolution. This is typically done using large-eddy simulation (LES), where the large energetic structures of the flow are resolved, while the smaller scales are modeled. This approach results in detailed time histories of three-dimensional wind fields on the chosen computational domain. Although it is much more computationally intensive, LES is now routinely used to study turbulent boundary layers, and it has been recently proposed as a high-fidelity strategy to produce physics-based low-altitude atmospheric velocity fields to support aeroelastic vehicle design (Deskos et al., 2020).

As one would expect, knowledge of the wind fields in the ABL is of interest mainly in aircraft takeoff and landing operations. They occur at low speeds and with full deployments of flaps and landing gear, which defines fundamentally different simulation and analysis challenges to those found at cruise conditions. Two aspects are of relevance here for aircraft design: the first one is the turbulence found within the ABL, which has been found to be critical on the operation of extremely low wing loading vehicles, such as solar-powered aircraft (del Carre and Palacios, 2019; Noll et al., 2004). Most often, however, the concern is mainly about the mean wind speeds. As the aircraft is near the ground, flight path control is a key driver, which brings severe restrictions to maximum *crosswinds* in takeoff and landing. Maximum demonstrated crosswind is a characteristic of the aircraft configuration, and it is one of the performance limits in its *Pilot Operating Handbook*. For civilian transport aircraft, FAA Part §25.237 requires that the aircraft demonstrates safe takeoff and landing operations with a 90° crosswind of at least 20 knots and not more than 25 knots.

2.4.2   Discrete Gusts

Discrete gusts can be seen as particular realizations of wind shear. In Section 2.3, the investigation on continuous turbulence was grounded on physical models of the atmosphere, but they also assumed isotropic turbulence conditions that often break down in practice (e.g., under severe weather conditions). Wind shear was introduced later to

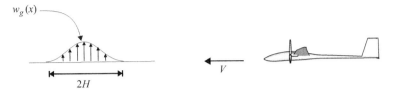

Figure 2.18 Some parameter definitions in a "1-cosine" gust encounter.

identify more deterministic situations associated with specific weather or geographical features. However, outside the ABL, they are hard to predict and, therefore, to use in design. To address that, the aircraft dynamic response is further investigated under deterministic gust profiles, known as *discrete gusts*, which correspond to idealized (and rather challenging) conditions.[9]

While several discrete gusts have been proposed, current analysis for vehicle certification is almost exclusively based on 1-D (i.e., spanwise constant) stationary gusts with a "1-cosine" (literally, "one minus cosine") spatial dependency; see Murrow et al. (1989) for a brief historical overview of this development. The velocity profile effectively recreates a "bump in the air," which is defined as

$$w_g = \frac{1}{2} w_0 \left(1 - \cos \frac{2\pi x}{2H} \right) \quad \text{if } 0 \leq x \leq 2H,$$

$$w_g = 0 \quad \text{elsewhere,} \tag{2.54}$$

where w_g corresponds to either a vertical or lateral velocity. Figure 2.18 sketches an encounter with a vertical gust. Its profile is represented by two constants, namely the gust intensity, w_0, and the gust *gradient* (or gust length), H. Commercial aircraft regulations (FAA, 2011) for gust load analysis specify how both values need to be chosen depending on the aircraft characteristics, altitude, and flight speed.

For large transport passenger aircraft, the gust gradient, H, needs to be investigated for values between 30 ft. and 350 ft. (9.14–106.68 m). For a given aircraft and flight condition, the gust length that creates the maximum loads on the airframe needs to be found. It is known as the critical gust length, for which the resulting gust is the *discrete tuned gust*. Typically, critical or tuned gust gradients are between 5 and 15 mean wing chords, depending on the aircraft. An example of this can be found in Section 3.6.2.

The gust intensity, w_0, is defined in the regulations in terms of equivalent air speed, Equation (2.2), by the formula

$$w_0 = w_{\text{ref}} F_g \left(H/H_{\text{max}} \right)^{1/6}, \tag{2.55}$$

---

[9] They are often justified as particularly challenging events that can be found in continuous turbulent signals, but this is similar to Borges's imaginary library, which contains all the books that could be written by infinite permutations of the letters of the alphabet. Every possible discrete gust (every book ever written, or that will ever be written!) has to be there by definition; but also by definition we have a zero probability of finding it. We think it is more appropriate to think of discrete gusts as standardized testing conditions that have been agreed to define a common benchmark to compare designs.

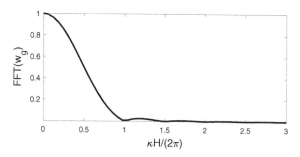

Figure 2.19 Normalized wavenumber content of a discrete "1-cosine" gust.

where $H_{max} = 350$ ft., F_g is the flight profile alleviation factor that depends on the aircraft weight and maximum operating altitude, typically taking values between 0.8 and 0.9, and w_{ref} is the reference gust velocity (EAS) defined as a function of altitude. Aircraft regulations define w_{ref} to be 56 ft./sec. (17.07 m/s) at sea level, 44 ft./sec. (13.41 m/s) at 15,000 ft., and 20.86 ft./sec. (6.36 m/s) at 60,000 ft. (18,288 m). Linear interpolation is used between them. This is for aircraft flying at its design cruise speed, V_C, or lower. In fact, regulations specify a design speed for maximum gust intensity, V_B, linked to vehicle stall characteristics, which is below V_C. For aircraft at its maximum design speed (the dive speed, V_D), the reference gust velocity is halved to account for the unlikely scenario of the aircraft encountering those conditions at that very large speed.

It is also interesting to investigate the frequency content of the discrete gusts. Figure 2.19 shows the normalized discrete Fourier transform of Equation (2.54) in terms of the wavenumber, κ, which has units of rad/m. The bandwidth of excitation scales with the gust gradient, H, and it is almost negligible for $\kappa > 2\pi/H$. As with continuous turbulence, the excitation on an aircraft flying at velocity V through the gust will be at angular velocities $\omega = V\kappa$.

For aircraft with wings of a very high aspect ratio, the assumption of spanwise homogeneous gust may be no longer appropriate, as we have seen in Section 2.3.4. For those situations, a modified discrete gust, known as the DARPA gust, has been proposed that includes a harmonic modulation along the spanwise direction of the "1-cosine" gust (Dillsaver et al., 2012; Hesse and Palacios, 2014).

2.5   Aircraft Wakes

Aircraft themselves produce a disturbance in the atmosphere in the form of wakes. The dominant features in aircraft wakes are the wingtip vortices that are generated due to the pressure difference between suction and pressure sides required for lift. Near the wingtip, that pressure jump generates a strong suction effect that manifests itself on a rotational motion on the fluid in the vertical plane. As a result, a pair of counter-rotating trailing vortices is created whose strength depends on the weight and aspect ratio of the generating aircraft. They may persist in the atmosphere for several

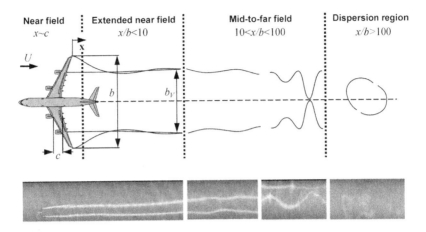

Figure 2.20 Evolution of a wake vortex (reprinted from Breitsamter (2011) with permission from Elsevier).

minutes before decaying due to the combination of their internal dynamics and the local atmospheric conditions (Spalart, 1998), as discussed in Section 2.4.1. Wingtip vortices are effectively a man-made form of wind shear.

The evolution of a vortex wake is typically divided in to four stages as shown in Figure 2.20. They are identified by their downstream location with respect to the aircraft, measured in wingspans, b, and given approximately as (Breitsamter, 2011): (1) the near field, $x \sim c < b$, dominated by the local geometry of the aircraft; (2) the extended near field, $0.5 \leq x/b \leq 10$, in which the two counter-rotating vortices are formed; (3) the mid-to-far field, $10 \leq x/b \leq 100$, in which the vortices persist and move downward; and (4) the dispersion region $x/b > 100$, in which the vortices eventually disappear. As discussed above, however, details depend on the local atmospheric conditions and control towers may often produce real-time estimates of such distances (Ahmad and Proctor, 2013).

A *wake vortex encounter* (WVE) occurs when an aircraft (the follower aircraft) passes through the trailing vortices of another (the leader aircraft), although there has been reported incidents of aircraft crossing their own wake during maneuvers (Claverias et al., 2013). A typical WVE event is schematically shown in Figure 2.21. The flow velocities in the mid-to-far field of the wake of the leader aircraft (which is not shown), where WVE are more likely to occur, can still be rather high and pose severe danger on the handling and/or integrity of the follower aircraft. This is exacerbated when the follower aircraft is much smaller than the leader aircraft. However, both the wake location and the decay rate can be reasonably estimated, as opposed to naturally occurring wind shear, and, as a result, criteria for minimum separation distances among aircraft have long been established. They become particularly important in managing congested aerospace and in determining landing slots on airports. The International Civil Aviation Organization (ICAO) currently specifies a minimum aircraft separation of between 5.6 and 11.1 km, depending on the weight of both aircraft.

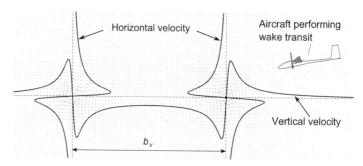

Figure 2.21 Idealized velocity field in the normal plane to an aircraft wake. Continuous lines show the horizontal and vertical velocity components along the axes going through the vortices.

Since the separation criteria are routinely enforced, loads on the airframe due to WVE is not specifically considered in aircraft certification. The underlying assumption is that any unexpected instance of WVE has already been conservatively considered by the analysis under the discrete gusts defined in Section 2.4.2. However, WVEs have sharper gradients than discrete gusts, which may result in a broader spectrum of loading on the follower aircraft. They also have a strong directionality, which results in wake fronts not necessarily perpendicular to the follower aircraft's flight path and that generate asymmetric loading. There is indeed some evidence that WVE can define critical loading conditions on some substructures of large military aircraft and UAVs (Claverias et al., 2013; Kier, 2013), which have less operational restrictions than civilian ones.

Analysis of WVE in the mid-to-far field is typically carried out with the wake modeled as the potential flow field generated by two counter-rotating vortex filaments of equal strength that run parallel to the aircraft flight path and at a distance $b_v \leq b$ (Figure 2.20). This is to account for the effect that rolled-up tip vortices move inward to achieve equilibrium. For an elliptic lift distribution, $b_v/b = \pi/4$ (Breitsamter, 2011). The circulation strength, Γ_v, on each of the vortices can be estimated from Joukowski's theorem as

$$\Gamma_v = \frac{L}{\rho V b_v},$$ (2.56)

where the lift is equal to the weight of the leader aircraft (for steady 1-g flight). The resulting flow field encountered by the follower aircraft can be obtained using the Biot–Savart law for a vortex line (Katz and Plotkin, 2001). To avoid nonphysical values near the vortex filaments, a vortex core model is typically introduced. A simple one often used was proposed by Burnham and Hallock (1982), such that the modulus of the azimuthal velocity in the plane of the vortex is

$$v_\theta(r) = \frac{\Gamma_v}{2\pi} \frac{r}{r^2 + r_c^2},$$ (2.57)

where r_c is the vortex core radius and r is the distance to the center of the vortex. A typical value for r_c in the extended near wake is around 3% of the wingspan. The

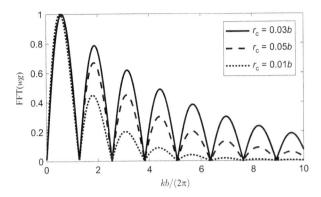

Figure 2.22 Normalized wavenumber content of the normal flow velocity in a wake vortex with core radius r_c. [`02NonSteadyAtmosphere/wakefft.m`]

resulting velocity distribution is illustrated in Figure 2.21. This defines a nonstationary background velocity field for the follower aircraft.

Figure 2.21 also includes the flow velocities along the vertical and horizontal axes passing through the origin of both vortices. Note that along the horizontal axes, this wake model only has vertical velocity and vice versa. While, in general, a 3-D velocity field would need to be considered, particularly with oblique crossing during ascent/descent trajectories, the vertical component is often the only one relevant when both aircraft are along horizontal trajectories. As with the discrete gust, the frequency content of the excitation is of interest for the aircraft dynamics. Figure 2.22 shows this in terms of the wavenumber, κ, normalized by the wingspan for a 90° encounter. It assumes an elliptic wing distribution and compares three different vortex core radii. For $\kappa < 2\pi/b_v$, which is the wavelength defined by the distance between the vortices, the amplitude is highest and mostly independent of the vortex radius, r_c. At higher wavenumbers, the amplitude strongly depends on the vortex radius, and it can be very high through a relatively large bandwidth if there is little dissipation.

We can also compare the frequency content of a wake encounter, Figure 2.22, with that of the "1-cosine" gust, Figure 2.19. Note that, although the discrete gust has a narrower spatial bandwidth, it is normalized by H, whose minimum value in regulations is around 10 m. This is much smaller than the span of airliners, typically between 40 and 80 m. Finally, it should be remarked that, as saturated airports demand shorter landing slots, more refined models have also been developed to consider vortex kinematics and decay rates (Bieniek and Luckner, 2014), which reduce the impact of the WVE on the follower aircraft and better correlate with the empirical evidence.

2.6 Summary

This chapter has described the mathematical models most commonly used to characterize nonstationary atmospheric conditions. Depending on the deterministic or nondeterministic nature of the atmospheric disturbances, we have divided natural phenomena in either wind shear or continuous turbulence. For the latter, we have seen that the description can only be done in a statistical manner, and we have introduced the mathematical apparatus necessary to describe the frequency content on turbulent flows. We have also introduced the analysis framework used to determine the statistics of the response of a linear system (a flexible aircraft) to a random excitation (atmospheric turbulence). This has led to a process for the generation of synthetic, or simulated, turbulence, by means of linear filters, which in later chapters will enable the simulation of the expected response of flexible aircraft and wings to atmospheric turbulence.

Wind shear, on the other hand, is described by a spatially varying *time-averaged* velocity field. Two important situations have been discussed: First, we have outlined the characteristics of the wind profile found near the ground, known as the ABL. Its features are critical in takeoff and landing, and we have discussed the main criteria for establishing atmospheric stability conditions. There are many available models for the ABL, and we have only introduced the most common ones used in wind engineering and outlined the potential for direct simulation of atmospheric flows using state-of-the-art high-performance computing. Second, we have discussed discrete "1-cosine" gusts that, despite being ad hoc velocity distributions (and therefore, extremely unlikely to ever occur in real flight), play a critical role in aircraft certification and analysis.

Finally, in Section 2.5, we have discussed a different situation in which the background velocity field is defined not by naturally occurring phenomena but through the aerodynamic interaction between two aircraft sharing a common airspace. We have then outlined the key characteristics of aircraft wakes, identifying four stages for the evolution and eventual dispersion of a vortex wake. A simple model based on a set of counter-rotating vortices has finally been utilized to approximate the background velocity field in WVEs.

2.7   Problems

Exercise 2.1. Show that the integral timescale for a scalar random variable $w(t)$ satisfies $T = (\pi/2)\left(\Phi_w(0)/\sigma_w^2\right)$.

Exercise 2.2. Noting that the definitions in this chapter also apply to deterministic processes, consider a signal given by the analytical expression $w(t) = 1 + \sin t$. Show that:

(i) its mean value is $\bar{w} = 1$;
(ii) its autocorrelation function $\phi_w(\tau) = 1 + \frac{1}{2}\cos\tau$;
(iii) its spectral density is $\Phi_w(\omega) = \delta(\omega) + \frac{1}{2}\delta(\omega - 1)$, for $\omega \geq 0$, with $\delta(\omega - \omega_0)$ being Dirac's delta function;
(iv) its mean square value is $\sigma_w = \frac{3}{2}$ and it satisfies Equation (2.18).

Exercise 2.3. Show that $\frac{\partial \phi_w(\tau)}{\partial \tau} = -\phi_{\dot{w}}(\tau)$, where we have used the standard definition, namely $\phi_{\dot{w}}(\tau) = \mathrm{E}[\dot{\mathbf{w}}(t + \tau)\dot{\mathbf{w}}^\top(t)]$.

Exercise 2.4. Consider again Example 2.1, for which the source codes are available in the companion site. Include a shift of -0.5 s on the second time signal, $w_2(t)$, and compute again the autocorrelation function. Investigate the asymmetry on the resulting cross-correlation functions by changing the integral timescale in both signals. Note that, for this, you will need to modify the transfer function used to generate each signal.

Exercise 2.5. Consider the Dryden turbulence model, Equation (2.45).

(i) Build a turbulence filter for the three components of the velocity field.
(ii) Show that the longitudinal and vertical velocity components satisfy Equation 2.42.

Exercise 2.6. Starting from a general nonrational FRF, a strategy to obtain a minimum-phase state-space model in Matlab goes as follows: first, a frequency-response data (FRD) model is generated by sampling the function in the frequency range of interest (Matlab function `frd`); second, the resulting FRD model is fitted into a minimum-phase state-space description using log-Chebyshev magnitude design (Matlab function `fitmagfrd`).

Using this strategy, approximate the synthetic turbulence filter to the von Kármán model, Equation (2.48), with state-space models of orders 4, 8, and 12, and show the convergence of their corresponding PSDs against von Kármán's expression.

Exercise 2.7. The short-period dynamics of an aircraft under atmospheric gusts can be described in the first approximation by the second-order system

$$\dot{\mathbf{x}} = \begin{bmatrix} 0 & 1 \\ -\omega_o^2 & -2\xi\omega_0 \end{bmatrix} \mathbf{x} + \begin{bmatrix} 0 \\ \beta \end{bmatrix} w_g,$$

where the state vector includes the instantaneous pitch angle and pitch rate, $\mathbf{x}(t) = \begin{bmatrix} \theta & \dot{\theta} \end{bmatrix}^\top$, and β includes the aerodynamic derivatives for the quasi-steady loading on the aircraft that flies with velocity V due to the instantaneous vertical gust velocity, $w_g(t)$.

(i) Compute the FRF of the aircraft pitch to gust vertical velocity.
(ii) If the gust profile corresponds to a continuous turbulence spectrum described by the Dryden model with intensity σ, obtain the PSD of the pitch response.

Exercise 2.8. Consider the response to seismic loading of the 2-DoF system of the figure. It is composed of two masses, m_1 and m_2, two extensional springs, k_1 and k_2, and two viscous dampers, c_1 and c_2. If z_1 and z_2 are the relative vertical coordinates of the masses with respect to the moving ground, its equations of motion are

$$m_1\ddot{z}_1 + (k_1 + k_2)z_1 - k_2z_2 = -m_1\ddot{z}_0,$$
$$m_2\ddot{z}_2 - k_2z_1 + k_2z_2 = -m_2\ddot{z}_0.$$

The spectral density of the ground acceleration is assumed to be a low-pass noise with a cutoff frequency ω_c, that is,

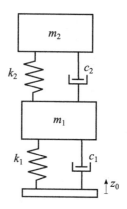

$$\Phi_{\ddot{z}_0} = \begin{cases} S_0 & \text{if } |\omega| < \omega_c \\ S_0/2 & \text{if } |\omega| = \omega_c \\ 0 & \text{if } |\omega| > \omega_c. \end{cases}$$

(i) Compute the admittance between the acceleration at the base and the displacements of the masses.

(ii) Compute the RMS value of the displacements of both masses.

Exercise 2.9. A large transport aircraft has a wingspan of 80 m, a mean wing chord of 10 m, and a maximum landing weight of 385 tons. If the aircraft lands at 75 m/s, compare the spatial distribution of the velocities in its wake to the vertical discrete gusts. Use $F_g = 0.9$ for the flight profile alleviation factor.

3 Dynamics of Elastically Supported Rigid Airfoils

3.1 Introduction

In Chapter 1, we have introduced the classical scope of aeroelasticity defined as a feedback interconnection (a coupled system) between the aerodynamic and structural response of streamlined deformable bodies in an airstream. We have also highlighted its relevance in aircraft analysis and design. This chapter presents an overview of the basic concepts of linear aeroelasticity using the now classical example of a rigid airfoil with a trailing-edge flap mounted on linear springs and in a low-speed nonstationary flow. Under the assumptions of small-amplitude oscillations and potential flow aerodynamics, this configuration can be studied using linear models for which there are well-known analytical solutions. They serve to illustrate most of the coupled physics found in the dynamic response of flexible wings and aircraft at subsonic speeds and under attached flow conditions, which will be the focus of subsequent chapters in this book.

First, the chapter presents the equations of motion describing the dynamics of an elastically supported airfoil. Then it proceeds to review, in some detail, the main results in linear unsteady aerodynamics of thin airfoils using the closed-form solutions in the frequency domain of Theodorsen and Garrick (1934) and Sears (1941). While most of the linear analyses of aeroelastic systems can be done in the frequency domain, particularly to assess their stability characteristics, the focus here is on time-domain state-space representations (Figure 1.10). For that purpose, we introduce rational-function approximations (RFAs) of the analytical frequency-response functions, which are then used to obtain state-space realizations of the unsteady aerodynamics. Using this representation, we finally investigate the static and dynamic stability character-istics of the resulting coupled aeroelastic system (leading to the definition of flutter and divergence), as well as their response to external disturbances, defined by either discrete gusts or continuous turbulence, as introduced in Chapter 2.

Most of the material in this chapter is covered in various forms in most textbooks on aeroelasticity. Bisplinghoff et al. (1955) and Fung (2008), in particular, are clas-sical references that include full analytical derivations of the unsteady aerodynamic theory of thin airfoils oscillating in an incompressible flow, as well as the frequency-domain methods to estimate flutter onset and continuous turbulence response. A more succinct description is included in Dowell et al. (2004), which also includes an intro-duction to state-space methods for time-domain analysis and control design, albeit built on a vortex-lattice solution of the airfoil aerodynamics (which is discussed in Chapter 7). Additional details on state-space methods for aeroelasticity can be found in the monograph by Lind and Brenner (2012). Finally, Hodges and Pierce (2014) and Wright and Cooper (2007) include useful summaries of the main results for thin-airfoil unsteady aerodynamics in the frequency domain, as well as the most common frequency-matching numerical methods used to determine flutter onset conditions.

3.2   Equations of Motion of a 2-DoF Elastically Supported Airfoil

As was mentioned in Section 3.1, in this section, we consider a thin rigid airfoil of chord c and unit span that is allowed to move within a freestream at a constant speed V and density ρ. The airfoil is supported by two massless springs attached to a single location, as schematically shown in Figure 3.1. This results in two rigid-body degrees of freedom (DoF): the rotation of the airfoil about the attachment point (or *pitch*, $\alpha(t)$, defined as positive nose up) and the vertical displacement (known as *plunge* or *heave*, $h(t)$, positive down). They are assumed to be always small. In particular, we assume that $\tan \alpha \approx \alpha$ and $\tan \frac{h}{V} \approx \frac{h}{V}$. Many physical realizations of this basic system have been built for wind-tunnel testing and are often referred to as *pitch-and-plunge apparatus* (Block and Strganac, 1998).

The airfoil may also have a trailing-edge control surface with a (known) commanded input, $\delta(t)$, also assumed to be small. The flap hinge is located at a distance x_{fh} from the leading edge, as shown in Figure 3.1. The hinge is defined as a point discontinuity along the airfoil camber line. Unless otherwise stated, x is the horizontal coordinate from the leading edge of the airfoil.

The feedback mechanism between aerodynamics and structural dynamics for this problem is shown in the block diagram (Figure 3.2). The aerodynamics of the airfoil determines the instantaneous lift and moment for the given kinematics of the airfoil and commanded flap deflections, while the governing equations of the underlying

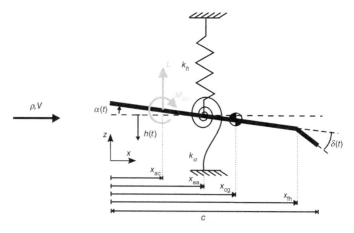

Figure 3.1 Physical system description of a 2-DoF airfoil with a trailing-edge control surface.

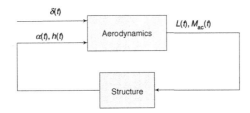

Figure 3.2 Feedback mechanism for a 2-DoF airfoil with the control surface.

mass-spring structural system determine the airfoil dynamics in response to the lift and moment.

From an *aerodynamic* point of view, the problem is therefore to evaluate the (instantaneous) forces exercised by the fluid on the airfoil as a function of (h, α, δ), which may vary with time. In this chapter, the airfoil is assumed to have zero thickness, that is, to be a rigid flat plate, for which analytical solutions are readily available. It is also set to have zero initial angle of attack, and therefore when $\alpha = 0$ and $\delta = 0$, it is a horizontal line on a horizontal freestream, which produces no aerodynamic forces. Let $\Delta p(x,t) = p_u - p_l$ be the instantaneous pressure jump between the suction (upper) and pressure (lower) sides of the airfoil, with x begin the chordwise coordinate. If the freestream dynamic pressure is given as $\frac{1}{2}\rho V^2$, we define the pressure coefficient along the chord as $c_p(x) = \frac{p}{\frac{1}{2}\rho V^2}$.

As the airfoil is assumed to be rigid, the interest is in the resultant forces and moments obtained from that pressure distribution. The reference point for their evaluation is chosen at the *aerodynamic center*, which is defined as the point along the chord for which a change of steady angle of incidence generates no change in pitch moment. Note that this is a steady-state definition, and, as we will see in Section 3.3.3, fast enough changes in pitch create an instantaneous aerodynamic moment. The chordwise coordinate of the aerodynamic center, measured from the airfoil leading edge, is x_{ac}. The resulting aerodynamic lift, L, and pitching moment about the aerodynamic center, M_{ac}, both defined per unit span length, are obtained as

$$L = -\frac{1}{2}\rho V^2 \int_0^c \Delta c_p(x)\,dx,$$

$$M_{ac} = \frac{1}{2}\rho V^2 \int_0^c (x - x_{ac}) \Delta c_p(x)\,dx. \qquad (3.1)$$

Note that the sign convention used in the definition of Δc_p implies that a negative pressure jump generates positive lift. We finally define the usual (dimensionless) sectional aerodynamic coefficients[1]

$$c_l = \frac{L}{\frac{1}{2}\rho V^2 c} \quad \text{and} \quad c_m = \frac{M_{ac}}{\frac{1}{2}\rho V^2 c^2}. \qquad (3.2)$$

---

[1] The convention in this book is to use lowercase symbols for the sectional lift and moment coefficients and uppercase for integrated lift and moments coefficients over a wing or a full aircraft. We also use the airfoil chord to obtain the usual scaling of the aerodynamic coefficients, but we use the semichord instead, as it is traditionally done in aeroelasticity, as the characteristic distance for nondimensionalization elsewhere in this chapter.

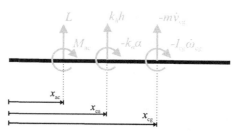

Figure 3.3 Forces and moments on a 2-DoF airfoil.

From a *structural dynamics* point of view, the problem is to evaluate the response of a mass-spring system to those aerodynamic forces. Viscous dampers can be easily included in this description (Ko et al., 1998), but this is not done here since, as we will see later, the aerodynamic loading typically adds (at sufficiently low speeds) high damping to the coupled system. The elastic properties are the stiffness constants for the rotational, k_α, and translational, k_h, springs, as well as the chordwise location where those springs are attached to the airfoil, which are referred to as the *elastic axis*. Although it may seem more appropriate to refer to it as the elastic center, this naming reflects the fact that realizations of this system on a finite span wing are typically built with elastic supports at both ends. The coordinate of the elastic axis from the leading edge is x_{ea} (see Figure 3.1). Finally, the inertial properties are the total mass of the airfoil per unit span, m, and its moment of inertia, I_{cg}, about the section center of mass, x_{cg}. The inertia of the flap is assumed to be negligible in this chapter for simplicity. For large flaps, however, that assumption is no longer valid and a 3-DoF structural system needs to be considered (Li et al., 2010).

In general, the elastic axis, aerodynamic center, and center of mass will be different points along the chord of the airfoil. The equations of motion for the airfoil can be easily obtained from a free-body diagram, including inertia forces (as defined by the D'Alembert principle), restoring forces and moments by the springs, and aerodynamic forces. Under the assumptions given above, and defining v_{cg} as the vertical velocity and ω_{cg} as the angular velocity around the center of mass, one has the forces and moments shown in Figure 3.3. Note that the control surface does not explicitly appear in the force balance, as this is a commanded input whose only action (within the assumptions of the current model) is to modify the lift and aerodynamic moment, as shown in Figure 3.2. Summing all vertical forces gives

$$-m\dot{v}_{cg} + k_h h + L = 0, \tag{3.3}$$

while the sum of moments about the elastic axis results in

$$-I_{cg}\dot{\omega}_{cg} + \left(x_{cg} - x_{ea}\right) m\dot{v}_{cg} - k_\alpha \alpha + M_{ac} + L\left(x_{ea} - x_{ac}\right) = 0. \tag{3.4}$$

Finally, the velocities at the center of mass are obtained from the time derivatives of the pitch and plunge degrees of freedom, as $v_{cg} = -\dot{h} - (x_{cg} - x_{ea})\dot{\alpha}$ and $\omega_{cg} = \dot{\alpha}$. After reordering terms, this results in the following equations of motion for the system:

$$m\ddot{h} + m\left(x_{\text{cg}} - x_{\text{ea}}\right)\ddot{\alpha} + k_h h = -L,$$
$$I_\alpha \ddot{\alpha} + m\left(x_{\text{cg}} - x_{\text{ea}}\right)\ddot{h} + k_\alpha \alpha = M_{\text{ac}} + \left(x_{\text{ea}} - x_{\text{ac}}\right)L,$$
(3.5)

where $I_\alpha = I_{\text{cg}} + m\left(x_{\text{cg}} - x_{\text{ea}}\right)^2$ is the moment of inertia about the elastic axis, which has been evaluated using the parallel axis theorem.

3.2.1 Nondimensional Equations of Motion

Introducing the nondimensional plunge position $\eta = \frac{h}{c/2}$, these equations can now be written in matrix form as in Equation (1.4), that is,

$$\mathbf{M}\ddot{\mathbf{q}} + \mathbf{K}\mathbf{q} = \mathbf{f}_q,$$
(3.6)

with the column vectors of generalized coordinates and forces given, respectively, as

$$\mathbf{q} = \begin{Bmatrix} \eta \\ \alpha \end{Bmatrix} \quad \text{and} \quad \mathbf{f}_q = \tfrac{1}{2}\rho V^2 \boldsymbol{\kappa}_g \begin{Bmatrix} c_l \\ c_m \end{Bmatrix}.$$
(3.7)

Here, we have introduced a matrix of airfoil geometric coefficients, defined as

$$\boldsymbol{\kappa}_g = 2 \begin{bmatrix} -1 & 0 \\ \frac{x_{\text{ea}} - x_{\text{ac}}}{c/2} & 2 \end{bmatrix},$$
(3.8)

and the mass and stiffness matrices for this 2-DoF system, defined as

$$\mathbf{M} = \begin{bmatrix} m & m\frac{x_{\text{cg}} - x_{\text{ea}}}{c/2} \\ m\frac{x_{\text{cg}} - x_{\text{ea}}}{c/2} & \frac{I_\alpha}{(c/2)^2} \end{bmatrix} \quad \text{and} \quad \mathbf{K} = \begin{bmatrix} k_h & 0 \\ 0 & \frac{k_\alpha}{(c/2)^2} \end{bmatrix}.$$
(3.9)

The second equation has been rescaled with the semichord to keep the symmetry of the mass matrix. Equation (3.6) defines the structural dynamics of the 2-DoF airfoil, that is, they correspond to the lower block in Figure 3.2. The solution needs to be made in conjunction with some initial conditions at $t = 0$ and an aerodynamic model that determines the lift and moment given airfoil kinematics, which is discussed in detail in the following section. Note that, in aerodynamic design, the main concern would be evaluating the (nondimensional) lift and moment (and also the drag) coefficients for a range of Mach and Reynolds numbers. Here, however, the forcing terms scale with the dynamic pressure, $\frac{1}{2}\rho V^2$, that is, aeroelasticity is concerned with the *absolute values of the aerodynamic forces* since they define the actual loading that deforms the flexible wing structure.

The 2-DoF airfoil problem considered here has the same structure as the one that would be obtained for a straight cantilever wing at zero angle of attack (see Example 6.3 on page 257). The translational and rotational springs can be associated with the bending and torsional stiffness of the wing, respectively, which occur around a spanwise reference line (the elastic axis) that does not necessarily coincide with the local sectional center of mass as in the current model. However, the aerodynamic forces as defined on the airfoil problem here would correspond to rigid-body motions (in plunge and pitch) of a finite wing, where the aerodynamic forces are generated from the change of geometry of the wing as it deforms in, for example, bending and torsion. It should be finally noted that the planar kinematics of the airfoil would also include the displacements along the horizontal axis (the x axis defined in Figure 3.1). In the problem definition, we have implicitly assumed that they are negligible. The main reason for this is that such motions would be the response to the drag forces on the airfoil, which are much smaller than the lift and moment at small angles of attack. Moreover, that motion would correspond to in-plane bending deformations of a 3-D wing, which typically have a much higher stiffness than the out-of-plane bending.

3.2.2   Flutter with a Steady Aerodynamic Model

Before we consider a more detailed aerodynamic model, it can be very illustrative to introduce the basic characteristics of the flutter instability using the previous equations of motion with simple aerodynamics. In particular, we consider the case in which the aerodynamic forces are $c_l = c_{l_\alpha}\alpha$ and $c_m = 0$, namely those on a stationary symmetric airfoil at low angles of attack. In that case, Equation (3.6) results in

$$\mathbf{M\ddot{q}} + \left(\mathbf{K} - \tfrac{1}{2}\rho V^2 \mathcal{A}\right)\mathbf{q} = \mathbf{0}, \tag{3.10}$$

where we have defined the *aerodynamic influence coefficients* matrix as

$$\mathcal{A} = 2c_{l_\alpha} \begin{bmatrix} 0 & -1 \\ 0 & \frac{x_{ea}-x_{ac}}{c/2} \end{bmatrix}. \tag{3.11}$$

Equation (3.10) is a linear and homogeneous ordinary differential equation in the generalized coordinates of the structural problem, $\mathbf{q}$. Their stability is determined by seeking nontrivial solutions of the form $\mathbf{q}(t) = \bar{\mathbf{q}}e^{pt}$, with $\bar{\mathbf{q}}$ and p being complex numbers that need to be determined. This results in the *characteristic polynomial* being written as

$$\det\left(p^2\mathbf{M} + \left(\mathbf{K} - \tfrac{1}{2}\rho V^2 \mathcal{A}\right)\right) = 0, \tag{3.12}$$

which results in a biquadratic equation (a quadratic equation on p^2) of the form

$$a_2 p^4 + a_1 p^2 + a_0 = 0, \tag{3.13}$$

with

$$\begin{aligned}
a_2 &= m\left[I_\alpha - m\left(x_{cg} - x_{ea}\right)^2\right], \\
a_1 &= k_h I_\alpha + k_\alpha m + \tfrac{1}{2}\rho V^2 c c_{l_\alpha} m \left(x_{ac} - x_{cg}\right), \\
a_0 &= k_h \left[k_\alpha + \tfrac{1}{2}\rho V^2 c c_{l_\alpha}\left(x_{ac} - x_{ea}\right)\right],
\end{aligned} \tag{3.14}$$

where we have multiplied all the coefficients by $(c/2)^2$ for convenience. We know that if $V = 0$, the roots of Equation (3.12) correspond to the natural frequencies of the structure, that is, two pairs of purely imaginary solutions (determined by the eigenvalues of the $(\mathbf{K}, \mathbf{M})$ pair). They are denoted here as ω_α and ω_h to indicate whether the mode shape is dominated by the pitching or plunging component, respectively. Next, we

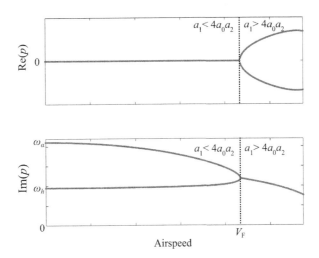

Figure 3.4 Typical evolution of the real and imaginary parts of the roots of Equation (3.12) with the airspeed.

want to assess the system stability as we increase the airspeed, which can be achieved by investigating the sign of the a_i coefficients. First, using the parallel axis theorem for the moment of inertia of the airfoil, it is easy to prove that $x_{cg} - x_{ea} < \sqrt{\frac{I_\alpha}{m}}$, and therefore $a_2 > 0$ always. Second, $a_0 > 0$ if $V < \sqrt{\frac{2k_\alpha}{c\rho c_{l_\alpha}(x_{ea}-x_{ac})}}$. In Section 3.5.1, we will see that this limit value corresponds to *aeroelastic torsional divergence*, which is an upper bound for the admissible velocities (and exists if and only if $x_{ea} > x_{ac}$). Finally, for sufficiently small values of the airspeed, $a_1 > 0$. As the solutions of Equation (3.13) are of the form

$$p^2 = \frac{-a_1 \pm \sqrt{a_1^2 - 4a_0a_2}}{2a_0},$$

(3.15)

we have that p^2 are negative real numbers if $a_1 > 4a_0a_2$, and complex numbers otherwise. As one of the square roots of any imaginary number always has a positive real part, the condition $a_1 = 4a_0a_2$ (noting that this condition always occurs before $a_1 < 0$) defines the transition from the airfoil switching from being (neutrally) stable to unstable. This bifurcation point is known as the *flutter onset* condition. The condition $a_1 = 4a_0a_2$, with a_0, a_1, and a_2 given by Equation (3.14), defines a quadratic equation in the airspeed V. The smallest nonzero solution of that equation, if exists, determines the flutter speed, V_f.

The typical evolution of the real and imaginary parts of p with the airspeed is shown in Figure 3.4. The solutions are two pairs of complex conjugate numbers, but only those with positive imaginary part are shown. For airspeeds below the flutter speed, all roots are purely imaginary, but they approach each other as the airspeed increases. At the flutter speed, V_F, both roots coalesce on a single frequency, the flutter frequency. This is known as *frequency-coalescence flutter*, the result of a nonsymmetric aeroelastic stiffness matrix. Finally, above the flutter speed, that is, when $a_1 < 4a_0a_2$, one of the roots has a positive real part and the system becomes dynamically unstable.

3.3   Unsteady Aerodynamics of Thin Airfoils

The focus shifts now to the evaluation of the nonstationary aerodynamic forces on the moving airfoil, that is, the lift and aerodynamic moment per unit length needed in Equation (3.5). For that purpose, a general overview of the relevant theoretical concepts is first necessary. The aerodynamic forces on thin airfoils are obtained here using 2-D potential theory for incompressible flows, which provides a quite accurate approximation of the pressure distribution in attached flows at low speeds. An excellent description of the underlying theory can be found in the book of Katz and Plotkin (2001); only the main features of the theory and the most relevant solutions for aeroelastic analysis are described here. For detailed derivations of the analytical solutions for thin airfoils, the reader is referred to the classical works of Bisplinghoff et al. (1955, Ch. 5), Fung (2008, Ch. 13), and Johnson (1980, Ch. 10).

In potential theory, the flow is assumed to be inviscid and irrotational. Here it is additionally assumed that it is incompressible.[2] However, it is allowed to be nonstationary. In general, we consider three sources of unsteadiness on airfoil aerodynamics: (1) the rigid-body motions of the body, given here by nonzero pitch and plunge rates and accelerations; (2) nonstationary conditions in the incoming flow, due to the presence of gust fields in the airstream (see Chapter 2), which are assumed here to be along the vertical axis; (3) changes of body geometry, which are restricted here to trailing-edge flaps, although we present first below the more general solution, which is also valid for compliant (but thin) airfoils. A fourth source on unsteadiness is flow separation, which is not captured in potential flow theory and is not discussed in this book. Note finally that very thin airfoils are subject to leading-edge separation at rather low angles at attack, which limits the range of direct applicability of the linear models used in this chapter.

3.3.1   Incompressible Potential Flow Equations

The assumptions of *thin-airfoil theory* are considered. First, it is assumed that the airfoil has very small thickness and camber; and second, it moves in the x–z plane (Figure 3.1) with small amplitudes with respect to an airstream of density ρ and constant horizontal velocity V (a small vertical gust will be added later). In such case, the boundary conditions on the flow can be approximately enforced at all times at $z = 0$, as shown in Figure 3.5a.

---

[2] The compressible theory for oscillating airfoils was first developed by Possio (1938). However, unlike for incompressible conditions, no analytical solutions have been found to that problem (Kier, 2013), and it is treated in this book as a particular case of the 3-D numerical solution of Section 6.5.

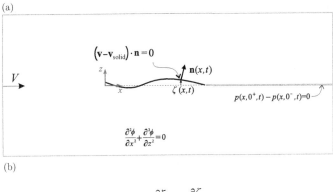

Figure 3.5 Definition of the thin-airfoil problem. (a) Boundary conditions in the velocity field (the dotted and gray lines represent the reference airfoil geometry and the far field, respectively). (b) Vortex-sheet representation.

The instantaneous airfoil shape is $\zeta(x,t)$, with $|\zeta| \ll c$ being the vertical coordinate along the airfoil. It is later defined in terms of the DoFs introduced in Section 3.2, although this is not necessary at this stage. The local velocity along the chordline of the airfoil is then $\mathbf{v}_{\text{solid}}(x,t) = \left(0, \frac{\partial \zeta}{\partial t}\right)$, and the local normal at that point can be approximated as

$$\mathbf{n}(x,t) = \left(-\sin\frac{\partial \zeta}{\partial x}, \cos\frac{\partial \zeta}{\partial x}\right) \approx \left(-\frac{\partial \zeta}{\partial x}, 1\right). \tag{3.16}$$

The 2-D flow velocity vector, $\mathbf{v}$, is written as a perturbation on the freestream velocity, that is,

$$\mathbf{v}(x,z,t) = \left(V + v_x(x,z,t), v_z(x,z,t)\right). \tag{3.17}$$

It is evaluated by enforcing conservation of mass, which for an incompressible fluid implies the zero divergence of the velocity, that is, $\nabla \cdot \mathbf{v} = 0$, or in terms of velocity components, $\frac{\partial v_x}{\partial x} + \frac{\partial v_z}{\partial z} = 0$. From the irrotationality condition ($\nabla \times \mathbf{v} = \mathbf{0}$, i.e., $\frac{\partial v_z}{\partial x} - \frac{\partial v_x}{\partial z} = 0$), there is also a velocity potential corresponding to the velocity field, that is, $\mathbf{v} = \nabla\phi$. As a result, incompressible potential flows are described by the Laplace equation on the velocity potential, namely

$$\nabla \cdot \nabla\phi = \frac{\partial^2 \phi}{\partial x^2} + \frac{\partial^2 \phi}{\partial z^2} = 0. \tag{3.18}$$

The difference in tangential-flow velocities between the suction and pressure sides of the airfoil at its trailing edge generates a downstream thin-shear layer that is known as the *wake* of the airfoil. For 2-D inviscid flows, this is represented by a discontinuity in the velocity field, which is further assumed to lay along the x axis in (linear) thin-airfoil theory, as shown in Figure 3.5a. As a result, we have the following boundary conditions on the fluid:

1. A far-field condition that enforces that the fluid moves with the freestream velocity far enough from the body,

$$v_x(x,z,t)\big|_{x^2+z^2\to\infty} = 0. \tag{3.19}$$

2. The nonpenetrating boundary condition on the airfoil, that is, zero normal relative velocity in the solid/fluid interface, $(\mathbf{v} - \mathbf{v}_{\text{solid}}) \cdot \mathbf{n} = 0$. This can be written as $\nabla\phi \cdot \mathbf{n} = \mathbf{v}_{\text{solid}} \cdot \mathbf{n}$, which is a *Neumann boundary condition* on the velocity potential. Under linear assumptions, the local velocity of the fluid is given by Equation (3.16), while the normal is that of Equation (3.17), such that the nonpenetrating condition becomes

$$v_z(x,0^{\pm},t) = \frac{\partial\zeta}{\partial t} + V\frac{\partial\zeta}{\partial x} \quad \text{for } 0 \leq x \leq c. \tag{3.20}$$

3. A zero pressure jump on the airfoil wake along the x axis. This can be written as

$$\Delta p(x,0,t) = 0 \quad \text{for } c < x < \infty. \tag{3.21}$$

The pressure field is obtained from the general form of Bernoulli's principle for unsteady flows.[3] If p_∞ is the static pressure, then

$$p + \frac{1}{2}\rho\|\mathbf{v}\|^2 + \rho\frac{\partial\phi}{\partial t} = p_\infty + \frac{1}{2}\rho V^2, \tag{3.22}$$

along streamlines. Note that, while the conservation of mass defined a time-independent problem in Equation (3.18), Bernoulli's equation includes the rate of change of the velocity potential. As a result, unsteady flows produce forces on the airfoil not only by the changes in the magnitude of the velocity along the airfoil but also from the local acceleration of the fluid as it follows the moving airfoil. This effectively is a reaction force (according to Newton's third law) that is known as *apparent mass* and is discussed in some detail in Section 3.3.3. It should also be noted that the time derivative in Equation (3.22) needs to be evaluated with respect to a inertial reference frame (Drela, 2014, Ch. 7).

The interest here is, in particular, in the pressure jump along the airfoil and its wake (i.e., across the x axis for $x > 0$ in our infinitesimally thin-airfoil model). In this case, and recalling that $\Delta\bullet = \bullet_u - \bullet_l$, the difference between the values of any given magnitude between the suction (upper) and pressure (lower) sides of the airfoil can be given by

---

[3] Bernoulli's equation is obtained from integration along streamlines of the momentum equation, which, for an inviscid fluid, can be written in terms of the velocity potential as $\frac{\partial}{\partial t}\nabla\phi + \nabla\phi \cdot \nabla(\nabla\phi) + \frac{1}{\rho}\nabla p = 0$. In a component form, the convective term can further be written as $\nabla\phi \cdot \nabla\frac{\partial\phi}{\partial x_i} = \nabla\phi \cdot \frac{\partial}{\partial x_i}\nabla\phi = \frac{1}{2}\frac{\partial}{\partial x_i}(\nabla\phi \cdot \nabla\phi)$. Therefore, $\nabla\left[\frac{\partial\phi}{\partial t} + \frac{\|\mathbf{v}\|^2}{2} + \frac{p}{\rho}\right] = 0$.

$$\Delta p = \frac{\rho}{2}(\mathbf{v}_l \cdot \mathbf{v}_l - \mathbf{v}_u \cdot \mathbf{v}_u) + \rho\frac{\partial}{\partial t}(\phi_l - \phi_u)$$

$$= \frac{\rho}{2}(\mathbf{v}_l + \mathbf{v}_u) \cdot (\mathbf{v}_l - \mathbf{v}_u) + \rho\frac{\partial}{\partial t}(\phi_l - \phi_u). \tag{3.23}$$

The first term includes the average velocity $\frac{\mathbf{v}_l + \mathbf{v}_u}{2}$ that can be approximated by the freestream velocity V under linear assumptions. As a result, the pressure jump depends uniquely on the derivatives of the potential jump as

$$\Delta p = -\rho\frac{\partial(\Delta\phi)}{\partial t} - \rho V\frac{\partial}{\partial x}(\Delta\phi). \tag{3.24}$$

Note finally that Equation (3.24) is valid both on the airfoil and its wake.

Solution with a Time-Varying Vortex Sheet

Solutions to the 2-D potential flow problem defined by Equations (3.18)–(3.21) are often sought using the boundary element method, and this is also the solution procedure followed here. For the thin-airfoil problem, this is done by introducing, as a fundamental solution, a time-varying vortex sheet (a vortex line since the problem is 2-D) along both the airfoil and the wake, as shown in Figure 3.5b. The instantaneous vortex strength distribution, $\gamma(x,t)$ for $0 \le x < \infty$, is then chosen to satisfy the boundary conditions (Katz and Plotkin, 2001, Ch. 5). By definition, a vortex segment generates a rotating velocity field with constant circulation, that is, the integral of the velocity about a closed circuit enclosing the vortex is constant. As a result, the vortex sheet automatically satisfies both Equation (3.18) (mass conservation) and the far-field equations. Therefore, we only need to enforce the boundary conditions on the airfoil and the wake.

Enforcing the nonpenetrating boundary condition on the airfoil, Equation (3.20), gives

$$V\frac{\partial\zeta}{\partial x} + \frac{\partial\zeta}{\partial t} = -\frac{1}{2\pi}\int_0^c \frac{\gamma_b(\xi,t)}{x-\xi}\,\mathrm{d}\xi - \frac{1}{2\pi}\int_c^\infty \frac{\gamma_w(\xi,t)}{x-\xi}\,\mathrm{d}\xi, \quad \text{for } 0 < x < c, \tag{3.25}$$

where we have distinguished between γ_b and γ_w, the unknown distributions of bound and wake vorticity, respectively. They are shown schematically in Figure 3.5b.

Before enforcing the boundary condition on the wake, we should note that, while there is no jump in the normal (vertical) velocity across the vortex sheet, the tangent (horizontal) velocity depends directly on the local vortex strength,[4] with $u_u(x,t) = \frac{\gamma(x,t)}{2}$ and $u_l(x,t) = -\frac{\gamma(x,t)}{2}$. Therefore, we have a direct relation between the vortex sheet strength and the corresponding potential velocity jump, written as

$$\gamma(x,t) = \frac{\partial(\Delta\phi(x,t))}{\partial x}. \tag{3.26}$$

---

[4] The horizontal velocity field generated by an arbitrary vortex sheet of length ℓ located along the x axis is $u(x,z) = \frac{1}{2\pi}\int_\ell \frac{z\gamma(\xi)}{(x-\xi)^2+z^2}\,\mathrm{d}\xi$. In the limit, $u(x,0^+) = \frac{\gamma(x)}{2\pi}\lim_{z\to 0+}\int_\ell \frac{z}{(x-\xi)^2+z^2}\,\mathrm{d}\xi$, which can be integrated with the change of variable $\lambda = \frac{x-\xi}{z}$ as $\lim_{z\to 0+}\int_\ell \frac{z^2}{\lambda^2 z^2 + z^2}\frac{\mathrm{d}\xi}{z} = \int_{-\infty}^\infty \frac{\mathrm{d}\lambda}{\lambda^2+1} = \tan^{-1}\lambda\Big|_{-\infty}^\infty = \pi$.

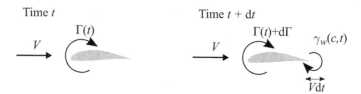

Figure 3.6 Kutta–Joukowski condition on thin-airfoil unsteady aerodynamics.

We can now consider the condition of zero static pressure jump across the wake, Equation (3.21). By differentiating Equation (3.24) along x and using Equation (3.26), one obtains

$$\frac{\partial \gamma_w}{\partial t} + V \frac{\partial \gamma_w}{\partial x} = 0, \qquad (3.27)$$

that is, the condition of no pressure jump along the wake under linear theory assumptions implies that the wake vorticity is convected downstream with constant velocity V. A more generic form of this expression is considered in Section 7.3, where vorticity is convected with the actual flow velocity.

Equation (3.27) needs boundary conditions at the onset of the wake, that is, at the trailing edge of the airfoil. They are obtained from the conservation of circulation around closed curves in an irrotational fluid (*Kelvin's theorem*). Defining the total circulation on the airfoil as $\Gamma(t) = \int_0^c \gamma_b(\xi, t)\, d\xi$, the instantaneous wake vorticity at the trailing edge is assumed to be equal and opposite to the instantaneous changes in circulation on the airfoil. This is schematically shown in Figure 3.6 and can be written as the following relation over an infinitesimal time step dt,

$$\gamma_w(c, t) V\, dt = -d\Gamma(t). \qquad (3.28)$$

This is a particular form of the *Kutta–Joukowski condition* at the trailing edge of nonstationary thin airfoils. As a result, the local vorticity at a point x downstream from the airfoil (with x as before, measured from the leading edge) and at time t can be obtained by tracking it back to the point when it was shed at the trailing edge, that is, $\gamma_w(x, t) = \gamma_w(c, t - \frac{x-c}{V})$, for $x > c$. We can obtain the instantaneous wake vorticity from the history of circulation changes on the airfoil as

$$\gamma_w(x, t) = -\frac{\dot{\Gamma}(t - \frac{x-c}{V})}{V}. \qquad (3.29)$$

Therefore, in nonstationary potential flows, Kelvin's theorem results in an airfoil wake that keeps the history of all past circulation changes.

Resultant Forces on the Airfoil

From Equation (3.29), we have a closed-form expression for the wake vorticity, $\gamma_w(x, t)$, as a function of the time history of the bound vorticity, $\gamma_b(x, t)$. With this, we can now solve Equation (3.25) for given airfoil motions, $\zeta(x, t)$. A numerical solution to this problem using a lump vortex discretization is presented in Section 7.2, where it will serve as an introduction to unsteady vortex lattice methods for aerodynamic force calculations on 3-D geometries. An analytical solution based on Chebyshev polynomials is used in the rest of this section.

Once the distribution of vorticity is known, the pressure difference between the upper and lower surfaces on the airfoil can be finally obtained from Equation (3.24). Note first that, integrating Equation (3.26) along the streamline at $z = 0$, we have $\Delta\phi(x) = \int_0^x \gamma_b(\xi,t)\,d\xi$. Consequently, the pressure jump from the pressure to the suction side of the airfoil is

$$\Delta p = -\rho V \gamma_b - \rho \frac{\partial}{\partial t} \int_0^x \gamma_b(\xi,t)\,d\xi, \tag{3.30}$$

or, in a nondimensional form,

$$\Delta c_p(x,t) = -\frac{2}{V^2}\left[V\gamma_b + \frac{\partial}{\partial t}\int_0^x \gamma_b(\xi,t)\,d\xi\right]. \tag{3.31}$$

Finally, substituting (3.31) into Equation (3.1) gives the lift and moment (about the aerodynamic center) as

$$L = \rho \int_0^c \left[V\gamma_b(x,t) + \frac{\partial}{\partial t}\int_0^x \gamma_b(\xi,t)\,d\xi\right]dx,$$
$$M_{ac} = -\rho \int_0^c \left[V\gamma_b(x,t) + \frac{\partial}{\partial t}\int_0^x \gamma_b(\xi,t)\,d\xi\right](x - x_{ac})\,dx. \tag{3.32}$$

The expressions above give the forcing terms in the equations of motion of the airfoil, Equation (3.6). If necessary, additional force/moment resultants, such as the aerodynamic moments at the flap hinge, can be obtained from the integration of the pressure distribution of Equation (3.31).

3.3.2   Quasi-Steady Approximation

The first approximation to unsteady effects in the aerodynamic forces comes from assuming that the airfoil moves slowly compared to the flow. This is assessed by scaling the time with the typical time that fluid particles take to cover the airfoil semichord as

$$s = \frac{Vt}{c/2}. \tag{3.33}$$

We refer to this nondimensional time as the *reduced time*.[5] It measures time in terms of semichords of freestream fluid motion, and, for example, $s = 2$ corresponds to the time that a flow particle moving at the freestream velocity takes to cover the chord of the airfoil. The assumption in quasi-steady aerodynamics is that the typical timescales in the airfoil motions correspond to very large reduced times. In that case, the flow particles effectively see the airfoil as stationary, and we can use the tools of steady aerodynamics, albeit with slowly moving boundary conditions,

---

[5] In this book, we use the symbol s to refer to the nondimensional time, as it is usually done in aeroelasticity, and we employ the nonitalicized s for the Laplace variable, which we only sparingly use.

Equation (3.20). Under quasi-steady assumptions, the pressure jump across the airfoil thickness, Equation (3.31), simplifies to

$$\Delta c_p(x) = -2\frac{\gamma_b(x)}{V}. \tag{3.34}$$

As a result, one obtains an integral equation that determines the pressure jump on the airfoil that results from a (given) normal velocity distribution,

$$\frac{\partial \zeta}{\partial x} + \frac{1}{V}\frac{\partial \zeta}{\partial t} = \frac{1}{4\pi}\int_0^c \frac{\Delta c_p(\xi)}{x-\xi}\,d\xi, \tag{3.35}$$

which is *Glauert's airfoil equation*. Once this integral equation is solved, the lift and moment are obtained from Equation (3.1). In classical aerodynamic theory, it is common to describe the flow field around the airfoil with respect to an observer moving with the airfoil itself. It is then convenient to introduce the airfoil–bound *upwash*, $w_b(x,t)$, as the instantaneous relative vertical fluid velocity with respect to the airfoil, which for linear theory simply implies a change of sign (that is, the fluid velocity with respect to the airfoil is equal and opposite to the airfoil velocity with respect to the fluid). Equally, the flow vertical velocity component v_z (which is opposite to the upwash) is referred to as "downwash." The upwash on the airfoil is then written as

$$w_b(x,t) = -v_z(x,0^\pm,t) = -V\frac{\partial \zeta}{\partial x} - \frac{\partial \zeta}{\partial t}, \quad \text{for } x \in [0,c]. \tag{3.36}$$

A closed-form solution to Equation (3.35) can be found using Chebyshev polynomials (Abramowitz and Stegun, 1972) that result in Fourier expansions under the change of variable $x = \frac{c}{2}(1 - \cos\theta)$, for $0 \le \theta \le \pi$, known as Glauert's transformation. Next, we write the bound vortex distribution using the enriched Fourier expansion

$$\frac{\gamma_b(\theta,t)}{V} = 2A_0\cot\frac{\theta}{2} - 4\sum_{n=1}^{\infty} A_n\sin n\theta, \tag{3.37}$$

where the first term (the *enrichment term*) is included to simplify the capture of the integrable singularity on the pressure field on the airfoil associated with the leading-edge suction (Katz and Plotkin, 2001, Ch. 5). Substituting this into Equation (3.35), we obtain the following expansion for the relative normal flow velocity (upwash) on the airfoil:

$$\frac{w_b(\theta,t)}{V} = A_0(t) + 2\sum_{n=1}^{\infty} A_n(t)\cos n\theta. \tag{3.38}$$

The coefficients of this expansion can be easily identified as $A_n = \frac{1}{\pi}\int_0^\pi \frac{w_b(\theta)}{V}\cos n\theta d\theta$, for $n = 0,1,2,\dots$. Note that the coefficients in Equation (3.37) have been chosen, as in Fung (2008), so that the resulting dimensionless upwash can be written as a Fourier cosine series. Different scaling of the series in Equation (3.37) is found in the literature (see, e.g., Ch. 5 of Katz and Plotkin (2001)), but we found this one to enable easier manipulation. The instantaneous pressure forces are now obtained from Equation (3.34), and after integration along the chord, the quasi-steady lift and moment per unit length can be obtained as

$$c_l^{qs}(t) = 2\pi \left[A_0(t) - A_1(t)\right],$$
$$c_m^{qs}(t) = \frac{\pi}{2}\left[A_1(t) - A_2(t)\right]. \tag{3.39}$$

Note that despite the use of an infinite series to represent arbitrary airfoil shapes and kinematics, Equation (3.38), the lift and moment depend only on the first three coefficients of the expansion. As a result, the lift on a stationary (thin) airfoil mainly depends on the angle of incidence and its camber, the moment at the aerodynamic center does not depend on the angle of attack (that is, of A_0), and a reflexed camber line (that mainly modifies A_2) can be used to modify aerodynamic moments with minimum impact on the airfoil lift.

Airfoil with Pitch-Plunge-Flap Degrees of Freedom

Back to the problem defined in Figure 3.1, the instantaneous airfoil shape is defined as

$$\zeta(x,t) = -h(t) - (x - x_{ea})\,\alpha(t) - \mathcal{H}(x - x_{fh})\delta(t), \tag{3.40}$$

where $\mathcal{H}(\bullet)$ is the Heaviside function, defined as $\mathcal{H}(x)=0$, for $x<0$, and $\mathcal{H}(x)=1$ for $x \geq 0$. Substituting Equation (3.40) into Equation (3.20) results in an instantaneous upwash that includes contributions from the pitch angle, $\alpha(t)$; the pitch and plunge rates, $\dot{h}(t)$ and $\dot{\alpha}(t)$, respectively; and the flap deflection, $\delta(t)$, and its rate $\dot{\delta}(t)$. The upwash becomes

$$\frac{w_b(x,t)}{V} = \alpha(t) + \frac{\dot{h}(t)}{V} + \frac{x - x_{ea}}{V}\dot{\alpha}(t) + \left[\delta(t) + \frac{x - x_{fh}}{V}\dot{\delta}(t)\right]\mathcal{H}(x - x_{fh}), \tag{3.41}$$

and its different contributions are shown in Figure 3.7. The left column in the figure shows the airfoil kinematics, while the right column includes the resulting upwash. The coefficients in Equation (3.38) can now be easily obtained for the downwash defined in Equation (3.41) using a computational algebra package. Introducing again the reduced time as described above, $s = \frac{Vt}{c/2}$ (and $\bullet' = \frac{d\bullet}{ds}$), and the nondimensional plunge of Section 3.2, $\eta = \frac{h}{c/2}$, the quasi-steady lift and moment coefficients are finally obtained as

$$c_l^{qs} = c_{l_\alpha}^{qs}\left(\alpha + \eta'\right) + c_{l_{\alpha'}}^{qs}\alpha' + c_{l_\delta}^{qs}\delta + c_{l_{\delta'}}^{qs}\delta',$$
$$c_m^{qs} = c_{m_{\alpha'}}^{qs}\alpha' + c_{m_\delta}^{qs}\delta + c_{m_{\delta'}}^{qs}\delta'. \tag{3.42}$$

Note that one can define an *effective angle of attack* from the combined effect of the instantaneous pitch angle, $\alpha(t)$, and the angle induced by the plunge rate, $\eta' = \frac{\dot{h}}{V}$. From the definition of the aerodynamic center, the moment coefficient about that point is independent of the angle of attack. For incompressible flows, it is found to be at the quarter-chord point of the airfoil from the leading edge, that is, $x_{ac} = c/4$. For the infinitesimally thin airfoil of Figure 3.1, we define next a nondimensional distance from the mid-chord, that is, $\nu(x) = \frac{x - c/2}{c/2}$. For the elastic axis (the point around which the rotation of the airfoil is defined), $\nu_{ea} = \frac{x_{ea} - c/2}{c/2}$, and the constant coefficients in

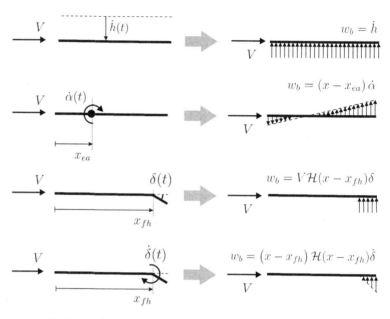

Figure 3.7 Contributions to the relative normal velocity (upwash) on a 2-DoF airfoil with a control surface.

Equation (3.42) can be written as

$$c_{l_\alpha}^{qs} = 2\pi,$$
$$c_{l_{\alpha'}}^{qs} = \left(\tfrac{1}{2} - \nu_{ea}\right) 2\pi,$$
$$c_{m_{\alpha'}}^{qs} = -\frac{1}{4}\pi,$$

(3.43)

for the dependencies with the pitch angle and its derivative. Note in particular that the moment coefficient is negative, which implies that c_m lags the airfoil pitching motion and produces an aerodynamic damping that tends to suppress pitching oscillations (McCroskey, 1982). It also does not depend on the position axis of rotation, that is, the moment due to the pitch rate, $c_{m_{\alpha'}}^{qs}$, is constant. On the contrary, when $\nu_{ea} = 1/2$, that is, when the airfoil is pitching around the 3/4-chord point ($x_{ea} = \frac{3c}{4}$), it is then $c_{l_{\alpha'}}^{qs} = 0$ and the instantaneous pitch rate does not produce lift.

The aerodynamic coefficients associated with the flap deflections are obtained in a similar manner. Defining $\nu_{fh} = \frac{x_{fh} - c/2}{c/2}$ and $\theta_{fh} = \cos^{-1}(-\nu_{fh})$, we have

$$c_{l_\delta}^{qs} = 2\pi - 2\theta_{fh} + 2\sin\theta_{fh},$$
$$c_{l_{\delta'}}^{qs} = \left(\tfrac{1}{2} - \nu_{fh}\right)(2\pi - 2\theta_{fh}) + (2 - \nu_{fh})\sin\theta_{fh},$$
$$c_{m_\delta}^{qs} = -\frac{1 + \nu_{fh}}{2}\sin\theta_{fh},$$
$$c_{m_{\delta'}}^{qs} = -\frac{1}{4}\left[\pi - \theta_{fh} + \tfrac{2}{3}\left(\tfrac{1}{2} - \nu_{fh}\right)(2 + \nu_{fh})\sin\theta_{fh}\right].$$

(3.44)

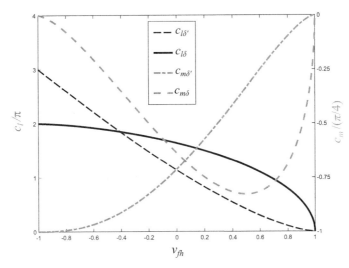

Figure 3.8 Normalized flap-related lift and moment (about the aerodynamic center) coefficients as a function of the nondimensional flap hinge location. [O3Airfoil/static_flap.m]

As defined above, ν_{fh} and ν_{ea} are nondimensional distances from the airfoil mid-chord point, which range from -1 at the leading edge to 1 at the trailing edge. Figure 3.8 shows the (linear-theory) lift and moment coefficients due to the trailing-edge flap deflection as a function of the nondimensional hinge location, ν_{fh}. As can be seen, lift is always positive for positive flap deflection and positive flap rates, while moments about the aerodynamic center are always negative. As $\nu_{fh} \to 1$, the hinge approaches the trailing edge and the flap chord (and therefore any aerodynamic forces) goes to zero. As the hinge moves toward the leading edge, both lift coefficients (c_{l_δ} and $c_{l_{\delta'}}$) and the moment due to the flap angular rate ($c_{m_{\delta'}}$) monotonically grow. However, the moment coefficient due to the static deflection, c_{m_δ}, has a maximum of $-\frac{\sqrt{27}}{8} \approx -0.65$, at $\nu_{fh} = \frac{1}{2}$, the 3/4-chord point. This negative moment has an important effect in the effectiveness of the flaps, as will be seen in Section 5.4. Note finally that the scale of the lift coefficient (the left vertical axis) is one order of magnitude larger than the scale of the moment coefficient (the right axis).

Equation (3.42) has been written in terms of aerodynamic derivatives, for which explicit solutions for a flat plate have been given in Equations (3.43) and (3.44). Note, however, that Equation (3.42) is a generic first-order expansion, and consequently it also holds for more complex flows and geometries undergoing small amplitude dynamics, provided the pitch angle is measured from the zero lifting line. The aerodynamic derivatives in those cases need to be computed by more appropriate aerodynamic models (that take account, for instance, of airfoil thickness), or through wind-tunnel experiments. For example, Brunton and Rowley (2013) have estimated the aerodynamic derivatives, including the unsteady effects discussed in the next section, in a low-Reynolds flow using the direct numerical simulation of the Navier–Stokes equations. Finally, while the derivatives with pitch and plunge of Equation (3.43) are very good approximations for thin airfoils, local separation and the presence of gaps around the flap hinge imply a much larger error in the estimation of flap derivatives using Equation (3.44).

3.3.3   Frequency-Domain Unsteady Forces

The assumption in Section 3.3.2 is that changes on the fluid momentum due to the airfoil motions convect downstream very quickly relatively to the characteristic timescale in the airfoil kinematics. Therefore, the fluid behaves as if the airfoil was stationary. As the typical frequency of the airfoil oscillations increases, the convecting wake of the airfoil has sufficient time to interact with the flow on the airfoil before any steady-state equilibrium is reached, and this needs to be included in the analysis. A closed-form analytical solution to the problem defined by Equations (3.25), (3.29), and (3.32) exists in the frequency domain (Theodorsen and Garrick, 1934). First, we define the *reduced frequency* as the frequency ω of the airfoil oscillations normalized with the convection timescale on the fluid (V/c), that is,

$$k = \frac{\omega c}{2V}, \tag{3.45}$$

where we have introduced a factor of 2 for consistency with classical formulations that use the semichord as typical length scale. Assuming harmonic oscillations of the airfoil, $w_b(x,t) = \overline{w}_b e^{i\omega t}$ (or, analogously, $w_b(x,s) = \overline{w}_b e^{iks}$), the problem can be solved using again a Chebyshev expansion of the upwash, Equation (3.38), but now with frequency-domain coefficients, $\overline{A}_n(ik)$, such that $A_n(s) = \overline{A}_n(ik)e^{iks}$. Closed-form expressions for the airfoil lift and moment about the aerodynamic center exist in terms of those coefficients and can be written as (Fung, 2008, Ch. 13)

$$\overline{c_l}(ik) = 2\pi \left[\left(\overline{A}_0 - \overline{A}_1 \right) \mathcal{C}(ik) + \left(\overline{A}_0 - \overline{A}_2 \right) \frac{ik}{2} \right],$$
$$\overline{c_m}(ik) = \frac{\pi}{2} \left[\left(\overline{A}_1 - \overline{A}_2 \right) + \left(\tfrac{1}{2}(\overline{A}_1 - \overline{A}_3) - (\overline{A}_0 - \overline{A}_2) \right) \frac{ik}{2} \right]. \tag{3.46}$$

These relations are known as the *Küssner–Schwartz formulas* in unsteady aerodynamics of thin airfoils. The complex function $\mathcal{C}(ik)$ is given as

$$\mathcal{C}(ik) = \frac{K_1(ik)}{K_0(ik) + K_1(ik)}, \tag{3.47}$$

where $K_\nu(ik)$ are the modified Bessel functions of the second kind[6] (Abramowitz and Stegun, 1972). $\mathcal{C}(ik)$ is known as Theodorsen's *lift-deficiency function* and characterizes the effect of the unsteady wake on the airfoil, which creates both a time delay and an amplitude reduction between the airfoil oscillations and the resulting aerodynamic forces. The amplitude and the phase of the lift-deficiency function are shown in Figure 3.13. Note that in the limit, we have $\mathcal{C}(0) = 1$. Equation (3.46) gives the

---

[6] In most textbooks, this expression is written in terms of Hankel functions of the second kind, $H_\nu^{(2)}(k)$, since $K_\nu(ik) = \frac{\pi}{2}(-i)^{\nu+1} H_\nu^{(2)}(k)$. We prefer to make explicit the dependency with ik for consistency with the notation used for other frequency-response functions.

lift and moment for a harmonic upwash with an arbitrary shape, which is defined by either airfoil motions or an incoming gust. Note that only the first four terms in the infinite expansion of the upwash, Equation (3.38), contribute to the lift and moment coefficients. Any higher order effects in the Glauert expansion, for example, in the deformations of a compliant airfoil (or in a flying carpet (Argentina et al., 2007)), would generate zero resultants. This is similar to what we have already found in the quasi-steady solution, Equation (3.39); however, a time-varying reflexed camber also contributes to the non stationary component of the lift.

The instantaneous lift and moment in (3.46) have two contributions. The first term is known as the *circulatory* part and comes from the same physical mechanism as the quasi-steady solution, Equation (3.39), with the difference that the unsteadiness in the wake introduces the modulation of the lift that is captured by $\mathcal{C}(ik)$. For $k \to 0$ (slow airfoil oscillations), $\mathcal{C}(0) = 1$, and the lift goes indeed to the steady lift. Interestingly, the lift due to the wake acts always at the quarter-chord point, and consequently, the circulatory part to the aerodynamic moments remains unchanged from the quasi-steady solution. The second term within the brackets grows linearly with the reduced frequency for both the lift and the moment in Equation (3.46), and it is known as the *apparent mass* (or noncirculatory) contribution. It is negligible when the frequency is small enough, but since the amplitude of the lift-deficiency function is never bigger than 1, it is the dominant term for large enough frequencies. Apparent mass effects result from the inertia of the flow that is displaced by the airfoil and becomes relevant when the ratio of fluid mass to airfoil mass is relatively high. This occurs when either a wing is moving on a dense fluid (i.e., hydrofoils in water) or when the airfoil mass is very small (i.e., the wings of a paraglider or a solar-powered aircraft).

Airfoil with Pitch-Plunge-Flap Degrees of Freedom

For the particular case of the 2-DoF airfoil with a control surface of Figure 3.1, the airfoil oscillations are described by $\eta(s) = \bar{\eta}e^{iks}$, $\alpha(s) = \bar{\alpha}e^{iks}$, and $\delta(s) = \bar{\delta}e^{iks}$. In that case, the lift and moment coefficients (with moments about the aerodynamic center) in Equation (3.46) correspond to the classical solutions of Theodorsen and Garrick (1934), namely

$$\bar{c}_l(ik) = \mathcal{C}(ik)\overline{c_l^{qs}} + \overline{c_l^{nc}},$$
$$\bar{c}_m(ik) = \overline{c_m^{qs}} + \overline{c_m^{nc}}, \tag{3.48}$$

where the superindex indicates the quasi-steady and noncirculatory components. Expressions for the circulatory terms are those obtained in the quasi-steady problem, Equation (3.42), only that now we assume the harmonic motion of the DoFs, that is,

$$\overline{c_l^{qs}}(ik) = \left(c_{l_\alpha}^{qs} + ikc_{l_{\alpha'}}^{qs}\right)\bar{\alpha} + ikc_{l_\alpha}^{qs}\bar{\eta} + \left(c_{l_\delta}^{qs} + ikc_{l_{\delta'}}^{qs}\right)\bar{\delta},$$
$$\overline{c_m^{qs}}(ik) = ikc_{m_{\alpha'}}^{qs}\bar{\alpha} + \left(c_{m_\delta}^{qs} + ikc_{m_{\delta'}}^{qs}\right)\bar{\delta}, \tag{3.49}$$

where all the coefficients have already been defined. The noncirculatory terms are directly computed from the terms in Equation (3.46) that include $\frac{ik}{2}$. The linear dependency with the reduced frequency implies one additional time derivative in the expressions that determine the force and moment coefficients from the DoFs. As a result, and when written in the time domain (i.e., $c_x^{nc}(s) = \overline{c_x^{nc}}(ik)e^{iks}$), we have

$$
\begin{aligned}
c_l^{nc}(s) &= c_{l_{\alpha'}}^{nc}\left(\alpha'+\eta''\right) + c_{l_{\alpha''}}^{nc}\alpha'' + c_{l_{\delta'}}^{nc}\delta' + c_{l_{\delta''}}^{nc}\delta'', \\
c_m^{nc}(s) &= c_{m_{\alpha'}}^{nc}\left(\alpha'+\eta''\right) + c_{m_{\alpha''}}^{nc}\alpha'' + c_{m_{\delta'}}^{nc}\delta' + c_{m_{\delta''}}^{nc}\delta''.
\end{aligned}
\tag{3.50}
$$

In these expressions, the coefficients that depend on the pitch angle and its time derivatives are

$$
\begin{aligned}
c_{l_{\alpha'}}^{nc} &= \pi, & c_{m_{\alpha'}}^{nc} &= -\frac{\pi}{4}, \\
c_{l_{\alpha''}}^{nc} &= -\pi\nu_{\text{ea}}, & c_{m_{\alpha''}}^{nc} &= -\frac{\pi}{4}\left(\frac{1}{4}-\nu_{\text{ea}}\right),
\end{aligned}
\tag{3.51}
$$

while the terms that depend on the flap deflection angle and its time derivatives become

$$
\begin{aligned}
c_{l_{\delta'}}^{nc} &= \pi - \theta_{\text{fh}} - \nu_{\text{fh}}\sin\theta_{\text{fh}}, \\
c_{l_{\delta''}}^{nc} &= -\nu_{\text{fh}}\left(\pi-\theta_{\text{fh}}\right) + \frac{1}{3}\left(2+\nu_{\text{fh}}^2\right)\sin\theta_{\text{fh}}, \\
c_{m_{\delta'}}^{nc} &= -\frac{1}{4}\left[\pi - \theta_{\text{fh}} + \left(\frac{2}{3}-\nu_{\text{fh}}-\frac{2}{3}\nu_{\text{fh}}^2\right)\sin\theta_{\text{fh}}\right], \\
c_{m_{\delta''}}^{nc} &= -\frac{1}{4}\left[\left(\frac{1}{4}-\nu_{\text{fh}}\right)\left(\pi-\theta_{\text{fh}}\right) + \left(\frac{2}{3}-\frac{5\nu_{\text{fh}}}{12}+\frac{\nu_{\text{fh}}^2}{3}+\frac{\nu_{\text{fh}}^3}{6}\right)\sin\theta_{\text{fh}}\right].
\end{aligned}
\tag{3.52}
$$

The only geometric constants are again the nondimensional coordinates of the elastic axis, $\nu_{\text{ea}} = \frac{x_{\text{ea}}-c/2}{c/2}$, and the flap hinge, $\nu_{\text{fh}} = \frac{x_{\text{fh}}-c/2}{c/2}$, as introduced in Equation (3.42). Recall also that $\nu = -\cos\theta$. In particular, ν_{ea} is typically a negative number, which means that the elastic axis is closer to the leading than to the trailing edge. On the contrary, ν_{fh} is normally larger than zero, as the flap hinge is near the trailing edge. Note that all the previous coefficients are easily obtained with just a few lines of code in a computational algebra package, although Theodorsen and Garrick (1934) managed very well obtaining those coefficients without computers. In spite of the assumptions of potential flow theory, aerodynamic loads predicted by Equation (3.48), particularly with respect to pitch and plunge motions, have been seen to compare very well with wind-tunnel data (see, e.g., Krzysiak and Narkiewicz, 2006). There is a larger error on the forces and moments resulting from flap deflections, as the mechanical design of flaps typically results in local separation that is not captured in potential theory. Equation (3.48) gives important insights into the key effects in unsteady aerodynamics, as can be seen in the next example.

Example 3.1 Lift Due to Oscillatory Pitch Motions.   Consider a thin airfoil with harmonic pitch oscillations around a point with coordinate x_{ea} from the leading edge. The pitch is then $\alpha(s) = \bar{\alpha}e^{iks}$, whereas the lift is given as $c_l(s) = \bar{c_l}e^{iks}$. We can then define the frequency-response function (see Section 1.4.2) between both magnitudes, as shown in Figure 3.9, which is given by

$$
G_{l\alpha}(ik) = \frac{\overline{c_l}(ik)}{\overline{\alpha}(ik)} = C(ik)\left(c_{l\alpha}^{qs} + ikc_{l_{\alpha'}}^{qs}\right) + \left(ikc_{l_{\alpha'}}^{nc} - k^2 c_{l_{\alpha''}}^{nc}\right).
\tag{3.53}
$$

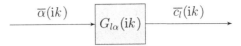

Figure 3.9 Frequency-response function between the pitch angle and the lift coefficient.

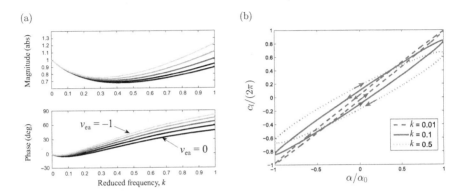

Figure 3.10 Frequency-response function between the pitch angle and the lift coefficient, $G_{l\alpha}$, and instantaneous values of both magnitudes. (a) $G_{l\alpha}/(2\pi)$ for $\nu_{ea} = 0, -0.25, -0.5, -0.75$ and -1. (b) Instantaneous lift for $\alpha = \alpha_0 \cos(ks)$ and $\nu_{ea} = -0.25$ (arrows indicate trajectory with time) [03Airfoil/GLalpha.m]

Substituting the analytical expressions for the coefficients for quasi-steady and noncirculatory contributions, Equations (3.43) and (3.51), respectively, yields

$$G_{l\alpha}(ik) = 2\pi\mathcal{C}(ik)\left(1 + ik\left(\tfrac{1}{2} - \nu_{ea}\right)\right) + \pi\left(ik + k^2\nu_{ea}\right). \tag{3.54}$$

Figure 3.10a shows the magnitude and phase of the frequency-response function for reduced frequencies between 0 and 1, and pitch axis (elastic axis) position varying between $\nu_{ea} = 0$ (mid-chord point) and $\nu_{ea} = -1$ (leading edge), in increments of $\Delta\nu_{ea} = -0.25$. At low reduced frequencies, the circulatory effects are dominant and, therefore, the amplitude of the lift initially drops. As the reduced frequency increases, the noncirculatory terms dominate and the amplitude goes up. This happens, however, at rather high reduced frequencies, since the typical situations in fixed-wing aircraft aeroelasticity rarely exceed $k = 0.5$ (where potential flow theory starts to fail anyway). Lift is initially delayed (negative phase) with respect to the pitch oscillations, again, due to the relative importance at low frequencies of the circulatory terms and the negative phase of the lift-deficiency function (see Figure 3.13). As the reduced frequency increases, the phase changes sign and becomes positive. Note finally that moving the pitch axis forward increases both the lift magnitude and its phase.

These results can be further understood by examining Figure 3.10b which shows the instantaneous values of the pitch angle and the lift coefficient for $\alpha = \alpha_0 \cos(ks)$ for rotations around the 37.5%-chord point ($\nu_{ea} = -0.25$) and for three values of the reduced frequency. Note the strong hysteresis between the pitch and lift during the cycle. It would not exist in steady flow conditions, that is, $k = 0$, where the solution is a straight line, but it is very significant even for very small values of the reduced frequency. For $k = 0.01$ and $k = 0.1$, the evolution along the plot with time is counterclockwise, which means that the maximum of the pitch angle is reached before the maximum of the lift coefficient. This situation is reversed for $k = 0.5$, where the frequency-response function in Figure 3.10a has a positive phase. As will be seen in Section 3.5.2, these hysteresis loops between airfoil kinematics and the corresponding aerodynamic forces are critical in determining whether the aeroelastic system extracts energy from the flow. In particular, flutter conditions occur when the energy balance in one cycle of oscillations is exactly zero and self-sustained oscillations are possible.

Compliant and Adaptive Airfoils

Although the focus of this book is on actuation using "conventional" trailing-edge control surfaces, Equation (3.32) can also be used to predict the unsteady aerodynamic loads for arbitrary geometry changes of the airfoil (provided the assumption of linear theory still holds). This only requires the computation of the coefficients in the Fourier expansion of the frequency-domain upwash for the airfoil kinematics of interest, which are then substituted into Equation (3.46). This approach has been used, for example, to investigate the nonstationary behavior of airfoils with changing camber deformations (Spielberg, 1953) and of compliant and adaptive trailing edges on rotor blades (Gaunaa, 2006; Kumar and Cesnik, 2015; Peters and Johnson, 1994). Alternative closed-form solutions to the same problem have been obtained using a polynomial expansion by Joseph and Mohan (2021). Compliant wing structures, however, also require the characterization of the sectional flexibility and increase the dimension of the problem (Murua et al., 2010; Su, 2017; Walker and Patil, 2014). Moreover, while adaptive concepts on nonload-bearing substructures, such as the compliant trailing edge of Lu and Kota (2003), are essentially 2-D, when the "morphing" concept affects the overall cross-sectional stiffness of the wing, the problem cannot be decoupled from the spanwise flexibility. In such cases, full 3-D analysis is necessary for a proper aeroelastic characterization (Sahoo and Cesnik, 2002).

3.3.4   Forces Generated by a Sinusoidal Gust

Unsteady thin-airfoil theory can also be used to estimate the forces created on airfoils interacting with a disturbance in the incident flow, that is, a gust encounter. Gusts mainly arise due to turbulence in the atmosphere and are characterized by their frequency content. As discussed in Section 2.3.3, for the length scales under consideration in aircraft applications, we can assume the frozen turbulence hypothesis (Taylor, 1938): The flow disturbance is a traveling gust that moves with the freestream velocity and is uniquely defined by its time history at a single measurement point (i.e., the leading edge of the airfoil).

Furthermore, while atmospheric gusts can occur in any direction, here we only consider those that are normal to the mean flow and, therefore, to the chordline of an airfoil at small angle of attack. Those are typically referred to as *upwash* and act on the airfoil by changing the effective angle of attack similarly to how plunge motion generates lift.

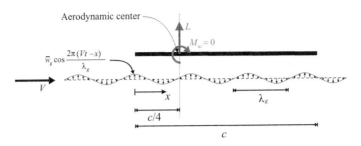

Figure 3.11 Thin airfoil in a sinusoidal vertical gust.

Consider an airfoil rigidly mounted at zero angle of attack and subject to a vertical sinusoidal gust, as shown in Figure 3.11. Under Taylor's assumption, the gust velocity profile can be represented as a traveling wave with horizontal velocity V, that is,

$$w_g(x,t) = \overline{w}_g \cos \frac{2\pi(Vt - x)}{\lambda_g}, \tag{3.55}$$

where λ_g is the wavelength of the gust. We assume again the reference for the spatial coordinates at the leading edge of the airfoil and, importantly, that the gust profile is not affected by the presence of the airfoil. Neglecting again airfoil thickness, that is, modeling the airfoil as a flat plate, Sears (1941) obtained a closed-form solution for the unsteady aerodynamic load. The angular frequency of the gust excitations is $\omega_g = \frac{2\pi V}{\lambda_g}$, and its associated reduced frequency can be defined as

$$k_g = \frac{\omega_g c}{2V} = \frac{2\pi V}{\lambda_g} \frac{c}{2V} = \frac{\pi c}{\lambda_g}. \tag{3.56}$$

Sears showed that the lift due to Equation (3.55) acts at the quarter-chord point of the airfoil at all times, that is, $M_{ac} = 0$, and it is given as

$$\overline{c}_l(ik_g) = c_{l_\alpha}^{qs} \mathcal{S}_0(ik_g) \frac{\overline{w}_g}{V}, \tag{3.57}$$

where $\mathcal{S}_0(ik_g)$ is a complex function, referred to here as *Sears's function with respect to the leading edge.*[7] It is given by

$$\mathcal{S}_0(ik_g) = e^{-ik_g} \mathcal{S}_{1/2}(ik_g), \tag{3.58}$$

with $\mathcal{S}_{1/2}(ik_g)$ being Sears's function about the mid-chord point defined in terms of modified Bessel functions as

$$\mathcal{S}_{1/2}(ik_g) = \frac{1}{ik_g\left(K_0(ik_g) + K_1(ik_g)\right)} \tag{3.59}$$

and, at the limit, $\mathcal{S}_{1/2}(0) = 1$. The input–output relation between gust disturbances and the resulting lift on the airfoil can be viewed as a single-input single-output (SISO) linear dynamical system with the frequency-response function of Figure 3.12, where $G_{lw}(ik_g) = c_{l_\alpha}^{qs} \mathcal{S}_0(ik_g)$.

---

[7] Sears (1941) uses the mid-chord point as the reference for x coordinates.

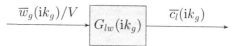

Figure 3.12 Lift due to gust represented as a frequency-response function (Equation (3.58)).

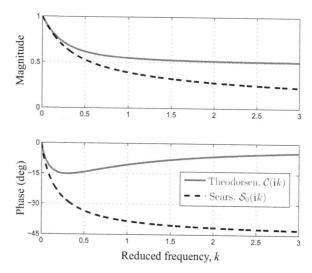

Figure 3.13 Theodorsen's lift-deficiency function and Sears's function vs. reduced frequency. [03Airfoil/theosears.m]

Figure 3.13 shows the amplitude and phase of Sears's function with respect to the leading edge, which is compared to Theodorsen's lift-deficiency function. The amplitude goes to zero in the limit $k_g \to \infty$. The phase lag is relative to the phase of the gust velocity at the leading edge of the airfoil and tends to $-45°$. Note also that Sears's and Theodorsen's functions coincide for small enough values of the reduced frequency, which indicates that the difference between lift due to changes of the pitch angle or gust velocities disappears under quasi-steady aerodynamic assumptions.

3.3.5   Induced Drag

As discussed in Section 3.2, the aerodynamic drag on the airfoil is typically not considered in aeroelastic analysis because the inplane component of the forces on airfoils or wings does not generate any significant deformations. Accordingly, the horizontal degree of freedom was constrained in the definition of the airfoil kinematics in Figure 3.1. This is a very important observation, and it is arguably the main reason to the widespread adoption of potential flow aerodynamics for the aeroelasticity of wings in nontransonic speeds: Potential flow theory gives an excellent approximation of a nonstationary pressure field with a very low computational cost. Moreover, pressure loads are the main contributor to lift and moments, which are the key magnitudes driving the aeroelastic response. Aerodynamic design, on the other hand, targets the average drag generated by the vehicle and, therefore, requires formulations that capture viscous effects on the fluid, which is typically achieved with numerical solutions to the Navier–Stokes equations.

Wing oscillations give a rather different scenario. The induced drag is the dominant component of the nonstationary contributions to the drag for moving airfoils in the absence of flow separation. This may have a significant effect for free-flying vehicles, particularly when they display strong couplings between their aeroelastic and flight dynamics characteristics (see Chapter 6). It is also the most common swimming locomotion mode in most aquatic animals who manage to cancel the drag (indeed generating thrust) by lateral motions, albeit in the form of a traveling wave (Wu, 1971).

The basic mechanism for drag modulation (or thrust generation) comes from the changes that heaving and pitching generate on the component of the pressure on the freestream direction. For steady-state problems, a well-known solution in 2-D potential flow aerodynamics is that the induced drag is identically zero for any airfoil shape. This is known as the d'Alembert paradox, and it is linked to the production of leading-edge suction, an integrable singularity in the pressure field at the leading edge. For a 2-D rigid plate under plunge/pitch motions, as defined in the sections 3.2, a closed-form analytical solution for the induced drag coefficient, c_{d_i}, was obtained by Garrick (1957) as

$$c_{d_i} = \alpha c_l - \sigma_0^2, \tag{3.60}$$

where c_l is the airfoil lift coefficient determined in Equation (3.48) and σ_0 is the leading-edge suction coefficient defined in the frequency domain (i.e., $\sigma_0(s) = \overline{\sigma}_0(ik)e^{iks}$) as

$$\overline{\sigma}_0 = \sqrt{2\pi} \left[C(ik) \left(\overline{\alpha} + ik\overline{\eta} + ik(\tfrac{1}{2} - \nu_{ea})\overline{\alpha} \right) - \tfrac{1}{2}ik\overline{\alpha} \right]. \tag{3.61}$$

It is easy to see that for stationary flows ($k = 0$), the induced drag is identically zero. Also, since the nonstationary-induced drag is a quadratic function of the pitch and plunge amplitudes, for oscillations around a reference with zero angle of attack, the linear approximation of the induced drag is identically zero. This is not the case, however, around a nonzero reference, a situation encountered by vibrations of wings in flight.

Example 3.2 Thrust Generated by Plunge Motions. Interestingly, it can also be seen that harmonic plunge motions at a zero initial angle of attack generate propulsive forces, that is, a negative drag. For that purpose, assume that $\eta(s) = \eta_0 \cos(ks)$ and $\alpha(s) = 0$ in Equation (3.60). If we separate the real and imaginary parts of the lift-deficiency function as $C(ik) = \mathcal{F}(k) + i\mathcal{G}(k)$, it results in a time history of the induced drag written as (Simpson et al., 2013)

$$c_{di}(s) = -2\pi k^2 \left[\mathcal{G}(k)\cos(ks) + \mathcal{F}(k)\sin(ks) \right]^2 \eta_0^2. \tag{3.62}$$

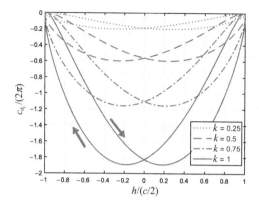

Figure 3.14 Instantaneous induced drag generated by harmonic plunge motions. The arrows indicate the direction of the trajectory for $k = 1$. [03Airfoil/garrick_plunge.m]

Note that the induced drag is always negative, and its amplitude increases with the frequency of oscillations. This can be further seen in Figure 3.14, which shows the instantaneous induced drag generated by the harmonic plunge motions for four different reduced frequencies. Note that the solution is symmetric with respect to the x axis ($\eta = 0$), as both up and downstroke motions generate the same forward forces. The maximum thrust is produced when the airfoil goes near the x axis, as this is the point with the maximum plunge rate and, therefore, an effective angle of attack. The arrows in Figure 3.14 indicate the direction of the evolution of the curve for $k = 1$. They show that the maximum in thrust is again delayed with respect to the crossings through the x axis, which is directly associated with the phase in the lift-deficiency function.

3.4   Finite-State Approximations in Unsteady Aerodynamics

The previous section has presented closed-form solutions, from linear potential flow theory, for the prediction of aerodynamic forces on moving airfoils and in nonstationary flows. They have been written as analytical expressions in the frequency domain, which gives a rich understanding of the parameter dependencies in (linear) unsteady aerodynamics. They can also be seen as frequency-response functions between inputs and outputs of the aerodynamic subsystem, that is, as an external description of the problem, as defined in Section 1.4.3. As shown in Figure 3.2, a coupled (i.e., closed-loop) system can then be assembled with the structural subsystem to describe the aeroelastic dynamics of the 2-DoF airfoil. However, the presence of Bessel functions in the aerodynamic transfer function complicates the analysis and limits the use of the standard tools of linear systems theory. For that purpose, it is convenient to approximate the analytical frequency-response functions by RFAs, which will allow, for example, the identification of poles and zeros on the coupled system or the description of the dynamics through state-space equations of finite dimension (see

Table 3.1 Zero-pole-gain coefficients of a second-order Padé RFA. [03Airfoil/theosears.m]

	Theodorsen	Sears
K	0.5	0.1159
z_1	−0.090	−4.034
z_2	−0.532	−0.229
p_1	−0.073	−0.140
p_2	−0.333	−0.793

Section 1.4.2). A more general introduction to the use of RFAs in unsteady aerodynamics can be found in Section 6.5.3. Here the focus is on the finite-state approximation of Theodorsen's and Sears's functions, as will be seen next.

3.4.1 Rational-Function Approximations

Any function of a single variable, $f(x)$, defined in a closed interval, $x_{min} \le x \le x_{max}$, can be approximated to any level of accuracy by a Padé approximant of sufficiently high order n, that is, a rational function of the form

$$f(x) \approx \frac{\sum_{j=0}^{n} \nu_n x^n}{1 + \sum_{j=1}^{n} \mu_n x^n}. \tag{3.63}$$

While, in general, it is not a requirement that the numerator and denominator be polynomials of the same order, this is typically assumed in aeroelasticity, and it is also done here. Equation (3.63) can be written in a zero-pole-gain form by computing the (in general, complex) roots of both the numerator and the denominator, that is,

$$\frac{\sum_{j=0}^{n} \nu_n x^n}{1 + \sum_{j=1}^{n} \mu_n x^n} = K \frac{(x - z_1)(x - z_2) \cdots (x - z_n)}{(x - p_1)(x - p_2) \cdots (x - p_n)}. \tag{3.64}$$

Padé approximants are used here to approximate the frequency-domain expressions for the lift and moment coefficients derived in Equation (3.48) for the response to harmonic motions of the airfoil and Equation (3.57) for the response to a sinusoidal gust. To obtain RFAs in those cases, we only need to approximate Theodorsen's and Sears's functions since all other terms in the equations already have polynomial dependency with the reduced frequency. As discussed in Section 1.4.3, the coefficients in this approximation, where $f(x)$ is now replaced by either $C(ik)$ or $S_0(ik)$, are obtained through a system identification procedure on a frequency sampling of Equations (3.47) and (3.58). As an example, a second-order RFA using a log-Chebyshev frequency-weighted approximation (Boyd and Vandenberghe, 2004, Ch. 6) with a sampling of 0.01 up to $k = 5$ for Theodorsen's function, and of 0.02 up to $k = 25$ for Sears's function, gives the coefficients shown in Table 3.1. Note in particular that all poles are in the negative real axis, and we have obtained a stable RFA.

For applications in unsteady aerodynamics, it is preferred to have RFAs that exactly preserve the static solutions at zero frequency (the static gains). This is enforced by

Table 3.2 Common RFA coefficients in 2-D
aerodynamics (Jones, 1938).

	Theodorsen	Sears
a_1	0.165	0.5
a_2	0.335	0.5
b_1	0.0455	0.13
b_2	0.3	1.0

setting explicitly that the coefficients of the RFA satisfy $C(0) = 1$ and $S_0(0) = 1$. A
second condition that is often enforced, although is of less practical consequence, is
to preserve the asymptotic behavior at very large frequencies, that is, $C(ik) \to 1/2$ and
$S_0(ik) \to 0$ as $k \to \infty$. This is enforced by imposing additional constraints on the identification of coefficients in Equation (3.63), which leads to approximations of the form

$$f(x) \approx 1 - \sum_{j=1}^{n} \frac{a_j x}{b_j + x}, \tag{3.65}$$

where a_j and b_j are a real coefficients on which the condition at high frequencies is
enforced as $\sum_{j=1}^{n} a_n = 1 - f(\infty)$. An often-cited approximation using Equation (3.65)
is the second-order RFA of the lift-deficiency function due to Jones (1938), whose
coefficients are included in Table 3.2. Many other approximations have been proposed
in the literature, and some of them have been reported by Brunton and Rowley (2013).

Figure 3.15a shows the Bode plot of the lift-deficiency function, including
Theodorsen's analytical solution (3.47), Jones's two-state approximation of Table 3.2,
the 2-state Padé approximation of Table 3.1, and a 4-state Padé approximation. The
4-state approximation is almost indistinguishable from the exact solution, while the
2-state one shows some errors in the phase component, but smaller than that in Jones's
approximation.

The same approximations are computed for Sears's function, Equation (3.58), and
are shown in Figure 3.15b. To achieve sufficient accuracy, the curve fitting of the Padé
approximants needs to be done here for a larger range of reduced frequencies (up to
$k = 25$ in the result in the figure) if no constraints on the asymptotic value as $k \to \infty$
are enforced. As before, the fourth-order Padé gives an excellent approximation, while
both the second-order Padé and Jones's approximations may be accurate enough, with
Jones's fit having slightly larger values in the low-frequency range.

RFAs can be used to transform the frequency-domain descriptions of unsteady aerodynamics into linear differential equations (*state-space* equations) in the time domain.
In this section, we consider only RFAs of the form of Equation (3.65). Their identification is more involved than for the general Padé approximants, as it requires
enforcing additional constraints, but they also lead to simpler expressions. The approximation introduced in Equation (3.63) will be considered in the more general setting of
Section 6.5.3, where it will be generalized into the so-called *minimum-state method*.

Lift and Moment Due to Airfoil Kinematics

We consider first the relation between the kinematics of the airfoil (determined by
its pitch, plunge and flap angles) and the lift and moment coefficients. Substituting

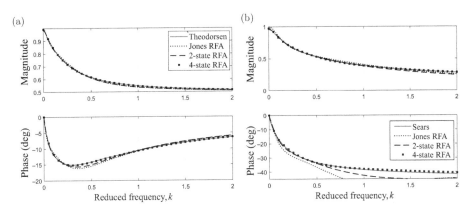

Figure 3.15 Analytical and RFA results of Theodorsen's and Sears's functions. (a) Theodorsen's function and (b) Sears's function. [`03Airfoil/theosears.m`]

Equation (3.65) into Equation (3.48) and collecting terms in powers of the reduced frequency, the frequency-domain forces and moments on the airfoil are

$$\left\{ \begin{matrix} \overline{c}_l \\ \overline{c}_m \end{matrix} \right\} = \left[\mathcal{A}_0 + \mathrm{i}k\mathcal{A}_1 + (\mathrm{i}k)^2 \mathcal{A}_2 - \sum_{j=1}^{n} \frac{\mathrm{i}k a_j}{\mathrm{i}k + b_j} \left(\mathcal{A}_3 + \mathrm{i}k\mathcal{A}_4 \right) \right] \overline{\mathbf{u}}_a, \qquad (3.66)$$

with

$$\mathbf{u}_a(s) = \left\{ \begin{matrix} \eta(s) \\ \alpha(s) \\ \delta(s) \end{matrix} \right\}, \qquad (3.67)$$

being the column vector of DoF and the following constant matrices of associated aerodynamic influence coefficients as

$$\mathcal{A}_0 = \begin{bmatrix} 0 & c_{l_\alpha}^{qs} & c_{l_\delta}^{qs} \\ 0 & 0 & c_{m_\delta}^{qs} \end{bmatrix}, \quad \mathcal{A}_1 = \begin{bmatrix} c_{l_\alpha}^{qs} & c_{l_{\alpha'}}^{qs+nc} & c_{l_{\delta'}}^{qs+nc} \\ 0 & c_{m_{\alpha'}}^{qs+nc} & c_{m_{\delta'}}^{qs+nc} \end{bmatrix}, \quad \mathcal{A}_2 = \begin{bmatrix} c_{l_{\alpha'}}^{nc} & c_{l_{\alpha''}}^{nc} & c_{l_{\delta''}}^{nc} \\ c_{m_{\alpha'}}^{nc} & c_{m_{\alpha''}}^{nc} & c_{m_{\delta''}}^{nc} \end{bmatrix},$$

$$\mathcal{A}_3 = \begin{bmatrix} 0 & c_{l_\alpha}^{qs} & c_{l_\delta}^{qs} \\ 0 & 0 & 0 \end{bmatrix}, \quad \mathcal{A}_4 = \begin{bmatrix} c_{l_\alpha}^{qs} & c_{l_{\alpha'}}^{qs} & c_{l_{\delta'}}^{qs} \\ 0 & 0 & 0 \end{bmatrix},$$

with the coefficients defined in Equations (3.43), (3.44), (3.51), and (3.52). The superindex $qs + nc$ refers to the sum of the corresponding quasi-steady and noncirculatory coefficients. Equation (3.66) can now be used to obtain a state-space realization of the aerodynamic subsystem. For that, we introduce an additional set of DoFs that define the internal states of the aerodynamic problem, and are referred to as *aerodynamic states* (or aerodynamic lags), as

$$\overline{\boldsymbol{\lambda}}_j(\mathrm{i}k) = \frac{\mathrm{i}k}{\mathrm{i}k + b_j} \overline{\mathbf{u}}_a(\mathrm{i}k), \qquad \text{for} \qquad j = 1, \ldots, n. \qquad (3.68)$$

Therefore, the RFA introduced in this section results in n column vectors whose dimension is the number of DoFs in the problem (three here). With this transformation, all terms in Equation (3.66) become polynomial functions of $\mathrm{i}k$, and transformation to the time domain is obtained from analytic continuation: multiplying by ik an analytic function in the frequency domain corresponds to computing its derivative $\frac{d}{ds}$ in the (nondimensional) time domain. As a result, the finite-state time-domain form of the unsteady aerodynamic equations on the airfoil is written as

$$\left\{ \begin{array}{c} c_l(s) \\ c_m(s) \end{array} \right\} = \mathcal{A}_0 \mathbf{u}_a + \mathcal{A}_1 \mathbf{u}_a' + \mathcal{A}_2 \mathbf{u}_a'' - \sum_{j=1}^{n} a_j \left(\mathcal{A}_3 \boldsymbol{\lambda}_j + \mathcal{A}_4 \boldsymbol{\lambda}_j' \right),$$

$$\boldsymbol{\lambda}_j' + b_j \boldsymbol{\lambda}_j = \mathbf{u}_a', \quad \text{with} \quad j = 1, \ldots, n. \tag{3.69}$$

Noting that the coefficients in Equation (3.65) are chosen such that $\sum_{j=1}^{n} a_j = 1 - \mathcal{C}(\infty) = 1/2$, the second equation in Equation (3.69) can be used to remove explicit dependencies of $\boldsymbol{\lambda}'$ in the first one. Consequently, Equation (3.69) becomes

$$\left\{ \begin{array}{c} c_l(s) \\ c_m(s) \end{array} \right\} = \mathcal{A}_0 \mathbf{u}_a + \left(\mathcal{A}_1 - \frac{1}{2}\mathcal{A}_4 \right) \mathbf{u}_a' + \mathcal{A}_2 \mathbf{u}_a'' - \sum_{j=1}^{n} a_j \left(\mathcal{A}_3 - b_j \mathcal{A}_4 \right) \boldsymbol{\lambda}_j,$$

$$\boldsymbol{\lambda}_j' + b_j \boldsymbol{\lambda}_j = \mathbf{u}_a', \quad \text{with} \quad j = 1, \ldots, n. \tag{3.70}$$

For a given initial time history of the airfoil kinematics, $\mathbf{u}_a(s)$, and initial conditions on the aerodynamic lags, $\boldsymbol{\lambda}(0)$ (typically zero, if the dynamics are computed around a static equilibrium), the linear time-invariant equations (3.70) can now be used to compute the forcing terms on the 2-DoF airfoil in the time domain. Note that rational-function RFAs trade complexity (there is no need to evaluate Bessel's functions in the lift-deficiency function) by problem size (we need to solve additional DoFs). It must be remarked that the aerodynamic lags are not physical magnitudes, and their definition is somehow arbitrary, as they rely first on the approximation introduced in Equation (3.65) and then on the change of variables defined by Equation (3.68), which is not unique. In the definition adopted here, they depend only on the rate of change of $\mathbf{u}_a$, while the effect of amplitudes, $\mathbf{u}_a$, and accelerations, $\mathbf{u}_a''$, only appear as feedthrough. This means that $\mathbf{u}_a$ and $\mathbf{u}_a''$ in this description produce only an instantaneous response with constant gains given by the aerodynamic stiffness and inertia matrices, that is, $\mathcal{A}_0$ and $\mathcal{A}_2$, respectively. This will be made even more apparent in the following section.

Lift and Moment Due to Gusts

The same procedure can be used to obtain a time-domain representation of the lift due to gust inputs. Substituting now Equation (3.65) for Sears's function into Equation (3.57), the frequency-domain forces and moments on the airfoil due to gust velocities are

$$\left\{ \begin{array}{c} \overline{c_l}(ik) \\ \overline{c_m}(ik) \end{array} \right\} = \left[\mathcal{A}_g - \sum_{j=1}^{\hat{n}} \frac{ik\hat{a}_j}{ik + \hat{b}_j} \mathcal{A}_g \right] \frac{\overline{w_g}}{V}, \tag{3.71}$$

The notation $\hat{\bullet}$ has been introduced to identify the RFA coefficients of Sears's problem (see Table 3.2). The only matrix of aerodynamic influence coefficients in Equation (3.71) is

$$\mathcal{A}_g = \begin{bmatrix} c_{l_\alpha}^{qs} \\ 0 \end{bmatrix}.$$

(3.72)

The corresponding finite-state form is then written as

$$\begin{Bmatrix} c_l(s) \\ c_m(s) \end{Bmatrix} = \mathcal{A}_g \frac{w_g}{V} - \mathcal{A}_g \sum_{j=1}^{\hat{n}} \hat{a}_j \hat{\lambda}_j',$$

$$\hat{\lambda}_j' + \hat{b}_j \hat{\lambda}_j = \frac{w_g}{V}, \quad \text{with} \quad j = 1, \ldots, \hat{n}.$$

(3.73)

In this case, the aerodynamic lags associated with the gust inputs, $\hat{\lambda}_j$, are scalar functions. They are defined as a function of the gust velocity (as opposed to the rate, as in Equation (3.68)), for convenience. Noting that $\sum_{j=1}^{\hat{n}} \hat{a}_j = 1$, which is derived from the condition that Sears's function tends to zero at high frequencies, it is possible to simplify this equation by eliminating $\hat{\lambda}_j'$ in the first equation using the second one, as before. As a result, the first equation in (3.73) can be written as

$$\begin{Bmatrix} c_l(s) \\ c_m(s) \end{Bmatrix} = \mathcal{A}_g \sum_{j=1}^{\hat{n}} \hat{a}_j \hat{b}_j \hat{\lambda}_j.$$

(3.74)

Therefore, there is no explicit dependency between the instantaneous gust velocities (the disturbance) and lift (the output). In other words, there is no feedthrough in the finite-state representation of Sears's problem, which is a feature that will reappear in 3-D problems (see Section 7.5).

3.4.2   State-Space Realization of the Finite-State Aerodynamics

The main advantage of introducing RFAs is that the instantaneous lift and moment can be obtained from the solution of a linear time-invariant state-space system (in nondimensional time), that is, RFAs enable the transformation from frequency to the time domain in Figure 1.10. Combining Equations (3.70) and (3.73), one obtains the state equations

$$\mathbf{x}_a' = \mathbf{A}_a \mathbf{x}_a + \mathbf{B}_a \mathbf{u}_a',$$

$$\mathbf{x}_g' = \mathbf{A}_g \mathbf{x}_g + \mathbf{B}_g \frac{w_g}{V},$$

(3.75)

and the combined output equation

$$\mathbf{y}_a = \mathcal{A}_0 \mathbf{u}_a + \left(\mathcal{A}_1 - \tfrac{1}{2} \mathcal{A}_4 \right) \mathbf{u}_a' + \mathcal{A}_2 \mathbf{u}_a'' + \mathbf{C}_a \mathbf{x}_a + \mathbf{C}_g \mathbf{x}_g,$$

(3.76)

where the inputs have been defined in Equation (3.67) as $\mathbf{u}_a(s) = \{\eta(s) \quad \alpha(s) \quad \delta(s)\}^\top$, and the aerodynamic states and outputs are defined, respectively, as

$$\mathbf{x}_a = \begin{Bmatrix} \lambda_1 \\ \vdots \\ \lambda_n \end{Bmatrix}, \quad \mathbf{x}_g = \begin{Bmatrix} \hat{\lambda}_1 \\ \vdots \\ \hat{\lambda}_{\hat{n}} \end{Bmatrix}, \quad \text{and} \quad \mathbf{y}_a = \begin{Bmatrix} c_l \\ c_m \end{Bmatrix}.$$

(3.77)

Note that while the internal states (the aerodynamic lags) of this problem depend only on $\mathbf{u}_a'$, the outputs (the lift and moment coefficients) also depend on $\mathbf{u}_a$ and $\mathbf{u}_a''$. If the unsteady aerodynamics were to be studied under prescribed airfoil kinematics, it would be convenient to define an augmented state vector that includes both $\mathbf{u}_a$ and $\mathbf{u}_a'$ so that the output equations would not have an explicit dependency of the time derivatives of the inputs (e.g., the acceleration, $\mathbf{u}_a''$). However, this is not done here since those terms are also the structural DoFs and can be combined in the aeroelastic coupling (see Section 3.5).

The aerodynamic system matrices are $\mathbf{A}_a = -\mathrm{diag}(b_1, b_1, b_1, \ldots, b_n, b_n, b_n)$ of dimension $3n$ and $\mathbf{A}_g = -\mathrm{diag}(\hat{b}_1, \ldots, \hat{b}_{\hat{n}})$ of dimension $\hat{n}$, where n and $\hat{n}$ are the order of the approximation of Theodorsen's and Sears's functions, respectively. The input matrices are simply

$$\mathbf{B}_a = \begin{bmatrix} \mathcal{I} \\ \vdots \\ \mathcal{I} \end{bmatrix} \quad \text{and} \quad \mathbf{B}_g = \begin{bmatrix} 1 \\ \vdots \\ 1 \end{bmatrix}, \tag{3.78}$$

where the unit matrices in $\mathbf{B}_a$ are of dimension 3, and the output matrices are

$$\mathbf{C}_a = \left[a_1 \left(\mathcal{A}_3 - b_1 \mathcal{A}_4 \right) \quad \cdots \quad a_n \left(\mathcal{A}_3 - b_n \mathcal{A}_4 \right) \right],$$
$$\mathbf{C}_g = \left[\hat{a}_1 \hat{b}_1 \mathcal{A}_g \quad \cdots \quad \hat{a}_{\hat{n}} \hat{b}_{\hat{n}} \mathcal{A}_g \right]. \tag{3.79}$$

The state-space description of Equations (3.75) and (3.76) provides a time-domain representation for the upper block in Figure 3.2, including the effect of atmospheric gusts. Note that, from an aerodynamic point of view, no fundamental distinction has been made between the flap angles, $\delta(s)$, and the airfoil motions, (α, η), which are all inputs into the system. It should be noted, however, that the inputs are not only the instantaneous values of the pitch, plunge, and flap DoFs but also their first and second derivatives with (nondimensional) time. However, those are readily available from the equations of motion of the structure.

3.4.3   Step Response: Impulsive Flows

It is interesting to consider a second approach, this time using step responses,[8] to construct time-domain solutions in dynamic aeroelasticity. We have seen in Section 1.4.2 that the dynamics of an LTI system can be uniquely described by its impulse response, which is the inverse Fourier transform of the transfer function between inputs and outputs. For applications in unsteady aerodynamics, the step response, which is the time integral of the impulse response, is often preferred instead, as it is easier to generate both experimentally and numerically. Just like the transfer function and the impulse

---

[8] In unsteady aerodynamics applications, step responses are often known as *indicial responses* (Ballhaus and Goorjian, 1978).

response, it is an external description of the system dynamics; that is, it only involves inputs (airfoil kinematics) and outputs (force and moment coefficients).

The starting point is again the analytical solutions in the frequency domain of Theodorsen and Sears, Equations (3.48) and (3.57), respectively. Next, using Fourier analysis (see Appendix A), we obtain the response to any arbitrary motion or disturbance and, in particular, to unit step functions (sudden changes in the pitch angle or upwash). Consider first the case

$$\alpha(s) = \alpha_0 \mathcal{H}(s), \tag{3.80}$$

with $\mathcal{H}(s)$ being the Heaviside function that has already appeared in Equation (3.40). Note, however, that the analysis that follows would equally apply to the step function on the plunge rate $\zeta'(s) = \zeta_0' \mathcal{H}(s)$. The dimensionless time $s = \frac{Vt}{c/2}$ becomes the distance in semichords traveled by the airfoil since the step change. The Fourier transform of Equation (3.80) is

$$\overline{\alpha}(ik) = \frac{1}{2\pi} \int_{-\infty}^{\infty} \alpha(s) e^{-iks} \, ds = \frac{1}{2\pi} \int_{0}^{\infty} \alpha_0 e^{-iks} \, ds = \frac{\alpha_0}{i2\pi k}. \tag{3.81}$$

Noting that the derivative of the unit step function is the unit impulse function (the Dirac delta function), that is, $\frac{d\mathcal{H}}{ds} = \delta^*(s),$[9] the pitch rate time history corresponding to Equation (3.80) is $\alpha'(s) = \alpha_0 \delta^*(s)$. The instantaneous lift is now obtained from the inverse Fourier transform of Equation (3.48), although this is not necessary for the apparent mass, which is already known in the time domain. As a result, the lift coefficient resulting from the step change on the pitch angle can be written as

$$c_l(s) = \int_{-\infty}^{\infty} c_{l\alpha}^{qs} C(ik) \overline{\alpha}(ik) e^{iks} \, dk + c_{l\alpha'}^{nc} \alpha_0 \delta^*(s), \tag{3.82}$$

for $s > 0$, which is often expressed as

$$c_l(s) = \left[c_{l\alpha}^{qs} \varphi(s) + c_{l\alpha'}^{nc} \delta^*(s) \right] \alpha_0, \tag{3.83}$$

with

$$\varphi(s) = \frac{1}{2\pi} \int_{-\infty}^{\infty} \frac{C(ik)}{ik} e^{iks} \, dk. \tag{3.84}$$

The first term in Equation (3.83) is the circulatory lift and the second the apparent mass. Note that this last term appears as a Dirac delta function, since noncirculatory terms in potential flow theory only appear while the airfoil accelerates. The function $\varphi(s)$ was first introduced by Wagner (1925), and it is therefore referred to as *Wagner's function*. The reciprocal relation between Wagner's and Theodorsen's functions, Equation (3.84), was first derived by Garrick (1938). Noting that $C(-ik) = C^*(ik)$, its complex conjugate, it follows that the real function $\varphi(s)$ can be computed from either the real or the imaginary part of the lift-deficiency function as

$$\varphi(s) = \frac{2}{\pi} \int_{0}^{\infty} \frac{\text{Re}(C(ik))}{k} \sin ks \, dk = 1 + \frac{2}{\pi} \int_{0}^{\infty} \frac{\text{Im}(C(ik))}{k} \cos ks \, dk. \tag{3.85}$$

---

[9] The asterisk in this section distinguishes between the Dirac delta function and flap deflection.

As a result, for a given time history of the pitch angle, $\alpha(s)$, for $s \geq 0$, and its Fourier transform, $\overline{\alpha}(ik)$, the circulatory lift can now be obtained by two alternative expressions:

$$c_l^{qs}(s) = 2\pi \int_{-\infty}^{\infty} \mathcal{C}(ik)\overline{\alpha}(ik)e^{iks}\,dk = 2\pi \int_0^s \varphi'(s-\sigma)\alpha(\sigma)\,d\sigma, \qquad (3.86)$$

where the first integral expression is the inverse Fourier transform of $\overline{c_l^{qs}}(ik)$, and the second one is the convolution integral, which was introduced in Equation (1.14), of the instantaneous angle of attack and the impulse response $\varphi'(\sigma)$. From integration by parts, the last expression can also be written as

$$c_l^{qs}(s) = 2\pi\varphi(s)\alpha(0) + 2\pi \int_0^s \varphi(s-\sigma)\alpha'(\sigma)\,d\sigma. \qquad (3.87)$$

Exact analytical solutions of Equation (3.84) are not available, but approximated ones to a given accuracy can be easily obtained using the RFAs to Theodorsen's lift-deficiency function introduced in Section 3.4.1. This effectively corresponds to going from the transfer function to the impulse response via the state-space model in Figure 1.10. In particular, using Equation (3.65) results in

$$\varphi(s) \approx \frac{1}{2\pi} \int_{-\infty}^{\infty} \left(\frac{1}{ik} - \sum_{j=1}^n \frac{a_j}{ik+b_j} \right) e^{iks}\,dk = 1 - \sum_{j=1}^n a_j e^{-b_j s}, \qquad (3.88)$$

where the coefficients for a second-order approximation are those of Table 3.2. An alternative approximation, using methods of nonlinear system identification, has been recently presented by Dawson and Brunton (2021). While it provides little practical advantage for the linear problems considered here, it uses a data-driven approach that can also be applied to generalizations of this problem, to include, for example, the effect of thickness or nonplanar wakes.

Similarly, we can obtain aerodynamic forces due to a step gust progressing over the airfoil at the freestream velocity V. The (vertical) gust velocity at the leading edge can be given as

$$w_g(s) = w_0 \mathcal{H}(s). \qquad (3.89)$$

Using the same procedure as before, we obtain

$$c_l^g(s) = c_{l_\alpha}^{qs} \frac{w_0}{V} \psi(s), \qquad (3.90)$$

with

$$\psi(s) = \frac{1}{2\pi} \int_{-\infty}^{\infty} \frac{\mathcal{S}_0(ik)}{ik} e^{iks}\,dk, \qquad (3.91)$$

where the function $\psi(s)$ is known as *Küssner's function*, for which an approximation of the form of Equation (3.88) can also be obtained (using the corresponding coefficients in Table 3.2). The reduced time, $s = \frac{Vt}{c/2}$, represents here the distance measured in semichords traveled by the gust front past the leading edge of the airfoil. Wagner's and Küssner's functions can also be computed numerically from the direct

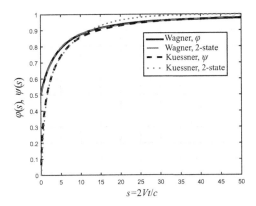

Figure 3.16 Wagner's and Küssner's step-response functions. Converged solutions and 2-state approximations. [03Airfoil/wagner_kuessner.m]

integration of Equations (3.84) and (3.91). Kayran (2006) has discussed some technical details on the numerical evaluation of Küssner's function directly from Sears's function, although they can be avoided by introducing first an RFA, as it is done here.

Converged solutions for Wagner's and Küssner's functions obtained from 10-state approximations are shown in Figure 3.16. Both step responses take about the same length of time (around 10 chord lengths, i.e., $s \approx 20$) to asymptote to around 5% of their steady states, but Wagner's function starts at exactly 50% of the final value (which corresponds to the limit of $C(ik)$ for $k \to \infty$), whereas Küssner's function starts at zero (as Sears's function tends to zero for high frequencies). This is due to the very different initial conditions in the problem: In the step change of angle of attack, that is, Wagner's problem, the boundary conditions are instantaneously changed, which have a sudden effect on the flow response,[10] while in the step gust change (Küssner's problem), the gust front is just outside the airfoil at $s = 0$, and then it sweeps through it until it reaches the trailing edge at $s = 2$. Figure 3.16 also includes the commonly used 2-state approximation using Equation (3.65) and the coefficients of Table 3.2. Two states clearly give already a very good approximation to capture Wagner's function, while they give slightly higher forces in the gust response for $s > 10$.

Step responses have been traditionally used in aeroelasticity to construct time-domain solutions to arbitrary inputs using convolution integrals (Bisplinghoff et al., 1955, Ch. 5), although this is not pursued here. Instead, state-space representations, such as those described in Section 3.4.2, are preferred, as they naturally relate to modern analysis methods on linear dynamical systems. Note finally that, while our discussion has focused only on lift, the step response of the moment can be obtained from Equation (3.48).

---

[10] Note that this is a limit solution that is only possible because this is an incompressible flow model, and therefore information "travels at infinite speed" (Hariharan and Leishman, 1996).

3.5 Aeroelastic Equations of Motion

In Section 3.2, we have presented the structural equations of motion for the airfoil with 2-DoF, Equation (3.6). Closed-form expressions for the lift and moment on the airfoil have been obtained in Equation (3.70) and Equation (3.73) in nondimensional time, $s = \frac{Vt}{c/2}$. The solution to the coupled problem is obtained by means of a feedback interconnection between both subsystems: The outputs of the structural solution become the inputs of the aerodynamic solution and vice versa. To simplify the description in what follows, we assume that there is no control surface. In that case, the finite-state approximations in Equation (3.66) are still valid after removing the columns of the matrices associated with the flap angle. The problem with control inputs is discussed, in a more general framework, in Section 6.6.1.

For more general problems, such as those defined later in this book, the aerodynamic and structural models are built on different numerical solutions and are coupled at every time step (typically, using subiterations). This approach is normally referred to as *co-simulation* (Bazilevs et al., 2013). Such a strategy, however natural, does not use the internal structure of the problem and, in particular, the feedthrough terms that were identified in Equations (3.75) and (3.76). As was seen there, the aerodynamic model effectively adds stiffness and inertia (as well as damping) to the structural DoFs, and it helps the analysis, as well as to any numerical schemes used to solve the problem, if those are aggregated in the formulation of the equations. This is done here.

Combining Equations (3.6), (3.75), and (3.76), one obtains

$$\mathbf{M}_{ae}\mathbf{q}'' + \mathbf{D}_{ae}\mathbf{q}' + \mathbf{K}_{ae}\mathbf{q} = \tfrac{\rho V^2}{2}\kappa_g\left(\mathbf{C}_a\mathbf{x}_a + \mathbf{C}_g\mathbf{x}_g\right),$$
$$\mathbf{x}'_a = \mathbf{A}_a\mathbf{x}_a + \mathbf{B}_a\mathbf{q}',$$
$$\mathbf{x}'_g = \mathbf{A}_g\mathbf{x}_g + \mathbf{B}_g\tfrac{w_g}{V}, \tag{3.92}$$

where we have introduced the aeroelastic mass, damping, and stiffness associated with the nondimensional time s as

$$\mathbf{M}_{ae} = \tfrac{4V^2}{c^2}\mathbf{M} - \tfrac{\rho V^2}{2}\kappa_g\mathcal{A}_2,$$
$$\mathbf{D}_{ae} = -\tfrac{\rho V^2}{2}\kappa_g\left(\mathcal{A}_1 - \tfrac{1}{2}\mathcal{A}_3\right),$$
$$\mathbf{K}_{ae} = \mathbf{K} - \tfrac{\rho V^2}{2}\kappa_g\mathcal{A}_0. \tag{3.93}$$

Note that, for $\tfrac{1}{2}\rho V^2 > 0$, neither matrix is symmetric, which reflects the fact that the aeroelastic system is nonconservative and cannot be associated with energy potentials.

Equation (3.92) can be written in a state-space form by defining the aeroelastic state variable

$$\mathbf{x}_{ae} = \left\{\mathbf{q}^\top \quad \mathbf{q}'^\top \quad \mathbf{x}_a^\top \quad \mathbf{x}_g^\top\right\}^\top. \tag{3.94}$$

The resulting first-order form of the aeroelastic equations of motion is

$$\frac{d\mathbf{x}_{ae}}{ds} = \mathbf{A}_{ae}\mathbf{x}_{ae} + \mathbf{B}_{ae}\tfrac{w_g}{V}, \tag{3.95}$$

with the system and input matrices defined, respectively, as

$$
\mathbf{A}_{ae} =
\begin{bmatrix}
\mathbf{0} & \mathbf{1} & \mathbf{0} & \mathbf{0} \\
-\mathbf{M}_{ae}^{-1}\mathbf{K}_{ae} & -\mathbf{M}_{ae}^{-1}\mathbf{D}_{ae} & \frac{\rho V^2}{2}\mathbf{M}_{ae}^{-1}\kappa_g\mathbf{C}_a & \frac{\rho V^2}{2}\mathbf{M}_{ae}^{-1}\kappa_g\mathbf{C}_g \\
\mathbf{0} & \mathbf{B}_a & \mathbf{A}_a & \mathbf{0} \\
\mathbf{0} & \mathbf{0} & \mathbf{0} & \mathbf{A}_g
\end{bmatrix}
, \quad
\mathbf{B}_{ae} =
\begin{bmatrix}
\mathbf{0} \\
\mathbf{0} \\
\mathbf{0} \\
\mathbf{B}_g
\end{bmatrix}.
$$

$$(3.96)$$

Note that the system matrix depends independently on both the airspeed and density (which is defined by the flight altitude), which means that in dynamic aeroelasticity solutions do not generally scale with the dynamic pressure. This is in opposition to the flight dynamics of rigid aircraft, where aerodynamic derivatives are tabulated with the dynamic pressure (or equivalent airspeed), as is shown in Chapter 4.

Moreover, for most practical cases, $\det(\mathbf{M}_{ae}) > 0$, and its inverse is well defined. In fact, for most vehicles, the mass of the airfoil is substantially higher than that of the displaced air, and the effect of the apparent mass can be neglected, that is, $\mathcal{A}_2 \approx \mathbf{0}$. In that case, $\mathbf{M}_{ae}$ depends only on the airspeed V due to the normalization of the time variable but not on the density ρ. This can be used to speed up the identification of flutter conditions (see Section 3.5.2) by fixing the airspeed and looking for the altitude at which the aeroelastic system becomes dynamically unstable. In that case, the matrix inverse in Equation (3.96) only needs to be computed once per flutter point evaluation.

Equation (3.95) describes the dynamics of a flapless airfoil under wind gusts. Time marching of the equations can be used to obtain the dynamics of the airfoil for given initial conditions, and this is described in Section 3.6. It can also be used to obtain the statistics of the airfoil response under continuous turbulence, which is exemplified in Section 3.7. Finally, the system matrix $\mathbf{A}_{ae}$ defined in Equation (3.96) determines the stability characteristics of the aeroelastic system, which will depend in this case on (V, ρ). The interest now is in the physical interpretation of the instability conditions, which for aeroelastic system result in either flutter or divergence events. Those are discussed next.

3.5.1   Divergence

Aeroelastic divergence is defined as the flight condition in which the feedback system between the structure and the aerodynamics, Figure 3.2, becomes statically unstable. Mathematically, this situation corresponds to the system matrix $\mathbf{A}_{ae}$ in Equation (3.95) having a null eigenvalue, or, equivalently, to finding the nontrivial solutions of Equation (3.95) (or Equation (3.92)) with $\mathbf{x}'_{ae} = \mathbf{0}$. Under those conditions, from the second equation in Equation (3.92), we have $\mathbf{x}_a = \mathbf{0}$, and the static stability is then uniquely determined by the properties of the aeroelastic stiffness matrix, $\mathbf{K}_{ae}$. Therefore, divergence will occur when

$$
\det(\mathbf{K}_{ae}) = \det\left(\mathbf{K} - \tfrac{\rho V^2}{2}\kappa_g\mathcal{A}_0\right) = 0.
\tag{3.97}
$$

Substituting the matrices in this expression by their definitions in Equations (3.6), (3.8), and (3.66), we have

$$\mathbf{K} - \frac{\rho V^2}{2} \kappa_g \mathcal{A}_0 = \begin{bmatrix} k_h & 2\frac{\rho V^2}{2} c_{l_\alpha}^{qs} \\ 0 & \frac{k_\alpha}{(c/2)^2} - \frac{\rho V^2}{2}\frac{x_{ea}-x_{ac}}{c/2} c_{l_\alpha}^{qs} \end{bmatrix}, \tag{3.98}$$

which gives the divergence dynamic pressure, $q_D = \frac{1}{2}\rho_D V_D^2$,

$$q_D = \frac{k_\alpha}{c\,(x_{ea} - x_{ac})\,c_{l_\alpha}^{qs}}. \tag{3.99}$$

It should be remarked that divergence on elastically supported airfoils is independent of the plunge stiffness. This is because aerodynamic forces are invariant with the vertical (plunge) location of the airfoil. Also, the stability in this case depends on the dynamic pressure, as opposed to the dynamic problems above (see Section 3.2.2), where density and airspeed appear independently in the equations. From Equation (3.99), it is clear that the onset of divergence is delayed by either increasing the torsional stiffness or by reducing the offset between the elastic axis and the aerodynamic center, which reduces the moment that the lift force generates on the elastic axis. Indeed, since the dynamic pressure can only be a positive number, divergence can only occur if that restoring moment is positive – that is, if the elastic axis is behind the aerodynamic center ($x_{ea} > x_{ac}$), according to the sign convention defined in Figure 3.1. Note that the static stability boundary determined by Equation (3.99) can also be obtained directly from the static equilibrium of moments on the airfoil as has been done, for instance, by Hodges and Pierce (2014, Ch. 3).

3.5.2   Flutter

Flutter occurs when the aeroelastic system defined by Equation (3.95) becomes dynamically unstable, that is, when the system matrix $\mathbf{A}_{ae}$ has a complex eigenvalue with positive real part. We can write this eigenvalue problem as a parametric problem in the flight condition, that is,

$$\mathbf{A}_{ae}(\rho, V)\phi = \phi\Lambda, \tag{3.100}$$

where Λ is the diagonal matrix whose terms are the system (complex) eigenvalues and ϕ are their corresponding eigenvectors (the also complex *aeroelastic modes*). For a given altitude (i.e., given ρ), the smallest airspeed for which one of the eigenvalues has a positive real part defines the *flutter speed*. Since a crossing along the real axis corresponds to aeroelastic divergence, flutter roots appear along the imaginary axis, and the frequency at the crossing point is known as the *flutter frequency*. Note that $\mathbf{A}_{ae}$ is the state matrix with system dynamics expressed in nondimensional time, $s = \frac{Vt}{c/2}$, and therefore the eigenvalues obtained in Equation (3.100) are also nondimensional. Note also that, even for the simple 2-DoF system considered in this chapter, the solution to this problem has to be obtained numerically (see Example 3.3).

Since the flutter onset point is a neutrally stable condition, that is, flutter corresponds to harmonic motions of the homogeneous system, the problem is often solved

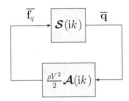

Figure 3.17 Feedback representation of the flutter problem.

directly in the (dimensional) frequency domain. For that, we consider again Equation (3.6), assume $\mathbf{q}(t) = \bar{\mathbf{q}}e^{i\omega t}$, and express the lift and moment coefficients using Theodorsen's solution, Equation (3.48), without including flap or gust inputs (as we are interested in the stability of the system). This results in an expression of the form

$$\left(-\omega^2 \mathbf{M} + \mathbf{K} - \tfrac{\rho V^2}{2}\mathcal{A}(ik)\right)\bar{\mathbf{q}} = \mathbf{0}, \tag{3.101}$$

where $\mathcal{A}(ik)$ is the frequency-domain matrix of *aerodynamic influence coefficients* associated with the generalized coordinates, which are usually known as generalized aerodynamic forces (GAFs). This allows the direct use of Theodorsen's closed-form solutions for the lift and moment on the airfoil, Equation (3.46), instead of their RFAs. The conventional approach in aeroelasticity is to consider Equation (3.101), which is known as the *flutter equation*, as a generalized nonlinear eigenvalue problem in the (unknown) flutter frequency. Iterative numerical schemes, such as the *p-k* method of Hassig (1971), have been developed to solve this problem (see Example 6.5.2).

When the unsteady aerodynamics is approximated by an RFA, Equation (3.66), this problem can also be analyzed as the (closed-loop) stability of the feedback system established between the structural frequency-response function $\mathcal{S}(ik) = \left(\mathbf{K} + \left(\frac{2ikV}{c}\right)^2 \mathbf{M}\right)^{-1}$ and the aerodynamic influence coefficients (AICs),

$$\mathcal{A}(ik) = \kappa_g \left[\mathcal{A}_0 + ik\mathcal{A}_1 + (ik)^2\mathcal{A}_2 - \sum_{j=1}^{n} \frac{ika_j}{ik+b_j}(\mathcal{A}_3 + ik\mathcal{A}_4)\right], \tag{3.102}$$

with the dynamic pressure $\tfrac{1}{2}\rho V^2$ acting as a feedback gain, as shown in Figure 3.17. The rescaling of the time variable in the structural admittance implies that we first need to fix the airspeed V and then evaluate aeroelastic stability with varying air density. In that case, the flutter onset condition is sought by increasing the value of ρ (decreasing flight altitude) starting from very small values (very high altitude) at constant speed. Note that $\rho = 0$ corresponds to the structural system in open loop (there is no aerodynamic feedback in Figure 3.17), and the poles are the normal modes of the structure.

It is more common, however, to assess dynamic stability on physical time. In that case, the air density is kept constant, and the aerodynamic transfer function is updated for either an increasing value of the airspeed or, more commonly, the corresponding dynamic pressure, up to the flutter conditions. Given a dynamically stable flight condition, that is, a pair (ρ, V), the *flutter margin* is determined as the largest perturbation on the flight speed for which the closed-loop aeroelastic system (Figure 3.17) remains stable. A detailed discussion of aeroelastic stability margins using the tools of robust control can be found in Lind and Brenner (2012, Ch. 5).

Finally, recall that the reduced frequency is the frequency of the structural vibrations normalized by the convective timescale. When the reduced frequency of the self-excited oscillations at the flutter point is very low (a typically threshold is $k_F < 0.05$), the aerodynamics can be simplified using the quasi-steady assumptions of Section 3.3.2. In that case, the characteristic polynomial that determines the dimensionless eigenvalues of the aeroelastic system with quasi-steady aerodynamics is $\det(p^2 \mathbf{M}_{ae} + p\mathbf{D}_{ae} + \mathbf{K}_{ae}) = 0$, that is, the same equation that one would consider to study damped vibrations in structural dynamics, but with the mass, damping, and stiffness modified to include the effect of the surrounding fluid. As mentioned above, those matrices are no longer symmetric, and therefore the resulting characteristic polynomial may have roots with positive real part. A demonstration of this approach is shown in Section 3.2.2.

Example 3.3 Effect of the Center-of-Mass Location in Flutter of a 2-DoF Airfoil.

As a numerical example, consider a 2-DoF airfoil with mass ratio $\mu = \frac{m}{\pi \rho c^2} = 5$, coordinates $x_{ac} = 0.25c$ and $x_{ea} = 0.35c$, and the radius of gyration about the elastic axis, $r_\alpha = \sqrt{I_\alpha/m} = 0.25c$. Four different locations of the center of mass are considered, $x_{cg}/c = \{0.35, 0.375, 0.40, 0.45\}$, all at or behind the elastic axis, that is, $x_{cg} \geq x_{ea}$. This case has been studied by Zeiler (2000) and Murua et al. (2010), among others.

In order to parameterize the system dynamic characteristics, we first identify the characteristic pitch and plunge angular frequencies as $\omega_\alpha = \sqrt{k_\alpha/I_\alpha}$ and $\omega_h = \sqrt{k_h/m}$, respectively. Note however that, in general, they are not the natural vibration modes of the structure (they are when $x_{cg} = x_{ea}$). All results in this section are normalized by the chord c, the pitch frequency ω_α, and the resulting six nondimensional parameters, namely $\mu, \omega_h/\omega_\alpha, x_{ac}/c, x_{ea}/c, x_{cg}/c$, and r_α, a uniquely defined dimensionless response of the 2-DoF airfoil problem. Moreover, if the reference is taken with respect to either the elastic axis, the aerodynamic center, or the CM, then the problem can be reduced to five independent parameters without loss of generality.

First of all, the divergence speed can be directly obtained from Equation (3.99) as

$$\frac{V_D}{c\omega_\alpha} = \frac{r_\alpha}{c}\sqrt{\frac{\mu}{\frac{x_{ea}}{c} - \frac{x_{ac}}{c}}} \approx 1.77,$$

that is, since aeroelastic divergence does not depend on the inertia properties, it is constant as x_{cg} is varied. Next, the dynamic stability is obtained by investigating the eigenvalues of Equation (3.95) as the airspeed increases. Three aerodynamic lags are considered, which yields a system of dimension 10 (four structural and six aerodynamic states). The predicted flutter speed, V_F, as a function of the range of frequencies and the location of x_{cg} is shown in Figure 3.18, while Figure 3.19a shows the *root locus* of the nondimensional eigenvalues, λ, for $\omega_h/\omega_\alpha = 0.5$ and $x_{cg} = 0.45c$. The parameter in the curves in Figure 3.19a is the nondimensional airspeed. For each speed, the 10 eigenvalues of the system are plotted. They include here two pairs of complex roots

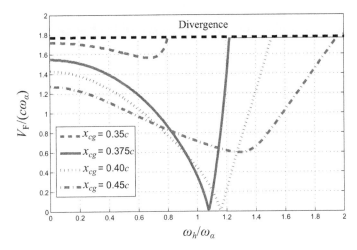

Figure 3.18 Nondimensional flutter speed of the 2-DoF airfoil as a function of frequency ratio for varying locations of the center of mass. [03Airfoil/flutter.m]

associated with the structural DoFs, and six roots along the negative real axis associated with the aerodynamic lags. Note that $\frac{\lambda V}{c/2}$ are the eigenvalues in physical time and, therefore, the plots show the eigenvalues normalized by the characteristic frequency in pitch, ω_α.

The eigenvalues in Figure 3.19a correspond to nondimensional airspeeds $\left(\frac{V}{\omega_\alpha c}\right)$ between 0.1 and 2.0, in increments of 0.1. The eigenvalues for $\frac{V}{\omega_\alpha c} = 0.1$ have been marked with a square, while the eigenvalues for nondimensional airspeeds 1.0 and 2.0 are marked by a circle. Two complex pairs can be seen starting from the imaginary axis for $V = 0$, where they correspond to the pitching and plunging modes in vacuum (the pitching mode is the one with higher frequency), while the poles at the real axis correspond to the aerodynamic lags. The unsteady aerodynamics initially increases the damping on the pitch and plunge modes. However, the softening of the pitching DoF due to the steady aerodynamics, which has been described in Section 3.2.2, is still a major destabilizing factor, as it brings the frequencies of both modes closer to each other as the airspeed increases. Eventually, one of the roots becomes unstable for $\frac{V_F}{\omega_\alpha c} = 1.04$, which is therefore the nondimensional flutter speed for this configuration. The corresponding dimensionless frequency of the neutrally stable oscillations at the flutter point is $\frac{\omega_F}{\omega_h} = 0.74$. Note also that at the divergence speed, one of the roots in the real axis crosses the right-hand plane in Figure 3.19a. Table 3.3 shows the structural components of the eigenvectors at both the flutter and divergence points. Divergence is a static process and the eigenvector is a real number, while flutter is dynamic and described by complex numbers. It is also easy to see that the time derivatives in the flutter mode are 90° out of phase with respect to the corresponding primitive variable. Importantly, there is a significant phase delay between the pitch and plunge DoFs (here, pitch is at −45° with respect to plunge), which is critical in enabling the self-sustained oscillations in flutter, as it staggers the cycles of power extraction from the fluid for each DoF (Patil, 2003). Note finally that the flutter eigenvector also includes

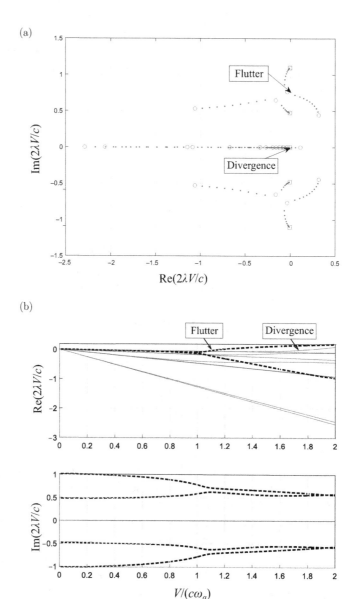

Figure 3.19 Stability diagrams for the 2-DoF airfoil with $x_{\mathrm{cg}} = 0.45c$ and $\omega_h/\omega_\alpha = 0.5$. Flutter and divergence occur at $\frac{V}{\omega_\alpha c}=1.04$ and 1.77, respectively. (a) Root locus. $\frac{V}{c\omega_\alpha}$ increasing from 0.1 (red square) in increments of 0.1 and (b) V–g diagram (dash: structural modes; continuous: aerodynamic modes)

nonzero components of the aerodynamic lags, but they do not have an obvious physical meaning and are rarely monitored. The divergence mode, on the other hand, is a static mode with only real components and zero value of both structural velocities and aerodynamic lags.

The dynamic response needs to be investigated in some detail at speeds near the flutter onset instability, both to identify warning mechanisms and to ensure that numerical

Table 3.3 Components of the unstable eigenvector for flutter and divergence for $\frac{\omega_h}{\omega_\alpha} = 0.5$.

	Flutter	Divergence
η	$0.46 \pm 0.44i$	1
α	1	-0.20
$\dot{\eta}$	$-0.15 \pm 0.16i$	0
$\dot{\alpha}$	$0.00 \pm 0.36i$	0

predictions are not very sensitive to modeling errors. In particular, a quick drop in damping when approaching the flutter speed means that the flutter will be a sudden, and therefore often destructive, event. To aid with this, it is customary to plot independently the real and imaginary parts of the eigenvalues against the nondimensional freestream speed, as is done in Figure 3.19b. This representation is known as the *V–g diagram*. It shows the same information as the root locus plot, but it clearly shows the flutter and divergence speed, as well as the evolution of the damping of the unstable mode near the instability points.

Repeating this analysis for different values of the frequency ratio and locations of the CM, one can obtain the flutter envelope as shown in Figure 3.18. Note first that the divergence speed is independent of the moving parameters in this problem, and it defines an upper bound in the stability analysis at $\frac{V}{c\omega_\alpha} = 1.77$. This has been marked as a gray region in Figure 3.18. Next, it can be seen that the flutter speed is strongly affected by the ratio $\frac{\omega_h}{\omega_\alpha}$, that is, the separation between the plunging and pitching frequencies of the airfoil. As mentioned in this example, the characteristic frequencies ω_h and ω_α are the natural frequencies of the structural subsystem only when the elastic axis and CM coincide (which corresponds to the condition $x_{cg} = 0.35$), but they are still close to the natural frequencies of the problem in all other cases. As can be seen, when both frequencies are close to each other, the flutter speed markedly decreases. The location of the center of mass also has a substantial effect on the flutter behavior, and small variations (of the order of 5% of the chord) can move the flutter boundary by a large amount. This property has been used as a passive mechanism for wing flutter suppression, known as *mass balancing*. In a mass-balanced wing (or aileron), a small mass is added at the leading edge in an outboard position to delay the flutter onset so that it occurs at a flight speed outside the flight envelope.

3.6 Discrete Gust Response of a 1-DoF Airfoil

Aircraft response to atmospheric gusts and turbulence is of critical importance from a structural design perspective, but also from a performance point of view. The loads appearing when the aircraft encounters gusts can be very high, and its knowledge is essential in airframe sizing. In general, gust loads analysis is carried out from the time-domain aeroelastic equations of motion with a forcing term to include the gust

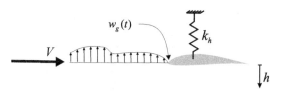

Figure 3.20 Airfoil encountering a vertical stationary gust. The airfoil can move only along the vertical axis, and $w_g(t)$ is the instantaneous vertical gust speed at its leading edge.

disturbance, Equation (3.92), together with a given profile of the velocity fluctuations of the incoming flow (such as those introduced in Chapter 2). To introduce the problem, this section focuses on the vertical motions of a rigid airfoil supported by a spring and subject to a discrete vertical gust, updating the classical presentation of this problem by Bisplinghoff et al. (1955, Section 10.6) with a modern linear systems theory description.

Consider a rigid airfoil of chord c moving horizontally at constant velocity V in a flow with density ρ. The airfoil has a mass m, and it is restricted by a spring of stiffness k_h to changing only its vertical position, denoted by the coordinate h (positive down). At time $t = 0$, it encounters a stationary vertical gust with an arbitrary (but known) spatial distribution of velocity as shown in Figure 3.20.

The corresponding time history of the vertical gust speed at the leading edge is $w_g(t)$. If the resultant instantaneous lift on the airfoil is $L(t)$, Equation (3.5) simplifies to

$$m\ddot{h} + k_h h = -L(t). \tag{3.103}$$

The instantaneous lift has two components: one coming from the gust disturbance and another coming from the motion of the airfoil, such as $L = \frac{1}{2}\rho V^2 c \left(c_l^g + c_l^a \right)$. As before, we introduce now the nondimensional plunge coordinate $\eta = \frac{h}{c/2}$ and the nondimensional time $s = \frac{2Vt}{c}$, resulting in

$$\frac{4V^2 m}{c^2}\eta'' + k_h \eta = -\rho V^2 (c_l^g + c_l^a). \tag{3.104}$$

The basic structure of this problem is shown in Figure 3.21. There are two independent inputs to the aerodynamics: one coming from the airfoil motion and another coming from the gust profile. The instantaneous resultant lift is then the input in Equation (3.104), which after integration updates the airfoil position, velocity, and acceleration.

Lift Due to Gust

In Section 3.3.4, we introduced Sears's solution to the lift on an airfoil encountering the sinusoidal gust, which was obtained in the frequency domain, Equation (3.57). A transfer function was then defined in Figure 3.12, for which an RFA was later introduced to facilitate the state-space formulation to compute the lift in the time domain, Equation (3.73), due to an arbitrary gust time history, $w_g(s)$. This is the first contri-

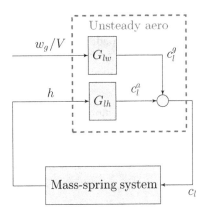

Figure 3.21 Block diagram for the dynamics of a 1-DoF airfoil subjected to a gust.

bution (G_{lw}) to the unsteady aerodynamics in Figure 3.21. From Equation (3.73), and using that $c_{l_\alpha}^{qs} = 2\pi$, we have

$$c_l^g(s) = 2\pi \sum_{j=1}^{\hat{n}} \hat{a}_j \hat{b}_j \hat{\lambda}_j,$$

$$\hat{\lambda}_j' + \hat{b}_j \hat{\lambda}_j = \frac{w_g}{V}, \quad \text{with } j = 1,\dots,\hat{n}. \tag{3.105}$$

Alternatively, the lift can be obtained from a convolution integral, Equation (1.14), starting from Küssner's solution to the step gust,

$$c_l^g(s) = 2\pi \left[\frac{w_g(0)}{V} \psi(s) + \int_0^s \frac{w_g'(\sigma)}{V} \psi(s-\sigma)\,d\sigma \right]$$

$$= 2\pi \int_0^s \frac{w_g(\sigma)}{V} \psi'(s-\sigma)\,d\sigma, \tag{3.106}$$

where $\psi(s)$ is Küssner's function, Equation (3.90), and the last expression has been obtained by integration by parts.

Lift Due to Airfoil Motions

This is obtained in the time domain using the RFA introduced in Equation (3.70), that is,

$$c_l^a(s) = \pi \left(\eta'' + \eta' + 2 \sum_{j=1}^{n} a_j b_j \lambda_j \right), \tag{3.107}$$

$$\lambda_j' + b_j \lambda_j = \eta'.$$

As before, derivatives are with respect to the nondimensional time, $s = \frac{Vt}{c/2}$, and it is $c_{l_\alpha}^{qs} = 2\pi$. The first term in Equation (3.107) is the apparent mass effect and the second relates to the effective angle of attack in the plunge motions. Only the latter affects the aerodynamic states, since they result from the circulatory effects.

Mass Ratio and Instantaneous Load Factor

The equations of motion in nondimensional time can be finally written as

$$\left(\frac{4V^2 m}{c^2} + \pi\rho V^2\right)\eta'' + \pi\rho V^2\eta' + k_h\eta + 2\pi\rho V^2\sum_{j=1}^{n} a_j b_j \lambda_j = -\rho V^2 c_l^g(s), \qquad (3.108)$$

with c_l^g given by either Equation (3.105) or 3.106, and the aerodynamic states λ_a defined by the second equation in Equation (3.107). If we divide this expression by $\pi\rho V^2$, we obtain the following dimensionless form:

$$(4\mu + 1)\eta'' + \eta' + 4\mu\bar\omega_h^2\eta + 2\sum_{j=1}^{n} a_j b_j \lambda_j = -\frac{1}{\pi}c_l^g(s), \qquad (3.109)$$

where $\mu = \frac{m}{\pi\rho c^2}$ is known as the *mass ratio* (which has been already introduced in Example 3.3) and $\bar\omega_h = \frac{c}{2V}\sqrt{\frac{k_h}{m}}$ is the natural frequency of the mass-spring system nondimensionalized by the convective timescale of the flow. The mass ratio compares the mass per unit span of the wing with the mass of an equivalent fluid disk of diameter c. When $m \gg \pi\rho c^2$, the apparent mass is negligible and $4\mu + 1 \approx 4\mu$ in the first term of Equation (3.109).

For a given gust profile, $w_g(s)$, the time history of the lift coefficient, $c_l^g(s)$, is uniquely defined, and the solution to Equation (3.109) determines the time history of the vertical position of the airfoil, $h(s) = \frac{c}{2}\eta$. The interest from a ride-comfort perspective is not in the amplitude of the airfoil (i.e., the aircraft) excursions in the vertical plane, but rather in the inertial forces, which are directly proportional to the instantaneous vertical acceleration. When written in terms of dimensional time, this is typically normalized by the acceleration of gravity to define the *instantaneous incremental load factor*,

$$\Delta n(t) = -\frac{\ddot h(t)}{g} = -\frac{2V^2}{gc}\eta''. \qquad (3.110)$$

Gust Alleviation Factor

A good metric of the impact of the gust loads on a vehicle is the maximum value achieved by the instantaneous load factor in the transient response after the gust encounter. The resulting maximum load factor, Δn_{max}, however, depends on the amplitude of the gust profile and, therefore, it is convenient to normalize by its maximum amplitude, w_0. This is typically done by considering the limit case of a steady gust on the quasi-steady response of a free-flying heavy airfoil, that is, without stiffness, damping, apparent mass, and aerodynamic lags. Since $\max\left|c_l^g\right| \sim \frac{|w_0|}{V}$, then $\max\left|\eta''\right| \sim \frac{|w_0|}{\mu V}$, and, from Equation (3.110), we can then rewrite the maximum load factor as

$$\Delta n_{max} = \frac{w_0 V}{\mu g c}\beta, \qquad (3.111)$$

where β is known as the *gust alleviation factor*, although this name can be somehow misleading as higher values do not imply higher alleviation (but higher loads).

Figure 3.22 1-DoF airfoil encountering a sharp-edged gust.

Nevertheless, if w_0 is the maximum amplitude of the gust velocity profile, then, from its definition, $|\beta| \leq 1$ always, and β effectively gives a measure of the load reduction (or alleviation) due to dynamic effects in the gust response. For a given vehicle, flight condition, and gust profile, the gust alleviation factor can be computed from the nondimensional acceleration by combining Equations (3.110) and (3.111) as

$$\beta = \frac{2\mu}{w_0/V} \max |\eta''|. \tag{3.112}$$

The final stage in the analysis process is the definition of the gust profile (the background nonstationary flow conditions), which can be any of those considered in Chapter 2. In the following subsections, the dynamics of 1-DoF airfoils under three typical conditions are investigated. First, we consider unsupported (free-flying) airfoils under a sharp-edged gust, that is, whose lift due to the gust is given by Küssner's result. Then we consider the same airfoil under a "1-cosine" discrete gust, as defined in Section 2.4.2. The third example is the stochastic analysis of the response of an elastically supported airfoil, subjected to continuous turbulence, using the von Kármán model of Section 2.3.3.

3.6.1 Response of an Unsupported Airfoil to a Sharp-Edged Gust

Consider now the particular case of the open-loop gust response of an unsupported airfoil ($k_h = 0$) as represented in Figure 3.22. Since the result only depends on the derivatives of $\eta(s)$, we solve in the nondimensional plunge speed $\nu = \frac{-\eta'}{w_0/V} = -\frac{\dot{h}}{w_0}$, that is, the ratio of the plunge speed over the gust amplitude. The equations of motion can then be written as

$$(4\mu + 1)\nu' = -\nu + 2\sum_{j=1}^{n} a_j b_j z_j + \frac{c_l^g(s)}{\pi w_0/V}, \tag{3.113}$$

$$z_j' = -\nu - b_j z_j,$$

with $z_j = \frac{\lambda_j}{w_0/V}$ corresponding to the rescaled aerodynamic lags. These equations can be written in a state-space form as

$$\mathbf{x}' = \mathbf{A}\mathbf{x} + \mathbf{B}u, \tag{3.114}$$

with $\mathbf{x}^\top = (\nu, z_1, \ldots, z_n)$, $u(s) = \frac{c_l^g(s)}{2\pi w_0/V}$, and

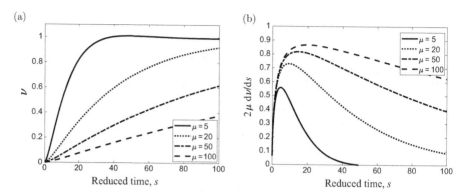

Figure 3.23 Response of unsupported 1-DoF airfoils to a unit sharp-edged gust for the varying mass ratio. (a) Nondimensional velocity, $\nu = -\dot{h}/w_0$ and (b) nondimensional acceleration, $2\mu\nu'$. [03Airfoil/gust_step.m]

$$\mathbf{A} = \begin{bmatrix} \frac{-1}{4\mu+1} & \frac{2a_1b_1}{4\mu+1} & \cdots & \frac{2a_nb_n}{4\mu+1} \\ -1 & -b_1 & \cdots & 0 \\ \cdots & \cdots & \cdots & \cdots \\ -1 & 0 & \cdots & -b_n \end{bmatrix} \quad \text{and} \quad \mathbf{B} = \begin{bmatrix} \frac{2}{4\mu+1} \\ 0 \\ \cdots \\ 0 \end{bmatrix}. \quad (3.115)$$

The input in Equation (3.114) has been normalized by the quasi-steady lift due to a constant gust velocity w_0. The only remaining variable coefficient in the equation is the mass ratio, which will determine the evolution of the vertical acceleration (and consequently the instantaneous load factor). To investigate this, the airfoil is subjected first to a sharp-edged gust of amplitude $0 < w_0 \ll V$, as shown in Figure 3.22. In this case, the input to the system is $u(s) = \psi(s)$, that is, Küssner's function, which has been defined in Equation (3.90). This problem can be solved by numerical integration for given initial conditions (here, $\nu(0) = 0$ and $z_j(0) = 0$), and the solutions for various mass ratios are shown in Figure 3.23. The time history of the nondimensional plunge speed $\nu = \frac{\dot{h}}{w_0}$ is shown in Figure 3.23a. Regardless of the mass of the vehicle, the solution goes to the steady state $\nu = 1$. This means that the vehicle moves upward with a steady gust, and the rate at which it reaches this asymptotic value strongly depends on the mass ratio, μ. The lower the mass of the vehicle, the sooner the asymptotic value is reached. This is further explored in Figure 3.23b, which shows the dimensionless acceleration scaled by the apparent mass, so the maximum value of each curve corresponds to the gust alleviation factor, β, defined in Equation (3.112). As can be seen, the higher the mass ratio, the higher is the value of the gust alleviation factor (while always being smaller than 1). Therefore, even though the maximum acceleration grows as the mass decreases, this growth is modulated by the dynamic effects (mainly aerodynamic damping), which become more important as the mass of the aircraft becomes smaller. For very heavy aircraft (very large μ), most of the gust energy goes into accelerating the aircraft; yet, as its mass is large, the amplitude of the acceleration is low and nearly constant for a long period. The speed raises nearly linearly with time to the steady-state value ($\nu = 1$).

3.6.2 Response of an Unsupported Airfoil to a "1-Cosine" Gust

The "1-cosine" gust was introduced in Section 2.4.2, where it was argued that it is a standarized scenario that is used in the certification process to demonstrate the appropriate dynamic response of the aircraft. Consider here the response of an unsupported airfoil to a "1-cosine" gust, as shown in Figure 3.24. The gust profile is defined as in Equation (2.54). As the aircraft moves through the gust at constant speed, V, $x = Vt$, and the vertical gust speed at the leading edge as a function of the nondimensional time, $s = \frac{Vt}{c/2} = \frac{2x}{c}$, becomes

$$w_g(s) = \frac{1}{2}w_0\left(1 - \cos\frac{\pi s}{2H/c}\right), \quad \text{if } 0 \le s \le 4H/c,$$

$$w_g(s) = 0 \qquad\qquad\qquad \text{elsewhere.} \tag{3.116}$$

Figure 3.25 shows the lift coefficient c_l^g, given by Equation (3.106), for "1-cosine" gusts of increasing length (the gust gradient varies between 10 and 50 airfoil semi-chords with constant increments). Even though they all have the same gust intensity w_0, shorter gusts will have smaller amplitude, since they are more affected by the aerodynamic lags (the delay in the production of lift). Note also that there is a small time delay between the point of the maximum gust speed (which corresponds to $s = 2H/c$) and the point of the maximum lift in the figure. Figure 3.25 also includes the time history of the airfoil nondimensional accelerations for $\mu = 5$, obtained from the numerical integration of Equation (3.109). The points of maximum and minimum accelerations are of particular relevance since they define the highest loads that the vehicle may experience during operation. Note that they correspond to two different values of the gust gradient ($H = 5c$ for the maximum instantaneous acceleration and $H = 12.5c$ for the minimum one). Identification of the worst loading conditions is a critical design consideration for airframe sizing, and this is investigated in detail in the following section.

Critical Gust Length

The normalized response of the airfoil to the "1-cosine" gust is, therefore, a function of two parameters: the mass ratio μ and the gust gradient H. The gust alleviation factor defined in Equation (3.112) is also a function of both parameters, that is, $\beta = \beta(\mu, H)$.

Figure 3.26a shows the gust alleviation factor as a function of the gust gradient, H, for mass ratios ranging from $\mu = 5$ to 100 in constant increments. As can be seen, the contour lines with fixed μ all have a maximum value. Furthermore, as one considers heavier aircraft (larger μ), this maximum gust alleviation factor, β, grows. For each value of μ, the gust gradient H at which β has a maximum is known as the *critical*

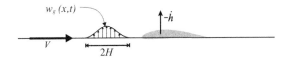

Figure 3.24 1-DoF airfoil encountering a "1-cosine" gust.

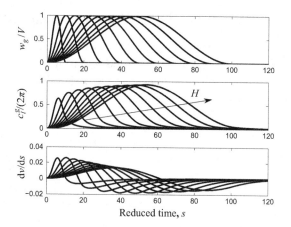

Figure 3.25 Gust profile, lift due to gust, c_l^g, and nondimensional acceleration for increasing gust gradients. $\mu = 5$ and H from $2.5c$ to $25c$ in increments of $2.5c$. [03Airfoil/gust_1cos.m]

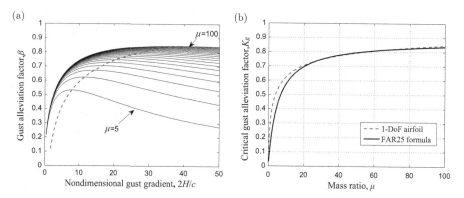

Figure 3.26 Gust alleviation factor and critical gust alleviation factor for a 1-DoF airfoil. (a) Gust alleviation factor, β (continuous lines: mass ratio μ from 5 to 100 in increments of 5; dotted line through maxima) (b) Critical gust alleviation factor, K_g. [03Airfoil/gust_beta.m]

gust length. The process of determining the critical gust length for a given aircraft and flight condition is typically referred to as *gust tuning*, and the corresponding gust as the *tuned gust*. We finally define the critical gust alleviation factor as the maximum value of β over all possible gust lengths, that is,

$$K_g = \max_H(\beta(\mu, H)), \qquad (3.117)$$

which is solely a function of the mass ratio, $\mu = \frac{m}{\pi \rho c^2}$. The locus of the critical gusts for each value of the mass ratio is also shown with a dotted line in Figure 3.26a.

In summary, the process goes as follows. For a given aircraft and flight condition, we determine the time history of load factors over all possible gust lengths. The point of the maximum load factor determines the critical gust length, and the corresponding gust alleviation factor is the *critical gust alleviation factor*. If we repeat this over all flight conditions (including changes on vehicle weight), we obtain the curve of

Figure 3.26b of critical gust alleviation factors as a function of the mass ratio. The critical factor K_g approximately satisfies the following expression, which is also included in Figure 3.26b:

$$K_g = \frac{0.88\mu}{5.3 + \mu}. \qquad (3.118)$$

This expression was originally proposed by Pratt (1953) and has since been adopted for airworthiness certification to estimate the dynamic alleviation in aircraft gust loading. It is typically referred to as the *FAR gust load formula* (Hoblit, 1988, Ch. 2), and it is used for preliminary sizing as follows. Given an aircraft with mass ratio μ, a set of static-equivalent loads are determined by scaling the lift distribution in forward flight with the load factor, Equation (3.111), where the gust alleviation factor β is replaced by the critical value, K_g, determined in Equation (3.118). Large aircraft, for which μ is large, have values of $K_g \approx 0.88$, and their dynamic loading is very similar to those of static loads. Nevertheless, in relative terms, aircraft mass (and therefore the mass ratio μ) increases much faster than flight velocity V, as aircraft become larger. Therefore, from Equation (3.111), it also follows that gusts with larger intensity w_0 are needed to generate equivalent levels of static loading as aircraft mass increases. Further discussion on this problem can be found in Ricciardi et al. (2013).

3.7   Continuous Turbulence Response of an Elastically Supported Airfoil

Section 2.3.2 presented the response of a linear dynamical system to random excitations. This was done in an statistical manner by using the spectral density to obtain the expected value (i.e., the value over long enough measurements) of the frequency content in the system response. In particular, it was seen in Equation (2.33) that the spectral density of the measurement outputs can be determined from the system admittance and the spectral density of the excitation. This is used now to investigate the uncontrolled vibrations of an elastically supported 2-DoF airfoil in the presence of a continuous turbulent airstream. The equations of motion of the aeroelastic system in the state-space form were given in Equation (3.95), and the turbulent flow velocities are assumed to be well described by the 1-D von Kármán spectrum, Equation (2.46). Only nonzero vertical gust velocities are considered.

Figure 3.27 describes the structure of this system. The turbulence filter generates a synthetic gust profile w_g/V at the airfoil leading edge from a Gaussian white noise of given intensity. This defines the disturbance input to the linear aeroelastic model defined in Equation (3.95). The system outputs need to be defined and can be chosen as any linear combination of the states, such as instantaneous aerodynamic forces, or the local velocities (or even accelerations), at any point on the airfoil. This defines the output matrix $\mathbf{C}_{ae}$. There is, however, no feedthrough, that is, no direct measurements of the gust velocity are included in the output vector, as assumed in Equation (2.19). The state-space equations of the aeroelastic system linking the gust profile and the chosen outputs become

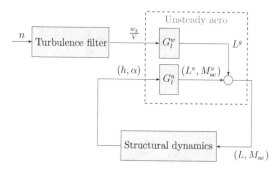

Figure 3.27 Block diagram for the dynamics of a 2-DoF airfoil in atmospheric turbulence.

$$\mathbf{x}'_{ae} = \mathbf{A}_{ae}\mathbf{x}_{ae} + \mathbf{B}_{ae}\frac{w_g}{V},$$
$$\mathbf{y}_{ae} = \mathbf{C}_{ae}\mathbf{x}_{ae}, \tag{3.119}$$

where $\mathbf{A}_{ae}$ and $\mathbf{B}_{ae}$ have been defined in Equation (3.96). This allows computing the frequency-response function (or admittance, Equation (1.19)) of the aeroelastic system, in terms of the reduced frequency, as the (column) matrix function

$$\mathbf{G}_{yw}(ik) = \mathbf{C}_{ae}(ik\mathcal{I} - \mathbf{A}_{ae})^{-1}\mathbf{B}_{ae}. \tag{3.120}$$

The input for Equation (3.120) is the nondimensional gust velocity, w_g/V, which is itself the output of the turbulence filter, as shown in Figure 3.27. The outputs are chosen in the definition of $\mathbf{C}_{ae}$. Let σ_w and ℓ be the (dimensional) RMS and integral length scale of the turbulent velocity, respectively. Substituting the vertical component of Equation (2.46) into Equation (2.33), this results in a spectral density that, in terms of the reduced frequency, is written as

$$\Phi_y(k) = \frac{\sigma_w^2}{V^2}\frac{2\ell}{\pi c}\frac{1+\frac{8}{3}\left(2ak\frac{\ell}{c}\right)^2}{\left[1+\left(2ak\frac{\ell}{c}\right)^2\right]^{11/6}}\mathbf{G}_{yw}(ik)\mathbf{G}_{yw}^*(ik). \tag{3.121}$$

As can be seen, the bandwidth of the excitation when written in terms of the reduced frequency is proportional to the ratio between the integral length scale ℓ of the continuous turbulence and the airfoil chord. Note that the spectral density in terms of the physical angular velocity is $\Phi_y(\omega) = \frac{c}{2V}\Phi_y(k)$.

Recall that the diagonal terms of Equation (3.121) give the PSD. The off-diagonal terms define the coherence between each pair of output signals, Equation (2.14), which is identically one for all cases here as all outputs are obtained from a single noise input. It is also interesting to obtain the output correlation matrix, or covariance, Σ_y, as a measure of the overall loading and coupling levels. The covariance can be obtained directly through the solution of the Lyapunov equation, Equation (2.27), with state and input matrices given by the series connection of an RFA of the von Kármán turbulence filter, as given by Equation (2.49), and the aeroelastic model of Equation (3.119). It can alternatively be obtained from the integration of the corresponding spectral density, as

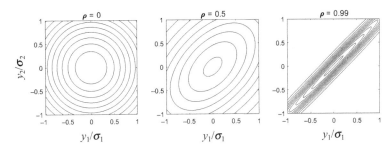

Figure 3.28 Equal probability ellipses for bivariate normal distributions for three values of the correlation coefficient, ρ. [03Airfoil/correlation.m]

in Equation (2.16). If we consider a problem with two outputs (e.g., plunge and pitch rates), the resulting covariance matrix can be written as

$$\Sigma_y = \frac{1}{2} \int_0^\infty \left(\Phi_y(k) + \Phi_y^\top(k) \right) dk, = \begin{bmatrix} \sigma_1^2 & \rho\sigma_1\sigma_2 \\ \rho\sigma_1\sigma_2 & \sigma_2^2 \end{bmatrix}, \qquad (3.122)$$

where σ_1 and σ_2 are the RMS values of each output and ρ is the *correlation coefficient*. The correlation coefficient always satisfies $-1 \leq \rho \leq 1$. Since we have assumed that atmospheric turbulence is a Gaussian random process, knowledge of the correlation function is sufficient to describe the joint probability of both outputs. This is obtained directly from the probability density function (PDF) of the bivariate normal distribution (Tong, 1990), given as

$$f(y_1, y_2) = \frac{1}{2\pi\sigma_1\sigma_2\sqrt{1-\rho^2}} \exp\left(-\frac{1}{2(1-\rho^2)} \left[\frac{y_1^2}{\sigma_1^2} + \frac{y_2^2}{\sigma_2^2} - 2\rho\frac{y_1}{\sigma_1}\frac{y_2}{\sigma_2} \right] \right). \quad (3.123)$$

The PDF gives the relativel likelihood of the output to take the simultaneous values y_1 and y_2. As can be easily seen, constant values of this joint probability define ellipses in the y_1 and y_2 planes, which are known as *equal probability ellipses*. Figure 3.28 shows the axis-normalized equal probability ellipses corresponding to three different values of the correlation coefficient. Integration of the PDF gives the likelihood of a pair (y_1, y_2) to be above or below a given value. This is very useful in load analysis, as it allows comparing the expected load levels with the admissible ones. For that purpose, it is translated into a *frequency of exceedance* that determines the probability (per unit of flight time) that a certain output (typically load) level is exceeded (Hoblit, 1988, Ch. 4). Even though equal probability ellipses do not provide spectral information about the loads on the aircraft, they are extensively used to identify worst-case scenarios for each pair of interesting quantities (here, plunge and pitch, but this also applies to the more general case of forces and moments on a wing discussed). The resulting loading conditions are often assessed using static analysis methods for airframe structural sizing.

Example 3.4 2-DoF Airfoil under von Kármán Turbulence. Consider again the 2-DoF airfoil analyzed in Example 3.3. It has a mass ratio $\mu = \frac{m}{\pi \rho c^2} = 5$; coordinates $x_{ac} = 0.25c$, $x_{ea} = 0.35c$, and $x_{cg} = 0.45c$; and radius of gyration $r_\alpha = 0.25c$. We consider first the gust response when the ratio of characteristic frequencies is $\omega_h/\omega_\alpha = 0.5$. From the stability analysis in Figure 3.19a, the nondimensional flutter speed is $\frac{V_F}{c\omega_\alpha} = 1.04$.

For this problem, a state-space description is built using four lags to approximate the lift-deficiency function, $C(ik)$, and Jones's two-lag approximation of Sears's function (Table 3.2). The system outputs are chosen as the instantaneous pitch angle and nondimensional plunge, which were already defined as the generalized coordinates **q** in Equation (3.7). Therefore, the output is extracted directly from the state vector, Equation (3.94), as

$$\mathbf{C}_{ae} = \begin{bmatrix} 1 & 0 & \cdots & 0 \\ 0 & 1 & \cdots & 0 \end{bmatrix}. \tag{3.124}$$

This results in a state-space description with 14 states: one input and two outputs. Equation (3.120) determines the corresponding admittance from the gust vertical velocity component at the airfoil leading edge to the airfoil DoFs (pitch and plunge). Its magnitude and phase are shown in Figure 3.29a for four freestream velocities, $\frac{V}{c\omega_\alpha} = \{0.2, 0.4, 0.6, 0.8\}$. As the response is in all cases to a unit amplitude of w_g/V, increasing the freestream velocity generates larger aerodynamic forces and therefore larger amplitude in the airfoil excursions. Note that at zero frequencies, the response to a constant upward gust is a positive pitch and a negative plunge (in agreement with the sign convention in Figure 3.1). However, there are significant phase changes as the frequency of the gust input increases. Two clearly distinct peaks can also be seen in the magnitude plot, corresponding to the two aeroelastic eigenvalues associated with the structural vibrations. The first peak (near $\omega = 0.5/\omega_\alpha$) is linked to the resonance mode in the plunge and the second (near $\omega = \omega_\alpha$) to the resonance mode in the pitch. Their amplitude and frequency are determined by the real and imaginary parts of the complex eigenvalues in Equation (3.94), respectively. As the freestream velocity increases and approaches the flutter speed, the frequencies of both peaks get closer.

In the results so far, we have considered the case where the plunge resonant frequency (always defined in vacuum) is below that of the pitch vibrations. This is the usual situation on most aeroelastic structures, but it is also interesting to consider the response if the opposite was true. Figure 3.29b shows the frequency-response functions between gust inputs and pitch and plunge DoFs for $\omega_h/\omega_\alpha = 1.63$ at the same four speeds. As can be seen from Figure 3.18, the nondimensional flutter speed is again $\frac{V_F}{c\omega_\alpha} = 1.04$, but the response is markedly different. The oscillations at low frequencies are much more damped than in the previous case, including those near the pitch-resonance frequency (near $\omega = \omega_\alpha$). However, there is a large amplitude response near the plunge mode (near $\omega = 1.63\omega_\alpha$), with a much higher impact on the structure.

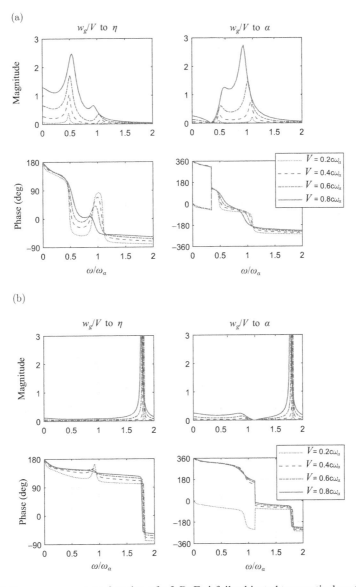

Figure 3.29 Frequency-response function of a 2-DoF airfoil subjected to a vertical gust ($\mu = 5$, $x_{ac} = 0.25c$, $x_{ea} = 0.35c$, $x_{cg} = 0.45c$). (a) $\omega_h/\omega_\alpha = 0.5$ and (b) $\omega_h/\omega_\alpha = 1.63$. [03Airfoil/turb2dof.m]

Finally, the frequency content in the response to atmospheric turbulence with integral length scale $\ell = c$, and the unit RMS of σ_w/V is shown in Figure 3.30a, for an airfoil with $\omega_h/\omega_\alpha = 0.5$. The figure includes the normalized PSD of the von Kármán turbulence spectrum for each of the considered speeds, as well as the PSD of the resulting nondimensional plunge and pitch oscillations. As can be seen in the figure, the PSDs of both outputs show rather similar features to the admittance function from which they are derived.

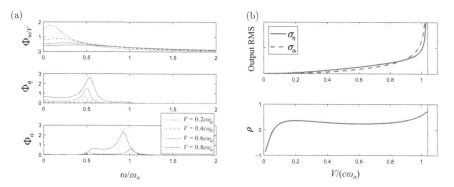

Figure 3.30 Response of a 2-DoF airfoil with $\omega_h/\omega_\alpha = 0.5$ to atmospheric turbulence. The limit $\frac{V}{c\omega_\alpha} = 1.04$ is the flutter speed. (a) PSD of the gust input (w_g/V) and of the plunge and pitch response. (b) RMS and correlation coefficient of the output magnitudes. [03Airfoil/turb2dof.m]

Integration over all frequencies of the spectral density matrix gives the covariance, Equation (3.122). Figure 3.30b shows the RMS of the plunge and pitch response as a function of the nondimensional forward flight speed, as well as the correlation factor between both signals. Note that the amplitude of the oscillations increases exponentially as the speed approaches the flutter speed (marked with a dotted vertical line in the plots). Finally, the correlation is seen here to change very rapidly at low speeds, while it remains mainly constant for flight speeds between 0.1 and 0.8 $c\omega_\alpha$. For a given probability level (i.e., 99.9%), the values in Figure 3.30b are inputs into Equation (3.123) to determine the associated equal probability ellipses at each flight speed.

3.8   Summary

This chapter has presented an overview of the key concepts in linear aeroelasticity, which form the basis on which we will develop many of the more general ideas in this book. The presentation has been done for the particular case of an elastically supported rigid airfoil with a trailing-edge flap, which defines a representative system of many aeroelastic phenomena typically encountered on aircraft wings. By restricting the study to airfoils at low speeds, classical analytical solutions in unsteady aerodynamics are readily available, and the coupled system dynamics have been investigated in detail using relatively few parameters. This model, however, provides only a qualitative description of the actual physics in more complex 3-D configurations, and it can rarely be used for predictive purposes.

We have seen that linear aeroelasticity can be described as a feedback interconnection between the structural dynamics and aerodynamic characteristics of the physical system under study. For the 2-DoF airfoil problem, the equations of motion of the structure have been defined through a lumped-parameter model for the airfoil mass and stiffness, while the classical frequency-domain solutions of Theodorsen and Sears have provided a representation of the unsteady aerodynamics under airfoil motions and external gusts, respectively. While frequency-domain solutions have been traditionally used in dynamic aeroelasticity, particularly for stability analysis, the focus here has been on time-domain formulations, which allows unrestricted access to the many powerful tools of linear systems and control theory. This also links better to the more general nonlinear formulations introduced later in this book, which are naturally written in the time domain.

As the classical frequency-domain solutions in unsteady aerodynamics are given in terms of Bessel functions, approximations in terms of rational functions are necessary to avoid costly Fourier transforms. This has been carried out using the now standard approach in aeroservoelasticity that builds an approximation to any order of accuracy using Padé approximants. As a result, linear-time-invariant state-space descriptions are obtained to describe the aeroelastic response of 2-DoF airfoils to gust disturbance.

System static and dynamic stability have been first established from the eigenvalues of the system matrix, which define the flight conditions, leading to aeroelastic divergence and flutter, respectively. The open-loop dynamics of the general system to vertical gusts was then investigated as an introduction to dynamic load analysis. Two specific situations have been of particular interest, namely the response to discrete and continuous gusts that are required in aircraft certification. For the discrete gust response, the response to "1-cosine" gust profiles has been analyzed and used to introduce the concepts of load factor and critical gust length, which are extensively used in aircraft design. For the continuous gust, the statistical models for the atmospheric turbulence of Section 2.3.3 have been used (in particular, von Kármán's model) for which the expected response of the airfoil was obtained using the methods of Section 2.3.2.

3.9 Problems

Exercise 3.1. An airfoil of chord c is mounted in a wind tunnel such that it can only move in pitch against a spring of stiffness k_α. The airfoil is fitted with a flap with a spring restraint of stiffness k_δ that creates a restoring moment on the hinge proportional to the incremental deflection from an initial value, δ_0, as shown in the following figure. Assume known (nonzero) values of the following aerodynamic derivatives, c_{l_α}, c_{l_δ}, c_{m_δ}, c_{h_α}, and c_{h_δ}, where $H = \frac{1}{2}\rho V c^2 c_h$ is the aerodynamic moment at the hinge.

Derive its static equilibrium equations and compute the divergence speed as a function of the hinge stiffness, k_δ.

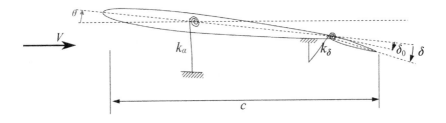

Exercise 3.2. Consider the civil structure of the following figure. It is modeled as a rigid plate of chord c, unit span, and constant mass per unit area ρ_s, which is supported by two linear springs with stiffness k_1 and k_2, such that $k_1 < k_2$. Both supports have the same natural length measured from the ground, z_0, and a maximum admissible elongation $\pm z_{max}$. We will consider the aeroelastic design of this structure to environmental loading:

(i) *Snow loads.* Including the weight of the structure, compute the maximum thickness of snow that can be accumulated on the roof before the springs fail, assuming that it covers the roof uniformly and that there is zero wind velocity.

(ii) *Wind loads.* Assuming that there is no snow, write down the equations of the static aeroelastic equilibrium as a function of the wind velocity, V, and use them to compute the maximum admissible wind velocity before structural failure.

(iii) Compute the maximum admissible wind velocity before the roof fails due to aeroelastic divergence.

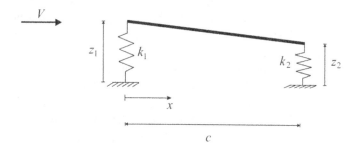

Exercise 3.3. A rigid airfoil of chord c is mounted in a wind tunnel such that its only DoF is the vertical translation against a spring of stiffness k_h. The airfoil is fitted with a trailing-edge control surface that is free to rotate with no spring restraint and has nonnegligible inertia. This results in the 2-DoF system of the following figure. Let m_c be the mass of the control surface and m the total mass, including the control surface, and let x_c be the location of the center of mass of the control surface from the hinge location and r_c its radius of gyration about its center of gravity. Under quasi-steady aerodynamic assumptions, the lift and aerodynamic hinge moment are approximated as

$$L = \tfrac{1}{2}\rho V^2 c \left(c_{l_\alpha} \frac{\dot{h}}{V_\infty} + c_{l_\delta}\delta\right),$$

$$H = \tfrac{1}{2}\rho V^2 c^2 \left(c_{h_\alpha} \frac{\dot{h}}{V_\infty} + c_{h_\delta}\delta\right).$$

Derive the equations of motion of the airfoil as a function of the lift, L, and the aerodynamic hinge moment, H, and introducing then the aerodynamic model given above, determine the flutter speed and frequency in a nondimensional form. Use the following constants: $\frac{m}{\rho c^2} = 4$, $x_c = 0.1c$, $m_c = 0.1m$, $r_c = 0.1c$, $c_{l_\alpha} = 2$, $c_{l_\delta} = 1$, $c_{h_\alpha} = -0.04$, and $c_{h_\delta} = -0.05$.

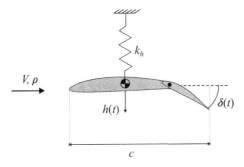

Exercise 3.4. A flat plate of chord c, mass m, and unit span is suspended by a spring of stiffness k_ξ in an airstream with velocity V and density ρ. The plate can only move vertically, with coordinate $\xi(t)$ positive up, and is fitted with a trailing-edge flap, with deflection $\delta(t)$ positive down. The full-span flap covers 25% of the airfoil chord. After neglecting the apparent mass, the frequency-domain lift coefficient can be assumed to be $\overline{c_l} = 2\pi\mathcal{C}(ik)\overline{\alpha_{qs}}$, with $c_l(t) = \overline{c_l}e^{iks}$, and $\alpha_{qs}(t) = \overline{\alpha_{qs}}e^{iks}$ being the equivalent angle of attack. Assume a two-term RFA to the lift-deficiency function, that is,

$$\mathcal{C}(ik) = 1 - \sum_{j=1}^{2} \frac{ika_j}{ik + b_j},$$

with known coefficients a_j and b_j.

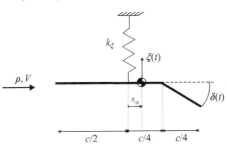

(i) Introducing aerodynamic states, obtain the differential equations that determine the nonstationary lift in dimensional time, $c_l(t)$, for the arbitrary time histories of $\xi(t)$ and $\delta(t)$.

(ii) If a sharp-edged vertical gust with speed w_0 hits the airfoil at time $t = 0$, write the equations of motion of its subsequent dynamics.

(iii) Consider a proportional feedback control of the form $\delta = -K_0\dot{\xi}$, with $K_0 > 0$. Discuss its effect on the damping and frequency in the response of the plate to the pervious gust.

Exercise 3.5. Consider a flat plate of unit span with a chord c and mass m, which can only move vertically against a spring of stiffness k_ξ, and it is positioned at the zero angle of attack in an airstream with velocity V and density ρ. The plate has an externally actuated trailing-edge flap with a very small chord εc. Assume that any motions are sufficiently slow for the lift on the plate to be well approximated by a quasi-steady model, that is,

$$c_l^a(s) = -\frac{2\pi}{c}\frac{d^2\xi}{ds^2} - \frac{4\pi}{c}\frac{d\xi}{ds} + 8\sqrt{\varepsilon}\delta,$$

with the nondimensional time $s = \frac{2Vt}{c}$.

(i) Determine the steady-state response of the plate for sinusoidal excitations of the flap, $\delta(s) = \delta_0 \sin(k_0 s)$, where k_0 is a reference-reduced frequency.

(ii) The plate has now its flap locked at $\delta = 0$ and encounters a gust with the known time history $w_g(s)$. A closed-form solution for the resulting lift is known in the frequency domain (assuming $w_g(s) = \overline{w_g}e^{iks}$) as

$$\overline{c_l^g}(ik) = 2\pi \mathcal{S}_0(ik)\frac{\overline{w_g}}{V},$$

where Sears's function can be approximated as $\mathcal{S}_0(ik) = 1 - \frac{ika}{ik+b}$, with known a and b. Write the differential equations that give the response of the plate to the gust.

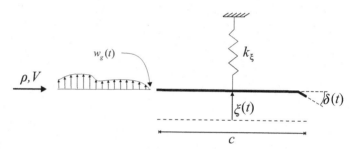

Exercise 3.6. Consider a linear aeroelastic system with N structural DoFs $\mathbf{q}$, mass $\mathbf{M}$, damping $\mathbf{D}$, and stiffness $\mathbf{K}$, and flow conditions defined by the airspeed V and density ρ. From a linear frequency-domain fluid solver, GAFs are obtained and expressed as $\frac{1}{2}\rho V^2 \mathbf{A}(ik)\overline{\mathbf{q}}$, with $\mathbf{q} = e^{i\omega t}\overline{\mathbf{q}}$ and $k = \frac{\omega c}{2V}$. Assume an RFA for the GAF matrix given with n lags and negligible apparent mass as $\mathbf{A}(ik) = \mathbf{A}_0 + ik\mathbf{A}_1 + \sum_{j=1}^{n}\frac{ik}{\gamma_j + ik}\mathbf{A}_{j+1}$.

(i) Write the structural and aerodynamic equations in a state-space form.

(ii) Write the coupled time-domain equations for the resulting feedback system, as shown in the following figure. Identify its state vector and state matrix.

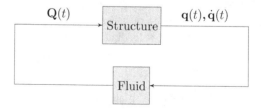

(iii) Explain how you would determine the flutter onset conditions for this system.

4 Dynamics of Rigid Aircraft in a Nonstationary Atmosphere

4.1 Introduction

In the previous two chapters, we have described the atmospheric conditions that need to be considered for the design and operation of most aircraft, as well as the linear models that describe the low-speed response of airfoils (both supported and unsupported) to some of those conditions. We now expand our scope to study the dynamics of an aircraft flying in a nonstationary atmosphere. This is done in this chapter using the standard framework of flight dynamics, in which the aircraft is idealized as a rigid body, while subsequent chapters will consider more general situations where aeroelastic effects also need to be considered. The rigid aircraft model is still the basic foundation on which all our more general flexible aircraft models are built, and therefore it is very convenient to review first the classical theory while introducing a suitable notation that facilitates subsequent studies.

As we have discussed in Section 1.2, flight dynamics is concerned with the study of the response of the vehicle to its aerodynamic environment and (auto)pilot commands, in order to assess its performance and stability or to design control systems. It also provides the framework to describe the evolution of the aircraft with respect to an Earth-based observer, to optimize trajectories and to design maneuvers. The starting point is a suitable description of the kinematics of rigid aircraft in 3-D space, for which we define multiple coordinate systems whose relative position and orientation are sought. Spatial orientations are described in this chapter using Euler angles. Although alternative parameterizations with better numerical performance also exist, Euler angles bring more intuition into our descriptions, and we will use them extensively in this book. They are then used to derive the general (nonlinear) equations of motion (EoMs) of the rigid aircraft, which are presented in this chapter under the assumption of quasi-steady aerodynamics. That implies that the aerodynamic loading can be written as a static map (e.g., a lookup table) between the aircraft instantaneous state and the resultant forces and moments at its center of mass (CM). To consider nonstationary atmospheric conditions, aerodynamic forces are more generally written here in terms of the instantaneous relative velocity between the aircraft and a moving air mass. Stationary solutions of the EoMs determine steady-state flight conditions, which are used to define steady maneuvers and the maneuvering envelope. Finally, the linearized response of the aircraft is studied, and the response to continuous turbulence is investigated on a simplified configuration using the methods for the stochastic analysis of Section 2.3.

It is assumed throughout this book that any flight segment under consideration is of relatively short duration (typically, of the order of seconds or a few minutes), for example, a maneuver or an encounter with a thermal gust. In that case, it is appropriate to assume the "flat-Earth" approximation, thus treating the Earth surface as flat and stationary. A frame of reference on the Earth will, therefore, define an inertial reference frame in most situations in this book. Furthermore, in that frame, gravity can be assumed to be a constant vertical force that is independent of the altitude changes during the analysis. Finally, we also assume that the total mass and moment of inertia of the vehicle do not change within a given flight segment (e.g., mass of burnt fuel is negligible).

The presentation of the material in this chapter follows the structure that can be found in most classical books of flight dynamics, in particular those of Ashley (1974), Etkin (1972), and Stengel (2004). The discussion is, however, often more succinct, as our objective is to establish the basic framework that can be slowly expanded as the book progresses. In particular, most flight dynamic books include rather extensive discussions on the estimation of aerodynamic derivatives and the effect of configuration parameters on aircraft stability. Those are only superficially considered here, as we present later in this book a more general strategy in which forces on the vehicle, including unsteady and aeroelastic effects, are solved for from first principles.

4.2   Kinematics of Rigid Aircraft

4.2.1   Frames of Reference

It is necessary first to establish a convention for the description of aircraft kinematics. In this chapter, the aircraft is considered to be a rigid body, and therefore its kinematic state is uniquely determined by a measure of its position and orientation, as a function of time, from a known reference. It is then convenient to define multiple spatial reference frames whose relative positions and orientations, and velocities are sought. All reference frames in this book are defined as right-handed Cartesian bases, defined by three orthogonal unit vectors with a common origin.

Earth axes. Under the "flat-Earth" assumption, the aircraft kinematics are described with respect to an observer in an *inertial frame of reference*, that is, one that moves at most with constant translational velocity. In flight dynamics, we normally refer to this global frame of reference as the *Earth axes*.

Principal axes. The orientation of the aircraft with respect to that inertial frame of reference is given by means of a *body-attached frame of reference*, or *body axes*, that is rigidly linked to the aircraft and with origin at its CM. Most aircraft have a plane of symmetry, and it is customary to define the body-attached frame of reference such that the y axis is normal to the longitudinal symmetry plane and positive on the starboard (right) wing. In that case, y becomes a principal axis of inertia. A natural choice for the x axis is to define it as a second principal axis of inertia, as shown in Figure 4.1. Finally, the z axis completes the right-hand-rule triad, lying in the plane of

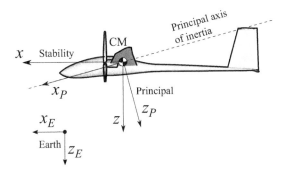

Figure 4.1 Frames of reference for a rigid aircraft: principal and stability axes (side view).

symmetry and positive downward in forward flight. The resulting body-attached frame of reference is known as the *principal axes* of the vehicle. They are uniquely defined given a vehicle geometry and mass (including payload) distribution. This frame of reference simplifies the inertia tensor, which becomes diagonal, as we will see in the following section. However, adopting this reference would unnecessarily complicate the EoMs, as this axis system would normally be tilted downward with respect to the aircraft velocity vector. This is the situation displayed in Figure 4.1.

Stability axes. A more intuitive and convenient choice for flight dynamics purposes results from aligning the x axis with the projection, on the symmetry plane, of the velocity vector of the aircraft in some nominal conditions. This defines the *stability axes*. The y–z is still the plane of symmetry, but in the reference condition, the velocity vector has no z component. In forward-flight conditions with no sideslip, which is the most common reference condition, the velocity is along the x axis. Note that the orientation of the frame with respect to the airframe changes with the point in the flight envelope (as, e.g., the aircraft may fly at different angles of incidence). Also, since the axes are body attached, the instantaneous aircraft velocity vector will not necessarily be along the x axis during maneuvers.

Wind axes. The kinematic description of both rigid and flexible aircraft dynamics requires the prior selection of a suitable frame of reference. Following the convention in flight dynamics, most derivations in this book are done using the vehicle's stability axes. However, the description of the (steady) aerodynamic characteristics of the aircraft is more easily done on a reference frame in which the x axis is always aligned with the velocity vector. This defines the *wind axes*. This frame of reference is defined such that x is always tangent to the flight path and positive forward, and z is in the plane of symmetry and positive downward as before. Wind axes are also useful when the aircraft is flying in crosswinds, as shown in Figure 4.2. Importantly, the wind axes are *not* a body-attached frame of reference.

Regarding notation, all Euclidean vectors (velocities, forces, moments, etc.) are written in this book as the column vector of their three components in a given reference frame. They are written in lowercase bold, with an uppercase subindex that indicates the frame of reference used in the projection. When necessary, we add a superscript to identify the point of application of the vector. For example, $\mathbf{v}_E^{CM}$ denotes the column

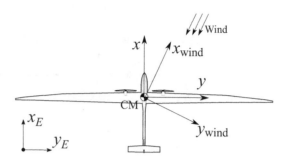

Figure 4.2 Frames of reference for a rigid aircraft: wind and stability axes (plan view).

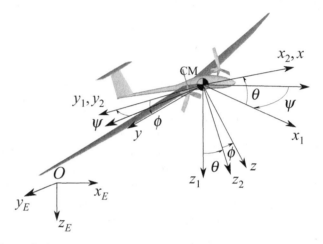

Figure 4.3 Euler angles between the Earth (origin O) and the body axes (origin at the CM).

matrix with the three components in the Earth frame of reference of the inertial velocity at the CM of the aircraft. In this chapter, most vector magnitudes are defined at the
aircraft CM, however, and the CM superindex is often removed to simplify notation,
such that $\mathbf{v}_E = \mathbf{v}_E^{CM}$.

4.2.2   Euler Angles

Once the reference frames are selected, one needs to define their relative orientation as
well as its evalution with time. Multiple alternative parameterizations (Euler or Bryant
angles, Rodrigues parameters, quaternions, etc.) have been proposed for the description of finite rotations in space. A comprehensive review can be found in Bauchau
(2011) and will not be repeated here. In this book, the orientation between two frames
of reference and, in particular, between the Earth and the body axes is parameterized using Euler angles. As we have discussed in the introduction, this provides some
insights into the mathematical formalism necessary to describe the rotation field while
still using rather simple and intuitive definitions. We note, however, that quaternions
have many computational advantages (as outlined, e.g., in Géradin and Cardona (2001,
Ch. 4)) and are often preferred in practice.

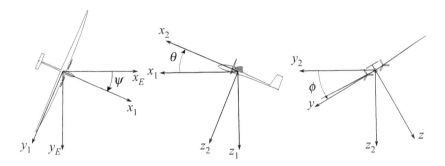

Figure 4.4 The three successive rotations defining local aircraft orientation. (a) First rotation: yaw; (b) second rotation: pitch; and (c) third rotation: roll.

Euler angles are defined by three successive rotations, starting from the Earth frame defined as in Figure 4.3, that is,

1. A rotation ψ about the vertical direction z_E of the Earth frame. ψ is the vehicle *heading* (or azimuth) and is chosen such that $-\pi \leq \psi \leq \pi$. The resulting frame of reference after the rotation is called 1 in Figure 4.3.
2. A rotation θ about the horizontal direction y_1. θ is the vehicle *pitch* (or elevation) and is chosen such that $-\frac{\pi}{2} \leq \theta \leq \frac{\pi}{2}$. The new frame of reference is called 2.
3. A rotation ϕ about the longitudinal direction x_2. ϕ, with $-\pi \leq \phi \leq \pi$, is the vehicle *bank* (or roll), and the final frame of reference defines the body-attached stability axis.

Once the order and the range of the rotations are defined, the three Euler angles that define the relative orientation between two frames of reference are unique. The convention above (yaw first, pitch second, and roll third, also known as 3-2-1), and shown in detail in Figure 4.4, is the one typically used in flight dynamics. The Euler angles allow now the transformation of coordinates of vectors in 3-D space. As an example, consider again the inertial velocity at the CM, written in terms of its components in the Earth frame, $\mathbf{v}_E$. In its components in the body-attached frame of reference, it will be $\mathbf{v}_B = \mathbf{R}_{BE}\mathbf{v}_E$, with $\mathbf{R}_{BE}$ being the *coordinate transformation matrix* from the Earth to body axes. This is obtained by the three successive rotations defined in Figure 4.4, that is,

$$
\mathbf{R}_{BE} = \begin{bmatrix} 1 & 0 & 0 \\ 0 & \cos\phi & \sin\phi \\ 0 & -\sin\phi & \cos\phi \end{bmatrix} \begin{bmatrix} \cos\theta & 0 & -\sin\theta \\ 0 & 1 & 0 \\ \sin\theta & 0 & \cos\theta \end{bmatrix} \begin{bmatrix} \cos\psi & \sin\psi & 0 \\ -\sin\psi & \cos\psi & 0 \\ 0 & 0 & 1 \end{bmatrix}. \quad (4.1)
$$

We can write it in shorthand form as $\mathbf{R}_{BE} = \boldsymbol{\tau}_x(\phi)\boldsymbol{\tau}_y(\theta)\boldsymbol{\tau}_z(\psi)$, where $\boldsymbol{\tau}_j$ identifies a rotation about the coordinate axis j. The coordinate transformation matrix from a body to the Earth can now be obtained simply by reversing the order of rotations, that is, first a rotation about x, then about y, and finally about z, with angles being equal an opposite to those defined above. Noting that each rotation is a matrix multiplication,

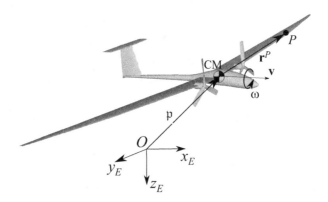

Figure 4.5 Kinematic description of a rigid aircraft.

they appear right to left in the expression as

$$\mathbf{R}_{EB} = \boldsymbol{\tau}_z(-\psi)\boldsymbol{\tau}_y(-\theta)\boldsymbol{\tau}_x(-\phi). \tag{4.2}$$

It is easy to see that $\boldsymbol{\tau}_x^\top(\phi) = \boldsymbol{\tau}_x(-\phi)$ and similarly for the other two individual rotations. As a result, the matrix defined in Equation (4.2) is the transpose of that defined in Equation (4.1), that is, $\mathbf{R}_{EB} = \mathbf{R}_{BE}^\top$. This means that by reversing the rotations, that is, applying them in the opposite order in which they were first defined, we can "undo" then. As a result, the coordinate transformation matrix is an *orthonormal matrix*, that is, its inverse is its transpose. In what follows, the inverse transformation will be simply indicated by a change of order of subindexes, as

$$\mathbf{R}_{EB} = \mathbf{R}_{BE}^{-1} = \mathbf{R}_{BE}^\top. \tag{4.3}$$

The transpose of the coordinate transformation matrix is often referred to as the *rotation matrix*. Rotations in 3-D space also play a key role in the description of the deformed shape of very flexible aircraft using geometrically nonlinear beam theory, and they will therefore appear again in Chapter 8. Further details on the properties of rotation matrices and the parameterization of rotations can be found in Appendix C.

4.2.3   Angular Velocity

Consider now the rigid aircraft of Figure 4.5. The position vector of its CM, in components in the Earth frame, is p_E. We are interested in describing the rotational motion of the aircraft about its CM. For that purpose, consider also an arbitrary point P in the aircraft, as shown in the figure. Its position vector with respect to the CM, expressed in its components in a certain body-attached frame of reference, B, is given by

$$\mathbf{r}_B^P = \left\{ \begin{matrix} x \\ y \\ z \end{matrix} \right\}, \tag{4.4}$$

which, since the aircraft is considered rigid, is constant with time. When expressed in its components in the Earth reference frame, it is $\mathbf{r}_E^P(t) = \mathbf{R}_{EB}(t)\mathbf{r}_B^P$, where the dependence on time has been made explicit as the orientation of the aircraft changes with time. Its inertial velocity, that is, the velocity of point P with respect to an inertial observer (in particular, one in the Earth frame), can be written, in its components in the Earth reference frame, as

$$\mathbf{v}_E^P = \dot{\mathbf{p}}_E + \dot{\mathbf{r}}_E^P = \mathbf{v}_E + \dot{\mathbf{r}}_E^P = \mathbf{v}_E + \dot{\mathbf{R}}_{EB}\mathbf{r}_B^P, \tag{4.5}$$

where $\mathbf{v}_E = \dot{\mathbf{p}}_E$ is the inertial velocity of the CM, written in terms of its components in the Earth frame. As mentioned above, the instantaneous direction of the inertial velocity vector defines the x axis of the wind axes. An important magnitude in flight dynamics is the angle between the velocity vector and the horizontal, which is known as the *flight-path angle* (or the climb angle), γ, positive upward. Defining the unitary column vector $\mathbf{e}_3 = \begin{Bmatrix} 0 & 0 & 1 \end{Bmatrix}^{\top}$, we can write

$$\gamma = -\sin^{-1}\frac{\mathbf{v}_E^{\top}\mathbf{e}_3}{V} \quad \text{and} \quad -\frac{\pi}{2} \le \gamma \le \frac{\pi}{2}, \tag{4.6}$$

where $V(t) = \left(\mathbf{v}_E^{\top}\mathbf{v}_E\right)^{1/2} = \left(\mathbf{v}_B^{\top}\mathbf{v}_B\right)^{1/2}$ is the magnitude of the instantaneous inertial velocity at the aircraft CM. In terms of its components in the body-attached frame of reference and recalling that $\mathbf{v}_E = \mathbf{R}_{EB}\mathbf{v}_B$, the inertial velocity of point P is given as

$$\mathbf{v}_B^P = \mathbf{v}_B + \mathbf{R}_{BE}\dot{\mathbf{R}}_{EB}\mathbf{r}_B^P. \tag{4.7}$$

From the properties given in Equation (4.3), $\mathbf{R}_{BE}\mathbf{R}_{EB} = \mathcal{I}$ is a unit matrix. This implies that $\frac{d}{dt}(\mathbf{R}_{BE}\mathbf{R}_{EB}) = \frac{d}{dt}(\mathcal{I}) = 0$, and

$$\frac{d}{dt}(\mathbf{R}_{BE}\mathbf{R}_{EB}) = \mathbf{R}_{BE}\dot{\mathbf{R}}_{EB} + \dot{\mathbf{R}}_{BE}\mathbf{R}_{EB} = \mathbf{R}_{BE}\dot{\mathbf{R}}_{EB} + \left(\mathbf{R}_{EB}^{\top}\dot{\mathbf{R}}_{BE}^{\top}\right)^{\top}, \tag{4.8}$$

which, using again Equation (4.3), leads to

$$\mathbf{R}_{BE}\dot{\mathbf{R}}_{EB} + \left(\mathbf{R}_{BE}\dot{\mathbf{R}}_{EB}\right)^{\top} = 0. \tag{4.9}$$

This means that $\mathbf{R}_{BE}\dot{\mathbf{R}}_{EB}$ is a skew-symmetric matrix, and consequently that the rate of change of the rotation matrix, after being *pulled back* to B by premultiplication with $\mathbf{R}_{BE}$, is uniquely given by three independent coefficients. We then choose them as $(\omega_x, \omega_y, \omega_z)$ such that

$$\mathbf{R}_{BE}\dot{\mathbf{R}}_{EB} = \begin{bmatrix} 0 & -\omega_z & \omega_y \\ \omega_z & 0 & -\omega_x \\ -\omega_y & \omega_x & 0 \end{bmatrix}. \tag{4.10}$$

In this expression, we have identified the nonzero entries of $\mathbf{R}_{BE}\dot{\mathbf{R}}_{EB}$ as the three components in body axes of a vector, which we define as the instantaneous *angular velocity* of the rigid body, $\boldsymbol{\omega}_B$. Moreover, it is very convenient to write Equation (4.10) as $\mathbf{R}_{BE}\dot{\mathbf{R}}_{EB} = \tilde{\boldsymbol{\omega}}_B$, where the operator $(\tilde{\bullet})$ is known as the *cross-product operator*, whose main properties are given in Appendix C.4. As a result, Equation (4.7) results in the well-known relation

$$\mathbf{v}_B^P = \mathbf{v}_B + \tilde{\boldsymbol{\omega}}_B\mathbf{r}_B^P. \tag{4.11}$$

Note that in our description, we identify the components of the translational and angular velocity vectors in the body-attached frame of reference as[1]

$$\mathbf{v}_B = \begin{Bmatrix} v_x \\ v_y \\ v_z \end{Bmatrix} \quad \text{and} \quad \boldsymbol{\omega}_B = \begin{Bmatrix} \omega_x \\ \omega_y \\ \omega_z \end{Bmatrix}. \tag{4.12}$$

Finally, we can write the angular velocity from the time derivative of the Euler angles by combining Equations (4.1) and (4.10). The algebra is rather tedious, but it is also straightforward. After going through it, the inertial angular velocity, expressed in its components in the body-attached frame of reference, can be shown as

$$\boldsymbol{\omega}_B = \begin{Bmatrix} \omega_x \\ \omega_y \\ \omega_z \end{Bmatrix} = \begin{Bmatrix} \dot{\phi} \\ 0 \\ 0 \end{Bmatrix} + \boldsymbol{\tau}_x(\phi) \begin{Bmatrix} 0 \\ \dot{\theta} \\ 0 \end{Bmatrix} + \boldsymbol{\tau}_x(\phi)\boldsymbol{\tau}_y(\theta) \begin{Bmatrix} 0 \\ 0 \\ \dot{\psi} \end{Bmatrix}. \tag{4.13}$$

Note that this result can also be obtained by the superposition of three independent angular velocities defined by the rate of change of the Euler angles around their respective axes of rotation. Those are the axes perpendicular to the planes in each of Figures 4.4a–4.4c, that is (1) an angular velocity of magnitude $\dot{\psi}$ along the Earth z axis, (2) an angular velocity of magnitude $\dot{\theta}$ along the pitching axis (axis $y_1 \equiv y_2$ in Figure 4.3), and (3) an angular velocity of magnitude $\dot{\phi}$ along the x axis of the body-attached frame of reference. This is known as the *addition theorem* (Bauchau, 2011, Ch. 4), which states that the angular velocity of a reference frame B with respect to a second frame E is the sum of the angular velocity of an arbitrary frame A with respect to E and of B with respect to A. A very important implication of the addition theorem is that angular velocities, contrary to Euler angles, are Euclidean vectors and define a vector space.

While the EoMs, could, in principle, be written using the time derivatives of the Euler angles as unknowns, the angular velocity makes for a much more compact and intuitive formulation. The transformation between both, Equation (4.13), can be written in a matrix form as

$$\boldsymbol{\omega}_B = \mathbf{T}(\phi, \theta, \psi) \begin{Bmatrix} \dot{\phi} \\ \dot{\theta} \\ \dot{\psi} \end{Bmatrix}, \tag{4.14}$$

where $\mathbf{T}(\phi, \theta, \psi)$ is known as the *tangential operator* corresponding to the Euler angles (see Appendix C), which is given by

$$\mathbf{T}(\phi, \theta, \psi) = \begin{bmatrix} 1 & 0 & -\sin\theta \\ 0 & \cos\phi & \sin\phi\cos\theta \\ 0 & -\sin\phi & \cos\phi\cos\theta \end{bmatrix}. \tag{4.15}$$

---

[1] Most textbooks define the velocity components as $\mathbf{v}_B = (u, v, w)$ and $\boldsymbol{\omega}_B = (p, q, r)$. This component description can be, however, quite restrictive to extend it to flexible vehicles and it has been found convenient here to use a notation based on subindices. Note also that, unless otherwise stated, coordinates x, y, z will correspond to the instantaneous (body-attached) stability axis, as shown in Figure 4.3.

Inversion of this expression results in a set of nonlinear ordinary differential equations (ODEs) to obtain the evolution of the Euler angles given a time history of angular velocity, that is,

$$\begin{Bmatrix} \dot{\phi} \\ \dot{\theta} \\ \dot{\psi} \end{Bmatrix} = \mathbf{T}^{-1}\boldsymbol{\omega}_B = \begin{bmatrix} 1 & \sin\phi\tan\theta & \cos\phi\tan\theta \\ 0 & \cos\phi & -\sin\phi \\ 0 & \sin\phi\sec\theta & \cos\phi\sec\theta \end{bmatrix} \begin{Bmatrix} \omega_x \\ \omega_y \\ \omega_z \end{Bmatrix}. \tag{4.16}$$

Given initial values of the Euler angles, integration in time of Equation (4.16) gives the instantaneous orientation of the rigid aircraft with respect to the Earth frame.

We have seen that any Euclidean vector, expressed in its components in an inertial reference frame, can be written in body axes by means of the coordinate transformation matrix defined in Equation (4.2). Introducing the angular velocity facilitates the computation of time derivatives. Take, for example, the velocity vector. It satisfies, $\mathbf{v}_E = \mathbf{R}_{EB}\mathbf{v}_B$, and therefore its first derivative is $\dot{\mathbf{v}}_E = \dot{\mathbf{R}}_{EB}\mathbf{v}_B + \mathbf{R}_{EB}\dot{\mathbf{v}}_B$. Premultiplying by $\mathbf{R}_{BE}$, we have finally,

$$\mathbf{R}_{BE}\dot{\mathbf{v}}_E = \dot{\mathbf{v}}_B + \tilde{\boldsymbol{\omega}}_B\mathbf{v}_B. \tag{4.17}$$

This result is known as the *transport theorem* (Greenwood, 1988). It hugely simplifies the derivation of EoMs on body-attached frames of reference, as we will see next.

4.3 Flight Dynamic Equations

4.3.1 Inertia Characteristics

Each material particle P of the vehicle, whose position vector is $\mathbf{r}_B^P$ as defined in Equation (4.4), can be associated with an infinitesimal mass dm. If $\mathcal{V}$ is the material volume of the aircraft, the total mass of the vehicle is simply

$$m = \int_{\mathcal{V}} dm. \tag{4.18}$$

From the definition of the CM, we also have that $\int_{\mathcal{V}} \mathbf{r}_B^P \, dm = 0$. The aircraft CM moves with some (inertial) linear and angular velocities, which in their components in body axes have been defined as $\mathbf{v}_B$ and $\boldsymbol{\omega}_B$, respectively. We define the corresponding total *linear momentum*, $\mathbf{p}_B$, as

$$\begin{aligned} \mathbf{p}_B &= \int_{\mathcal{V}} \mathbf{v}_B^P \, dm = \int_{\mathcal{V}} \left(\mathbf{v}_B + \tilde{\boldsymbol{\omega}}_B \mathbf{r}_B^P \right) dm \\ &= \mathbf{v}_B \int_{\mathcal{V}} dm + \tilde{\boldsymbol{\omega}}_B \int_{\mathcal{V}} \mathbf{r}_B^P \, dm = m\mathbf{v}_B. \end{aligned} \tag{4.19}$$

Similarly, the total *angular momentum* about the CM, $\mathbf{h}_B$, is

$$\begin{aligned} \mathbf{h}_B &= \int_{\mathcal{V}} \tilde{\mathbf{r}}_B^P \mathbf{v}_B^P \, dm \\ &= \int_{\mathcal{V}} \tilde{\mathbf{r}}_B^P \left(\mathbf{v}_B + \tilde{\boldsymbol{\omega}}_B \mathbf{r}_B^P \right) dm \\ &= -\tilde{\mathbf{v}}_B \int_{\mathcal{V}} \mathbf{r}_B^P \, dm - \int_{\mathcal{V}} \tilde{\mathbf{r}}_B^P \tilde{\mathbf{r}}_B^P \boldsymbol{\omega}_B \, dm = \mathbf{I}_B \boldsymbol{\omega}_B, \end{aligned} \tag{4.20}$$

where we have used some of the properties given in Equation (C.23) and again that $\int_{\mathcal{V}} \mathbf{r}_B^P \, dm = 0$ from the definition of the CM. We have also defined here the *inertia tensor* in body axes, $\mathbf{I}_B$, which, using the cross-product operator introduced in Equation (4.10), can be written as

$$\mathbf{I}_B = -\int_{\mathcal{V}} \tilde{\mathbf{r}}_B^P \tilde{\mathbf{r}}_B^P \, dm = \begin{bmatrix} I_{xx} & -I_{xy} & -I_{xz} \\ -I_{yx} & I_{yy} & -I_{yz} \\ -I_{zx} & -I_{zy} & I_{zz} \end{bmatrix}. \qquad (4.21)$$

The terms in the diagonal are the moments of inertia of the body, for example, $I_{xx} = \int_{\mathcal{V}} (y^2 + z^2) \, dm$. The off-diagonal terms depend on the products of inertia, for example, $I_{yz} = I_{zy} = \int_{\mathcal{V}} yz \, dm$. Note that when the aircraft is symmetric about the y axis, it is $I_{xy} = 0$ and $I_{yz} = 0$.

If expressed in its components in the Earth frame, the angular momentum becomes

$$\mathbf{h}_E = \mathbf{R}_{EB} \mathbf{h}_B = \mathbf{R}_{EB} \mathbf{I}_B \mathbf{R}_{BE} \boldsymbol{\omega}_E. \qquad (4.22)$$

Since by definition $\mathbf{h}_E = \mathbf{I}_E \boldsymbol{\omega}_E$, the coordinate transformation on the inertia tensor is finally given by the *congruent transformation* $\mathbf{I}_E = \mathbf{R}_{EB} \mathbf{I}_B \mathbf{R}_{BE}$. Moreover, we assume in what follows that the distribution of masses in the rigid aircraft does not change within the timescales of interest, which implies, for example, that the mass of the burnt fuel during that interval is negligible. As a result, both mass m and inertia $\mathbf{I}_B$ can be considered to be constant magnitudes with time, while $\mathbf{I}_E$ depends on the instantaneous aircraft orientation.

4.3.2   Equations of Motion of a Rigid Body

The EoM$_s$ are obtained from the Lagrange's equations, Equation (1.2), applied to a rigid body under external forces and moments. The kinetic energy is defined as $\mathcal{T} = \frac{1}{2} \int_{\mathcal{V}} \mathbf{v}_B^{P\top} \mathbf{v}_B^P \, dm$, which can be written in terms of the translational and angular velocities at the CM as

$$\mathcal{T} = \frac{1}{2} m \mathbf{v}_E^\top \mathbf{v}_E + \frac{1}{2} \boldsymbol{\omega}_E^\top \mathbf{I}_E \boldsymbol{\omega}_E. \qquad (4.23)$$

Substituting Equation (4.23) into Equation (1.2) results in the total applied forces equating the rate of change of linear momentum, and the applied moments about the CM equating the rate of change of angular momentum. In particular, the aircraft is subject to time-dependent external forces from which we can compute the total resultant forces and the resultant moments around the CM. Using the notation in this book, those vectors of resultant forces and moments are written in their components in the Earth frame as the column vectors $\mathbf{f}_E(t)$ and $\mathbf{m}_E(t)$, respectively. As a result, the EoMs are

$$\begin{aligned} \frac{d\mathbf{p}_E}{dt} &= \frac{d}{dt}(m\mathbf{v}_E) = \mathbf{f}_E, \\ \frac{d\mathbf{h}_E}{dt} &= \frac{d}{dt}(\mathbf{I}_E \boldsymbol{\omega}_E) = \mathbf{m}_E. \end{aligned} \qquad (4.24)$$

The first set of equations derives directly from Newton's second law, of motion while the second set is known as Euler's equations (or Euler's rotation equations). As a result, it is common to refer to them jointly as the *Newton–Euler equations*. As has just been mentioned, the inertia tensor in the Earth axes, $\mathbf{I}_E$, is a function of time, and it is therefore more convenient to write the equations in body axes (in particular, the stability axes). Since $\mathbf{h}_E = \mathbf{R}_{EB}\mathbf{h}_B$, we have

$$\dot{\mathbf{h}}_E = \dot{\mathbf{R}}_{EB}\mathbf{h}_B + \mathbf{R}_{EB}\dot{\mathbf{h}}_B, \tag{4.25}$$

which results again in the transport theorem, Equation (4.17). Using this relation and the transformations in Equation (4.22), we finally obtain

$$m\dot{\mathbf{v}}_B + m\tilde{\omega}_B\mathbf{v}_B = \mathbf{f}_{aB} + \mathbf{f}_{pB} + \mathbf{f}_{gB},$$
$$\mathbf{I}_B\dot{\omega}_B + \tilde{\omega}_B\mathbf{I}_B\omega_B = \mathbf{m}_{aB} + \mathbf{m}_{pB}, \tag{4.26}$$

where the external forces and moments have been split among aerodynamic, propulsive, and gravitational loads. They are force and moment resultants acting on the CM and have been written in their components in the body-attached frame of reference. They are discussed next.

4.3.3 Gravitational Forces

For a given aircraft mass, m, and under the flat-Earth assumptions, the gravitational forces in body axes will solely depend on the instantaneous orientation of the vehicle. Noting that the z_E axis in the Earth frame of reference is always defined to be vertical and positive downward, the resultant gravitational force (at the CM) is

$$\mathbf{f}_{gB} = mg\mathbf{R}_{BE}\mathbf{e}_3, \tag{4.27}$$

where g is the gravitational acceleration assumed here to be constant during the time of interest. In terms of the Euler angles introduced in Equation (4.1), the components of the gravitational force in body axes are

$$\mathbf{f}_{gB} = \left\{ \begin{array}{c} -mg\sin\theta \\ mg\cos\theta\sin\phi \\ mg\cos\theta\cos\phi \end{array} \right\}. \tag{4.28}$$

Therefore, while the equations of motion for a rigid vehicle, Equation (4.26), could be written in terms of linear and angular velocities and their time derivatives, the gravitational forces introduce an additional explicit dependency on the vehicle orientation.[2] This means that we have to solve simultaneously Equation (4.26), which determines the six components of the linear and angular velocities, and Equation (4.16), which defines a differential equation on the Euler angles given the instantaneous angular velocity.

---

[2] A second potential dependency of the EoMs on the absolute spatial orientation of the vehicle occurs when the aircraft flies through a nonstationary atmosphere as is discussed in Section 4.3.5.

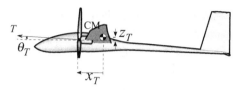

Figure 4.6 Propulsion system geometry definition.

4.3.4   Propulsive Forces

The propulsion system generates a thrust force of magnitude T on the vehicle. Thrust is typically expressed in terms of a nondimensional *thrust coefficient*, C_T, which is assumed to be a known function of the airspeed, V, and a throttle setting, δ_T (see section 2.5 of Stengel (2004) for details on typical power plants). It is defined as

$$T = \tfrac{1}{2}\rho V^2 SC_T(V, \delta_T), \qquad (4.29)$$

where ρ is the local air density. In power plant selection, S is either the propeller disk area or the engine exhaust area; however, in flight dynamics modeling, it is typically the main wing reference area. For a given mass configuration of the aircraft, the instantaneous thrust will generate a resultant force and moment at the CM, which will depend on the position and orientation of the propulsive devices in the body-attached frame of reference. Assuming that the propulsion system is symmetric with respect to the aircraft x–z plane, the geometric parameters become the coordinates x_T and z_T, as well as the pitch orientation θ_T, as shown in Figure 4.6. That finally defines the propulsive forces in Equation (4.26) as

$$\mathbf{f}_{pB} = \begin{Bmatrix} T\cos\theta_T \\ 0 \\ -T\sin\theta_T \end{Bmatrix} \text{ and } \mathbf{m}_{pB} = \begin{Bmatrix} 0 \\ T(z_T\cos\theta_T + x_T\sin\theta_T) \\ 0 \end{Bmatrix}. \qquad (4.30)$$

4.3.5   Quasi-Steady Aerodynamic Forces

Regarding the aerodynamic loads, specifically for this chapter, we assume that the timescales in the aircraft maneuvers are large enough such that quasi-steady aerodynamics, as defined in Section 3.3.2, can be considered. It is also assumed, as everywhere in this chapter, that the effect of any structural deformations on the aerodynamic forces can be neglected. In such a case, the aerodynamic forces on the rigid aircraft depend on the instantaneous relative velocity between the vehicle's CM and the atmosphere. We consider, in general, nonstationary atmospheric conditions, which define a varying distribution of wind velocities on the aircraft. Similar to Section 2.3.4, we, however, assume a "frozen" wind velocity field that remains unperturbed by the presence of the aircraft and that has a typical wavelength sufficiently large for the instantaneous distribution of wind velocities to vary at most linearly over the aircraft. This situation was sketched in Figure 2.17. The instantaneous wind velocity field on the moving aircraft can then be described by the time histories of two vectors of linear and angular gust velocities at the CM of the vehicle. They are naturally given with respect to the Earth frame of reference and will be referred to as $\mathbf{v}_{gE}(t)$ and $\boldsymbol{\omega}_{gE}(t)$, respectively (MIL-HDBK-1797, 1997; Richardson et al., 2013). Note that, as has already been discussed in Section 2.3.3, their particular time histories depend on both the spatial distribution of the wind field and the instantaneous orientation of the aircraft with respect to this field. The combination of both effects defines a time history of the components of both gust velocity vectors on the body-attached frame of reference (typically, the stability axes) as

$$\mathbf{v}_{gB} = \begin{Bmatrix} v_{gx}(t) \\ v_{gy}(t) \\ v_{gz}(t) \end{Bmatrix} \quad \text{and} \quad \boldsymbol{\omega}_{gB} = \begin{Bmatrix} \omega_{gx}(t) \\ \omega_{gy}(t) \\ \omega_{gz}(t) \end{Bmatrix}. \tag{4.31}$$

The aircraft CM is then selected to compute the instantaneous relative mean velocity between the vehicle and the atmosphere, which defines a nonstationary *background velocity field*. This is done for consistency with the aircraft EoMs, but any other reference point could have been equally chosen. Next, we define the aircraft instantaneous *angle of attack*, α, as the ratio of the components of the relative velocity vector in the plane of symmetry of the aircraft,

$$\alpha(t) = \tan^{-1} \frac{v_z(t) - v_{gz}(t)}{v_x(t) - v_{gx}(t)} \quad \text{and} \quad -\pi \leq \alpha \leq \pi. \tag{4.32}$$

Therefore, changes of the instantaneous angle of attack occur not only under changes of the CM inertial velocity but also under changes of the gust velocity. Note also that on the stability axis in calm air, the reference velocity goes along the x axis, which implies $\alpha = 0$ in forward flight using this definition.

The lateral component of the relative velocity defines the *sideslip angle*, β, that is,

$$\beta = \tan^{-1} \frac{v_y - v_{gy}}{\sqrt{\left(v_x - v_{gx}\right)^2 + \left(v_z - v_{gz}\right)^2}} \quad \text{and} \quad -\frac{\pi}{2} \leq \beta \leq \frac{\pi}{2}. \tag{4.33}$$

The sideslip angle measures the lateral misalignment of the instantaneous relative velocity vector with respect to the symmetry plane of the aircraft, and it is used in the description of the aircraft lateral dynamics and when flying in crosswind. Both the angle of attack and the sideslip angle are shown in Figure 4.7.

The relative velocity at the CM between the aircraft and the atmosphere can, therefore, be alternatively defined by either the three components of the velocity vector on the (body-attached) stability axis or by its magnitude, the angle of attack, and sideslip, that is,

$$\mathbf{v}_B - \mathbf{v}_{gB} = \begin{Bmatrix} v_x - v_{gx} \\ v_y - v_{gy} \\ v_z - v_{gz} \end{Bmatrix} = V_{\text{TAS}} \begin{Bmatrix} \cos\alpha\cos\beta \\ -\sin\beta \\ \sin\alpha\cos\beta \end{Bmatrix}, \tag{4.34}$$

where $V_{\text{TAS}}(t) = \|\mathbf{v}_B - \mathbf{v}_{gB}\|$ is the true airspeed, that is, the instantaneous magnitude of the relative velocity between the aircraft and the wind. The aerodynamic forces

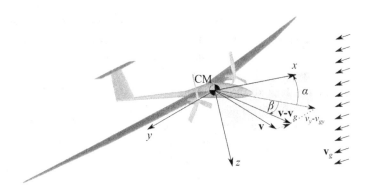

Figure 4.7 Angle of attack, α, and sideslip angle, β, for an aircraft in the presence of a constant wind velocity (not aligned with the x direction).

have, in general, nonzero resultants on all three components of forces and moments in Equation (4.26). The models used to obtain them vary quite substantially depending on the flight conditions and the characteristics of the vehicle. In general, we can state that they depend on three different sets of variables as follows.

First, they depend on the *vehicle geometry*, that is, magnitudes such as the wingspan, typical chord c, reference wing area S, and airfoil shape. For most aircraft, those are fixed parameters for each flight segment, while having take-off and landing configurations that differ from their cruise geometry.

Second, they depend on the *flight condition*, that is, the point of operation in the flight envelope, which determines the air density, as well as the Mach and Reynolds numbers of operation.

Third, the instantaneous aerodynamic forces depend on the relative linear and angular velocities between the vehicle and the atmosphere, and on the instantaneous deflections of any control surfaces (typically, elevators, ailerons, and rudder). If any of them varies sufficiently fast for the flow unsteadiness to become significant, then their rates (time derivatives) may need to be considered as well, as was done in Equation (3.42). However, this is not considered here, and in this chapter we use quasi-steady assumptions on the aerodynamic forces, that is, that changes in all magnitudes that define their values are sufficiently slow for the associated nondimensional frequency $\omega c/(2V)$ to be very small. This is the reduced frequency defined in Equation (3.45), which becomes again relevant here.

Consequently, for a given geometrical configuration on a vehicle, we write the aerodynamic forces and moments at the aircraft CM, expressed in their components in body axes, as

$$\mathbf{f}_{aB} = \tfrac{1}{2}\rho V_{\text{TAS}}^2 S\mathbf{c_{f_B}}\left(\mathbf{v}_B - \mathbf{v}_{gB}, \boldsymbol{\omega}_B - \boldsymbol{\omega}_{gB}, \boldsymbol{\delta}_c; \text{Ma}, \text{Re}\right),$$
$$\mathbf{m}_{aB} = \tfrac{1}{2}\rho V_{\text{TAS}}^2 S\Lambda_{\text{ref}}\mathbf{c_{m_B}}\left(\mathbf{v}_B - \mathbf{v}_{gB}, \boldsymbol{\omega}_B - \boldsymbol{\omega}_{gB}, \boldsymbol{\delta}_c; \text{Ma}, \text{Re}\right),$$

(4.35)

where $\boldsymbol{\delta}_c(t)$ is the vector with the instantaneous deflection angles in each of the control surfaces, Ma and Re are the Mach and Reynolds numbers, respectively, at the nominal conditions, $\Lambda_{\text{ref}} = \text{diag}\left(b_{\text{ref}}, c_{\text{ref}}, b_{\text{ref}}\right)$, and $\mathbf{c_{f_B}}$ and $\mathbf{c_{m_B}}$ are the nondimensional forces and moments after normalizing with the dynamic pressure, a reference area S, and reference lengths c_{ref} and b_{ref}. Typically, those are chosen so that S is the planform wing area, c_{ref} is the mean aerodynamic chord of the wing, and b_{ref} is the wingspan. Subindex B has been retained in both force and moment coefficients to indicate that their three components correspond to a projection on the stability axes.

Aerodynamic Derivatives

In many situations, such as at the time of stability analysis, control design, and estimation of performance, the aerodynamic forces and moments need to be known only for small changes, around a reference condition, of the control inputs, as well as the aircraft and wind velocities. The reference point is defined here as a trimmed-level flight in still air of the aircraft with velocity V, for which the reference force and moment coefficients are known and constant. Let the reference force and moment coefficients be $\mathbf{c}_{f0}$ and $\mathbf{c}_{m0}$, respectively. In such a case, the expressions above can be linearized about that reference, and the resulting incremental aerodynamic forces and moments are

$$\Delta \mathbf{f}_{aB} = \tfrac{1}{2}\rho V^2 S \left(\Delta \mathbf{c}_{f_B} + 2\mathbf{c}_{f0}\frac{\Delta v_x - \Delta v_{gx}}{V} \right),$$

$$\Delta \mathbf{m}_{aB} = \tfrac{1}{2}\rho V^2 S \Lambda_{ref} \left(\Delta \mathbf{c}_{m_B} + 2\mathbf{c}_{m0}\frac{\Delta v_x - \Delta v_{gx}}{V} \right).$$

(4.36)

The last term in both equations comes from perturbations of the dynamic pressure about its reference value $\tfrac{1}{2}\rho V^2$ since V goes along the x axis in the reference condition. The incremental force and moment are also given in their components in the body-attached stability axes B in the reference condition. Next, we need to evaluate the constant matrices that determine the linear dependency between the incremental aerodynamic force and moment coefficients, $\Delta \mathbf{c}_{f_B}$ and $\Delta \mathbf{c}_{m_B}$, and the perturbation of $\mathbf{v}_B$, $\mathbf{v}_{gB}$, ω_B, ω_{gB}, and δ_c from their reference values. They are known as *aerodynamic derivatives*. Noting, however, that under the quasi-steady aerodynamics assumption, forces on the aircraft only depend on its relative velocity with respect to the air mass, all dependencies with the gust velocities are equal and opposite to those of the corresponding body velocities. Therefore, only the latter ones need to be evaluated.

The convention followed in the definition of the tabulated aerodynamic derivatives comes from the classical setup in wind-tunnel testing. There, the vehicle model is positioned at an angle with respect to a constant airstream, and forces are then measured in relation to the tangent and normal directions to the flow. As a result, the quasi-steady aerodynamic characteristics of an aircraft are normally given in the wind axes, which result in the usual definitions of lift and drag as the force resultants on the negative z and x axes, respectively.[3]

---

[3] This is the most common convention, although not a universally accepted one. Recall that the "actual" aerodynamic forces on the aircraft are the distributed pressure and shear stress on the wet surfaces. Lift and drag are integral magnitudes that can only be measured in a wind-tunnel test through the reaction forces on the model supports.

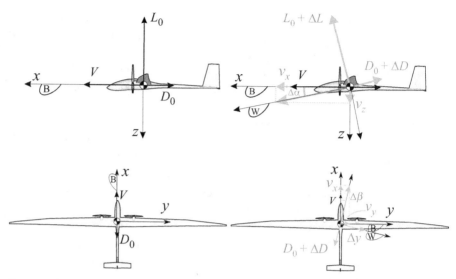

Figure 4.8 Instantaneous wind (W) and stability (B) axes in the evaluation of aerodynamic derivatives. (a) Reference condition in the $x-z$ plane, (b) perturbations in the $x-z$ plane, (c) reference condition in the $x-y$ plane, and (d) perturbations in the $x-y$ plane.

The resulting scenario is sketched in Figure 4.8. The aircraft is initially in forward flight, which defines the orientation of the frame of reference B, that is, the body-attached (stability) axes on which the EoMs are derived. A small perturbation is then considered, which results in an instantaneous velocity vector $\mathbf{v}_B(t) = \left(V + \Delta v_x(t), \Delta v_y(t), \Delta v_z(t)\right)^\top$. This defines the (incremental) angle of attack and the sideslip angle, respectively, as

$$
\begin{aligned}
\Delta \alpha &= \tan^{-1}\left(\frac{\Delta v_z}{V + \Delta v_x}\right) \approx \frac{\Delta v_z}{V}, \\
\Delta \beta &= \tan^{-1}\left(\frac{\Delta v_y}{V + \Delta v_x}\right) \approx \frac{\Delta v_y}{V}.
\end{aligned}
\tag{4.37}
$$

It is assumed that the angle of attack and the sideslip angle are sufficiently small such that they can, respectively, be defined in the $x-z$ and $x-y$ planes in the stability axes, as shown in Figure 4.8. Remember also that Δv_z is positive downward following the convention used in the definition of the stability axes (see Section 4.2.1)

The instantaneous wind axes are finally defined by means of a rotation with the (incremental) angle of attack from the reference condition, which gives the transformation matrix

$$
\mathbf{T}_{WB} = \begin{bmatrix} \cos\Delta\alpha & 0 & \sin\Delta\alpha \\ 0 & 1 & 0 \\ -\sin\Delta\alpha & 0 & \cos\Delta\alpha \end{bmatrix} \begin{bmatrix} \cos\Delta\beta & \sin\Delta\beta & 0 \\ -\sin\Delta\beta & \cos\Delta\beta & 0 \\ 0 & 0 & 1 \end{bmatrix} \approx \begin{bmatrix} 1 & \Delta\beta & \Delta\alpha \\ -\Delta\beta & 1 & 0 \\ -\Delta\alpha & 0 & 1 \end{bmatrix}.
\tag{4.38}
$$

In that reference, aerodynamic forces and moments are obtained as

$$\mathbf{c_{f_W}} = \mathbf{T}_{WB}\mathbf{c_{f_B}},$$
$$\mathbf{c_{m_W}} = \mathbf{T}_{WB}\mathbf{c_{m_B}}, \tag{4.39}$$

where the components of both vectors have the usual definitions (Drela, 2014), that is, $\mathbf{c_{f_W}} = (-C_D, C_Y, -C_L)$, the coefficients of drag, lateral force, and lift, respectively, and $\mathbf{c_{m_W}} = (C_{\mathcal{L}}, C_{\mathcal{M}}, C_{\mathcal{N}})$, the coefficients of roll, pitch, and yaw aerodynamic moment, respectively. They are tabulated with respect to changes on the angle of attack, sideslip angle, flap deflections, and nondimensional angular rates. The first three have been already defined in Equations (4.32) and (4.33), and Equation (4.35), respectively, and the angular rates are also given with respect to the wind axes, that is, as three components the column vector $\omega_W = \mathbf{T}_{WB}\omega_B$. The most common definition is

$$\omega_W = \frac{V}{2}\Lambda_{\text{ref}}\begin{Bmatrix} \bar{p} \\ \bar{q} \\ \bar{r} \end{Bmatrix}, \tag{4.40}$$

where Λ_{ref} have been defined in Equation (4.35), and $\bar{p}$, $\bar{q}$, and $\bar{r}$ are the nondimensional roll, pitch, and yaw rates, respectively. Typical linear dependencies go as

$$C_D \approx C_{D0} + C_{D\alpha}\Delta\alpha + C_{Dq}\Delta\bar{q} + C_{D\delta_e}\Delta\delta_e,$$
$$C_L \approx C_{L0} + C_{L\alpha}\Delta\alpha + C_{Lq}\Delta\bar{q} + C_{L\delta_e}\Delta\delta_e, \tag{4.41}$$
$$C_{\mathcal{M}} \approx C_{\mathcal{M}0} + C_{\mathcal{M}\alpha}\Delta\alpha + C_{\mathcal{M}q}\Delta\bar{q} + C_{\mathcal{M}\delta_e}\Delta\delta_e,$$

for the longitudinal coefficients, while the lateral force and moment coefficients are written as

$$C_Y \approx C_{Y0} + C_{Y\beta}\Delta\beta + C_{Yp}\Delta\bar{p} + C_{Yr}\Delta\bar{r} + C_{Y\delta_a}\Delta\delta_a + C_{Y\delta_r}\Delta\delta_r,$$
$$C_{\mathcal{L}} \approx C_{\mathcal{L}0} + C_{\mathcal{L}\beta}\Delta\beta + C_{\mathcal{L}p}\Delta\bar{p} + C_{\mathcal{L}r}\Delta\bar{r} + C_{\mathcal{L}\delta_a}\Delta\delta_a + C_{\mathcal{L}\delta_r}\Delta\delta_r, \tag{4.42}$$
$$C_{\mathcal{N}} \approx C_{\mathcal{N}0} + C_{\mathcal{N}\beta}\Delta\beta + C_{\mathcal{N}p}\Delta\bar{p} + C_{\mathcal{N}r}\Delta\bar{r} + C_{\mathcal{N}\delta_a}\Delta\delta_a + C_{\mathcal{N}\delta_r}\Delta\delta_r,$$

where $C_{L\alpha} = \frac{\partial C_L}{\partial\alpha}$, $C_{Lq} = \frac{\partial C_L}{\partial\bar{q}}$, etc., with the derivatives evaluated around the reference conditions, are known as *stability derivatives*, and $C_{L\delta_e}$, $C_{Y\delta_a}$, etc. are analogously defined and known as the *control derivatives*. They are normally tabulated in terms of the Mach number and the angle of attack. It is also common to distinguish between stiffness derivatives, which are related to changes with α, β, and δ_c, and damping derivatives, associated with their rates. An excellent discussion of each of the coefficients in Equations (4.41) and (4.42) and their typical values on some conventional configurations can be found in Stevens et al. (2016, Ch. 2).

For sufficiently fast maneuvers, it is often also necessary to add to the linear expansions above the stability derivatives associated with the rates of angle of attack and control surface deflection, but those are not explicitly included here. Those are instead part of the more general formulation that we present in Chapter 7. Note that, as the angle of attack is defined as the ratio of velocities in Equation (4.32), the aerodynamic forces associated with $\Delta\dot{\alpha}$ correspond to the plunging rates defined in Section 3.3 and not to the pitch rates, in spite of the notation used. As a result, $C_{L\dot{\alpha}}$ represents the apparent mass of the fluid displaced by the acceleration of a plunging aircraft. Finally, for sufficiently fast maneuvers at transonic speeds, one needs to consider derivatives with the Mach number. Two particular cases are important. First, the changes of moment due to the shift of center of pressure that occurs in the transition between subsonic and supersonic flow, which results in the so-called *tuck derivative*, $\frac{\partial C_M}{\partial \text{Ma}}$. Second, the increase in the drag with the Mach number in that regime, $\frac{\partial C_D}{\partial \text{Ma}}$, known as the *speed-damping derivative*. As the independent variable in the state vector is typically the airspeed, using Equation (4.34), the tuck and speed-damping derivatives are written in terms of V using the chain rule when added to Equation (4.41).

With this, we have that the reference force and aerodynamic coefficients in Equation (4.36) are

$$
\mathbf{c}_{\mathbf{f0}} = \begin{bmatrix} -C_{D0} \\ C_{Y0} \\ -C_{L0} \end{bmatrix} \quad \text{and} \quad \mathbf{c}_{\mathbf{m0}} = \begin{bmatrix} C_{\mathcal{L}0} \\ C_{\mathcal{M}0} \\ C_{\mathcal{N}0} \end{bmatrix}, \tag{4.43}
$$

and the incremental force and moment coefficients are finally obtained from small perturbations to Equation (4.39) as

$$
\begin{aligned}
\Delta \mathbf{c}_{\mathbf{f}_B} &= \tfrac{\partial}{\partial \Delta \alpha} \left(\mathbf{T}_{BW} \right) \mathbf{c}_{\mathbf{f0}} \Delta \alpha + \tfrac{\partial}{\partial \Delta \beta} \left(\mathbf{T}_{BW} \right) \mathbf{c}_{\mathbf{f0}} \Delta \beta + \Delta \mathbf{c}_{\mathbf{f}_B}, \\
\Delta \mathbf{c}_{\mathbf{m}_B} &= \tfrac{\partial}{\partial \Delta \alpha} \left(\mathbf{T}_{BW} \right) \mathbf{c}_{\mathbf{m0}} \Delta \alpha + \tfrac{\partial}{\partial \Delta \beta} \left(\mathbf{T}_{BW} \right) \mathbf{c}_{\mathbf{m0}} \Delta \beta + \Delta \mathbf{c}_{\mathbf{m}_B},
\end{aligned} \tag{4.44}
$$

where $\Delta \mathbf{c}_{\mathbf{f}_B}$ and $\Delta \mathbf{c}_{\mathbf{m}_B}$ are given by the first-order approximations defined in Equations (4.41) and (4.42), respectively, and the partial derivatives of the rotation matrix with the incremental angle of attack and sideslip are given by

$$
\frac{\partial}{\partial \Delta \alpha} \left(\mathbf{T}_{BW} \right) = \begin{bmatrix} -\sin \Delta \alpha & 0 & -\cos \Delta \alpha \\ 0 & 0 & 0 \\ \cos \Delta \alpha & 0 & -\sin \Delta \alpha \end{bmatrix}_{\Delta \alpha = 0} = \begin{bmatrix} 0 & 0 & -1 \\ 0 & 0 & 0 \\ 1 & 0 & 0 \end{bmatrix}, \tag{4.45}
$$

$$
\frac{\partial}{\partial \Delta \beta} \left(\mathbf{T}_{BW} \right) = \begin{bmatrix} -\sin \Delta \beta & -\cos \Delta \beta & 0 \\ \cos \Delta \beta & -\sin \Delta \beta & 0 \\ 0 & 0 & 0 \end{bmatrix}_{\Delta \beta = 0} = \begin{bmatrix} 0 & -1 & 0 \\ 1 & 0 & 0 \\ 0 & 0 & 0 \end{bmatrix}. \tag{4.46}
$$

Substituting Equation (4.44) back into Equation (4.36), the incremental force coefficients can be explicitly obtained in terms of the incremental linear and angular velocities of the CM and the control inputs. We will write them here as

$$
\begin{Bmatrix} \Delta \mathbf{f}_{aB} \\ \Delta \mathbf{m}_{aB} \end{Bmatrix} = \rho V \mathcal{A}_{rr} \begin{Bmatrix} \Delta \mathbf{v}_B - \Delta \mathbf{v}_{gB} \\ \Delta \boldsymbol{\omega}_B - \Delta \boldsymbol{\omega}_{gB} \end{Bmatrix} + \frac{1}{2} \rho V^2 \mathcal{A}_{ru} \delta_c, \tag{4.47}
$$

where $\mathcal{A}_{rr}$ and $\mathcal{A}_{ru}$ are the constant matrices of *rigid-body aerodynamic influence coefficients* associated with the states and inputs in the problem. They depend on the geometry, the stability and control derivatives, and the flight regime (reference Reynolds and Mach numbers). The estimation of stability and control derivatives for rigid aircraft, Equations (4.41) and (4.42), is extensively described in most textbooks on flight dynamics (e.g., Stengel, 2004, Ch. 3) and will not be discussed here. Computationally efficient methods are also now available to construct them from CFD simulations at the vehicle level, such as the one proposed by Da Ronch et al. (2013) using harmonic balance methods. Section 4.6 discusses an example of the derivation of the aerodynamic influence coefficients in Equation (4.47) under simplified conditions, while more general linear aerodynamic models, valid also for flexible aircraft, will be described in Section 6.5.1.

4.4   State-Space Description

We have so far derived the EoMs that describe the flight dynamics of a rigid aircraft and the main features of the external forces on the vehicle. From a mathematical point of view, the aircraft can be seen as a dynamical system; we are concerned about its equilibrium conditions (trimmed flight, static maneuvers), stability (its ability to return to the equilibrium when it is perturbed), and maneuverability (response to commanded inputs). Those problems are presented in the final sections of this chapter. To facilitate those studies, we first "repackage" in this section the EoMs into a state-space formulation.

4.4.1   State and Input Vectors

First, we define the *state vector* for the rigid aircraft dynamics, $\mathbf{x}_r(t)$, as

$$\mathbf{x}_r = \left\{ v_x \quad v_y \quad v_z \quad \omega_x \quad \omega_y \quad \omega_z \quad \phi \quad \theta \right\}^\top. \tag{4.48}$$

The state vector includes all the necessary information to track future vehicle kinematics given its current values and the future input time histories. We have not included the heading, ψ, as under the flat-Earth assumption, this value has no effect in the vehicle dynamics – we say that the dynamics is *invariant* with respect to ψ.

Next, we proceed to identify the possible actions on the system. From Section 4.3.4, we have seen that the throttle setting, δ_T, is the command associated with the power plant. In general, we may have multiple, independently actuated engines, but for simplicity, we use a single input signal in our discussion. Equally, the aerodynamic forces given by Equation (4.35) include commanded inputs by means of the control surface deflections, δ_c. We assume a single command for elevators, δ_e, ailerons. With this, we can define the *input vector*, $\mathbf{u}(t)$, as

$$\mathbf{u} = \left\{ \delta_e \quad \delta_a \quad \delta_r \quad \delta_T \right\}^\top. \tag{4.49}$$

Finally, the nonstationary atmospheric conditions, as seen by the aircraft at any given instant, define the *disturbance vector*, $\mathbf{w}(t)$, which is defined for convenience as

$$\mathbf{w} = \left\{ -v_{gx} \quad -v_{gy} \quad -v_{gz} \quad -\omega_{gx} \quad -\omega_{gy} \quad -\omega_{gz} \right\}^\top. \tag{4.50}$$

With this, we can write the dynamics of the system, given by Equations (4.16) and (4.26), in their most generic form as

$$\begin{aligned} \dot{\mathbf{x}}_r &= \mathfrak{f}(\mathbf{x}_r, \mathbf{u}, \mathbf{w}) \\ &= \mathfrak{f}_{\mathrm{gyr}}(\mathbf{x}_r) + \mathfrak{f}_{\mathrm{aero}}(\mathbf{x}_r, \mathbf{u}, \mathbf{w}) + \mathfrak{f}_{\mathrm{prop}}(\mathbf{x}_r, \mathbf{u}, \mathbf{w}) + \mathfrak{f}_{\mathrm{grav}}(\mathbf{x}_r). \end{aligned} \tag{4.51}$$

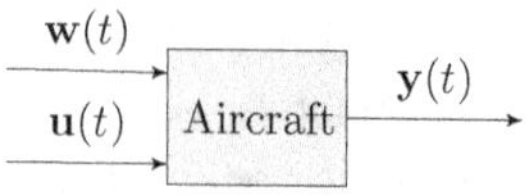

Figure 4.9 Basic block diagram for the aircraft state equations.

The term f_{gyr} includes the gyroscopic forces in the problem, which are the terms with the cross products of the angular velocity in Equation (4.26). For convenience, we also include there the right-hand side of the kinematic equations, Equation (4.16). In Equation (4.51), we have also made explicit that the gyroscopic terms only depend on the state of the system. The aerodynamic forcing terms are $f_{aero} = \left\{ \mathbf{f}_{aB}^\top \quad \mathbf{m}_{aB}^\top \quad 0 \quad 0 \right\}^\top$, with the forces and moments that have been defined in Equation (4.35). Note that this brings a dependency of the equations on the flight condition (altitude and flight Mach number). Similar expressions define the propulsion and gravitational terms, that is, f_{prop} and f_{grav}, from Equations (4.28) and (4.30), respectively. Equation (4.51) is normally complemented by an output equation that extracts magnitudes of interest into a generic *output vector*, $\mathbf{y}$ (for instance, to interface with a flight simulator). In its most general form, the resulting system is written as

$$\dot{\mathbf{x}}_r = f(\mathbf{x}_r, \mathbf{u}, \mathbf{w}),$$
$$\mathbf{y} = g(\mathbf{x}_r, \mathbf{u}, \mathbf{w}). \tag{4.52}$$

The output function, g, is often defined as a linear combination of the states, and often the values of certain states themselves, but it could also include magnitudes defined by nonlinear relations, such as the instantaneous airspeed, $V = \sqrt{v_x^2 + v_y^2 + v_z^2}$.

Equation (4.52) needs to be solved together with the initial condition, that is, a set of inputs, $\mathbf{u}(0) = \mathbf{u}_0$, states, $\mathbf{x}_r(0) = \mathbf{x}_{r0}$, and wind conditions, $\mathbf{w}(0) = \mathbf{w}_0$, at time $t = 0$. The flow diagram describing Equation (4.52) is the simple input–output relation, as shown in Figure 4.9.

4.4.2   Small-Perturbation Equations

We have seen that the EoMs for rigid aircraft can be written as a set of nonlinear state-space equations. Most analyses use as a reference an equilibrium point of the aircraft (*trim conditions*), defined by combinations of inputs $\mathbf{u}_0$, states $\mathbf{x}_{r0}$, and constant wind conditions $\mathbf{w}_0$ such that

$$f(\mathbf{x}_{r0}, \mathbf{u}_0, \mathbf{w}_0) = \mathbf{0}. \tag{4.53}$$

This equilibrium could be in straight-level flight or in a steady maneuver, which are discussed, for the particular case $\mathbf{w}_0 = \mathbf{0}$, in Section 4.5. Small-perturbation analysis on the dynamic equations around this reference condition is useful to analyze the stability characteristics of this equilibrium point, as well as to define (linear) feedback control strategies. For that purpose, we approximate the time histories around an equilibrium condition as

$$\mathbf{x}_r(t) \approx \mathbf{x}_{r0} + \Delta \mathbf{x}_r(t),$$

$$\mathbf{u}(t) \approx \mathbf{u}_0 + \Delta \mathbf{u}(t), \tag{4.54}$$

$$\mathbf{w}(t) \approx \mathbf{w}_0 + \Delta \mathbf{w}(t),$$

where $\Delta\mathbf{x}_r(t)$, $\Delta\mathbf{u}(t)$, and $\Delta\mathbf{w}(t)$ are assumed to be sufficiently small. Substituting Equation (4.54) into the first line in Equation (4.52), and subtracting the reference condition, given in Equation (4.53), result in

$$\Delta \dot{\mathbf{x}}_r = \mathfrak{f}(\mathbf{x}_{r0} + \Delta\mathbf{x}_r, \mathbf{u}_0 + \Delta\mathbf{u}, \mathbf{w}_0 + \Delta\mathbf{w}) - \mathfrak{f}(\mathbf{x}_{r0}, \mathbf{u}_0, \mathbf{w}_0). \tag{4.55}$$

The right-hand side of this equation gives the incremental forcing terms corresponding to the small perturbations defined above. In particular, if we invoke the form of Equation (4.51), it can be written as $\Delta\mathfrak{f} = \Delta\mathfrak{f}_{gyr} + \Delta\mathfrak{f}_{aero} + \Delta\mathfrak{f}_{prop} + \Delta\mathfrak{f}_{grav}$, although this is not explicitly used now, and it is left instead for the next sections. Using its first-order Taylor expansion, Equation (4.55) can be approximated as

$$\Delta \dot{\mathbf{x}}_r = \left.\frac{\partial \mathfrak{f}}{\partial \mathbf{x}_r}\right|_0 \Delta\mathbf{x}_r + \left.\frac{\partial \mathfrak{f}}{\partial \mathbf{u}}\right|_0 \Delta\mathbf{u} + \left.\frac{\partial \mathfrak{f}}{\partial \mathbf{w}}\right|_0 \Delta\mathbf{w}$$

$$= \mathbf{A}_{rr}\Delta\mathbf{x}_r + \mathbf{B}_r\Delta\mathbf{u} + \mathbf{E}_r\Delta\mathbf{w}, \tag{4.56}$$

which is a continuous-time LTI state-space representation of the rigid aircraft dynamics, with the system dynamics given as in Section 1.4.2 but now also with disturbance inputs coming from the wind velocity. In Equation (4.56), we have defined the rigid aircraft state matrix $\mathbf{A}_{rr}$, the input matrix $\mathbf{B}_r$, and the disturbance input matrix $\mathbf{E}_r$ at the trim condition. Detailed expressions for each of them are obtained in Section 4.6 around a steady climb. Contributions of the aerodynamic forces and moments are of particular importance. From Equation (4.47), it can be seen that the three terms in the linearized aerodynamics define contributions to the state, input, and disturbance matrices in Equation (4.56).

This linearization is only valid locally around a reference condition, and therefore the system matrices are typically parameterized by four independent variables that define steady-state flight conditions for a given aircraft configuration (Stengel, 2004), namely altitude (which determines air density), true airspeed (the velocity of the aircraft relative to the mean wind), which is often given in terms of the Mach number, bank angle (in rolling maneuvers), and climb angle (in climbing maneuvers).

4.5   Steady-State Flight

We have by now built the basic framework to investigate the aircraft flight dynamics, which can now be explored. We are first concerned with the *maneuverability* of rigid-wing aircraft, that is, their ability to follow a desired flight path, and then with the *maneuver loads*, that is, the internal stresses that appear in the airframe due to the external forces necessary to alter the vehicle trajectory. In general, we distinguish between *steady* and *dynamic maneuvers* depending on whether the aircraft linear and angular velocity components in the body-attached reference frame are constant with time or are time dependent. We restrict ourselves here to steady maneuvers that are further assumed to be performed in calm air. Steady maneuvers, or *steady-steady flight*, are used to define the nominal maneuvering envelope of a given vehicle (Section 4.5.2), while also determining the reference conditions for dynamic stability analysis and the initial conditions for dynamic events, such as the response of the aircraft to nonstationary atmospheric conditions. Those are studied in subsequent sections in this chapter.

4.5.1   Steady Maneuvers

Steady maneuvers correspond to solutions of Equation (4.26) with constant translational and angular velocity in the body-attached reference frame, that is,

$$\mathbf{f}_{aB} + \mathbf{f}_{pB} + \mathbf{f}_{gB} - m\tilde{\boldsymbol{\omega}}_B \mathbf{v}_B = 0,$$
$$\mathbf{m}_{aB} + \mathbf{m}_{pB} - \tilde{\boldsymbol{\omega}}_B \mathbf{I}_B \boldsymbol{\omega}_B = 0,$$
(4.57)

which are solved together with the kinematic relations defined by Equation (4.16). The first set of equations here represents the equilibrium of forces along each of the three body-fixed axes. The second set, likewise, represents the moment equilibrium about each of the body-fixed axes. All six equations must be satisfied simultaneously for the aircraft to be considered in equilibrium (or trim). The problem to be solved is then finding the control inputs, for which Equations (4.57) are identically satisfied. This results in five different steady maneuvers that can be achieved by an air vehicle, namely straight-level flight ($\dot{\phi} = \dot{\theta} = \dot{\psi} = 0, \phi = \theta = 0$), steady climb ($\dot{\phi} = \dot{\theta} = \dot{\psi} = 0, \phi = 0$), steady pull-up ($\dot{\phi} = \dot{\psi} = 0, \phi = 0$), steady turn ($\dot{\phi} = \dot{\theta} = 0$), and steady roll ($\dot{\theta} = \dot{\psi} = 0$). The first four, shown schematically in Figure 4.10, are considered in vehicle design and are discussed next. Note that straight-level flight is a particular case of a steady climb.

Steady Climb

In a *steady climb*, the aircraft follows a rectilinear trajectory at constant speed, that is, the nonzero terms in the state vector are $v_x = V$ and $\theta = \theta_0$, which are both constant with time. Note also that in this case, $\theta = \theta_0 = \gamma$, which is the flight-path angle (see Equation (4.6)). Considering only the set of force equilibrium equations now, we can use that the external forces in the body-fixed frame based on the stability axes are

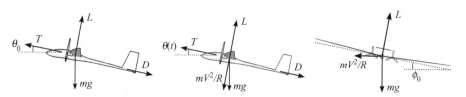

Figure 4.10 Steady Maneuvers. (a) Steady climb, (b) steady pull-up, and (c) steady turn.

$$\mathbf{f}_{aB} = \left\{ \begin{array}{c} -D \\ 0 \\ -L \end{array} \right\}, \ \mathbf{f}_{pB} = \left\{ \begin{array}{c} T\cos\theta_T \\ 0 \\ -T\sin\theta_T \end{array} \right\}, \ \text{and} \ \mathbf{f}_{gB} = \left\{ \begin{array}{c} -mg\sin\theta_0 \\ 0 \\ mg\cos\theta_0 \end{array} \right\}. \quad (4.58)$$

The first force equilibrium equation from the set in Equation (4.57) results in

$$D - T\cos\theta_T + mg\sin\theta_0 = 0, \quad (4.59)$$

where we used the fact that $\omega_B = 0$. The second force equilibrium equation is trivially satisfied, while the third force equation is given by

$$L + T\sin\theta_T - mg\cos\theta_0 = 0. \quad (4.60)$$

The characteristics of the climb are closely linked to the available thrust. The key metric used for this is the *specific excess power*, which measures the difference between the maximum available power and the power necessary to maintain level flight at a given airspeed and altitude (Stengel, 2004).

Steady Pull-Ups and Turns

In steady pull-up and turn maneuvers, the aircraft follows a circular trajectory in either a vertical (pull-up) or a horizontal (turn) plane. Let R be the radius of the trajectory in either case. The corresponding centripetal forces have been included in the free-body diagrams of Figure 4.10b and 4.10c. In the case of *steady pull-up*, the nonzero states are $v_x = V$ and $\omega_y = \frac{V}{R}$, both constant. Since $\omega_y = \dot{\theta}$, the pitch angle varies with time as $\theta(t) = \theta_0 + \frac{V}{R}t$. The nontrivial force equilibrium equations in this case along the x and z directions, respectively, are

$$L + T\sin\theta_T - mg\cos\theta(t) - m\frac{V^2}{R} = 0, \quad (4.61)$$
$$T\cos\theta_T - D - mg\sin\theta(t) = 0.$$

This clearly requires that lift (and therefore drag) and thrust be constantly updated to sustain the maneuver. Finally, the *steady turn* is similarly defined with a constant angular rate $\frac{V}{R}$ in the horizontal plane and a constant roll angle ϕ_0. As the stability axes are body attached, this results in an angular velocity with two nonzero components, $\omega_y = \frac{V}{R}\sin\phi_0$ and $\omega_z = \frac{V}{R}\cos\phi_0$ (both positive if we assume that the aircraft is turning clockwise in the horizontal plane). By definition, the x component of the stability axis is chosen to coincide with the translational velocity in the maneuver, and therefore it is $v_x = V$. This leads to the following force equilibrium equations along the x, y, and z stability axes, respectively:

$$T\cos\theta_T - D = 0,$$
$$mg\sin\phi_0 - m\frac{V^2}{R}\cos\phi_0 = 0, \quad (4.62)$$
$$mg\cos\phi_0 - (L - T\sin\theta_T) + m\frac{V^2}{R}\sin\phi_0 = 0.$$

In this expression, the engine mount angle, θ_T, is given with respect to the x–y stability axis, whose definition may vary at different trim conditions. Contrary to the pull-up conditions, the equilibrium equations in a steady turn are more naturally written in vertical and horizontal axes, such that the last two equations in (4.62) become

$$(L - T\sin\theta_T)\sin\phi_0 - m\frac{V^2}{R} = 0,$$

$$(L - T\sin\theta_T)\cos\phi_0 - mg = 0.$$

(4.63)

Steady turns are, therefore, achieved with a fixed commanded input on the vehicle controls. Euler angles are $\phi(t) = \phi_0$, $\theta(t) = 0$, and $\psi = \psi_0 + \frac{V}{R}t$, where ψ_0 is the initial heading at $t = 0$. Note that the components of the angular velocity in the body-attached stability axes satisfy Equation (4.13). From the previous expressions, it is clear that the maneuverability characteristics of an aircraft in both pull-up and turn (in particular, the achievable turning radius R) will be closely linked to the maximum available lift. The metric used here is the *load factor*, which is discussed next.

4.5.2   Load Factor and Maneuvering Envelope

We define the *load factor*, n, as the ratio between lift and weight, that is, $n = \frac{L}{mg}$ during a steady maneuver.[4] Steady maneuvers are normally defined by their load factor, and the maximum maneuvering load factor that a vehicle can sustain is a key structural sizing parameter. According to the category of the aircraft, the regulations define the load factor range, varying from values as low as 2 or 3 in the case of transport aircraft, to values as high as 8 or 9 for fighter planes. It is, however, customary to express load factors with respect to gravity, so one would talk about a "$6g$" maneuver. For the three conditions shown in Figure 4.10, it is easy to show that the load factors are

$$n_{\text{climb}} = \cos\theta_0,$$

$$n_{\text{pull-up}} = \cos\theta + \frac{V^2}{gR},$$

$$n_{\text{turn}} = \frac{1}{\cos\phi_0}.$$

(4.64)

Note that $n_{\text{pull-up}}$ is only defined as an instantaneous load factor, as the pitch orientation of the aircraft will vary during the maneuver. In fact, pull-up maneuvers for airframe design are defined on level flight conditions, $\theta = 0$, since this defines the maximum value of $n_{\text{pull-up}}$. Commercial transport aircraft regulations require the airframe to be able to sustain a 2.5-g pull-up from level flight. On the other extreme, a very small, or even zero, load factor can be obtained if the aircraft performs this maneuver with a negative angular velocity (*pull-down* maneuver). This is the situation for parabolic flights (zero-gravity maneuver), which generate the perception of weightlessness inside the cabin.

The maximum and minimum load factors that an aircraft can sustain, for given gross weight, configuration – e.g., clean or landing – and altitude, will depend on the airspeed. This determines the vehicle's *maneuvering envelope*, which is given as a plot of the available load factor over airspeed. This results in a V–n diagram, which

---

[4] For transient processes (i.e., dynamic maneuvers), we use instead the instantaneous load factor, which is defined in Equation (3.110).

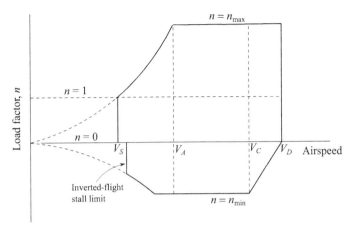

Figure 4.11 Typical maneuvering envelope.

typically looks like the one shown in Figure 4.11. Noting that a gust encounter results in a change of lift, and therefore a change of load factor, additional flight restrictions may occur when the design gust loads of Chapter 2 are also included in the V–n diagram (Niu, 1999, Ch. 3), but they are not shown in the figure. The upper and lower bounds in the load factor, n_{max} and n_{min}, respectively, are structural limits; the minimum and maximum airspeeds are operational limits, that is, the stall V_S and dive V_D speed, respectively. At the lower speeds ($V < V_A$, where V_A is known as the *design maneuvering speed*, the maximum achievable load factor is determined by stall conditions (C_{Lmax}) and similarly for the negative load factor. Note that in the horizontal axis, the speed is typically the equivalent airspeed (EAS), Equation (2.2), which makes the V-n diagram altitude independent.

The lowest speed at which the aircraft can achieve the maximum load factor is the so-called *corner speed*, V_{co}. For the clean configuration, $V_{co} = V_A$ in Figure 4.11. At this speed, one obtains the minimum turn radius, R_{min} (and the maximum attainable turn rate) at the given configuration. The minimum turn radius is derived using Equation (4.63), that is,

$$(L - T\sin\theta_T)\cos\phi_{max} = mg,$$
$$(L - T\sin\theta_T)\sin\phi_{max} = \frac{mV_{co}^2}{R_{min}}, \tag{4.65}$$

where the maximum roll angle is obtained from the maximum load factor in Equation (4.64), that is, $\cos\phi_{max} = 1/n_{max}$. As a result, we have

$$\begin{aligned}
\frac{mV_{co}^2}{R_{min}} &= \frac{mg}{\cos\phi_{max}}\sin\phi_{max} \\
&= mg\frac{\sqrt{1 - \cos^2\phi_{max}}}{\cos\phi_{max}} \\
&= mg\sqrt{n_{max}^2 - 1},
\end{aligned} \tag{4.66}$$

and, finally,

$$R_{\min} = \frac{V_{co}^2}{g\sqrt{n_{\max}^2 - 1}}. \tag{4.67}$$

Note that the results above assume that we have enough thrust and lift available to keep altitude and airspeed during the turning maneuver at the maximum load factor.

4.6   Linearized Vehicle Dynamics about a Steady Climb

Consider a trimmed rigid aircraft with leveled wings ($\phi_0 = 0$) that follows a rectilinear flight path with constant speed V and inclined at an angle θ_0 above the horizon. The aircraft is also at a certain altitude, which defines a local, and known, air density ρ. This situation is illustrated in Figure 4.12. We assume, as it is typically the case, that the aircraft is symmetric about its x–z plane, which implies that $I_{xy} = 0$ and $I_{yz} = 0$ in the inertia tensor, Equation (4.21), that is,

$$\mathbf{I}_B = \begin{bmatrix} I_{xx} & 0 & -I_{xz} \\ 0 & I_{yy} & 0 \\ -I_{zx} & 0 & I_{zz} \end{bmatrix}. \tag{4.68}$$

We also assume that the reference thrust, T_0, is oriented along the x axis (i.e., $\theta_T = 0$), and it does not generate pitching moments about the aircraft CM ($x_T = z_T = 0$ in Equation (4.30)), as we have already done in Section 4.5.

We are interested in investigating the small-amplitude dynamics of the aircraft about this reference trajectory. In order to first characterize the reference conditions, the trim equations of motion for this problem in the stability axes are obtained from Equations (4.59) and (4.60), yielding

$$\begin{aligned} L_0 &= mg\cos\theta_0, \\ D_0 &= T_0 - mg\sin\theta_0. \end{aligned} \tag{4.69}$$

This is the particular case of steady climb for the general equilibrium condition (4.53). The lift, drag, and thrust necessary to sustain the climb are obtained by a particular setting of the control inputs and vehicle states. Specific details of these settings, however, are not necessary for the discussion that follows. They also enforce that the resultant moment components at the CM are also zero in the reference condition.

In order to investigate the subsequent dynamic response of the aircraft to either pilot commands or atmospheric disturbance, we use the general EoM (4.26) expressed in components along the body-fixed frame (stability axes). Introducing the inertia tensor from Equation (4.68), we end up with the following six equations:

$$\begin{aligned} m(\dot{v}_x - \omega_z v_y + \omega_y v_z) &= -mg\sin\theta + f_x, \\ m(\dot{v}_y + \omega_z v_x - \omega_x v_z) &= mg\cos\theta\sin\phi + f_y, \\ m(\dot{v}_z - \omega_y v_x + \omega_x v_y) &= mg\cos\theta\cos\phi + f_z, \end{aligned} \tag{4.70}$$

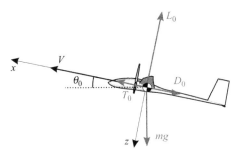

Figure 4.12 Reference conditions: steady climb.

$$I_{xx}\dot{\omega}_x - I_{xz}\dot{\omega}_z + (I_{zz} - I_{yy})\omega_z\omega_y - I_{xz}\omega_x\omega_y = m_x,$$
$$I_{yy}\dot{\omega}_y + (I_{xx} + I_{zz})\omega_z\omega_x + I_{xz}\omega_x(\omega_x^2 - \omega_z^2) = m_y,$$
$$I_{zz}\dot{\omega}_z - I_{xz}\dot{\omega}_x + (I_{yy} - I_{xx})\omega_x\omega_y + I_{xz}\omega_z\omega_y = m_z,$$

where the gravitational effects have been brought forward explicitly, while the aerodynamic and propulsive ones are left in terms of vector components in the stability axes in the reference condition, f_i and m_i, with $i = x, y, z$. They are solved together with the kinematic relations given by Equation (4.16), which can be rewritten as

$$\dot{\phi} = \omega_x + \omega_y \sin\phi \tan\theta + \omega_z \cos\phi \tan\theta,$$
$$\dot{\theta} = \omega_y \cos\phi - \omega_z \sin\phi, \tag{4.71}$$
$$\dot{\psi} = \omega_y \sin\phi \sec\theta + \omega_z \cos\phi \sec\theta.$$

Since we are interested in the small perturbations about a reference trajectory, we can consider incremental loading from the steady-state conditions as

$$\begin{array}{ll} f_x = T_0 - D_0 + \Delta f_x, & m_x = \Delta m_x, \\ f_y = \Delta f_y, & m_y = \Delta m_y, \\ f_z = -L_0 + \Delta f_z, & m_z = \Delta m_z, \end{array} \tag{4.72}$$

which results in small perturbations on the aircraft state variables from their reference values, that is,

$$\begin{array}{lll} v_x = V + \Delta v_x, & \omega_x = \Delta\omega_x, & \phi = \Delta\phi, \\ v_y = \Delta v_y, & \omega_y = \Delta\omega_y, & \theta = \theta_0 + \Delta\theta, \\ v_z = \Delta v_z, & \omega_z = \Delta\omega_z, \end{array} \tag{4.73}$$

where, as before, $\Delta\bullet$ refers to small incremental changes in magnitude, for example, $\|\frac{\Delta v_x}{V}\| \ll 1$. Angles $\theta_0 + \Delta\theta$ and $\Delta\phi$ are the instantaneous climb and bank flight-path angles, respectively. Moreover, changes on the azimuth angle, $\Delta\psi$, have no effect on the aircraft dynamics and do not need to be included. The relation between sufficiently small incremental forces in Equation (4.72) and the corresponding (small) response is given by the *small-perturbation EoMs*. They are obtained by substituting Equations (4.72) and (4.73) into Equations (4.70) and (4.71), respectively, and neglecting nonlinear terms. The linearized EoMs become

$$
\begin{aligned}
m\Delta\dot{v}_x &= \Delta f_x - (mg\cos\theta_0)\,\Delta\theta, \\
m\Delta\dot{v}_y &= \Delta f_y + (mg\cos\theta_0)\,\Delta\phi - mV\Delta\omega_z, \\
m\Delta\dot{v}_z &= \Delta f_z - (mg\sin\theta_0)\,\Delta\theta + mV\Delta\omega_y, \\
I_{xx}\Delta\dot{\omega}_x - I_{xz}\Delta\dot{\omega}_z &= \Delta m_x, \\
I_{yy}\Delta\dot{\omega}_y &= \Delta m_y, \\
-I_{xz}\Delta\dot{\omega}_x + I_{zz}\Delta\dot{\omega}_z &= \Delta m_z,
\end{aligned}
\tag{4.74}
$$

where we have used the following trigonometric relations:

$$
\begin{aligned}
\sin\theta &= \sin(\theta_0 + \Delta\theta) = \sin\theta_0\cos\Delta\theta + \sin\Delta\theta\cos\theta_0 \approx \sin\theta_0 + \Delta\theta\cos\theta_0, \\
\cos\theta &= \cos(\theta_0 + \Delta\theta) = \cos\theta_0\cos\Delta\theta - \sin\theta_0\sin\Delta\theta \approx \cos\theta_0 - \Delta\theta\sin\theta_0.
\end{aligned}
\tag{4.75}
$$

Similarly, the linearized kinematic relations are obtained as

$$
\begin{aligned}
\Delta\dot{\phi} &= \Delta\omega_x + \Delta\omega_z\tan\theta_0, \\
\Delta\dot{\theta} &= \Delta\omega_y.
\end{aligned}
\tag{4.76}
$$

For given external forces and moments, Equations (4.74) and (4.76) define a linear system with eight equations and eight unknowns. A noticeable feature of the small-perturbation equations is that, under typical aerodynamic and propulsive features (i.e., displaying symmetry about the x–z plane), they can be split into two independent sets of equations, corresponding to the longitudinal and lateral aircraft dynamics. However, while the longitudinal dynamics are only dependent on longitudinal states, the lateral-directional dynamics are also dependent on the trim pitch and flight-path angles. They are considered in more detail below.

4.6.1   Longitudinal Problem

The longitudinal problem corresponds to aircraft kinematics in the x–z plane. It involves the translations along both axes, as well as the pitch rotation and the pitch rate, which are given by

$$
\begin{aligned}
m\Delta\dot{v}_x &= \Delta f_x - (mg\cos\theta_0)\,\Delta\theta, \\
m\Delta\dot{v}_z &= \Delta f_z - (mg\sin\theta_0)\,\Delta\theta + mV\Delta\omega_y, \\
I_{yy}\Delta\dot{\omega}_y &= \Delta m_y, \\
\Delta\dot{\theta} &= \Delta\omega_y.
\end{aligned}
\tag{4.77}
$$

The incremental forces Δf_x and Δf_z are the tangent and normal forces, respectively, to the forward-flight velocity after perturbations. To simplify this description, we assume that the thrust force has constant magnitude and its direction is always aligned with the x axis. This assumption implies that Δf_x, Δf_z, and Δm_y only depend in our model on the aerodynamic forces. The incremental aerodynamic forces resulting from perturbations of the aircraft and wind velocities have been computed in

Equations (4.36) and (4.44). For the longitudinal problem, they can be written as

$$\Delta f_x = \frac{1}{2}\rho V^2 S \left[C_{L0}\Delta\alpha - \Delta C_D - 2C_{D0}\frac{\Delta v_x - \Delta v_{gx}}{V} \right],$$

$$\Delta f_z = \frac{1}{2}\rho V^2 S \left[-C_{D0}\Delta\alpha - \Delta C_L - 2C_{L0}\frac{\Delta v_x - \Delta v_{gx}}{V} \right], \qquad (4.78)$$

$$\Delta m_y = \frac{1}{2}\rho V^2 S c_{\text{ref}} \left[\Delta C_M + 2C_{M0}\frac{\Delta v_x - \Delta v_{gx}}{V} \right].$$

The incremental aerodynamic coefficients, ΔC_D, ΔC_L, and ΔC_M, can be obtained by Equations (4.41) and (4.42), with $\Delta\alpha = \frac{\Delta v_z - \Delta v_{gz}}{V}$ and $\Delta\bar{q} = \Delta\omega_y - \Delta\omega_{gy}$. Collecting the results above, we can write the incremental aerodynamic forces in the form of Equation (4.47), which becomes, for the longitudinal problem,

$$\left\{ \begin{array}{c} \Delta f_x \\ \Delta f_z \\ \Delta m_y \end{array} \right\} = \rho V \boldsymbol{A}_{rr} \left\{ \begin{array}{c} \Delta v_x \\ \Delta v_z \\ \Delta\omega_y \end{array} \right\} - \rho V \boldsymbol{A}_{rr} \left\{ \begin{array}{c} \Delta v_{gx} \\ \Delta v_{gz} \\ \Delta\omega_{gy} \end{array} \right\} + \frac{1}{2}\rho V^2 \boldsymbol{A}_{ru}\delta_e, \qquad (4.79)$$

with aerodynamic influence coefficient matrices defined as

$$\boldsymbol{A}_{rr} = S \begin{bmatrix} -C_{D0} & \frac{1}{2}(C_{L0} - C_{D\alpha}) & -\frac{1}{2}VC_{Dq} \\ -C_{L0} & -\frac{1}{2}(C_{D0} + C_{L\alpha}) & -\frac{1}{2}VC_{Lq} \\ c_{\text{ref}}C_{M0} & \frac{1}{2}c_{\text{ref}}C_{M\alpha} & \frac{1}{2}Vc_{\text{ref}}C_{Mq} \end{bmatrix} \qquad (4.80)$$

and

$$\boldsymbol{A}_{ru} = S \begin{bmatrix} -C_{D\delta_e} & -C_{L\delta_e} & c_{\text{ref}}C_{M\delta_e} \end{bmatrix}^\top. \qquad (4.81)$$

We have established the linear relation between perturbations of the aircraft flight dynamic states (linear and angular velocities in the x–z plane) and inputs (elevator deflections), on the one side, and the incremental aerodynamic forces appearing on the vehicle, on the other side. They further depend on the aircraft configuration and geometry, as well as flight conditions. All the aircraft-level longitudinal aerodynamic coefficients are defined using S and c_{ref} as normalization constants, as before. So far, we have not made any assumptions on the flow regime, and indeed Equation (4.78) is valid for low- and high-speed flight. In general, the incremental aerodynamic coefficients, ΔC_L, ΔC_D, and ΔC_M, are obtained for a given aircraft and flight regime in terms of Equations (4.41) and (4.42), with tabulated values for the stability and control derivatives. Here, following Ashley (1974), we introduce next a simplified aerodynamic model to estimate those coefficients for a conventional wing–fuselage–tail aircraft.

A Simplified Quasi-Steady Aerodynamic Model for a Conventional Configuration

For a conventional aircraft configuration, and assuming quasi-steady aerodynamics, the lift is generated both on the wing and the tail, as illustrated in Figure 4.13. Let $C_{L\alpha}^w$ and $C_{L\alpha}^t$ be the lift–curve slope of the wing and the horizontal tail, respectively, and let $C_{L\delta}^t$ be the tail-lift increment to a unit elevator input. They are typically normalized

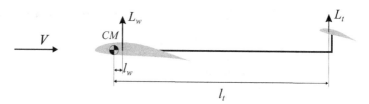

Figure 4.13 Wing–tail geometry definition.

with respect to their corresponding reference areas, that is, S_t for the tail coefficients and S for the wing coefficients. The incremental lift due to changes on the aircraft velocity becomes

$$\Delta C_L = \Delta C_L^w + \Delta C_L^t = \left(C_{L\alpha}^w + \frac{S_t}{S} C_{L\alpha}^t \right) \frac{\Delta v_z}{V} + \frac{S_t}{S} \left(C_{L\alpha}^t \frac{l_t \Delta \omega_y}{V} + C_{L\delta}^t \delta_e \right). \quad (4.82)$$

From here, we can directly identify some of the aircraft aerodynamic coefficients in Equations (4.80) and (4.81), in particular,

$$
\begin{aligned}
C_{L\alpha} &= C_{L\alpha}^w + \frac{S_t}{S} C_{L\alpha}^t, \\
C_{Lq} &= \frac{c_{\text{ref}}}{V} \mathbb{V}_h C_{L\alpha}^t, \\
C_{L\delta_e} &= \frac{c_{\text{ref}}}{l_t} \mathbb{V}_h C_{L\delta}^t,
\end{aligned}
\quad (4.83)
$$

where we have introduced the nondimensional parameter $\mathbb{V}_h = \frac{S_t l_t}{S c_{\text{ref}}}$, which is known as the *horizontal tail volume coefficient*, where l_t is the distance between the aerodynamic center of the tail and the CM. The tail volume coefficient is a design parameter used to assess tail effectiveness in improving longitudinal stability (through the negative moments generated against a perturbation on the pitch rate) or controllability (through the moments generated by elevator deflections). Typically, $\mathbb{V}_h$ takes values between 0.3 and 0.6 for conventional configurations. Note that the tail aerodynamics may also include the effect of the downwash of the main wing, although this has not been explicitly written. We have finally assumed that the lift on the tail is in the same direction that the lift on the main wing, which neglects the contributions of wing downwash and pitch rate onto the instantaneous wind velocity vector on the tail. See Etkin (1972, section 6.3) for further details.

The drag coefficient is mostly independent of the forward-flight velocity at subsonic speeds. It does change, however, under changes of the lift coefficient (and, therefore, the angle of attack) through the induced drag. As a result, changes in the drag coefficient can then be shown as

$$\Delta C_D = \frac{S}{\pi b_{\text{ref}}^2 e} \left(C_L^2 - C_{L0}^2 \right) \approx \frac{S}{\pi b_{\text{ref}}^2 e} 2 C_{L0} \Delta C_L, \quad (4.84)$$

where e is the Oswald's efficiency factor ($e = 0.8$ for a rectangular wing). It has also been assumed that both e and the aspect ratio have the same value for both the wing and the tail, leading to

$$\Delta C_D = \frac{2SC_{L0}}{\pi b_{\text{ref}}^2 e} \left[C_{L\alpha} \frac{\Delta v_z}{V} + \frac{S_t}{S} \left(C_{L\alpha}^t \frac{l_t \Delta \omega_y}{V} + C_{L\delta}^t \delta_e \right) \right], \quad (4.85)$$

and, therefore,

$$C_{D\alpha} = \frac{2SC_{L0}}{\pi b_{\text{ref}}^2 e} C_{L\alpha},$$

$$C_{Dq} = \frac{2SC_{L0}}{\pi b_{\text{ref}}^2 e} \frac{c_{\text{ref}}}{V} \mathbb{V}_h C_{L\alpha}^t, \quad (4.86)$$

$$C_{D\delta_e} = \frac{2SC_{L0}}{\pi b_{\text{ref}}^2 e} \frac{c_{\text{ref}}}{l_t} \mathbb{V}_h C_{L\delta}^t.$$

Finally, the pitch moment coefficient (positive nose up) at the aircraft CM is approximated by adding the contributions of the lift on both the wing and the tail, as

$$\Delta C_{\mathcal{M}} = -\frac{l_w}{c_{\text{ref}}} \Delta C_L^w - \frac{l_t}{c_{\text{ref}}} \Delta C_L^t$$

$$= -\left(\frac{l_w}{c_{\text{ref}}} C_{L\alpha}^w + \mathbb{V}_h C_{L\alpha}^t \right) \frac{\Delta v_z}{V} - \mathbb{V}_h C_{L\alpha}^t \frac{l_t \Delta \omega_y}{V} - \mathbb{V}_h C_{L\delta}^t \delta_e, \quad (4.87)$$

where l_w is the distance between the aerodynamic center of the wing and the CM, which is indicated by the positive sign in Figure 4.13. From this, we can again identify some additional aircraft aerodynamic coefficients in Equations (4.80) and (4.81), that is,

$$C_{\mathcal{M}\alpha} = \frac{l_w}{c_{\text{ref}}} C_{L\alpha}^w - \mathbb{V}_h C_{L\alpha}^t,$$

$$C_{\mathcal{M}q} = -\frac{l_t \mathbb{V}_h}{V} C_{L\alpha}^t, \quad (4.88)$$

$$C_{\mathcal{M}\delta_e} = -\mathbb{V}_h C_{L\delta}^t.$$

Collecting all the results above, we write them as in Equation (4.80), which for the longitudinal problem takes the form

$$\mathcal{A}_{rr} = S \begin{bmatrix} -C_{D0} & \left(\frac{1}{2} - \frac{SC_{L\alpha}}{\pi b_{\text{ref}}^2 e} \right) C_{L0} & -\frac{SC_{L0}}{\pi b_{\text{ref}}^2 e} \mathbb{V}_h c_{\text{ref}} C_{L\alpha}^t \\ -C_{L0} & -\frac{1}{2}(C_{D0} + C_{L\alpha}) & -\frac{1}{2} \mathbb{V}_h c_{\text{ref}} C_{L\alpha}^t \\ c_{\text{ref}} C_{\mathcal{M}0} & -\frac{1}{2}(l_w C_{L\alpha}^w + \mathbb{V}_h c_{\text{ref}} C_{L\alpha}^t) & -\frac{1}{2} \mathbb{V}_h c_{\text{ref}} l_t C_{L\alpha}^t \end{bmatrix} \quad (4.89)$$

and

$$\mathcal{A}_{ru} = S \begin{bmatrix} -\frac{2SC_{L0}}{\pi b_{\text{ref}}^2 e} \frac{\mathbb{V}_h c_{\text{ref}}}{l_t} C_{L\delta}^t & -\frac{\mathbb{V}_h c_{\text{ref}}}{l_t} C_{L\delta}^t & -\mathbb{V}_h c_{\text{ref}} C_{L\delta}^t \end{bmatrix}^\top. \quad (4.90)$$

We have established the linear relation between perturbations of the aircraft flight dynamic states (linear and angular velocities in the x–z plane) and inputs (elevator deflections), on the one side, and the incremental aerodynamic forces appearing on the vehicle, on the other side. They depend on the geometry of the aircraft

$(S, b_{\text{ref}}, l_w, S_t, l_t, e)$, stability and control derivatives, $(C_{L\alpha}^w, C_{L\alpha}^t, C_{L\delta}^t)$, the reference flight conditions (ρ, V) and steady forces at that reference (C_{L0}, C_{D0}, C_{M0}). Note that we have written the AICs in terms of both the volume coefficient $\mathbb{V}_t$ and the reference chord, c_{ref}, but they always appear as a product in Equations (4.80) and (4.81), which are uniquely defined with the geometry parameters listed above.

State Equation

We have obtained explicit expressions for the incremental aerodynamic forces, which can now be introduced in the linearized EoMs of the longitudinal problem, Equation (4.77). As we have established in Equation (4.47), the AICs related to the velocities of the aircraft are equal and opposite to those related to any wind gusts (measured at the aircraft CM). Therefore, those can also be added to the description of the aircraft dynamics, which becomes

$$\mathbf{M}\begin{Bmatrix} \Delta\dot{v}_x \\ \Delta\dot{v}_z \\ \Delta\dot{\omega}_y \end{Bmatrix} = (\mathbf{F}_1 + \rho V \boldsymbol{\mathcal{A}}_{rr})\begin{Bmatrix} \Delta v_x \\ \Delta v_z \\ \Delta\omega_y \end{Bmatrix} + \mathbf{F}_2\Delta\theta - \rho V\boldsymbol{\mathcal{A}}_{rr}\begin{Bmatrix} \Delta v_{gx} \\ \Delta v_{gz} \\ \Delta\omega_{gy} \end{Bmatrix} + \frac{1}{2}\rho V^2 \boldsymbol{\mathcal{A}}_{ru}\delta_e,$$

$$\Delta\dot{\theta} = \Delta\omega_y,$$

$$(4.91)$$

with

$$\mathbf{M} = \begin{bmatrix} m & 0 & 0 \\ 0 & m & 0 \\ 0 & 0 & I_{yy} \end{bmatrix}, \quad \mathbf{F}_1 = \begin{bmatrix} 0 & 0 & 0 \\ 0 & 0 & mV \\ 0 & 0 & 0 \end{bmatrix}, \quad \text{and} \quad \mathbf{F}_2 = \begin{bmatrix} -mg\cos\theta_0 \\ -mg\sin\theta_0 \\ 0 \end{bmatrix}. \quad (4.92)$$

For a given initial time history of elevator commands, $\delta_e(t)$, and gust disturbance velocities, $\mathbf{w}_r = -\begin{Bmatrix} \Delta v_{gx} & \Delta v_{gz} & \Delta\omega_{gy} \end{Bmatrix}^\top$, Equation (4.91) determines the (linearized) longitudinal response of the aircraft. The steady force aerodynamic coefficients, C_{L0} and C_{D0}, which have appeared in the aerodynamic influence coefficient matrices, are given from the trim state in the reference condition, Equation (4.69), and the trimmed pitching moment $C_{M0} = 0$. If those are explicitly substituted into Equation (4.91), the inversion of the (diagonal) mass matrix $\mathbf{M}$ yields, after aggregating terms, the state-space form of the EoMs for the small-perturbation longitudinal aircraft flight dynamics as

$$\dot{\mathbf{x}}_r = \mathbf{A}_{rr}\mathbf{x}_r + \mathbf{B}_r\delta_e + \mathbf{E}_r\mathbf{w}_r. \quad (4.93)$$

The state vector in this problem is $\mathbf{x}_r = \begin{Bmatrix} \Delta v_x & \Delta v_z & \Delta\omega_y & \Delta\theta \end{Bmatrix}^\top$, the constant-coefficient *state matrix* is

$$\mathbf{A}_{rr} = \begin{bmatrix} \dfrac{2(mg\sin\theta_0 - T_0)}{mV} & \dfrac{g\cos\theta_0}{V}\left(1 - \dfrac{2SC_{L\alpha}}{\pi b_{\text{ref}}^2 e}\right) & -\dfrac{2g\cos\theta S_t l_t C_{L\alpha}^t}{\pi b_{\text{ref}}^2 eV} & -g\cos\theta_0 \\[3mm] -\dfrac{2g\cos\theta_0}{V} & \dfrac{mg\sin\theta_0 - T_0}{mV} - \dfrac{\rho VSC_{L\alpha}}{2m} & V - \dfrac{\rho VS_t l_t C_{L\alpha}^t}{2m} & -g\sin\theta_0 \\[3mm] 0 & -\dfrac{\rho V\left(Sl_w C_{L\alpha}^w + S_t l_t C_{L\alpha}^t\right)}{2I_{yy}} & -\dfrac{\rho VS_t l_t^2 C_{L\alpha}^t}{2I_{yy}} & 0 \\[3mm] 0 & 0 & 1 & 0 \end{bmatrix}, \quad (4.94)$$

the *input matrix*, which in this case is a column matrix, is

$$\mathbf{B}_r = \tfrac{1}{2}\rho V^2 \tfrac{S_t}{m} C_{L\delta}^t \left[-\frac{4mg\cos\theta}{\rho V^2 \pi b_{\text{ref}}^2 e} \quad -1 \quad -\frac{l_t m}{I_{yy}} \quad 0 \right]^\top, \tag{4.95}$$

and, finally, the *disturbance matrix* is given by

$$\mathbf{E}_r = \begin{bmatrix} \frac{2(mg\sin\theta_0 - T_0)}{mV} & \frac{g\cos\theta_0}{V}\left(1 - \frac{2SC_{L\alpha}}{\pi b_{\text{ref}}^2 e}\right) & -\frac{2g\cos\theta S_t l_t C_{L\alpha}^t}{\pi b_{\text{ref}}^2 e V} \\ -\frac{2g\cos\theta_0}{V} & \frac{mg\sin\theta_0 - T_0}{mV} - \frac{\rho VSC_{L\alpha}}{2m} & -\frac{\rho VS_t l_t C_{L\alpha}^t}{2m} \\ 0 & -\frac{\rho V\left(Sl_w C_{L\alpha}^w + S_t l_t C_{L\alpha}^t\right)}{2I_{yy}} & -\frac{\rho VS_t l_t^2 C_{L\alpha}^t}{2I_{yy}} \\ 0 & 0 & 0 \end{bmatrix}. \tag{4.96}$$

We can assume finally that there are appropriate sensors on the vehicle to measure the instantaneous velocities of all states $\mathbf{x}_r$, which can support *full-state feedback*. The output equation, assuming that no feedthrough is necessary, is then the identity relation

$$\mathbf{y} = \mathbf{C}_r \mathbf{x}_r, \tag{4.97}$$

with the *output matrix* $\mathbf{C}_r = \mathcal{I}$. The eigenvalues of the state matrix determine the stability of the aircraft. In general, there will be two complex pairs, a low-frequency low-damping pair, known as the long-period (or *phugoid*) mode, and a higher frequency higher damping pair, known as the *short-period mode*, or short-period pitch oscillations (SPPOs). The values typically depend on the location of the aircraft CM (which enters the equations through l_w and l_t) and the horizontal tail area S_t. This is illustrated in Example 4.1.

Finally, we can use the methods of Section 2.3.2 to determine the statistical characteristics in the longitudinal response of an aircraft, defined by Equations (4.93) and (4.97), to atmospheric turbulence given by, for example, the von Kármán model, Equation (2.46). This is illustrated in Example 4.2. Conversely, such methods can also be used to estimate atmospheric turbulence levels from aircraft acceleration measurements during flight. This was used by Cornman et al. (1995) to propose a system in which commercial transport aircraft act as meteorological sensors by reporting load factor statistics in cruise flight, Equation (3.110), and cross-correlating them with detailed weather maps.

Example 4.1 Longitudinal Stability of a Rigid Glider. Consider a glider at an altitude of 1,000 m ($\rho = 1.112$ kg/m^3) on a descent trajectory with a glide angle $\theta_0 = -2°$ and velocity $V = 30$ m/s. The vehicle has a rectangular wing and a tail with symmetric airfoils. Furthermore, the wing is mounted at zero angle of incidence, while the tail is at $\alpha_{t0} = -5°$. Its linear dynamics is assumed to be well approximated by Equation (4.93), with the following geometrical, mass, and aerodynamic parameters,

$$
\begin{aligned}
m &= 318 \text{ kg} & S &= 6.11 \text{ m}^2 \\
I_{yy} &= 432 \text{ kg·m}^2 & S_t &= 1.14 \text{ m}^2 \\
C_{L\alpha}^w &= 5.55 & b_{\text{ref}} &= 15 \text{ m} \\
C_{L\alpha}^t &= 4.30 & l_t &= 4.63 \text{ m} \\
C_{L\delta}^t &= 0.45 & l_w &= -0.30 \text{ m}
\end{aligned}
$$

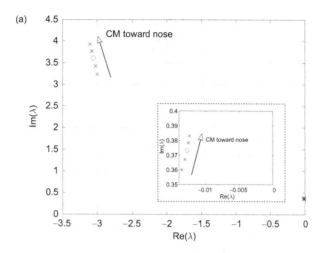

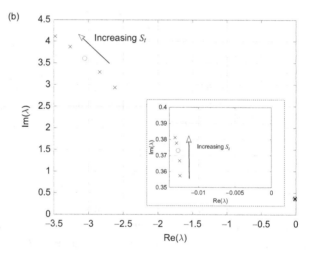

Figure 4.14 Evolution of the longitudinal eigenvalues with selected aircraft coefficients (the circle represents the baseline properties). Effects of (a) center-of-mass position and (b) the horizontal tail area, S_t. [04RigidAircraft/glider.m]

where, as before, the reference area S is the wing area. First, we can determine the trim condition. The load factor in trim is $n = \cos\theta_0 = 0.9994$. The value of the trim angle of attack, α_0, and the corresponding elevator deflection, δ_0, to keep the trajectory at the given path are given by the equilibrium of forces in the normal direction and the equilibrium of pitching moment, respectively, as

$$SC_{L\alpha}^w \alpha_0 + S_t C_{L\alpha}^t (\alpha_0 + \alpha_{t0}) + S_t C_{L\delta}^t \delta_0 = \frac{2nmg}{\rho V^2},$$

$$l_w SC_{L\alpha}^w \alpha_0 - l_t S_t C_{L\alpha}^t (\alpha_0 + \alpha_{t0}) - l_t S_t C_{L\delta}^t \delta_0 = 0.$$

Table 4.1 Eigenvectors in the reference conditions.
[04RigidAircraft/glider.m]

	Short period	Phugoid
v_x/V	$0.01 \mp i0.02$	0.87
v_z/V	1	-0.11
$\omega_y l_t/V$	$-0.16 \pm i0.58$	$0.06 \mp i0.01$
θ	$0.75 \mp i0.34$	$-0.12 \mp i0.99$

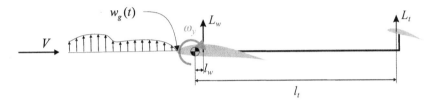

Figure 4.15 Wing–tail configuration under a vertical gust.

The numerical solution to this problem gives $\alpha_0 = 9.89°$ and $\delta_0 = -4.35°$. Next, we construct the state matrix, Equation (4.94), from which the eigenvalues (in rad/s) are obtained as $\lambda_1 = -0.013 \pm i0.373$ and $\lambda_2 = -3.05 \pm i3.60$. They correspond to a phugoid and a short-period mode, respectively, with their normalized eigenvectors, as shown in Table 4.1. They display the typical characteristics of the longitudinal modes: The phugoid is a slow exchange between kinetic (velocity) and potential energy (pitch), while the short period is a fast oscillation in pitch. The sensitivity of the eigenvalues to the location of the CM (which can be modified in the model by changing l_w while keeping $l_w + l_t$ constant) and the area of the stabilizer, S_t, are shown in Figure 4.14. The plot only shows the roots with a positive imaginary part, and their complex conjugates are also solutions. In all cases, results are for the nominal value of the parameter, marked with a circle, as well as 80%, 90%, 110%, and 120% of the reference value, which are identified with crosses. Moving the center-of-gravity aft (toward the rear of the plane, i.e., l_w growing and l_t decreasing) reduces the frequency of both flight dynamic modes, while increasing the area of the stabilizer increases both frequencies and introduces additional damping on the short-period mode. This behavior is typical of conventional wing–body–tail aircraft at subsonic speeds.

Example 4.2 Pitch Response of a Rigid Aircraft under Continuous Turbulence.   Consider a rigid aircraft with a conventional configuration of Figure 4.15 and moment of inertia I_{yy} about its CM. The aircraft flies with constant velocity V through vertical (positive up) atmospheric turbulence w_g, which is approximated using the von Kármán model of Section 2.3.3. The aircraft wing, with the reference area S, has its aerodynamic center at a distance l_w ahead of the CM. Its horizontal tail, with the reference area S_t, has the resultant lift applied at a distance l_t behind the CM.

We assume that the only relevant dynamics is the pitching of the plane,[5] with the instantaneous pitch rate $\omega_y(t)$ (positive nose up), and that the aerodynamic forces are quasi-stationary, with the lift–curve slope for the wing and tail known and given as $C_{L\alpha}^w$ and $C_{L\alpha}^t$, respectively. The EoMs for the aircraft around a certain steady-state reference condition are simply

$$I_{yy}\dot{\omega}_y = \Delta M_y = -l_w \Delta L_w - l_t \Delta L_t, \tag{4.98}$$

where ΔM_y are the instantaneous aerodynamic pitching moment at the CM during the transient dynamics, which are due to the variations of lift at both the wing and the tail. Assuming that the wing's aerodynamic center is close enough to the CM such that $l_w \|\omega_y\| \ll \|w_g\|$, the lift on the wing is simply due to the change of the effective angle of attack as the atmospheric turbulence impacts the wing, that is,

$$\Delta L_w = \tfrac{1}{2}\rho V^2 S C_{L\alpha}^w \frac{w_g(t)}{V}. \tag{4.99}$$

The aerodynamic forces on the tail have contributions from the atmospheric turbulence and from the instantaneous pitch rate, that is,

$$\Delta L_t = \tfrac{1}{2}\rho V^2 S_t C_{L\alpha}^t \frac{w_g(t) + l_t \omega_y(t)}{V}. \tag{4.100}$$

Here we have neglected the time that the fluid needs to move between the wing and the tail, $\frac{l_t - l_w}{V}$, which can result in a small delay if either V or the turbulent length scale is sufficiently small (as, e.g., in low-altitude flight). Substituting Equations (4.99) and (4.100) in the EoM yields,

$$I_{yy}\dot{\omega}_y + \tfrac{1}{2}\rho V^2 S_t l_t C_{L\alpha}^t \frac{l_t \omega_y}{V} = -\tfrac{1}{2}\rho V^2 \left(S l_w C_{L\alpha}^w + S_t l_t C_{L\alpha}^t \right) \frac{w_g(t)}{V}. \tag{4.101}$$

This equation can be nondimensionalized using $\hat{\omega}_y = \frac{l_t \omega_y}{V}$ and $\hat{w}_g = \frac{w_g}{V}$. The time can also be normalized by the characteristic period $T = a\ell/V$ in the atmospheric turbulence introduced in Equation (2.48). This results in

$$T\frac{d\hat{\omega}_y}{dt} + k_t \hat{\omega}_y = -\left(k_w + k_t \right) \hat{w}_g, \tag{4.102}$$

where

$$k_t = \mu_{yy}^{-1} \frac{S_t}{S} \frac{a\ell}{l_t} C_{L\alpha}^t \quad \text{and} \quad k_w = \mu_{yy}^{-1} \frac{l_w}{l_t} \frac{a\ell}{l_t} C_{L\alpha}^w, \tag{4.103}$$

with the pitch inertia parameter $\mu_{yy} = \frac{2 I_{yy}}{\rho S l_t^3}$. The corresponding frequency-response function is then

$$G_{\hat{\omega}_y \hat{w}_g}(i\omega) = -\frac{k_w + k_t}{k_t + i\omega T}, \tag{4.104}$$

which can be combined with the FRF of the von Kármán turbulence model, Equation (2.48), to determine the statistics of the vehicle response. This is the situation

---

[5] The solution to the general longitudinal problem can be found in Buck and Newman (2006).

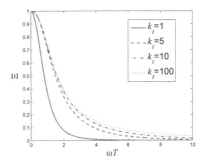

Figure 4.16 Normalized PSD of the pitch response to von Kármán turbulence, Equation (4.108). [04RigidAircraft/pitchpsd.m]

depicted in Figure 2.13, which results here in a combined FRF given by

$$G_{\omega_y n} = G_{\omega_y w_g} G_{w_g n} = \frac{1}{l_t} G_{\hat{\omega}_y \hat{w}_g} G_{w_g n} = -\frac{\sigma_w}{l_t} \sqrt{\frac{T}{\pi a}} \frac{k_w + k_t}{k_t + i\omega T} \frac{1 + \frac{2\sqrt{2}}{\sqrt{3}} i\omega T}{(1 + i\omega T)^{11/6}}. \quad (4.105)$$

The PSD of the pitch response is defined as $\Phi_{\omega_y}(\omega) = \left| G_{\omega_y n}(i\omega) \right|^2 \Phi_n$, where Φ_n has, as defined above, a unit amplitude. As a result, we finally have

$$\Phi_{\omega_y}(\omega) = \frac{\sigma_w^2 T}{\pi a l_t^2} \frac{(k_w + k_t)^2}{k_t^2 + \omega^2 T^2} \frac{1 + \frac{8}{3}\omega^2 T^2}{(1 + \omega^2 T^2)^{11/6}}, \quad (4.106)$$

which can also be written as

$$\Phi_{\omega_y}(\omega) = \frac{\sigma_w^2 T}{\pi a l_t^2} \left[\frac{k_w}{k_t} + 1 \right]^2 \Xi(\omega T), \quad (4.107)$$

with

$$\Xi(\omega T) = \frac{k_t^2}{k_t^2 + \omega^2 T^2} \frac{1 + \frac{8}{3}\omega^2 T^2}{(1 + \omega^2 T^2)^{11/6}}. \quad (4.108)$$

As a result, the intensity of the pitch disturbance grows with the intensity of the gust, σ_w, and with the ratio $\frac{k_w}{k_t} = \frac{S_l w C_{L\alpha}^w}{S_t l_t C_{L\alpha}^t}$, which depends only on the aircraft geometry. A large distance between the tail and the CM, l_t, results in a smaller amplitude of the pitch dynamics. The nondimensional frequency content is given by the term $\Xi(\omega T)$ in the equation and is shown in Figure 4.16 as a function of the constant k_t. Increasing the value of k_t, which can be achieved by either a lower aircraft pitch inertia or a smaller wing area, increases the bandwidth of the response.

4.6.2 Lateral Problem

Similar to the longitudinal problem, the lateral (also known as the *lateral-directional*) problem is defined by the equations in lateral translation, roll and yaw rotations, and their rates. From Equation (4.74), we obtain

$$m\Delta\dot{v}_y = \Delta f_y + (mg\cos\theta_0)\,\Delta\phi - mV\Delta\omega_z,$$
$$I_{xx}\Delta\dot{\omega}_x - I_{xz}\Delta\dot{\omega}_z = \Delta m_x, \tag{4.109}$$
$$-I_{xz}\Delta\dot{\omega}_x + I_{zz}\Delta\dot{\omega}_z = \Delta m_z.$$

The incremental force Δf_y corresponds to the lateral force normal to the forward-flight velocity after perturbation to the lateral translation and roll angle,[6] while the incremental Δm_x and Δm_z moments correspond to the roll and yaw moments, respectively, after the perturbations. Other than θ_0 be nonzero (reference condition associated with the longitudinal trim), the other relevant reference conditions (i.e., $C_{Y0}, C_{N0},$ and $C_{\mathcal{L}0}$) are zero (assuming symmetric aircraft) for the steady straight path case being considered here. Again, we aim to compute how the aerodynamic force and moments change with the perturbations of the states from the reference condition. The perturbations of the states of interest are a subset of Equation (4.73) and given by

$$v_y = \Delta v_y,$$
$$\omega_x = \Delta\omega_x, \qquad \phi = \Delta\phi, \tag{4.110}$$
$$\omega_z = \Delta\omega_z.$$

where the perturbation in lateral velocity leads to the sideslip angle $\Delta\beta = \frac{\Delta v_y}{V}$.

To evaluate the incremental lateral force, we refer again to the general case given in Equation (4.44), and from it, we obtain

$$
\begin{aligned}
\Delta f_y &= \frac{1}{2}\rho V^2 S \Delta C_Y \\
&= \frac{1}{2}\rho V^2 S \left(C_{Y\beta}\frac{\Delta v_y}{V} + C_{Yp}\Delta\omega_x + C_{Yr}\Delta\omega_z + C_{Y\delta_a}\Delta\delta_a + C_{Y\delta_r}\delta_r \right),
\end{aligned} \tag{4.111}
$$

where we used the lateral force coefficient as defined in Equation (4.42). Similarly, the incremental moments can be written as

$$
\begin{aligned}
\Delta m_x &= \frac{1}{2}\rho V^2 Sb_{\text{ref}}\Delta C_{\mathcal{N}} \\
&= \frac{1}{2}\rho V^2 Sb_{\text{ref}}\left(C_{\mathcal{N}\beta}\frac{\Delta v_y}{V} + C_{\mathcal{N}p}\Delta\omega_x + C_{\mathcal{N}r}\Delta\omega_z + C_{\mathcal{N}\delta_a}\delta_a + C_{\mathcal{N}\delta_r}\delta_r \right), \\
\Delta m_z &= \frac{1}{2}\rho V^2 Sb_{\text{ref}}\Delta C_{\mathcal{L}} \\
&= \frac{1}{2}\rho V^2 Sb_{\text{ref}}\left(C_{\mathcal{L}\beta}\frac{\Delta v_y}{V} + C_{\mathcal{L}p}\Delta\omega_x + C_{\mathcal{L}r}\Delta\omega_z + C_{\mathcal{L}\delta_a}\delta_a + C_{\mathcal{L}\delta_r}\delta_r \right).
\end{aligned}
$$
$$\tag{4.112}$$

From these results, we can write the incremental aerodynamic forces in the form of Equation (4.47) that becomes, for the lateral problem,

$$
\begin{Bmatrix} \Delta f_y \\ \Delta m_x \\ \Delta m_z \end{Bmatrix} = \rho V \boldsymbol{A}_{rr} \begin{Bmatrix} \Delta v_y \\ \Delta\omega_x \\ \Delta\omega_z \end{Bmatrix} + \frac{1}{2}\rho V^2 \boldsymbol{A}_{ru} \begin{Bmatrix} \delta_a \\ \delta_r \end{Bmatrix}, \tag{4.113}
$$

---

[6] The azimuth angle ψ is immaterial to the dynamic characterization of the vehicle as discussed in Section 4.6 and it does not need to be included here.

with aerodynamic influence coefficient matrices defined as

$$\mathcal{A}_{rr} = S \begin{bmatrix} \frac{1}{2}C_{Y\beta} & \frac{V}{2}C_{Yp} & \frac{V}{2}C_{Yr} \\ \frac{b_{\mathrm{ref}}}{2}C_{\mathcal{N}\beta} & \frac{Vb_{\mathrm{ref}}}{2}C_{\mathcal{N}p} & \frac{Vb_{\mathrm{ref}}}{2}C_{\mathcal{N}r} \\ \frac{b_{\mathrm{ref}}}{2}C_{\mathcal{L}\beta} & \frac{Vb_{\mathrm{ref}}}{2}C_{\mathcal{L}p} & \frac{Vb_{\mathrm{ref}}}{2}C_{\mathcal{L}r} \end{bmatrix} \tag{4.114}$$

and

$$\mathcal{A}_{ru} = S \begin{bmatrix} C_{Y\delta_a} & C_{Y\delta_r} \\ b_{\mathrm{ref}}C_{\mathcal{N}\delta_a} & b_{\mathrm{ref}}C_{\mathcal{N}\delta_r} \\ b_{\mathrm{ref}}C_{\mathcal{L}\delta_a} & b_{\mathrm{ref}}C_{\mathcal{L}\delta_r} \end{bmatrix}. \tag{4.115}$$

Corresponding state equations can be obtained here, similar to what was done for the longitudinal problem, with respect to the state vector $\mathbf{x}_r = \{\Delta v_y \ \Delta \omega_x \ \Delta \omega_z \ \Delta \phi\}^\top$.

4.7   Summary

This chapter has reviewed the basic theories and fundamental engineering concepts that describe the flight dynamics of rigid aircraft. This forms the basis for the more general formulations for flexible aircraft that will be introduced in the following chapters. Importantly, the traditional descriptions in flight dynamics have been slightly modified such that it can be easily expanded to include the additional DoFs on a flexible aircraft. The focus has also been on state-space formulations, and we have heavily used notations of matrix algebra whenever possible – this not only will easy our way in future chapters but also is a good practice to integrate traditional flight dynamics analysis into techniques of linear (and nonlinear) systems theory. Finally, we have put special emphasis, as we do throughout the book, on flight under nonstationary atmospheric conditions.

The chapter has started with the definitions of the reference frames that are needed to describe the kinematics of a rigid aircraft, and Euler angles were used to parameterize its rotations. Other parameterizations are possible; some of them will be later introduced in Chapter 9 to describe the large deformations on flexible vehicles. The EoMs have then been written in the stability axes, as it is customary in flight dynamics. The main features of gravitational, propulsive, and aerodynamic forces acting on the aircraft have also been outlined.

Once the general framework to describe and analyze the aircraft dynamics has been assembled, we have looked into two relevant situations. First, the constant-velocity solutions of the differential equations have defined steady (or equilibrium) maneuvers. Those are the basic building blocks to define the flight performance of a vehicle and have generalized the definition of the load factor that we had already introduced in Section 3.6. Finally, we have looked into the small-perturbation dynamics of a rigid aircraft about a steady straight path. This problem will reappear later in this book with additional physics. A glider example has been to investigate longitudinal response to atmospheric turbulence, using the tools of Section 2.3.

4.8   Problems

Exercise 4.1. The instantaneous angular velocity can be written in terms of its components in the inertial reference frame as $\omega_E = \mathbf{R}_{BE}\omega_B$.

(i) Show that $\tilde{\omega}_E = \dot{\mathbf{R}}_{EB}\mathbf{R}_{BE}$.
(ii) Using the previous result, show then that $\tilde{\omega}_E = \mathbf{R}_{EB}\tilde{\omega}_B\mathbf{R}_{BE}$.

Exercise 4.2. Derive the expression of Newton–Euler equations in body axes, Equation (4.26), from the conservation properties in the Earth axes, Equation (4.24). How do they change if the body-attached frame of reference is not at the CM?

Exercise 4.3. An aircraft carrier is heading north at a certain constant speed V_s. We define a frame of reference linked to the ship, with x_s pointing north and z_s vertical up. An aircraft is also flying north with a constant forward-flight velocity, V_∞. From radar measurements, it is known that at $t = 0$, the aircraft CM is at $[x_s(0) = x_{s0}, y_s(0) = y_{s0}, z_s(0) = z_{s0}]$ with respect to the ship. The subsequent history of CM velocities is then recorded from on-board equipment (body-fixed axes), and it is known to be $\mathbf{V}_B(t)$ and $\mathbf{\Omega}_B(t)$.

(i) Write the equations that determine the position of the aircraft in the ship frame at a certain time $t = T$, with $T > 0$.
(ii) Determine a_y and a_z and the landing trajectory if $\mathbf{\Omega}_B = 0$ and the components of the translational velocity, in the B frame, follow the profile

$$V_x = V_s + V_\infty \cos\frac{\pi t}{2T},$$

$$V_y = a_y \cos\frac{\pi t}{2T},$$

$$V_z = a_z \cos\frac{\pi t}{2T}.$$

Exercise 4.4. After Ashley (1974), the weathervane of the figure rotates about an axis perpendicular to the plane with a large angle, ψ. It has a moment of inertia I_{zz} about that axis. Assume that the drag is negligible and that the lift is always perpendicular to the constant wind velocity and proportional to $\sin\Psi$. Rotations around the hinge of the weathervane have a viscous damping moment proportional to $\dot{\Psi}$.

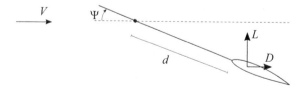

(i) Show that the EoMs have the following structure:

$$\ddot{\Psi} + a_1\dot{\Psi} + a_2\sin 2\Psi = 0.$$

(ii) Write the problem in a nonlinear state-space form as a function on a_1 and a_2.

(iii) Using a forward-Euler discretization, that is, $\dot{\mathbf{x}}(t_n) = \frac{\mathbf{x}(t_{n+1}) - \mathbf{x}(t_n)}{t_{n+1} - t_n}$, write a numerical algorithm to solve the state-space equations for the given initial conditions $\Psi(0) = \Psi_0$ and $\dot{\Psi}(0) = 0$.

(iv) Choose values for the parameters and explore the solutions to this problem numerically.

Exercise 4.5. Consider an aircraft with the maneuver envelope of Figure 4.11.

 (i) For leveled turns, plot its maximum turn rate $\dot{\Psi}_{max}$ against the flight speed.
(ii) A rectangular wing with chord c and semispan b is modeled using lifting line theory and airfoils with the known $c_{l\alpha}$ and $c_{l\delta}$. If the maximum trailing-edge flap deflection is δ_{max}, determine the size and location of the best flaps to achieve a 2% increment in lift.

Exercise 4.6. Introduce the nondimensional time $\bar{t} = \frac{gt}{V_\infty}$ and the nondimensional state variables $\bar{v}_x = \frac{v_x}{V_\infty}$, $\bar{v}_z = \frac{v_z}{V_\infty}$, and $\bar{\omega}_y = \frac{\omega_y l_t}{V_\infty}$ in the longitudinal EoMs for a rigid aircraft and show that the corresponding state matrix in a nondimensional form can be written as

$$\bar{A}_r = \begin{bmatrix} 2\left(\sin\Theta_0 - \bar{T}\right) & \cos\Theta_0\left(1 - \frac{2SC_{L\alpha}}{\pi b^2 e}\right) & 0 & -\cos\Theta_0 \\ -2\cos\Theta_0 & \sin\Theta_0 - \bar{q}C_{L\alpha} - \bar{T} & \frac{V_\infty^2}{gl_t} - \bar{q}\bar{S}_t C_{L\alpha}^t & -\sin\Theta_0 \\ 0 & -\bar{q}\bar{l}_t\left[\bar{l}_w C_{L\alpha}^w + \bar{S}_t \bar{l}_t C_{L\alpha}^t\right] & -\bar{q}\bar{l}_t^2 \bar{S}_t C_{L\alpha}^t & 0 \\ 0 & 0 & \frac{V_\infty^2}{gl_t} & 0 \end{bmatrix},$$

with $\bar{T} = \frac{T_0}{mg}$, $\bar{q} = \frac{q_\infty S}{mg}$, $\bar{S}_t = \frac{S_t}{S}$, $\bar{l}_x = \frac{l_x}{r_y}$, and $r_y = \sqrt{\frac{I_{yy}}{m}}$.

5 Dynamics of Flexible Aircraft with Quasi-Steady Deformations

5.1 Introduction

The previous chapter has described the dynamics of a rigid aircraft in a nonstationary atmosphere and introduced alongside the standard flight mechanics terminology. While that approach is sufficient to investigate the flight performance and to design control systems for many air vehicles, an increasing number of aerial platforms are now built on more flexible airframes for which the rigid-body assumption is no longer applicable. This and subsequent chapters, thus, consider more general cases in which the aircraft may have nonnegligible elastic deformations under the forces required to maintain flight or to perform a maneuver. To introduce the problem, we restrict ourselves in this chapter to vehicles for which the characteristic timescales in their flight dynamics are much larger than those of the aircraft structural vibrations. This implies the existence of a frequency separation between the aircraft flight mechanics and any structural vibrations, and we refer to the resulting problem as the study of aircraft dynamics with quasi-steady flexible effects. This is the typical situation during maneuvers of conventional (wing–body–tail) aircraft with high-aspect-ratio wings and results, as we will see in this chapter, in a relatively simple description of the aircraft dynamics. In particular, under those conditions, the structural (or, more precisely, static aeroelastic) response acts as a constant feedback gain on the rigid-body flight-dynamics equations. This preserves the mathematical structure of the description of the vehicle response discussed in Chapter 4, but with modified aerodynamic (now aeroelastic) derivatives. The next level of approximation corresponds to the inclusion of inertia effects on the structural response, and possibly as well as the unsteadiness in the aerodynamics, which was already discussed in Chapter 3. This is first done under linear elastic assumptions (Chapter 6) and later, in Chapter 9, for aircraft undergoing arbitrarily large changes on their external geometry during flight. We note finally that the assumption of quasi-steady flexible effects on an aircraft does not imply that the aircraft may not display structural vibrations in, for example, a gust encounter. It is simply that its high-frequency response can be independently analyzed (i.e., it is decoupled) from the lower frequency flight dynamics. The focus of this chapter is on the relatively slow processes involved in the flight dynamics of such vehicles.

This chapter discusses, in some detail, the different physical approximations in flexible aircraft dynamics (Section 5.2) before introducing the general quasi-steady problem in Section 5.3. Critically, the description of the aircraft dynamics with quasi-static structural response builds directly on the rigid aircraft models of Chapter 4, and the general structure of the resulting *elastified* flight-dynamics equations is described within that context. The elastified equations also serve to introduce classic linear static aeroelastic concepts, such as divergence (which was already introduced for 2-D problems in Section 3.5.1) and control reversal, and this has been done in Section 5.4. The last sections of this chapter exemplify this general formulation on two simplified but representative problems: First, the dynamic response to the elevator input of an aircraft with a flexible fuselage and second the roll control of a rigid vehicle with a flexible wing in torsion. They facilitate some detailed analyses of the key effects of airframe flexibility in longitudinal and lateral aircraft dynamic characteristics.

5.2 Physical Approximations to Flexible Aircraft Dynamics

Aircraft achieve flight by balancing (and unbalancing) aerodynamic and propulsion forces, Earth's gravity, and their own inertia. The instantaneous force and moment resultants at a reference point (typically chosen as the center of mass) determine the aircraft flight path, maneuverability, and general flight dynamics. This is the description of Figure 5.1a and has already been studied in Chapter 4. However, as shown schematically in Figure 5.1b, all those forces are at source distributed loads. They come from pressures and viscous shear stresses on the wetted surfaces, as well as body forces proportional to the local acceleration, including gravity, and reaction forces at the engine mounts, although only the aerodynamic forces are sketched in the figure. The CM is an extremely useful and ingenious mathematical construction, but it is not *the aircraft*. The forces on the aircraft that enable flight may also deform its airframe, which further complicates the analysis as the deformation modifies the forces themselves. This is the essential feature of aeroelastic problems, which has been introduced in Chapter 3. Not all those couplings are active for a given configuration, and different physical approximations may be used to describe the dynamics of a flexible aircraft, depending on both the magnitude and the rate of change of the elastic displacements and rotations.

(a)   (b)

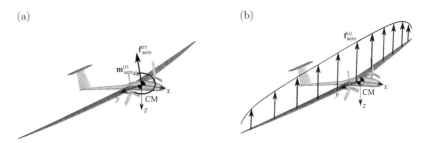

Figure 5.1 Typical spanwise distribution of aerodynamic loading on an aircraft wing and its resultant force and moment at the CM. (a) Resultant forces determine rigid-body dynamics and (b) distributed forces determine airframe structural deformations.

Most primary structures (i.e., wing box, fuselage, and tail) on a conventional aircraft can be analyzed in a first approximation using beam models and, as will be seen in Chapter 9, this approximation can suffice for many aeroelastic applications. From elementary strength of materials, the out-of-plane displacements of a cantilever beam of span b under a constant force distribution per unit span of magnitude f go as $\frac{fb^3}{EI}$, where EI is the effective bending stiffness that depends on the material stiffness (given by its characteristic Young's modulus, E) and the cross-sectional area moment of inertia, I. In particular, the forcing on each wing is proportional to the total lift of the aircraft divided by its span. Since the lift is itself proportional to the total weight of the aircraft, the nondimensional parameter that determines wing bending flexibility for an aircraft of mass m is, therefore,

$$\Pi_1 = \frac{mgb^2}{EI}. \tag{5.1}$$

The rigid aircraft assumptions of Chapter 4 implied that Π_1 is negligible, which is achieved by either very high stiffness EI, very small wingspan b, or very light construction (small m). The nondimensional parameter defined in Equation (5.1) is related to the *Cauchy number* between a fluid and a homogeneous solid, which is defined as the ratio between the magnitude of the fluid forces on the solid boundary (proportional to the dynamic pressure) and the elastic forces in the material (given by its bulk modulus, K), namely $\mathrm{Ca} = \frac{\rho V^2}{K}$.

Wing deformations on aircraft with high-aspect-ratio wings are often reduced by means of *span loading*. This is the result of redistributing some nonstructural masses along the wings so that the local weight partially cancels the local sectional lift in steady flight. This is schematically shown in Figure 5.2, where for both cases, the total aircraft weight (and the integral of the lift forces) is the same. However, in Figure 5.2b, some nonstructural components (e.g., fuel tank and batteries) have been moved from the fuselage to the outer portions of the wings. In this situation, $M^*_{\mathrm{fus}} < M_{\mathrm{fus}}$ and $m^*_{\mathrm{wing}} > m_{\mathrm{wing}}$. The additional weight on the wing produces a negative moment that reduces its bending. However, wing deformations in such case may still be significant during dynamic events that produce large changes on the instantaneous lift, for instance, during maneuvers or when encountering atmospheric gusts (Niu, 1999, Ch. 3).

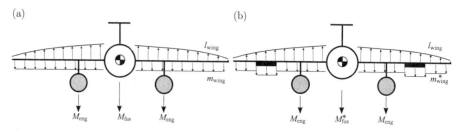

Figure 5.2 Locating nonstructural mass on the wing for the span loading effect. (a) Reference configuration and (b) nonstructural mass is moved to the outer wing.

When $\Pi_1 = \mathcal{O}(1)$, that is, when the displacements of the wing are comparable to its span, geometry changes due to airframe flexibility may alter the instantaneous values of both the first and the second moments of inertia of the aircraft. This results in different properties of the aircraft in its *jig shape*, that is, the shape that it has when stationary on the ground and in its *trim shape* during steady forward flight. Furthermore, during dynamic events, the CM is no longer a material point of the vehicle as its position changes with respect to any body-attached reference frame.[1] Vehicle deformations also alter its external shape and, therefore, the instantaneous aerodynamic force distribution, resulting in aeroelastic couplings between elastic displacements and forces. For a given aircraft, inertial changes and aeroelastic couplings do not necessarily occur simultaneously. For example, small wing torsional deformations often generate rather large changes of lift (an aeroelastic effect) but negligible changes to the overall inertial properties (in particular, the moment of inertia about the pitch axis) of the aircraft.

Figure 5.3 outlines the modeling approaches that reflect the increasingly complex physics on flexible aircraft, most of which is discussed in detail in the rest of this book. They are depicted within axes representing the maximum amplitude of the structural displacements and the highest frequency in the aircraft dynamics. As the structural displacements become larger, the order in which the different physical effects become significant is configuration dependent. Figure 5.3 shows some typical situations in the authors' own experience. The upper-left corner of the figure is the static equilibrium, or trim, of a rigid aircraft. For negligible structural displacements and slow aircraft dynamics, we are in the remit of the flight dynamics of rigid vehicles (Chapter 4). When structural deformations are still very small but no longer negligible, one needs first to consider quasi-steady aeroelastic derivatives, as discussed in Section 5.1, and then to update the inertia (which results in the center of gravity moving (floating) on

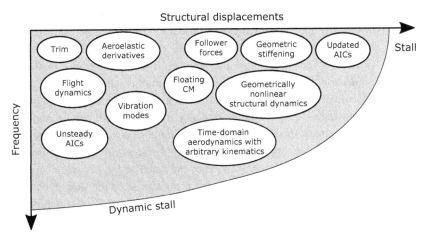

Figure 5.3 Modeling requirements as a function of the amplitude and frequencies of the elastic displacements.

---

[1] This situation will be analyzed in some detail in Section 6.3.

a reference attached to a material point on the vehicle). Inertial changes due to elastic deformations is discussed in Chapter 6. As deflections increase, geometrically nonlinear structural effects start to appear: First, one finds the rotation of the aerodynamic and propulsion force vectors as they follow the structure and subsequently the effective stiffening of the structure as it deforms. They are known as the *follower force* and *geometric stiffening* effects, respectively, and are studied in Chapter 9. Sufficiently large deformations on the airframe eventually need to be reflected on relevant changes on the wing geometry that determines the aerodynamic forces. As a result, one needs to update the aerodynamic influence coefficients, or AICs, as the structure deforms. This will be discussed in Chapter 7. Finally, under sufficiently large shape changes, flow separation is bound to occur and the aerodynamic surfaces *stall*. Depending on the aircraft, stall may occur before any of the previous physical effects become relevant (e.g., wings may stall before the geometric stiffening of their structure becomes noticeable).

The rate at which geometry changes on a flexible air vehicle occur is also critically important in establishing the relevant physical process in its dynamic description. This defines the vertical axis (frequency) in Figure 5.3. The key nondimensional parameter here is the ratio between the higher frequency, ω_{max}, associated with the typical timescales in the flight dynamics (e.g., the short-period mode) and the lowest natural frequency of the (unsupported) structure, ω_1, that is,

$$\Pi_2 = \frac{\omega_{max}}{\omega_1}. \tag{5.2}$$

Very low values of Π_2 correspond to quasi-steady approximations of the structural response, which is the subject of the rest of this chapter. On the contrary, if $\Pi_2 = \mathcal{O}(1)$, then one needs to consider the distributed inertia that leads to the vibration of the airframe, as indicated in Figure 5.3. This typically results in aircraft with much more complex dynamics, for example, when structural vibrations appear as a result of fast actions of the pilot and may result in a dangerous situation of *pilot-induced oscillations* (Drewiacki et al., 2019). The mathematical framework needed to describe this situation is introduced in some detail in Chapter 6. Large-amplitude dynamics of the airframe needs a nonlinear theory for both aerodynamics and structures. For the aerodynamics, this results in time-domain unsteady aerodynamics with arbitrary kinematics of the lifting surfaces, while for the structure, this is known as geometrically nonlinear structural dynamics. They are discussed in Chapters 7 and 8, respectively.

A final aspect to consider is the unsteadiness in the flow dynamics. Flows around stationary streamlined bodies at sufficient low angles of attack can be well approximated using steady aerodynamic assumptions. This is no longer the case if the characteristic timescale of any changes of airframe geometry is comparable to the convective timescale in the flow dynamics (i.e., the time that air particles take to cover the wing), which defines another important nondimensional parameter, namely

$$\Pi_3 = \frac{\omega_{max} c}{V}, \tag{5.3}$$

where c is a typical chord and V is the reference forward-flight speed. This magnitude is effectively (twice) the reduced frequency that was introduced in Chapter 3. If the frequency is given in Hertz, it is also known as the Strouhal number, St. The coupling between the flight dynamics and the structural vibrations may appear at smaller frequencies than the unsteady aerodynamic effects, but all three effects often need to be considered together. At sufficiently high frequencies, flow separation occurs, resulting in dynamic stall conditions. Stall and dynamic stall define a limiting region within which most aircraft normally operate, as shown in Figure 5.3. They are not discussed any further in this book, which assumes throughout attached flow conditions. An excellent introduction to the physics and modeling of dynamic stall can be found in Leishman (2006).

5.3   Elastified Aircraft Dynamics

5.3.1   Degrees of Freedom and Equations of Motion

Chapter 4 has described the dynamics of a rigid aircraft in terms of the evolution of translational/angular velocities and rotations at the vehicle's CM. Describing the dynamics of a flexible aircraft further requires additional DoFs to capture the instantaneous deformed state and, consequently, additional equations that determine their values. In general, this is an infinite-dimensional problem in which the instantaneous elastic displacement field is determined by partial differential equations (Meirovitch, 1991). However, and for nearly all practical purposes, that displacement field can be approximated by a finite-dimensional representation, that is, a description in terms of a finite number of DoFs with interpolating assumptions that allow recovering the solution anywhere in the aircraft structure to a given accuracy.

Consequently, the instantaneous deformed state of the aircraft is described here by a column vector $\mathbf{x}_s(t)$ of elastic DoFs that are associated with a discrete representation of the airframe. It can be chosen as a vector of nodal displacements in a finite-element model or, most commonly, of the amplitude of the linear normal modes of the structure *in vacuo* (see Section 6.4.1). Its definition is problem dependent and will be made specific in subsequent sections; the only condition we need now is that each elastic DoFs adds "new information" to the description of the vehicle shape – in particular, that they are not rigid-body modes.

The displacement field described by the elastic DoFs is given on the reference frame used to describe the rigid-body dynamics. For consistency with the description in Chapter 4, the origin of this frame is chosen to be at the CM of the aircraft, although this is not strictly necessary. As an example, Figure 5.4 shows the first six elastic linear normal modes of a finite-element model of a typical wide-body aircraft. In a flexible aircraft model built with those modes, $\mathbf{x}_s(t)$ would be the time history of the amplitude of each mode in the dynamic response of the aircraft whose instantaneous shape is determined by the superposition of all modes. Importantly, only a few elastic modes have been shown to give good approximations on practical problems (Karpel and Brainin, 1995). As will be seen in Section 6.4, the linear normal modes are

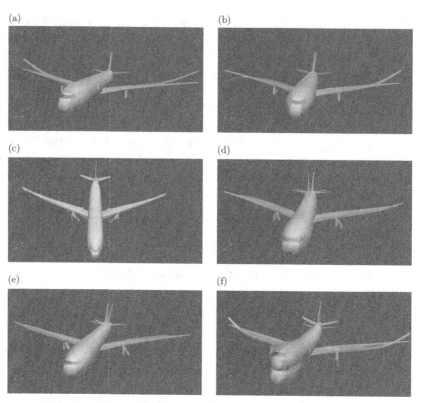

Figure 5.4 First six elastic vibration modes of the finite-element model of a typical long-range commercial transport aircraft (Cea and Palacios, 2022). (a) Mode 7: first symmetric wing bending; (b) Mode 8: first antisymmetric wing bending; (c) Mode 9: symmetric wing-pylon; (d) Mode 10: antisymmetric wing-pylon; (e) Mode 11: antisymmetric wing-fuselage; and (f) Mode 12: symmetric wing-fuselage.

described by both a displacement field (the mode shapes or eigenvectors) and a corresponding natural frequency (the associated eigenvalue). The assumption in this chapter is that the lowest natural frequency is still much higher than the typical frequencies in the aircraft dynamics, and only the mode shapes are used to construct a quasi-steady model for the aircraft elastic deformations. Information about their natural frequencies is not used.

Next, and with reference to Figure 5.3, we assume: (1) small amplitudes of those elastic displacements, so that the structure remains in the linear elastic regime, and (2) with slow enough changes for the structure to have a quasi-steady response. We further assume that the elastic displacements are (3) too small to change the instantaneous inertia characteristics of the aircraft, which are therefore considered to be constant, but (4) large enough to change the instantaneous aerodynamic and propulsion forces. This means that the position of the CM is approximately constant as the aircraft deforms, which is a characteristic, in particular, of transport aircraft with a stiff fuselage and moderately flexible wings. Under those conditions, Equation (4.51) needs to be modified to include the new dependencies with the elastic DoFs as

$$\dot{\mathbf{x}}_r = \mathbf{f}^{(r)}(\mathbf{x}_r, \mathbf{x}_s, \mathbf{u})$$
$$= \mathbf{f}^{(r)}_{\text{gyr}}(\mathbf{x}_r) + \mathbf{f}^{(r)}_{\text{grav}}(\mathbf{x}_r) + \mathbf{f}^{(r)}_{\text{aero}}(\mathbf{x}_r, \mathbf{x}_s, \mathbf{u}) + \mathbf{f}^{(r)}_{\text{prop}}(\mathbf{x}_r, \mathbf{x}_s, \mathbf{u}), \tag{5.4}$$

where we have added the superscript $\bullet^{(r)}$ to the forcing terms to indicate that they correspond to the force and moment resultants at the CM. The dependence of the resultant propulsive forces on the elastic DoF is included to consider the case of wing-mounted engines, where wing deformations may change the orientation of the thrust vector. The assumption on the inertia properties implies that the resultant gravitational forces in Equation (5.4) are still given by Equation (4.28) and that the location of the CM of the aircraft effectively remains unaltered by the deformations of the airframe. To simplify this description, we have not included the disturbance vector associated with the wind speed that appeared in Equation (4.51).

Note that the dimension of $\mathbf{x}_s$ can be much larger than that of $\mathbf{x}_r$, that is, $N_r \ll N_s$. This is because we only need $N_r = 8$ states, as we saw in Section 4.4.1, to describe the full (6-DoFs) rigid aircraft flight dynamics, but we may need to track the instantaneous position of thousands of nodes in a high-resolution finite-element model. To determine $\mathbf{x}_s$, we need the information of the spatial distribution of various external forces applied on the aircraft, as shown schematically in Figure 5.1b. This means that it is not sufficient to know the value of the total lift to estimate the bending of the wings, which needs instead more detailed information about the spanwise lift distribution (as well as about the stiffness of the wing structure). Under the small-amplitude assumption, the structural deformations under those distributed forces are determined by the stiffness matrix of the structure, $\mathbf{K}_{ss}$, which is assumed here to be known and invertible, that is, the structure has no rigid-body DoFs. In that case, the (instantaneous) static equilibrium on the structure can be written as

$$\mathbf{K}_{ss}\mathbf{x}_s = \mathbf{f}^{(s)}(\mathbf{x}_r, \mathbf{x}_s, \mathbf{u})$$
$$= \mathbf{f}^{(s)}_{\text{gyr}}(\mathbf{x}_r) + \mathbf{f}^{(s)}_{\text{grav}}(\mathbf{x}_r) + \mathbf{f}^{(s)}_{\text{aero}}(\mathbf{x}_r, \mathbf{x}_s, \mathbf{u}) + \mathbf{f}^{(s)}_{\text{prop}}(\mathbf{x}_r, \mathbf{x}_s, \mathbf{u}). \tag{5.5}$$

The terms on the right-hand side of Equation (5.5) are the discrete (either nodal or modal) forces generated by the gyroscopic, gravity, aerodynamic, and propulsion loading on the structure. This is sometimes referred to as the *static-elastic model* for the flexible effects (Schmidt, 2012, Ch. 7). In general, all the right-hand-side terms in Equation (5.5) may contribute to the deformation of the vehicle, but in many cases, only the aerodynamic forces are relevant (and the propulsion forces for wing-mounted engines). Equations (5.4) and (5.5) depend each on both $\mathbf{x}_r(t)$ and $\mathbf{x}_s(t)$ and, therefore, need to be solved simultaneously. This results in a coupled process that can schematically be represented by the feedback system shown in Figure 5.5. The structure and rigid-body blocks correspond to Equations (5.4) and (5.5), respectively. Note that if the focus is on the flight dynamic response of the flexible aircraft, the dotted line in Figure 5.5 defines a system whose inputs are the classical control surface deflections and engine inputs and whose outputs are the rigid-body states, that is, the same inputs and outputs that would be monitored in the flight dynamics of a rigid aircraft. However, the process of determining one from the other needs now to incorporate the structural properties of the flexible aircraft into the internal dynamics model.

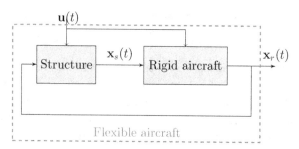

Figure 5.5 Rigid-elastic interactions in a flexible aircraft in still air.

The description so far has been generic, and Sections 5.5 and 5.7 show two particular but representative realizations of the coupled system defined by Equations (5.4) and (5.5). Before that, we will study the equilibrium condition (aeroelastic trim) and the linearized dynamics of a flexible aircraft that can be described by an elastified model.

5.3.2   Flexible Aircraft Trim

Trim equilibrium of a flexible aircraft is obtained from any set of inputs $\mathbf{u}_0$ and flight conditions for which the vehicle achieves steady-state values on all rigid and elastic DoFs, $\mathbf{x}_r(t) = \mathbf{x}_{r0}$ and $\mathbf{x}_s(t) = \mathbf{x}_{s0}$, respectively. This is referred to as *aeroelastic trim* and corresponds to identifying the fixed points of the coupled system defined by Equations (5.4) and (5.5), that is,

$$
\begin{aligned}
\mathbf{0} &= \mathfrak{f}^{(r)}(\mathbf{x}_{r0}, \mathbf{x}_{s0}, \mathbf{u}_0), \\
\mathbf{K}_{ss}\mathbf{x}_{s0} &= \mathfrak{f}^{(s)}(\mathbf{x}_{r0}, \mathbf{x}_{s0}, \mathbf{u}_0).
\end{aligned}
\tag{5.6}
$$

This expression is a generalization of Equation (4.53) for a flexible aircraft. As was done there, a nonzero (but constant) wind velocity can be considered in the trim equations, but this is not explicitly included here to simplify the description. The forcing terms on the right-hand side of Equation (5.6) include, in general, gyroscopic, gravity, aerodynamic, and propulsion forces. Similar to the rigid aircraft equilibrium, the gyroscopic terms are zero in forward-flight conditions. An example of flexible aircraft trim in forward flight is included in Section 5.6.

Stability axes on the trimmed flexible aircraft in forward flight are defined in the same manner as for rigid aircraft (Section 4.2.1), that is, with origin at the CM, the *x* axis along the inertial velocity vector at the CM, and the *y* axis perpendicular to the symmetry plane and positive toward the right wing. The *z* axis is perpendicular to the other two to form a right-handed frame of reference. While we have assumed that the location of the CM remains unaffected by flexibility effects, the vehicle angle of attack to achieve a certain lift is, in general, different between a rigid and flexible model of a given aircraft. This is exemplified in Figure 5.6, which shows the rigid (undeformed) aircraft and the flexible wing shapes calculated for a representative high-aspect-ratio-wing transport aircraft under three levels of deformation, which may be

Figure 5.6 Simulated equilibrium shapes for a transport aircraft in three different operation modes.

achieved with different mass cases. As the total mass increases, additional lift needs to be generated, which results in larger deformations in the wing. Indeed, for the highest loaded case in the figure, the assumption of linear structural deformations may no longer be valid. As the wing deforms, the aircraft geometry changes and so does, in general, the orientation of the stability axis in the symmetry plane of the aircraft with respect to the vehicle's undeformed geometry.

5.3.3 Linearized Equations

If the small-perturbation problem around a trimmed forward-flight condition is now considered, Equations (5.4) and (5.5) become the linear relations

$$\Delta \dot{\mathbf{x}}_r = \mathbf{A}_{rr} \Delta \mathbf{x}_r + \mathbf{A}_{rs} \Delta \mathbf{x}_s + \mathbf{B}_r \Delta \mathbf{u},$$
$$\mathbf{K}_{ss} \Delta \mathbf{x}_s = \mathbf{A}_{sr} \Delta \mathbf{x}_r + \mathbf{A}_{ss} \Delta \mathbf{x}_s + \mathbf{B}_s \Delta \mathbf{u}, \tag{5.7}$$

where the state and input matrices are obtained from the partitioned Jacobian of the nonlinear forcing terms, that is,

$$\mathbf{A}_{rr} = \left. \frac{\partial \mathbf{f}^{(r)}}{\partial \mathbf{x}_r} \right|_{(\mathbf{x}_{r0}, \mathbf{x}_{s0}, \mathbf{u}_0)} \qquad \mathbf{A}_{ss} = \left. \frac{\partial \mathbf{f}^{(s)}}{\partial \mathbf{x}_s} \right|_{(\mathbf{x}_{r0}, \mathbf{x}_{s0}, \mathbf{u}_0)}$$

$$\mathbf{A}_{rs} = \left. \frac{\partial \mathbf{f}^{(r)}}{\partial \mathbf{x}_s} \right|_{(\mathbf{x}_{r0}, \mathbf{x}_{s0}, \mathbf{u}_0)} \qquad \mathbf{A}_{sr} = \left. \frac{\partial \mathbf{f}^{(s)}}{\partial \mathbf{x}_r} \right|_{(\mathbf{x}_{r0}, \mathbf{x}_{s0}, \mathbf{u}_0)} \tag{5.8}$$

$$\mathbf{B}_r = \left. \frac{\partial \mathbf{f}^{(r)}}{\partial \mathbf{u}} \right|_{(\mathbf{x}_{r0}, \mathbf{x}_{s0}, \mathbf{u}_0)} \qquad \mathbf{B}_s = \left. \frac{\partial \mathbf{f}^{(s)}}{\partial \mathbf{u}} \right|_{(\mathbf{x}_{r0}, \mathbf{x}_{s0}, \mathbf{u}_0)}.$$

Matrices $\mathbf{A}_{rr}$ and $\mathbf{B}_r$ are defined in Equation (4.56) for the rigid aircraft, and they remain unaltered. All other matrices are associated with the structural effects on the aircraft dynamics. Equation (5.7) is a differential-algebraic equation that solves the incremental structural and rigid-body DoFs. Since the equations are linear, we can easily find a solution by first solving for $\Delta \mathbf{x}_s$ in the structural equation

$$\Delta \mathbf{x}_s = (\mathbf{K}_{ss} - \mathbf{A}_{ss})^{-1} (\mathbf{A}_{sr} \Delta \mathbf{x}_r + \mathbf{B}_s \Delta \mathbf{u}), \tag{5.9}$$

which is then substituted back into the first part of Equation (5.7). This process is known as the *residualization* of the coupled equations, and it is equivalent to including the structural effects as a constant feedback (a static gain) on the rigid aircraft equations, as has been represented in the block diagram of Figure 5.5. The result is a set of modified system dynamics equations of the form

$$\Delta \dot{\mathbf{x}}_r = \check{\mathbf{A}}_r \Delta \mathbf{x}_r + \check{\mathbf{B}}_r \Delta \mathbf{u}, \tag{5.10}$$

with

$$\begin{aligned}
\check{\mathbf{A}}_r &= \mathbf{A}_{rr} + \mathbf{A}_{rs} \left(\mathbf{K}_{ss} - \mathbf{A}_{ss} \right)^{-1} \mathbf{A}_{sr}, \\
\check{\mathbf{B}}_r &= \mathbf{B}_r + \mathbf{A}_{rs} \left(\mathbf{K}_{ss} - \mathbf{A}_{ss} \right)^{-1} \mathbf{B}_s.
\end{aligned} \tag{5.11}$$

Matrices $\check{\mathbf{A}}_r$ and $\check{\mathbf{B}}_r$ are referred to as the *elastified* (sometimes also called as *flexibilized* (Karpel and Sheena, 1989)) system and input matrices and describe the linearized dynamics of a flexible aircraft with quasi-static deformations. Therefore, this elastified description retains the structure of the rigid aircraft flight-dynamics problem, but with modified coefficients that also depend on the flight condition. Equation (5.10) still has dimension N_r, and it is in effect the mathematical representation of the flexible aircraft shown by the dotted block in Figure 5.5. It is important to note that one actually only needs Equation (5.5) to be linear to be able to eliminate the structural DoFs in Equation (5.4), which means that the fully nonlinear flight-dynamics equations of a rigid aircraft can be updated in this manner. This situation typically occurs under small-displacement assumptions, attached-flow conditions, and sufficiently slow aerodynamics.

The dynamic stability of a flexible aircraft is determined by the (in general, complex) eigenvalues of $\check{\mathbf{A}}_r$, with the corresponding eigenvectors being the flexible-aircraft flight dynamic modes. Therefore, under the quasi-steady assumption of the elastic DoFs, there are still $N_r = 8$ internal states and, therefore, modes on the aircraft. Recall from Chapter 4 that three of those modes (phugoid, short period, and Dutch roll) normally appear as complex pairs, with the final two being the roll subsidence and spiral modes. They are still present on flexible aircraft, but their characteristics now depend on both the rigid-body and the aeroelastic properties of each vehicle. Finally, we can add an output equation to Equation (5.10), which is similarly written as

$$\Delta \mathbf{y} = \check{\mathbf{C}}_r \Delta \mathbf{x}_r + \check{\mathbf{D}}_r \Delta \mathbf{u}. \tag{5.12}$$

The information that is included in the output vector $\Delta \mathbf{y}$ depends on the available sensors for the control system or on the interests of the analyst. Since Equation (5.10) "internally" solves for the structural DoFs through the second part of Equation (5.7), the output vector can include structural (e.g., wing loads) as well as rigid-body information.

5.3.4 Aeroelastic Effects on the Aerodynamic Forces

The description in Sections 5.3.1 to 5.3.3 has kept a generic description of the distributed forces acting on the aircraft. As was seen for the rigid aircraft in Section 4.3.5, the aerodynamic forces have some basic dependencies that are convenient to introduce early in the analysis. In particular, and as for the rigid aircraft, the instantaneous distributed aerodynamic forces on the aircraft depend on the instantaneous vehicle geometry, the flight operating point (defined by the Mach and Reynolds numbers), and the instantaneous relative velocity and control deflection. The difference here is that the aircraft geometry also changes with time according to the structural state, $\mathbf{x}_s(t)$, and by extension as well as the position of the control surfaces in the deforming aircraft (noting that nonlinear couplings may arise from their combined effect). That dependency appears in the integral (lift and moment), and the distributed aerodynamic loads in still air can then be written as

$$\mathbf{f}_{aB} = \tfrac{1}{2}\rho V^2 S \mathbf{c}_{fB}\left(\mathbf{v}_B, \boldsymbol{\omega}_B, \boldsymbol{\delta}_c, \mathbf{x}_s; \mathrm{Ma}, \mathrm{Re}\right),$$

$$\mathbf{m}_{aB} = \tfrac{1}{2}\rho V^2 S \Lambda_{\mathrm{ref}} \mathbf{c}_{mB}\left(\mathbf{v}_B, \boldsymbol{\omega}_B, \boldsymbol{\delta}_c, \mathbf{x}_s; \mathrm{Ma}, \mathrm{Re}\right), \qquad (5.13)$$

$$\mathbf{f}_{\mathrm{aero}}^{(s)} = \tfrac{1}{2}\rho V^2 S \mathbf{c}_f\left(\mathbf{v}_B, \boldsymbol{\omega}_B, \boldsymbol{\delta}_c, \mathbf{x}_s; \mathrm{Ma}, \mathrm{Re}\right),$$

where we have normalized all coefficients with the reference dynamic pressure, $\tfrac{1}{2}\rho V^2$, and a reference area, S. The first equations in (5.13) have already appeared for the rigid aircraft in Equation (4.35), with the difference that here we are restricting the description to still air, for simplicity. The generalized forces resulting from the aerodynamic loading on the structural DoFs, $\mathbf{f}_{\mathrm{aero}}^{(s)}$, are also proportional to the dynamic pressure and, therefore, their coefficients have also been normalized in Equation (5.13). We have also normalized them here with the reference area for consistency, although this is not always done in practice as $\mathbf{c}_f$ may still be a dimensional coefficient. Now we consider perturbations around a reference condition, defined by the air density ρ and the forward-flight speed V, for which the force coefficients take, in general, nonnegative values $\mathbf{c}_{f0}$, $\mathbf{c}_{m0}$, and $\mathbf{c}_{f0}$. The corresponding linearized expressions now become

$$\Delta\mathbf{f}_{aB} = \tfrac{1}{2}\rho V^2 S\left(\Delta\mathbf{c}_{fB} + 2\mathbf{c}_{f0}\frac{\Delta v_x}{V}\right),$$

$$\Delta\mathbf{m}_{aB} = \tfrac{1}{2}\rho V^2 S\Lambda_{\mathrm{ref}}\left(\Delta\mathbf{c}_{mB} + 2\mathbf{c}_{m0}\frac{\Delta v_x}{V}\right), \qquad (5.14)$$

$$\Delta\mathbf{f}_{\mathrm{aero}}^{(s)} = \tfrac{1}{2}\rho V^2 S\left(\Delta\mathbf{c}_f + \mathbf{c}_{f0}\frac{\Delta v_x}{V}\right).$$

The incremental force coefficient can finally be written as a linear combination of the incremental rigid and elastic states and the inputs. This results in incremental forces of the form

$$\left\{\begin{array}{c}\Delta\mathbf{f}_{aB}\\\Delta\mathbf{m}_{aB}\end{array}\right\} = \rho V \mathcal{A}_{rr}\left\{\begin{array}{c}\Delta\mathbf{v}_B\\\Delta\boldsymbol{\omega}_B\end{array}\right\} + \tfrac{1}{2}\rho V^2 \mathcal{A}_{rs}\Delta\mathbf{x}_s + \tfrac{1}{2}\rho V^2 \mathcal{A}_{ru}\Delta\boldsymbol{\delta}_c,$$

$$\qquad (5.15)$$

$$\Delta\mathbf{f}^{(s)} = \rho V \mathcal{A}_{sr}\left\{\begin{array}{c}\Delta\mathbf{v}_B\\\Delta\boldsymbol{\omega}_B\end{array}\right\} + \tfrac{1}{2}\rho V^2 \mathcal{A}_{ss}\Delta\mathbf{x}_s + \tfrac{1}{2}\rho V^2 \mathcal{A}_{su}\Delta\boldsymbol{\delta}_c.$$

For rigid aircraft flight simulation and analysis, the contribution of the aerodynamic and propulsion forces to the linearized equations, $\mathcal{A}_{rr}$ and $\mathcal{A}_{ru}$, is known in the form of tabulated aerodynamic derivatives that depend on the flight condition. Their evaluation was shown in Equation (4.47). For a flexible aircraft, that database of derivatives is further *elastified* with the addition of the matrices $\mathcal{A}_{sr}$, $\mathcal{A}_{rs}$, $\mathcal{A}_{ss}$, and $\mathcal{A}_{su}$ that are the *AICs* related to the elastic DoFs. These aeroelastic AICs also depend, in general, on the trim condition that may vary at each point in the flight envelope (altitude and Mach number). Some general methods for evaluating these coefficients are discussed in Section 6.5.1, while simple models are used in the two examples discussed later in this chapter. When the EoMs can be residualized as in Equation (5.10), explicit knowledge of the aeroelastic AICs is not strictly required. In that case, it is more convenient instead to directly tabulate a set of *modified derivatives* of the rigid-body problem (Milne, 1962) that implicitly include aeroelastic dependencies.

5.4   Aeroelastic Divergence and Control Reversal

The elastified EoMs introduced in Equation (5.10) allow now for a more general definition of some static aeroelastic concepts that were introduced in Chapter 3 for the 2-DoF airfoil problem. In particular, a general definition of aeroelastic divergence and control reversal of a flexible aircraft can now be defined. They are derived, respectively, from the static stability and controllability of the linear system describing the flexible aircraft dynamics.

5.4.1   Divergence

Equation (5.11) describes the flexible aircraft dynamics under quasity-steady deformations using only the rigid-body DoFs. That has required the prior residualization of the elastic DoFs in Equation (5.9), which needs the inversion of $(\mathbf{K}_{ss} - \mathbf{A}_{ss})$. Note that in our model of the aircraft, only the aerodynamic and propulsion contributions to the distributed forces on the airframe $\mathfrak{f}^{(s)}$ depend on the structural states and, therefore, contribute to the Jacobian $\mathbf{A}_{ss}$, Equation (5.8). Propulsion derivatives have a minor effect, and they are often excluded from the analysis. Thus, the existence of aeroelastic equilibrium implies that matrix $\left(\mathbf{K}_{ss} - \frac{1}{2}\rho V^2 \mathcal{A}_{ss} \right)$ cannot be singular. A singular matrix is associated with a static instability (a loss of effective wing stiffness), and that condition is equivalent to zeroing the determinant of the matrix, that is,

$$\det \left(\mathbf{K}_{ss} - \tfrac{1}{2}\rho V_D^2 \mathcal{A}_{ss} \right) = 0. \tag{5.16}$$

The smallest positive root, if it exists, of the resulting characteristic polynomial defines the dynamic pressure $\frac{1}{2}\rho V_D^2$ at which the aircraft becomes statically unstable. This condition is known as aeroelastic *divergence* and sometimes also called *torsional divergence* because of its strong dependence on wing torsional stiffness. Moreover, since the structural DoFs have been defined such that they do not include any rigid-body motions, $\mathbf{K}_{ss}$ is not singular, and the loss of static stability needs to occur for a nonzero dynamic pressure. Note that real positive roots may not exist and, therefore, it is theoretically possible to find divergence-free aircraft configurations (although in practice this is only necessary within the flight envelope).

In general, the matrix $\mathcal{A}_{ss}$ changes with the flight condition (e.g., through the Mach number). Therefore, once a solution for the dynamic pressure is obtained from Equation (5.16), one needs to iterate this equation by updating the derivatives to the estimated divergence condition, which ensures that the system is linearized about the unstable flight condition. This is referred to, in general, as aircraft stability analysis with *matched flight conditions*. It is also sometimes convenient to seek *unmatched* solutions in which the stability conditions are sought at a fixed reference point without updating $\mathcal{A}_{ss}$. Unmatched solutions are usually found when divergence may occur at a different flow regime than the nominal condition (e.g., on transonic aircraft, for which the instability may occur at supersonic speeds) and would therefore require a different modeling approach. In that case, the difference between the reference dynamic pressure and the estimated divergence dynamic pressure should be interpreted as a safety margin on the static stability of the vehicle.

The definition of divergence introduced here is a generalization of that used for the 2-DoF aeroelastic system studied in Section 3.5.1. As was discussed there, the loss of effective stiffness associated with divergence typically brings catastrophic structural failure and vehicle loss. Consequently, current aircraft certifications require that all points in the flight envelope have at least a 15% safety margin with respect to any predicted divergence speed. The solutions for the 2-DoF problem are also relevant for the wing problem. In particular, the key design parameters that determine wing divergence are still the torsional stiffness of the wing and the distance between the *elastic axis* and the aerodynamic center of the wing sections. This is exemplified in a simplified wing model in Section 5.6.

Finally, it should be noted that, even though divergence has been defined from a singularity condition in the static solution of the coupled aeroelastic and flight-dynamics equations, its mathematical definition only depends on the interaction between structural stiffness, $\mathbf{K}_{ss}$, and the aerodynamic forces due to elastic deformations, $\mathcal{A}_{ss}$. Consequently, aircraft divergence analyses are routinely carried out on bespoke static aeroelastic models. They are used to identify divergence-free flight conditions for which more involved flexible aircraft dynamics analyses are then performed.

5.4.2 Aircraft Controllability and Control Reversal

We consider next a second relevant problem that arises from Equation (5.10). It is related to any potential changes to the *controllability* characteristics of the aircraft with the available control effectors, typically control surface deflections and thrust settings. In control theory, controllability of a linear dynamical system is achieved when the available inputs can drive the internal state of the system between any two finite values in a finite time. On aircraft this implies, for example, that any desired velocity at the

CM can be achieved by choosing sufficiently large control inputs during sufficiently long times. Theoretically, under linear assumptions, controllability would ensure that any desired state is achievable, while in reality this is, of course, limited by the actual physical constraints of the system.

Consider then states at two different times, that is, $\Delta \mathbf{x}_r(t_0)$ and $\Delta \mathbf{x}_r(t_f)$, of a system with dynamics as in Equation (5.10). The linear system is controllable if there is a set of inputs $\Delta \mathbf{u}^*(t)$ that can take the system from one state to the other in the given time interval. Using Equation (1.11) with initial conditions at $t = t_0$, the previous condition can be written as

$$\Delta \mathbf{x}_r(t_f) = e^{\breve{\mathbf{A}}_r(t_f - t_0)} \Delta \mathbf{x}_r(t_0) + \int_{t_0}^{t_f} e^{\breve{\mathbf{A}}_r(t_f - t)} \breve{\mathbf{B}}_r \Delta \mathbf{u}^*(t) \, dt. \tag{5.17}$$

Multiplying Equation (5.17) by $e^{-\breve{\mathbf{A}}_r t_f}$, we have

$$e^{-\breve{\mathbf{A}}_r t_f} \Delta \mathbf{x}_r(t_f) - e^{-\breve{\mathbf{A}}_r t_0} \Delta \mathbf{x}_r(t_0) = \int_{t_0}^{t_f} e^{-\breve{\mathbf{A}}_r t} \breve{\mathbf{B}}_r \Delta \mathbf{u}^*(t) \, dt. \tag{5.18}$$

Next, from the Cayley–Hamilton theorem (Horn and Johnson, 2012, Ch. 2), the state-transition matrix (the exponential map) can be written as the sum of N_r terms of the form $e^{-\breve{\mathbf{A}}_r t} = \sum_{n=0}^{N_r-1} a_n(t) \breve{\mathbf{A}}_r^n$, where $a_n(t)$ are known (scalar) functions. Substituting into the right-hand side of Equation (5.18) gives

$$e^{-\breve{\mathbf{A}}_r t_f} \Delta \mathbf{x}_r(t_f) - e^{-\breve{\mathbf{A}}_r t_0} \Delta \mathbf{x}_r(t_0) = \sum_{n=0}^{N_r-1} \breve{\mathbf{A}}_r^n \breve{\mathbf{B}}_r \mathbf{b}_n^*, \tag{5.19}$$

with $\mathbf{b}_n^* = \int_{t_0}^{t_f} a_n(t) \Delta \mathbf{u}^*(t) \, dt$. For a given initial and final state, and time interval, Equation (5.19) defines N_r linear equations with $N_r N_u$ unknowns: the set $\mathbf{b}_n^*$, for $n = 0, \dots, N_r - 1$. This problem can be solved if and only if the matrix of coefficients on the right-hand side of Equation (5.19) is of full row rank. This matrix is written as

$$\mathcal{C} = \begin{bmatrix} \breve{\mathbf{B}}_r & \breve{\mathbf{A}}_r \breve{\mathbf{B}}_r & \breve{\mathbf{A}}_r^2 \breve{\mathbf{B}}_r & \cdots & \breve{\mathbf{A}}_r^{N_r-1} \breve{\mathbf{B}}_r \end{bmatrix}, \tag{5.20}$$

which is known as the *controllability matrix* of the system defined by Equation (5.10). Therefore, a (flexible) aircraft is controllable if $\text{rank}(\mathcal{C}) = N_r$, which defines requirements on actuator selection. For rigid aircraft, it can be shown that controllability can be ensured with a conventional wing–fuselage–tail configuration using the standard aileron, rudder, elevator, and engine thrust inputs (Stengel, 2004, Ch. 5–6). The full-rank condition on the controllability matrix gives a necessary condition for acceptable aircraft flight performance, although it is not a sufficient one. It provides essential information about the type of actuation that may be required, but it does not provide a quantitative metric to size those actuators. This information is given by the *controllability Gramian*, which will be introduced in Section 7.4.1.

As we have seen in Figure 5.5, the dynamics of flexible aircraft under quasi-steady elastic assumptions include static aeroelastic effects as a static gain with the freestream dynamic pressure as a scaling parameter. This changes, first, the dynamic stability properties of the vehicle with respect to those of a rigid aircraft (i.e., the eigenvalues of the closed-loop system defined between the rigid and flexible aircraft character-istics as in Figure 5.5), and second, the controllability characteristics of the flexible aircraft, which are established through the analysis of Equation (5.20). Loss of aircraft controllability driven by the aeroelastic feedback implies that at a sufficiently high dynamic pressure, certain rigid-body DoFs (e.g., aircraft roll) are not affected by the control inputs (e.g., aileron deflection). This typically comes associated with a change of sign in the response of the uncontrollable states below and above the uncontrollable condition. In the example of aileron reversal, which is the most common situation (see Section 5.7), a positive aileron deflection generates positive roll below a certain dynamic pressure, and negative ones about it. Consequently, this situation is known as *control reversal*, and the dynamic pressure at which this sign switch occurs is the *control reversal dynamic pressure*.

5.5   Longitudinal Dynamics of an Aircraft with a Flexible Fuselage

As a first example, consider the longitudinal dynamics of an aircraft in a steady climb with a rigid wing and tail but with a flexible fuselage. The linear dynamics of this problem can be described in a first approximation as in Section 4.6, with the effect of the flexible fuselage modeled by a rotational spring at a distance l_β from the tail, as shown in Figure 5.7. This results in an additional DoF that is selected here as the rotation angle, β, of the tail section with respect to the main fuselage. This angle is assumed to be sufficiently small for a linear solution to this problem to be still valid.

If the natural frequency of the fuselage vibrations is much higher than the typical frequencies in the rigid-body dynamics, Equations (4.77) and (4.78) are still valid for this problem. However, the instantaneous lift on the tail is also affected now by the tail rotation, and consequently Equations (4.82) and (4.85), and (4.87) become

$$\Delta C_L = C_{L\alpha}\frac{\Delta v_z}{V} + \frac{S_t}{S}C_{L\alpha}^t\left(\frac{l_t\Delta\omega_y}{V} - \beta\right) + \frac{S_t}{S}C_{L\delta}^t\Delta\delta_e,$$

$$\Delta C_D = \frac{2SC_{L0}C_{L\alpha}}{\pi b^2 e}\frac{\Delta v_z}{V},$$

$$\Delta C_{M_y} = -\left(\frac{l_w}{c_{\text{ref}}}C_{L\alpha}^w + \frac{S_t l_t}{Sc_{\text{ref}}}C_{L\alpha}^t\right)\frac{\Delta v_z}{V} - \frac{S_t l_t}{Sc_{\text{ref}}}C_{L\alpha}^t\left(\frac{l_t\Delta\omega_y}{V} - \beta\right) - \frac{S_t l_t}{Sc_{\text{ref}}}C_{L\delta}^t\Delta\delta_e,$$

$$(5.21)$$

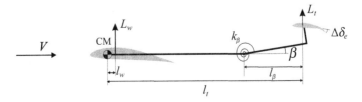

Figure 5.7 Geometry of an aircraft with a flexible hinge in the fuselage.

where, as in Section 4.6, v_z and ω_y are the normal and angular velocities of the CM, respectively, and $C_L = C_L^t + C_L^w$. If we further assume that $\beta = 0$ at the equilibrium condition (the aircraft trimmed for level flight or a steady climb), then the fuselage rotation in the subsequent aircraft dynamics is determined by the instantaneous equilibrium of moments at the hinge, that is,

$$k_\beta \beta = l_\beta \Delta L_t = \frac{1}{2} \rho V^2 S_t l_\beta \left[C_{L\alpha}^t \left(\frac{\Delta v_z}{V} + \frac{l_t \Delta \omega_y}{V} - \beta \right) + C_{L\delta}^t \Delta \delta_e \right]. \tag{5.22}$$

This results in the following EoMs for the flexible aircraft:

$$\dot{\mathbf{x}}_r = \mathbf{A}_r \mathbf{x}_r + \mathbf{A}_{rs} \beta + \mathbf{B}_r \Delta \delta_e, \tag{5.23a}$$

$$k_\beta \beta = \mathbf{A}_{sr} \mathbf{x}_r + A_s \beta + B_s \Delta \delta_e, \tag{5.23b}$$

with $\mathbf{x}_r = \left\{ \Delta v_x \quad \Delta v_z \quad \Delta \omega_y \quad \Delta \theta \right\}^{\mathsf{T}}$, $\mathbf{A}_r$ and $\mathbf{B}_r$ still given by Equations (4.94) and (4.95), respectively, and with the new coefficients being

$$A_s = -\frac{1}{2} \rho V^2 S_t l_\beta C_{L\alpha}^t,$$

$$\mathbf{A}_{rs} = \frac{\rho V^2 S_t C_{L\alpha}^t}{2m} \begin{bmatrix} 0 & 1 & \frac{l_t m}{I_{yy}} & 0 \end{bmatrix}^{\mathsf{T}},$$

$$\mathbf{A}_{sr} = \frac{1}{2} \rho V S_t l_\beta C_{L\alpha}^t \begin{bmatrix} 0 & 1 & l_t & 0 \end{bmatrix},$$

$$B_s = \frac{1}{2} \rho V S_t l_\beta C_{L\delta}^t. \tag{5.24}$$

As has been seen in Section 5.3.3, the *elastification* of the solution consists of implicitly solving the elastic DoFs and substituting into the rigid-body equations, as in Equation (5.10). Here, this means the elimination of β in Equation (5.23) by solving them as a function of the rigid-body states and inputs in Equation (5.23b). Equation (5.23a) finally results in

$$\dot{\mathbf{x}}_r = \left(\mathbf{A}_r + \frac{1}{k_\beta - A_s} \mathbf{A}_{rs} \mathbf{A}_{sr} \right) \mathbf{x}_r + \left(\mathbf{B}_r + \frac{B_s}{k_\beta - A_s} \mathbf{A}_{rs} \right) \Delta \delta_e. \tag{5.25}$$

Since $k_\beta - A_s = k_\beta + \frac{1}{2} \rho V^2 S_t l_\beta C_{L\alpha}^t > 0$, there is no divergence at any airspeed. Note, however, that if l_β were negative (e.g., the tail was in a canard configuration), then aeroelastic divergence could be found if the fuselage was sufficiently flexible. A numerical example is included next.

Example 5.1 Longitudinal Stability of a Glider with a Flexible Fuselage   Consider again the glider problem that was introduced in Example 4.1 to illustrate the longitudinal dynamics of a rigid aircraft. A flexible aircraft is considered now by including the bending compliance of the fuselage using a rotational spring as shown in Figure 5.7. The spring is assumed to be at the midpoint between the wing's and tail's aerodynamic centers, $l_\beta = \frac{l_w + l_t}{2}$, and the spring constant is nondimensionalized as $\sigma = \frac{k_\beta}{\frac{1}{2} \rho V^2 S_t l_t}$. Substituting the problem coefficients into Equation (5.25) gives an *elastified* state matrix

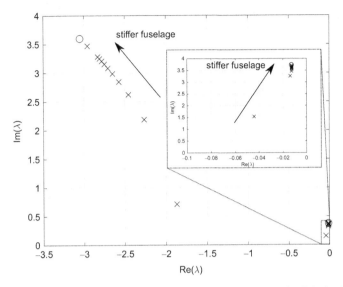

Figure 5.8 Longitudinal eigenvalues of the glider of Example 4.1 with a flexible fuselage, as a function of k_β. [05Quasisteady/gliderflex.m]

of the form

$$
\check{\mathbf{A}}_r =
\begin{bmatrix}
-0.02 & 0.28 & 0 & -9.81 \\
-0.65 & -2.05 & 28.81 & 0.34 \\
0 & -0.48 & -4.06 & 0 \\
0 & 0 & 1 & 0
\end{bmatrix}
+ \frac{1}{\sigma + 2.29}
\begin{bmatrix}
0 \\
0.15 \\
0.51 \\
0
\end{bmatrix}
\begin{bmatrix}
0 & 3.92 & 18.16 & 0
\end{bmatrix},
$$

that is, the longitudinal flexible aircraft response is still obtained from the solution of the dynamical system of dimension 4, with a modified state matrix that depends on the flexibility of the fuselage, σ. As expected for a rigid fuselage, we have $\sigma \to \infty$, and the second term in $\check{\mathbf{A}}_r$ vanishes.

The eigenvalues of the system with positive imaginary part[2] are shown in Figure 5.8 as a function of σ, with the circle representing the rigid aircraft modes (the short period and phugoid modes identified in Example 4.1) and the crosses the solutions for the values of σ between 2 and 20, in increments of 2, as well as $\sigma = 50$. As the stiffness of the fuselage decreases, the characteristic frequency of the short period and phugoid modes also drops. Moreover, the phugoid mode becomes more damped, while the short-period oscillations have an increasingly smaller damping. Note, however, that as the fuselage stiffness decreases, its fundamental resonant frequency, which is not included in this model, also lowers. Consequently, below a certain value of k_β, the assumption of quasi-steady deformations is no longer valid, and we would need to consider the more general formulation of Chapter 6.

---

[2] Their complex conjugates are also eigenvalues but are not shown.

5.6   Aeroelastic Trim of Aircraft with Flexible Straight Wings

Consider next the case of an aircraft with flexible wings of a high aspect ratio (i.e., the ratio between its semispan, b, and its mean aerodynamic chord) and no sweep (i.e., their spanwise direction is perpendicular to the flight speed, V). In this section, we are interested in its trim equilibrium, for which we first introduce simple models to represent the structure of the wing, as well as its loading.

5.6.1   Structural Model

A high-aspect-ratio wing structure can be modeled in a first approximation as a conventional (prismatic and isotropic) beam. Moreover, assuming the beam reference line to be the *elastic axis*, the torsion and bending responses of the wing become decoupled. The y axis is then chosen to coincide with the beam elastic axis to describe the wing kinematics, as shown in Figure 5.9. The resulting torsional model is known as a *torque tube*. Furthermore, and in order to take account of the fact that wings are normally stiffer at the root than at the tip, the torsional stiffness $GJ(y)$ is allowed to vary along the wing semispan, b. If the wing is subjected to a torsional moment per unit length $\tau(y)$, which will be due to the aerodynamic lift–moment distribution along the wing (and, in general, also due to the propulsion forces of wing-mounted engines), then the torsional equilibrium of the wing is determined by the differential equation (Hodges and Pierce, 2014, Ch. 2),

$$\frac{\mathrm{d}}{\mathrm{d}y}\left(GJ(y)\frac{\mathrm{d}\theta}{\mathrm{d}y}\right)+\tau(y)=0, \tag{5.26}$$

with $0 < y < b$ and $\theta(y)$ being the elastic torsional angle defined such that a positive elastic torsion pitches up the right wing. Equation (5.26) needs to be solved together with the boundary conditions of the problem, which, for a cantilever wing, are

$$\theta(0) = 0,$$
$$\theta'(b) = 0. \tag{5.27}$$

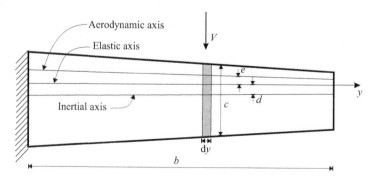

Figure 5.9 Planform view of a flexible cantilever wing.

5.6.2   Loads Model

Let $c(y)$ be the wing chord as a function of the span. We assume that aerodynamic forces are given by 2-D *strip theory* – that is, aerodynamic forces depend only on the local airfoil geometry and the local angle of incidence – and that the spanwise aerodynamic effects (in particular, wingtip loss) can be neglected.

The spanwise coordinate y is measured along the elastic axis defined in Section 3.2. Aerodynamic information on the airfoils is given at their aerodynamic center, which defines the aerodynamic axis at a distance $e(y)$ ahead of the elastic axis. The aerodynamic force distribution depends on the value of the local angle of attack along the wing, $\alpha(y)$. For a flexible wing, the total angle of attack comes from the superposition of the rigid (initial) angle of attack of each section, $\alpha_r(y)$, and the elastic twist angle along the wing axis $\theta(y)$, that is, $\alpha = \alpha_r + \theta$. Furthermore, the rigid angle of attack is given by the sum of the aircraft angle of attack, α_r^o, which depends on the pilot input, and the wing root mount angle and distributed pretwist along the span, $\alpha_r^p(y)$, which is a geometric property of the wing.

On the structural model, we consider the twist moment created by the aerodynamic forces on the wing sections, as well as the twist moment generated by the distributed mass, $m(y)$, of the wing subjected to the gravitational acceleration, g. The sectional centers of gravity are at a distance $d(y)$ behind the elastic axis and define the inertial axis in Figure 5.9. For a straight wing with the strip-theory assumption, and reference dynamic pressure $q = \frac{1}{2}\rho V^2$, we have

$$\tau(y) = q \left(e c c_l + c^2 c_{m_o} \right) + nmgd, \tag{5.28}$$

where $c_l(y) = c_{l_\alpha}(y)\left(\alpha_r(y) + \theta(y)\right)$ is the airfoil lift, and n is the load factor, which has been defined in Section 4.5.2. Equation (5.26) is then rewritten as

$$\frac{d}{dy}\left(GJ(y)\frac{d\theta}{dy} \right) + q e c c_{l_\alpha}\theta = -nmgd - qc\left(e c_{l_\alpha}\alpha_r + c c_{m_o} \right). \tag{5.29}$$

For given flight conditions n, α_r^o, and q, this differential equation determines the elastic twist $\theta(y)$ along the span. This defines the aeroelastic equilibrium conditions for the wing at the current flight conditions. Note that the equilibrium problem has been written in terms of a continuous function for the wing elastic twist, unlike the general solution in Equation (5.7). The discrete form of this equation is presented, for a slightly more general case, in Section 5.7.3.

5.6.3   Aeroelastic Equilibrium for Uniform Wings

If the wing has uniform cross sections, then all wing coefficients in Equation (5.29) have constant values. We also assume that $e > 0$, that is, the aerodynamic axis is ahead of the elastic axis, a typical situation in aircraft wings. Then Equation (5.29) can be written as

$$\frac{d^2\theta}{dy^2} + \lambda^2\theta = -\lambda^2\theta_0, \tag{5.30}$$

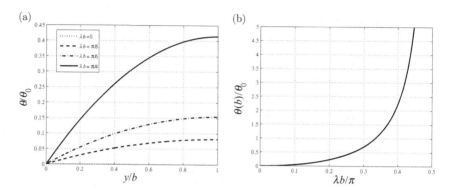

Figure 5.10 Twist for a straight wing with constant properties. (a) Twist along the span and (b) wingtip twist.

with

$$\lambda^2 = \frac{q e c c_{l_\alpha}}{GJ} \quad \text{and} \quad \theta_0 = \alpha_r + \frac{c_{m_o}}{c_{l_\alpha}} \frac{c}{e} + \frac{nmg}{q e c_{l_\alpha}} \frac{d}{c}. \tag{5.31}$$

Note that all terms in θ_0 are typically positive numbers. The solutions to this differential equation are of the form $\theta = -\theta_0 + A \sin \lambda y + B \cos \lambda y$, where the coefficients A and B are evaluated from the boundary conditions. This results in $A = \theta_0 \tan \lambda b$, $B = \theta_0$, and an elastic twist along a wing of a constant cross section given by

$$\theta = \theta_0 (\cos \lambda y + \tan \lambda b \sin \lambda y - 1). \tag{5.32}$$

This function is shown in Figure 5.10a for increasing λb. From Equation (5.31), the value of the nondimensional constant λb increases with the dynamic pressure, q. As is shown in Figure 5.10b, the wing-twist deformation grows with speed into an asymptotic value, in which there is no possible equilibrium between the elastic and aerodynamic forces. This defines aeroelastic *divergence* as the smallest positive dynamic pressure for which this occurs. From Equation (5.32), this is obtained with $\lambda b = \pi/2$, and thus

$$q_D = \frac{\pi^2}{4} \frac{GJ}{e c c_{l_\alpha} b^2}. \tag{5.33}$$

Wing divergence strongly depends on the offset between the aerodynamic and elastic axes, which should remain as small as possible if static aeroelastic stability is a concern.

Aeroelastic Derivatives

Assuming that lift is generated only at the wings, the total lift on the aircraft is

$$L = 2q \int_0^b c c_{l_\alpha} (\alpha_r + \theta) \, dy, \tag{5.34}$$

with $\theta(y)$ given by Equation (5.32) and, therefore, having linear dependency on the aircraft angle of attack, α_r^o. As always in this book, C_L denotes the lift coefficient of the aircraft, while $c_l(y)$ is the sectional coefficient of the wing airfoils. The lift derivative with respect to the aircraft angle of attack, that is, the change of lift coefficient of the flexible aircraft with changes on α_r^o, becomes

$$
\begin{aligned}
\frac{\mathrm{d}C_L}{\mathrm{d}\alpha_r^o} &= \frac{\partial C_L}{\partial \alpha_r^o} + \frac{\partial C_L}{\partial \theta}\frac{\partial \theta}{\partial \alpha_r^o} \\
&= c_{l_\alpha} + \frac{1}{b}\int_0^b c_{l_\alpha}\left(\cos\lambda y + \tan\lambda b\sin\lambda y - 1\right)\mathrm{d}y = c_{l_\alpha}\frac{\tan\lambda b}{\lambda b}.
\end{aligned}
\tag{5.35}
$$

At speeds sufficiently below the divergence speed, we can approximate this solution by a second-order Taylor expansion of the form $C_{L\alpha} \approx c_{l_\alpha}\left[1 + \frac{1}{3}(\lambda b)^2\right]$. From the definition of the dimensionless dynamic pressure λ in Equation (5.31), this relation implies that at sufficiently low speeds, the elastified derivative increases linearly with the dynamic pressure with respect to the rigid aircraft derivative, c_{l_α}. Note finally that, as the aircraft approaches the divergence speed, $\lambda b = \pi/2$, the aeroelastic derivative given by Equation (5.35) tends to infinity.

Relation between Flight Speed, Load Factor, and Aircraft Angle of Attack

We have assumed so far that, when the flight conditions are given, the dynamic pressure, q, the load factor, n, and the aircraft angle of attack, α_r^o, are known. However, only two of these magnitudes are actually independent in steady flight. To see this, consider that the weight of the aircraft is

$$
W = W_{\text{nonlift}} + 2\int_0^b mg\,\mathrm{d}y,
\tag{5.36}
$$

where W_{nonlift} is the weight of the nonlifting surfaces (fuselage, payload, engines, etc.). The lift on the aircraft, Equation (5.34), needs to satisfy $L = nW$ for a given load factor, as was seen in Section 4.5.2. As a result, the angle of attack of the aircraft needs to satisfy

$$
\alpha_r^o = \frac{nW}{2qcc_{l_\alpha}b} - \frac{1}{b}\int_0^b (\alpha_r^p + \theta)\,\mathrm{d}y.
\tag{5.37}
$$

Substituting the expression for the spanwise distribution of elastic twist given in Equation (5.32), we finally obtain that, for $q < q_D$,

$$
\alpha_r^o = \frac{nW}{2qcc_{l_\alpha}b} + \theta_0\left(\frac{\tan\lambda b}{\lambda b} - 1\right) - \frac{1}{b}\int_0^b \alpha_r^p\,\mathrm{d}y.
\tag{5.38}
$$

This expression determines the angle of attack, α_r^o, of an aircraft that enables it to fly at a given point in the flight envelope (altitude, speed) and on a given steady maneuver defined by a load factor n.

5.7 Roll Control of Flexible Straight Wings

Roll dynamics of aircraft with flexible wings is a typical situation in which aeroelasticity affects the vehicle's dynamic characteristics. This section investigates this scenario on a simple roll dynamics model of an aircraft with high-aspect-ratio straight wings. The wings are defined and modeled as in Section 5.6, with the addition of ailerons. The aircraft is initially in level flight with constant forward-flight speed, V. Its linearized dynamics about that reference are determined by the lateral dynamics in Equation (4.70) with $\theta_0 = 0$, that is,

$$
\begin{aligned}
m\dot{v}_y &= F_y + mg\phi - mV\omega_z, \\
I_{xx}\dot{\omega}_x - I_{xz}\dot{\omega}_z &= M_x, \\
-I_{xz}\dot{\omega}_x + I_{zz}\dot{\omega}_z &= M_z, \\
\dot{\phi} &= \omega_x,
\end{aligned}
\tag{5.39}
$$

where, to simplify the notation, the Δ symbol has been dropped from the perturbation variables. This description corresponds to Equation (5.4), with linearized expressions for the gravity and gyroscopic forces appearing in the first equation and a rigid-body state vector given by $\mathbf{x}_r = \left(v_x, \omega_x, \omega_y, \phi\right)^\top$. The symbols F_y, M_x, and M_z represent the aerodynamic and propulsive lateral force, roll moment, and yaw moment, respectively, which are the known functions of the aircraft rigid-body velocities, control and thrust inputs, and also of the structural deformations. Notice that only the antisymmetric components of the loads, control, and trim parameters play a role in the lateral response.

In what follows, the vehicle is further assumed to have very large and straight wings of semispan b, such that the fuselage aerodynamic effects become negligible. It also has outboard ailerons with antisymmetric command that start at a distance y_a from the wing root and cover up to the wingtip, as shown in Figure 5.11. We further assume, to simplify our description, that the elastic axis coincides with the spanwise vehicle stability axis, y. Finally, the effect of thrust forces in the lateral dynamics is assumed to be negligible.

For a positive roll (when there is no control reversal), the aileron deflection will be negative on the right (starboard) wing and positive on the left (port) wing, that is,

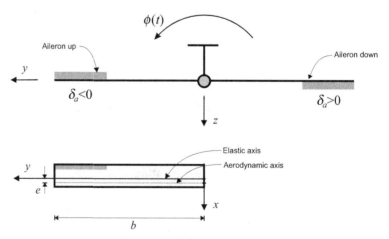

Figure 5.11 Simplified model for roll control of a flexible aircraft (front view of the full aircraft and plan view of the right wing).

$$\delta_a = -\delta_a^*, \quad \text{for } y > y_a,$$
$$\delta_a = \delta_a^*, \quad \text{for } y < -y_a. \tag{5.40}$$

To evaluate the forces and moments in Equation (5.39), we first need mathematical models for the structural response and for the distribution of aerodynamic loads on the wings (which are the cause of their deformations). From the aeroelastic equilibrium between them, we are now able to obtain the resultant aerodynamic forces that drive the aircraft dynamics. Each of those steps is discussed in detail next.

5.7.1 Structural Model

We use again the torque tube model introduced in Section 5.6.1. Equation (5.26) can be written in its weak form as

$$\int_0^b \left(\delta\theta' GJ\theta' - \delta\theta\tau \right) dy = 0, \tag{5.41}$$

with $\delta\theta$ being a virtual rotation field that satisfies the natural boundary conditions on the structure (i.e., $\delta\theta(0) = 0$ for a cantilever beam). This form of the equation is very suitable for numerical approximations in which the unknown variable (the rotation) and the weight functions (the virtual rotation) are approximated by finite expansions (Reddy, 2002, Ch. 7). In particular, we use here *Galerkin's method* (sometimes also known as the Ritz–Galerkin method), in which the rotation and the virtual rotation along the span are both approximated by the same set of *test functions*. These test functions need to be linearly independent, define a complete set, and satisfy the natural boundary conditions. They are assumed to be known and are given here by a row vector $\mathbf{N}(y)$ of dimension N_s, such that

$$\theta(y) \approx \mathbf{N}(y)\mathbf{x}_s,$$
$$\delta\theta(y) \approx \mathbf{N}(y)\delta\mathbf{x}_s. \tag{5.42}$$

Examples of test functions for this problem are half-range sine Fourier series, in which case $\mathbf{x}_s$ is the amplitude of the modes, or piecewise linear interpolating functions in a finite-element discretization, in which case $\mathbf{x}_s$ is the nodal rotations. Substituting into Equation (5.41) and canceling the $\delta\mathbf{x}_s$ term that multiplies both sides of the equation give

$$\mathbf{K}_{ss}\mathbf{x}_s = \mathbf{f}_{\text{aero}}^{(s)}, \tag{5.43}$$

with

$$\mathbf{K}_{ss} = \int_0^b \mathbf{N'}^\top(y)GJ(y)\mathbf{N'}(y)dy,$$
$$\mathbf{f}_{\text{aero}}^{(s)} = \int_0^b \mathbf{N}^\top(y)\tau(y)dy. \tag{5.44}$$

This is, therefore, the particular realization of Equation (5.5) for our model of the flexible aircraft response in roll.

5.7.2   Aerodynamic Model

As in Section 5.6.2, we use *strip theory*, such that the lift and moment coefficients (about the aerodynamic center) on each wing section are those of the local airfoil. We ignore the effect of the initial angle of attack of the aircraft, since it is a symmetric effect and only antisymmetric forces generate roll. For thin airfoils, general expressions for the aerodynamic coefficients have been obtained in Section 3.3.2, but only the static terms are retained here. Recall also that, for trailing-edge control surfaces, $c_{l_\delta} > 0$ and $c_{m_\delta} < 0$ (Equation (3.44)). The lift–curve slope of the airfoils, c_{l_α}, is also known.

If the aircraft has a rolling rate of $\dot{\phi}$ due to the ailerons, then at any station y, the normal velocity due to roll is $y\dot{\phi}$ (positive downward along the right wing). This effectively changes the angle of incidence, as shown in Figure 5.12.

If the local chord is $c(y)$, under the strip-theory assumption, we have the following lift and pitching moment about the local aerodynamic center (both per unit length) at a point y along the wingspan,

$$l(y) = qc \left[c_{l_\alpha} \left(\theta(y) + \frac{y\dot{\phi}}{V} \right) - c_{l_\delta} \delta_a^* \mathcal{H}(y - y_a) \right],$$

$$m_{\text{ac}}(y) = -qc^2 c_{m_\delta} \delta_a^* \mathcal{H}(y - y_a),$$

(5.45)

where $\mathcal{H}(y)$ is the Heaviside (step) function, which has been defined in Equation (3.40). Equation (5.45) corresponds to the right wing, that is, $y > 0$. On the left wing ($y < 0$), the distributions of lift and aerodynamic moment are equal but with opposite sign. The three different contributions to the wing lift distribution of Equation (5.45) are schematically shown in Figure 5.13. Note that a positive value of $\theta(y)$ implies that the elastic deformation of the wing reduces the effectiveness of the aileron.

Finally, the torsional moment due to the aerodynamic effects generated on the right wing during the rolling maneuver is

$$\tau(y) = el + m_{\text{ac}},$$

(5.46)

with the lift and moment per unit length given by Equation (5.45).

5.7.3   Aeroelastic Equilibrium

Once the torsional moment $\tau(y)$ generated by the aerodynamic loads is obtained, we can solve the aeroelastic equations. Substituting Equation (5.45) into Equation (5.43), we obtain

$$\mathbf{K}_{ss}\mathbf{x}_s = q \left(\mathcal{A}_{ss}\mathbf{x}_s + \mathcal{A}_{sr}\frac{b\dot{\phi}}{V} + \mathcal{A}_{su}\delta_a^* \right),$$

(5.47)

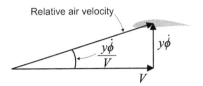

Figure 5.12 Increase in the incidence angle for the right wing with a positive roll rate.

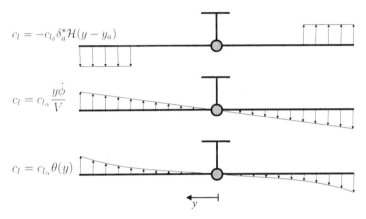

Figure 5.13 Contributions to the spanwise wing lift distribution in a flexible aircraft in roll.

which corresponds to the approximation defined in Equation (5.7) and with the matrices in the problem defined as

$$
\begin{aligned}
\mathcal{A}_{ss} &= \int_0^b \mathbf{N}^\top ecc_{l_\alpha} \mathbf{N}\, dy, \\
\mathcal{A}_{sr} &= \int_0^b \mathbf{N}^\top ecc_{l_\alpha} (y/b)\, dy, \\
\mathcal{A}_{su} &= -\int_{y_a}^b \mathbf{N}^\top (ecc_{l_\delta} + c^2 c_{m_\delta})\, dy.
\end{aligned}
\tag{5.48}
$$

These integrals depend on known parameters and can be numerically evaluated for a given problem. Solving now for $\mathbf{x}_s$ and replacing $\theta(y) = \mathbf{N}(y)\mathbf{x}_s$ in Equation (5.47), we obtain the torsion distribution along the flexible wing in terms of the rigid-body variables, δ_a^* and $\dot{\phi}$, as

$$
\theta(y) = q\mathbf{N}(y)\,(\mathbf{K}_{ss} - q\mathcal{A}_{ss})^{-1}\left(\mathcal{A}_{sr}\frac{b\dot{\phi}}{V} + \mathcal{A}_{su}\delta_a^* \right).
\tag{5.49}
$$

Note that for this equation to have a solution, $(\mathbf{K}_{ss} - q\mathcal{A}_{ss})$ must be invertible, that is, $\det(\mathbf{K}_{ss} - q\mathcal{A}_{ss}) \neq 0$. This means that the dynamic pressure must be below the vehicle's divergence dynamic pressure (if a positive value of q_D does exist). Finally, the total rolling moment, M_x, generated on the aircraft from both wings can now be obtained by the integration of the lift in Equation (5.45). Using the symmetry of the problem, we need to integrate only along one wing, which results in

$$M_x = -2 \int_0^b yl(y)\,dy$$

$$= 2q \int_{y_a}^b ycc_{l_\delta} \delta_a^*\,dy - 2q \int_0^b ycc_{l_\alpha} \left(\mathbf{N}\mathbf{x}_s + \frac{y\dot\phi}{V} \right) dy, \tag{5.50}$$

where, after introducing a reference chord c_0, we can explicitly carry out both integrals for a given configuration as

$$M_x = qc_0 b^2 \left[C_\delta \delta_a^* + C_{\dot\phi} \frac{b\dot\phi}{V} \right]. \tag{5.51}$$

Two dimensionless *elastified* aerodynamic derivatives, C_δ and $C_{\dot\phi}$, have been defined in this expression as

$$C_\delta = \frac{2}{b^2 c_0} \int_{y_a}^b ycc_{l_\delta}\,dy - \frac{2q}{b^2 c_0} \int_0^b ycc_{l_\alpha} \mathbf{N}(\mathbf{K}_{ss} - q\mathcal{A}_{ss})^{-1} \mathcal{A}_{su}\,dy,$$

$$C_{\dot\phi} = -\frac{2}{b^3 c_0} \int_0^b y^2 cc_{l_\alpha}\,dy - \frac{2}{b^2 c_0} \int_0^b ycc_{l_\alpha} q\mathbf{N}(\mathbf{K}_{ss} - q\mathcal{A}_{ss})^{-1} \mathcal{A}_{sr}\,dy. \tag{5.52}$$

The first integral in either expression is the contribution of the rigid aircraft aerodynamics, which depends solely on the aerodynamic properties of the wing airfoils and the wing geometry. It is positive in C_δ and negative in $C_{\dot\phi}$. Due to the normalization in Equation (5.51), they remain constant with changes in the dynamic pressure. The second integral includes the aeroelastic effects, which also depend on the structural characteristics of the wing and, importantly, the dynamic pressure. In aircraft wings, the elastic axis is typically behind the aerodynamic axis ($e > 0$ in Figure 5.11). In such a case, positive lift on a wing section generates a pitch-up moment and therefore positive values of the torsional angle. As a result, the usual situation is for the elastic contribution in both C_δ and $C_{\dot\phi}$ to be negative, and increasingly so with higher dynamic pressure.

Therefore, the usual situation is for C_δ to be positive at low dynamic pressures and negative for sufficiently high speeds. There is a dynamic pressure for which $C_\delta = 0$, for which aileron deflections do not generate rolling moments (a *control reversal* condition). This is illustrated in Example 5.2 on a wing with constant properties, for which analytical solutions of this problem can be obtained.

5.7.4   Elastic Effects in the Roll Response

In the previous section, we have derived expressions for the *elastified* aerodynamic derivatives associated with roll. They can now be introduced into Equation (5.39) to investigate the lateral dynamics of the flexible aircraft. For simplicity, we further assume that the inertia tensor is diagonal, in which case, the roll response of the aircraft becomes uncoupled from the rest of the lateral dynamics, and it is determined by

$$I_{xx}\dot\omega_x = M_x,$$

$$\dot\phi = \omega_x. \tag{5.53}$$

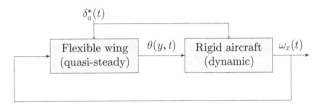

Figure 5.14 Block diagram for roll control with flexible wings.

Substituting Equation (5.51) into this expression and reordering terms, we can write the state-space form of the roll dynamics as

$$\begin{Bmatrix} \dot{\omega}_x \\ \dot{\phi} \end{Bmatrix} = \begin{bmatrix} \frac{qcb^3 C_{\dot{\phi}}}{VI_{xx}} & 0 \\ 1 & 0 \end{bmatrix} \begin{Bmatrix} \omega_x \\ \phi \end{Bmatrix} + \begin{Bmatrix} \frac{qcb^2 C_\delta}{I_{xx}} \\ 0 \end{Bmatrix} \delta_a^*. \qquad (5.54)$$

This equation corresponds to Equation (5.10) for this simplified problem. As before, the effect of flexibility in the response to aileron inputs can be seen as a constant gain feedback, with the block diagram of Figure 5.14.

The dynamic stability of the wing is determined by the eigenvalues of the system matrix in Equation (5.54). They are simply $\lambda_1 = \frac{qcb^3 C_{\dot{\phi}}}{VI_{xx}}$ and $\lambda_2 = 0$. The first one corresponds to the roll (subsidence) mode, while the second is an integrator. As was discussed in Section 5.7.3, both the flexible and rigid contributions to $C_{\dot{\phi}}$ are typically negative, which results in the roll dynamics having higher damping on a flexible than on a rigid wing. The controllability matrix, Equation (5.20), for this problem is

$$\mathcal{C} = \begin{bmatrix} \frac{qcb^2 C_\delta}{I_{xx}} & 0 \\ \frac{qcb^2 C_\delta}{I_{xx}} & \frac{q_\infty^2 c^2 b^5 C_\delta C_{\dot{\phi}}}{VI_{xx}^2} \end{bmatrix}, \qquad (5.55)$$

which has a full rank if $C_\delta \neq 0$, independently of the value of $C_{\dot{\phi}}$. As expected, the control reversal condition ($C_\delta = 0$) defines the loss of controllability in the roll dynamics.

Equation (5.54) can easily be integrated when written in terms of the roll angle and its derivatives as

$$I_{xx}\ddot{\phi} - q\frac{cb^3 C_{\dot{\phi}}}{V}\dot{\phi} = qcb^2 C_\delta \delta_a^*(t). \qquad (5.56)$$

Note that, since $C_{\dot{\phi}} < 0$, the damping term is always positive. Given the dynamic pressure, q, this equation can now be solved for any input on the aileron, which will be defined by either pilot or FCS commands. As an example, we consider the response – starting from level flight – to a sudden deflection of the ailerons, that is, a step response with $\delta_a^*(t) = \delta_0 \mathcal{H}(t)$. The solution is then

$$\dot{\phi}(t) = \frac{B}{A}\left(1 - e^{-At}\right)\delta_0, \qquad (5.57)$$

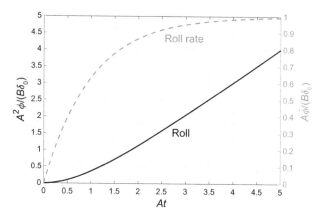

Figure 5.15 Nondimensional roll and roll rate on an aircraft with flexible high-aspect-ratio wings.

which has been given in terms of the constants

$$A = -\frac{qcb^3 C_{\dot{\phi}}}{V I_{xx}} \quad \text{and} \quad B = \frac{qcb^2 C_{\delta}}{I_{xx}}. \tag{5.58}$$

Integration of the roll rate gives the roll angle as

$$\phi(t) = \left[\frac{Bt}{A} + \frac{B}{A^2}\left(e^{-At} - 1\right)\right]\delta_0. \tag{5.59}$$

This is shown in Figure 5.15, which includes the time histories of the nondimensional roll and roll rate. The transient effects are driven by the value of A, after which the solution goes to a steady-state roll rate. This is also obtained from imposing $\ddot{\phi} = 0$ and results in $\dot{\phi}_{ss} = \frac{B}{A}\delta_0 = -\frac{VC_{\delta}}{bC_{\dot{\phi}}}\delta_0$.

Note that for dynamic pressures below reversal, $\dot{\phi}_{ss} > 0$. For the case of a rigid wing, the steady-state roll rate simplifies to $\frac{3}{2}\frac{c_{l\delta}}{c_{l\alpha}}\frac{V}{b}\delta_a^*$, which results in a roll effectiveness given as

$$\eta_{\text{roll}} = \frac{\dot{\phi}_{ss}\big|_{\text{flex}}}{\dot{\phi}_{ss}\big|_{\text{rigid}}} = -\frac{2}{3}\frac{C_{\delta}/c_{l\delta}}{C_{\dot{\phi}}/c_{l\alpha}}. \tag{5.60}$$

Example 5.2 Analytical Solution for a Wing with Constant Properties.   As in Section 5.6, closed-form solutions exist for wings with constant properties. Equation (5.26) for a flexible wing that rolls at a (still unknown) rate $\dot{\phi}(t)$ now becomes

$$GJ\frac{d^2\theta}{dy^2} + el + m_{ac} = 0. \tag{5.61}$$

Replacing the expressions for the aerodynamic loads, and dividing the resulting equation by GJ, we can write

$$\frac{d^2\theta}{dy^2} + \lambda^2\theta = \nu\lambda^2\delta_a^* - \lambda^2\frac{y\dot{\phi}}{V}, \tag{5.62}$$

with newly defined terms

$$\lambda^2 = \frac{qecc_{l_\alpha}}{GJ} \quad \text{and} \quad \nu = \frac{cc_{m_\delta} + ec_{l_\delta}}{ec_{l_\alpha}}. \tag{5.63}$$

Note that we have assumed implicitly that $e > 0$ in the definition of λ, which is by far the most typical situation in aircraft wings. We can then define the dimensionless parameter λb, which measures the ratio between aerodynamic and structural moments in wing twisting. The general solution[3] to this problem is of the form

$$\theta = A_1 \cos \lambda y + A_2 \sin \lambda y + \nu \delta_a^* - \frac{y\dot{\phi}}{V}. \tag{5.64}$$

Constants A_1 and A_2 are obtained from the boundary conditions (5.27). Enforcing $\theta(0) = 0$, we have $A_1 = -\nu\delta_a^*$, and then setting $\theta'(b) = 0$ gives

$$-\lambda A_1 \sin \lambda b + \lambda A_2 \cos \lambda b - \frac{\dot{\phi}}{V} = 0. \tag{5.65}$$

Thus,

$$A_2 = -\nu\delta_a^* \tan \lambda b + \frac{1}{\cos \lambda b} \frac{\dot{\phi}}{\lambda V}. \tag{5.66}$$

As a result, the torsional angle along the right wing becomes

$$\theta(y,t) = (1 - \tan \lambda b \sin \lambda y - \cos \lambda y)\nu\delta_a^*(t) - \left(\lambda y - \frac{\sin \lambda y}{\cos \lambda b}\right)\frac{\dot{\phi}(t)}{\lambda V}. \tag{5.67}$$

Note that, even though we have only considered the static equilibrium of the wing, the torsional angle will change with time as it depends on the instantaneous values of $\delta_a^*(t)$ and $\dot{\phi}(t)$. It can also be easily seen that for $\lambda b = \pi/2$, the wing twist goes to infinite. This defines the *divergence* condition for this wing, which can therefore be obtained analytically. Substituting this condition into Equation (5.63) defines the divergence dynamic pressure as

$$q_D = \frac{\pi^2}{4} \frac{GJ}{ecbc_{l_\alpha}}. \tag{5.68}$$

Divergence is, therefore, delayed with stiffer wings, which is a costly solution, as stiffness brings weight, or with a wing design with an elastic axis located near its airfoils' aerodynamic centers. See Section 3.2 of Hodges and Pierce (2014) for an extensive discussion on the divergence of cantilever wings.

The total rolling moment is obtained as in Equation (5.51) and results in an identical expression,

$$M_x = -2\int_0^b yl(y)\,\mathrm{d}y = qcb^2\left[C_\delta\delta_a^* + C_{\dot\phi}\frac{b\dot\phi}{V}\right]. \tag{5.69}$$

---

[3] If $\lambda^2 < 0$ as in the case when $e < 0$, the form of the solution changes from sin and cos to sinh and cosh.

The *elastified* aerodynamic derivatives can now be explicitly evaluated, after some algebra, as

$$C_\delta = c_{l_\delta} \left[1 + \left(1 - \frac{c}{e} \frac{c_{m_\delta}}{c_{l_\delta}} \right) \left(1 - \frac{2}{\lambda^2 b^2} \frac{1 - \cos \lambda b}{\cos \lambda b} \right) \right],$$

$$C_{\dot\phi} = 2 c_{l_\alpha} \frac{\lambda b - \tan \lambda b}{\lambda^3 b^3}. \tag{5.70}$$

The dynamic pressure, $q = \frac{1}{2}\rho V^2$, appears explicitly in Equation (5.69) and also in both C_δ and $C_{\dot\phi}$, through their dependence on λb. As discussed above, λb measures the relative importance between aerodynamic and structural effects. For very stiff wings, $GJ \to \infty$ and then $\lambda b \to 0$. Its upper limit is given by the divergence condition, $\lambda b = \pi/2$. Between those limits, the value of λb strongly affects the roll response.

Consider first C_δ, which measures the effectiveness of the ailerons to generate roll. From Section 3.3.2, an airfoil with a trailing-edge flap has $c_{l_\delta} > 0$, $c_{m_\delta} < 0$, and $c_{l_\alpha} > 0$. We have assumed above that $e > 0$ and, therefore, the sign of the second term in C_δ is determined by the sign of $f_1(\lambda b) = 1 - \frac{2}{\lambda^2 b^2} \frac{1 - \cos \lambda b}{\cos \lambda b}$. This expression has been plotted in Figure 5.16a, and it is seen to be increasingly negative with λb. Therefore, C_δ will be positive (positive aileron deflection, as defined in Equation (5.40), produces the positive roll moment) for small values of λb and negative (positive deflection gives the negative moment) for large enough values of this parameter. There is a point with $C_\delta = 0$, which defines the *control reversal* condition. At that point, the ailerons are ineffective.

As can be also seen in Figure 5.16a, the second derivative, $C_{\dot\phi}$, is always negative. The physical reason is the restoring effect that aerodynamics has on roll motion. At small speeds, $\tan \lambda b \approx \lambda b + \frac{1}{3}\lambda^3 b^3$, and then $C_{\dot\phi} \approx \frac{2}{3} c_{l_\alpha}$. From Figure 5.16a, we see that as λb grows, the negative moment becomes even larger, which slows the roll response of wings with either larger span (larger b), higher compliance (lower GJ), or higher sectional aerodynamic moment (high e).

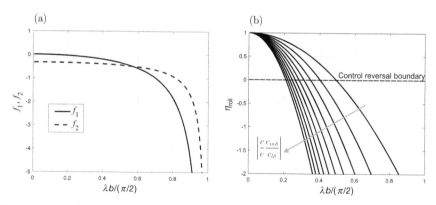

Figure 5.16 (a) Dimensionless functions in Equation (5.70) ($f_1(\lambda b) = 1 - \frac{2}{\lambda^2 b^2}\frac{1 - \cos \lambda b}{\cos \lambda b}$ and $f_2(\lambda b) = \frac{\lambda b - \tan \lambda b}{\lambda^3 b^3}$) and (b) resulting steady-state roll effectiveness of a flexible wing.

We can finally investigate the roll effectiveness defined in Equation (5.60). Substituting the coefficients obtained in Equation (5.70) into that expression gives

$$
\eta_{roll} = -\frac{1}{3}\frac{\lambda^3 b^3 + \left(1 - \frac{c}{e}\frac{c_{m\delta}}{c_{l\delta}}\right)\left(\lambda^3 b^3 - 2\lambda b\frac{1-\cos\lambda b}{\cos\lambda b}\right)}{\lambda b - \tan\lambda b}.
\tag{5.71}
$$

This function is shown in Figure 5.16b for the values of the dimensionless coefficient $-\frac{c}{e}\frac{c_{m\delta}}{c_{l\delta}} \in [2,20]$ in increments of 2. This is the ratio between the moments at the aerodynamic center due to the aileron deflection (which are negative and therefore act against roll) and the moments generated by the lift on the elastic axis (which are positive if $e > 0$). As can also be seen in Figure 5.16b, the roll rate achieved by a flexible wing is always smaller than that of the equivalent rigid wing. For large enough values of λb, the effectiveness becomes negative and the wing roll control operates on reversal.

The closed-form expressions in this example give some insights into the desired features on aircraft wings to avoid (or delay) aileron reversal. It is naturally beneficial to increase the wing torsional stiffness, GJ, although this has a weight penalty. A more feasible strategy is therefore to use ailerons with a large chord, which have a smaller value of $\left|\frac{c_{m\delta}}{c_{l\delta}}\right|$. Indeed, all moving wingtips would be the ideal actuation to avoid reversal, since they give no adverse torsional moments. Two final strategies, which are used in vehicles with very large or very flexible wings, are to move the ailerons inboard or to replace them altogether by spoilers that are designed to produce no adverse torsional moment.

5.8 Summary

After reviewing the basic concepts of flight dynamics and aeroelasticity in the previous two chapters, this chapter has introduced a first situation in which the dynamics of a flexible aircraft is dominated by the interactions between them. The focus has been on the dynamics of aircraft with nonnegligible elastic deformations under flight loads, but on which there is sufficient frequency separation between the structural vibration and rigid-body characteristics. In the time domain, this implies that the typical timescales in the flight dynamics are much larger than the natural vibration characteristics of the structure. This is a very common situation in the flight dynamics of modern air vehicles, which we have characterized as flexible aircraft with quasi-steady deformations.

The mathematical formulation of the dynamics of such vehicles builds directly on the rigid-body formulation presented in Chapter 4 with additional DoFs to describe the instantaneous shape of the deforming aircraft. However, because of the quasi-steady assumption of the elastic response, the structural equations become algebraic relations between the (potentially many) elastic DoFs and the rigid-body ones, and consequently the elastic DoFs can be residualized to the rigid-body equations. In effect, the flexibility of the airframe acts as a constant feedback gain on the rigid-body dynamics, with the instantaneous flight speed as input and the resulting deformed shape of the aircraft (which modifies its aerodynamic characteristics) as its output.

The resulting elastified EoMs for the flexible aircraft can be linearized and written in a standard space-state form. They depend internally on the aeroelastic equilibrium of the aircraft, which is itself affected by the dynamic pressure around which the linearization is made. The analysis of the flexible aircraft for stability was shown first, where the flight dynamic modes may be very strongly affected by the flight conditions. It was also shown that a sufficiently high dynamic pressure may produce a statically unstable condition that has been identified as aeroelastic divergence. Equally, controllability analysis of an aeroelastic vehicle shows that a loss of controllability is possible from the elastic effects canceling the incremental forces produced by control surfaces, thus leading to a situation known as control reversal.

5.9   Problems

Exercise 5.1. For an aircraft, such as that shown in Figure 5.11, show that if the elastic axis and the aerodynamic center coincide (that is, $e = 0$), then the dynamic pressure for control reversal is

$$q_R = -\frac{12}{5}\frac{GJ}{c^2 b^2}\frac{c_{l\delta}}{c_{l\alpha}c_{m\delta}}.$$

Exercise 5.2. Suppose an aircraft, with a layout such as that shown in Figure 5.11, is built with a wing of aspect ratio 12 and airfoils with $e/c = 0.1$, $c_{l\alpha} = 5$, $c_{m\delta} = -0.5$, $c_{l\delta} = 1.2$. Then

(i) Determine its nondimensional control reversal dynamic pressure.
(ii) Determine its nondimensional divergence dynamic pressure.

Exercise 5.3. Consider a high-aspect-ratio rectangular wing of semispan b and chord c in an airstream with velocity V. The wing flexibility is not negligible and its torsional stiffness is GJ, measured about an axis that is at a distance e aft the aerodynamic center (see the following figure). The wing is fitted with an aileron that covers half of the span and has a constant section with the known coefficients $c_{l\alpha}$, $c_{l\delta}$, and $c_{m\delta}$ about the aerodynamic center of the airfoil. δ is the deflection angle of the aileron, positive downward.

(i) Write the differential equations and boundary conditions that determine the initial twist along the span when the flap is deflected. It can be helpful to split the problem in two, namely $0 < y_1 < b/2$ and $0 < y_2 < b/2$, with y_1 and y_2 as defined in the figure.
(ii) Express the equations in a nondimensional form and find a solution that satisfies the boundary conditions.
(iii) Compute the control reversal speed.

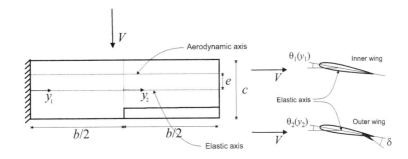

Exercise 5.4. Consider a reconnaissance UAV in a steady-level flight at speed V and at a certain altitude where the air density is ρ. The aircraft has prismatic straight wings of semispan b and constant chord c, and its aerodynamics can be approximated using lifting-line theory with lift–curve slope $c_{l\alpha}$ for all wing airfoils. All other aerodynamic constants can be neglected. The wing is flexible and its structure can be modeled as a torque tube with constant torsional stiffness GJ and elastic axis located at a distance e aft the aerodynamic axis.

(i) If the angle of incidence at the wing root is α_r^0, derive the equations that determine the elastic twist of the wings, $\theta(y)$, and the corresponding total lift on the aircraft.

(ii) If the total mass of the aircraft is m, establish the relation between the angle of incidence α_r^0 to sustain flight and the position e of the elastic axis.

Exercise 5.5. The static aeroelastic characteristics of a racing car's rear wing can be studied using the model of a wing. The wing has span $2l$ and chord c, and it is mounted at an angle α_0 (positive pitching downward) in an airstream with dynamic pressure $\frac{1}{2}\rho V^2$. The wing's structure is a flexible beam along the leading edge with constant bending and torsional stiffness, EI and GJ, respectively, which is clamped on two rigid supports on both ends. Consider thin-strip aerodynamics with symmetric airfoils and the known lift-curve slope $c_{l\alpha}$, including the ground effect.

(i) Derive the aeroelastic equilibrium equations that determine the elastic twist distribution, $\theta(y)$, along this wing.

(ii) Compute the vertical displacement at the center of the wing.

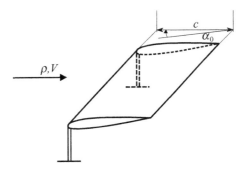

Exercise 5.6. Consider the aircraft of mass m that is initiating a roll maneuver as in the figure below. It has high-aspect-ratio rigid wings of semispan b and a negligible fuselage diameter in comparison with the wing semispan. Assuming also that the inertial coupling between the yaw and the roll is negligible, the transient dynamics in the turn can be approximated by $I_{xx}\ddot{\phi} = M_x$, where I_{xx} is the roll inertia, $\phi(t)$ is the instantaneous roll angle, and M_x is the total roll moment. Assume that the aircraft has full span trailing-edge flaps with known airfoil aerodynamic derivatives $(c_{l\alpha}, c_{l\delta}, c_{m\delta})$.

(i) Obtain the steady-state roll response to a step antisymmetric input of the flaps.
(ii) Repeat the problem if the rigid wings are mounted to the fuselage on a flexible hinge in torsion with know stiffness k_θ at a chordwise location e aft the aerodynamic axis.

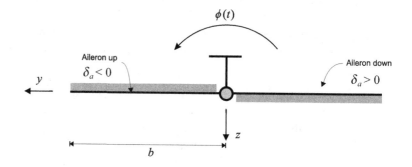

6   Dynamics of Flexible Aircraft with Small-Amplitude Vibrations

6.1   Introduction

Chapter 5 has introduced a first approximation to the dynamics of flexible aircraft, in which the rigid aircraft dynamics equations are augmented with the instantaneous static aeroelastic equilibrium (determined from a force balance between the distributed external loading on the aircraft and the internal forces in its deforming structure). As we have discussed in Section 5.2, it requires two main assumptions on the aircraft dynamics: first, that the distributed inertia of the vehicle does not participate in the structural response, that is, that structural dynamic effects are negligible; and second, that the elastic deformations are such that they do not appreciably change the instantaneous rigid-body inertia of the vehicle. The first assumption effectively requires that the airframe structural resonances (its natural vibration frequencies) occur at much higher frequencies than those characteristic of the vehicle flight dynamics. The second assumption is satisfied when the amplitudes of the elastic displacements are sufficiently small. Both approximations hold across most of the flight envelope in flight dynamics descriptions of a rather large number of air vehicles, from fighter jets to current-generation twin-aisle commercial transport aircraft. Consequently, the modeling framework of Chapter 5, which describes the vehicle dynamics using elastified aircraft EoMs, is widely used in the flight dynamics community.

In this chapter, we broaden our focus to consider both dynamic aeroelastic effects and the dynamic interactions between aeroelasticity and flight dynamics that may occur in flexible aircraft. This effectively implies removing both previous assumptions in order to consider a rather more general, fully coupled problem in the dynamics of flexible vehicles. While the current platforms in which such a theory is required are less numerous, it is increasingly found in at least two types of aircraft. The first group are vehicles with very high aspect ratio wing that are designed to achieve very large aerodynamic efficiency gains. This category includes vehicles such as flying wings (Britt et al., 2000), solar-powered aircraft (Klöckner et al., 2013), and, critically, many of the future-generation transport aircraft under development at the time of writing. An example of this last group is NASA's concept for a Transonic Truss-Braced Wing (TTBW) aircraft, shown in Figure 6.1a. The second group are the new generation of supersonic transport aircraft, such as NASA/Lockheed Martin X-59 Quiet Supersonic Transport (QueSST), shown in Figure 6.1b. They need a slender nose-cone to reduce

(a)   (b)

Figure 6.1 Two concept aircraft with relatively low-stiffness designs. (a) Transonic Truss-Braced Wing aircraft and (b) NASA X-59 QueSST (photos by NASA and Lockheed Martin).

(a)   (b)

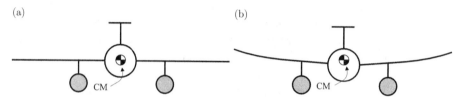

Figure 6.2 Wing deflections alter the location of the aircraft center of mass. (a) Rigid aircraft and (b) flexible aircraft.

the noise footprint, thus resulting in a flexible fuselage with very low natural vibration frequencies (Connolly et al., 2020).

In a vast majority of cases, we can still assume that the amplitudes of the elastic displacements in the airframe are sufficiently small for its stiffness to be well approximated by geometrically-linear structural theories. However, contrarily to what we have done in Chapter 5, we consider now the changes in the instantaneous global inertia properties of the vehicle that may result from the changes in geometry under deformation. This is because relatively large deflections of the wing (say, with wing bending deformations that result in tip displacements up to around 10% of the semispan) can be well captured by linear elasticity theory, but the corresponding shift in, for example, the location of the aircraft center of mass may have a nonnegligible effect on the vehicle dynamics. This situation is schematically shown in Figure 6.2. It is important to remark that changes in the global inertia of the aircraft are driven not only by the wing deformation but also, the weight fraction of the wings with respect to the total aircraft mass (i.e., very light wings with large deformations may not displace the aircraft CM). As a consequence of this, span loading the wings, which was discussed in Section 5.2 as a strategy to minimize wing bending by increasing the fraction of mass on the wings, may have the adverse effect of producing strong coupling between rigid-body and elastic degrees of freedom on dynamic events such as gust response (del Carre et al., 2019b).

While Chapter 5 has introduced the flexible aircraft dynamics as an extension to the classical rigid aircraft theory of Chapter 4, the mathematical description of the dynamics of a flexible vibrating aircraft needs the more general framework provided by flexible multibody dynamics (Bauchau, 2011; Géradin and Cardona, 2001; Shabana, 1997), although the problem can often be restricted to a single deformable body as it is done in the rest of this chapter. The flexible-body description needs to be further augmented with (in general, unsteady) aerodynamic models that give distributed forces on all wet surfaces, and also possibly a dynamic model of the propulsion system. The resulting flexible aircraft dynamic models are valid for a vast majority of existing air vehicles. In particular, the assumption of small-amplitude vibrations dramatically simplifies the description of the restoring elastic forces on the airframe, which can still be obtained from linear finite element analysis. Furthermore, it also has a very important knock-on effect on the description of the unsteady aerodynamics, which can then be well characterized by linearized models (for points in the flight envelope with attached-flow conditions). A more general nonlinear problem is discussed in Chapter 9.

The kinematic description of a deforming aircraft first needs some reevaluation of the frames of reference that have been introduced in Section 4.2.1 for a rigid aircraft. This is done in Section 6.2, where we further distinguish between body-attached and floating frames of reference. To make the description as general as possible, all inertial couplings between the elastic and rigid-body DoFs are initially retained, which results in the coupled formulation of Section 6.3. After that, we introduce restricted formulations under different assumptions, including fully linearized dynamics, and discuss their applicability. In particular, we will see in Section 6.4 that if elastic deformations are sufficiently small not to change the global (rigid-body) inertial characteristics, then the structural dynamics can be written in modal coordinates, which enormously reduces the size of the problem. We refer to this last problem as the nonlinear rigid-linear elastic EoMs of a flexible aircraft.

The EoMs are finally completed with a linear aerodynamic model in Section 6.5, which generalizes the aerodynamic derivatives used in Chapter 4 to consider flexible wings and unsteady effects. The general linearized formulation needed for both coupled stability analysis and linear control design is finally introduced in Section 6.6. If the effects of steady loading in the vehicle dynamics are not considered, this problem leads to the standard formulation used in dynamic aeroelastic analysis for flutter and gust response, which will also be discussed. The chapter concludes with examples of the effect of flexibility on flight dynamic stability and coupled flight dynamic-aeroelastic stability in Section 6.6.2.

6.2 Frames of Reference

As with the description of the kinematics of a rigid aircraft in Section 4.2.1, the starting point is a set of Earth axes, that is, an inertial frame of reference that is valid across a relatively short flight segment. As flexible aircraft can no longer be described as rigid bodies, there is no single aircraft-fixed reference frame that defines uniquely the instantaneous position and orientation everywhere in the vehicle. However, it is not practical to describe the kinematics of each individual material point on the aircraft

(a)   (b)

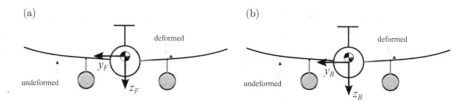

Figure 6.3 Floating vs. body-attached frames of reference under wing deformations. A floating frame tracks the instantaneous CM while a body-attached frame tracks a material point. (a) Floating frame and (b) body-attached frame.

with respect to the Earth frame.[1] An intermediate frame is therefore commonly used to separate, as much as possible, the flight dynamics, which describes the aircraft trajectory, from the structural dynamics, which gives its instantaneous shape. Two alternative definitions can be used here:

- *A floating frame of reference*, defined to track certain integral magnitudes (e.g., the instantaneous position of the center of mass) and, therefore, that moves with respect to an observer on the vehicle.
- *A body-attached frame of reference*, whose origin is rigidly linked to a given material point of the aircraft and with axes such that the local rotations around that point are zero.

Both situations are schematically shown in Figure 6.3 for an aircraft with a stiff fuselage and flexible wings. In either case, the instantaneous position of the origin of the reference frame and its instantaneous orientation with respect to the Earth frame are used to describe the aircraft flight dynamics state, while the relative position of the material points of the deforming aircraft with respect to that frame defines its elastic state. In practice, the instantaneous floating frame for flexible aircraft is typically defined such that the integral values of the linear and angular momenta of the relative motion are identically zero. Floating axes defined in this way are known as *mean axes* (Milne, 1962). Enforcing the condition of zero linear momentum implies that the origin of the mean axes is always at the CM, while the angular momentum condition determines its instantaneous orientation. Note that the mean axes is a unique frame for a given aircraft deformed state, regardless of the reference taken for the definition of the displacement field (Milne, 1968).

There is some more flexibility in the definition of the body-attached frame, although it is often convenient to define them such that they coincide with the stability axes (defined as in Section 4.2.1) in a reference condition, typically the trimmed aircraft in level flight. Note also that there may not be an actual material point at the reference CM, but one can still define the origin of a body-attached frame in relation to one such point.

---

[1] An exception to this occurs when velocity fields (as opposed to displacement fields) are used to describe the flexible vehicle kinematics, as it will be seen in the intrinsic description of beams in Section 8.6.

Floating frames, such as mean axes, are therefore more effective in decomposing the description of the aircraft kinematics between a rigid and an elastic part. Furthermore, as their evolution is defined by the resultant forces and moments on the vehicle, they have relatively "smooth" dynamics (that is, the total inertia of the aircraft acts as a low-pass filter on high-frequency loads). The kinematics of a poorly defined body-attached frame linked, for example, to a vibrating element may be harder to interpret. They may also require additional time resolution to track the vehicle dynamics to a given accuracy. However, unless additional assumptions about the vehicle kinematics are incorporated, consistent application of a mean-axes formulation implies an extra layer of mathematical complexity in the solution process (Guimaraes Neto et al., 2016; Saltari et al., 2017).

Body-attached axes are better associated with the actual sensors on the vehicle, naturally providing local information. The main advantage of body-attached axes, however, is that they simplify the mathematical description of the problem and the associated numerical methods and, therefore, are preferred here. They also provide a natural framework for the more general geometrically nonlinear problems that will be considered in Chapter 9.

We need to emphasize, however, that both definitions, when properly applied, give the same results, as is shown in Section 6.4.3. In particular, one can always retrieve the instantaneous location of the CM from the body-attached frame of reference, and one can always compute the instantaneous acceleration at a certain physical sensor location from a floating-frame description. As we discuss in Section 6.4.3, mean axes are typically used, together with additional kinematic assumptions, to forcibly decouple the aircraft's rigid and flexible dynamics. This is sometimes referred to as *practical mean axes* (Schmidt, 2012), but the additional assumptions need to be carefully justified case by case.

6.3 Dynamics of a Flexible Unsupported Body with Small Deformations

The EoMs of a generic deformable, unsupported body are derived here using variational methods, as described, for example, in Chapter 4 of Géradin and Rixen (1997) for general problems in mechanics, and by Meirovitch (1991) for specific application to the dynamics of a flexible structure with rigid-body DoFs. Its application to flexible aircraft dynamics, in a body-attached reference frame and under linear elastic assumptions, was later presented by Meirovitch and Tuzcu (2004) and, more succinctly, by Haghighat et al. (2012b). Following on them, the approach here assumes small deformations from the onset, but the equations can be more generally derived as the linearization of the problem with large structural displacements about an arbitrary equilibrium point. Such an approach, which is strictly necessary when the equilibrium point is geometrically nonlinear, has been followed by Shearer and Cesnik (2007) and Hesse and Palacios (2012), among others, and it is discussed in Chapter 9.

The starting point for describing the dynamics of a flexible aircraft is Hamilton's principle, which has been introduced in Section 1.4.1. However, before we can

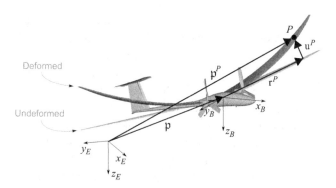

Figure 6.4 Position in the Earth frame of a material point P in a flexible body.

evaluate the kinetic and strain energy, we need to define the *configuration space*, that is, establish a description of the instantaneous shape resulting from the deformation. In that process, we need to make some choices. While those are conditioned by their application to flexible aircraft dynamics, the description that follows is still sufficiently generic.

6.3.1   Kinematics with Small Deformations

Consider the flexible aircraft schematically represented in Figure 6.4, whose instantaneous deformed geometry needs to be described with respect to an inertial (Earth) frame of reference. As it was done in Section 4.2.1 for a rigid aircraft, we refer to this frame as the E frame of reference. Also, as in the rigid aircraft problem, an aircraft-bound frame of reference needs to be chosen to define the vehicle orientation. A *body-attached* frame of reference, B, is introduced to track the trajectory and orientation of the body with respect to the Earth frame, E, as shown in Figure 6.4. Its origin and orientation are typically chosen as the center of mass and stability axes in a reference condition (e.g., the aircraft in steady straight flight), although this is not yet assumed here to keep the generality of the description.

Let P be an arbitrary material point in the aircraft. Its position vector with respect to the origin of the Earth frame, expressed in its components in that frame, is $\mathfrak{p}_E^P(t)$. The instantaneous position of the material point P is obtained (Figure 6.4) as the sum of the position of the origin of a body-attached reference frame B, $\mathfrak{p}_E(t)$, the position with respect to that reference of point P in the undeformed aircraft, $\mathbf{r}_E^P(t)$, and the effect of the instantaneous deformations on the vehicle, given by the local displacement, $\mathbf{u}_E^P$, that is,

$$\mathfrak{p}_E^P(t) = \mathfrak{p}_E(t) + \mathbf{r}_E^P(t) + \mathbf{u}_E^P(t). \tag{6.1}$$

Note that, as we have done in all mathematical derivations in Chapter 4, Euclidean vectors in 3-D space are always written here in terms of their components on the frame of reference identified by the subscript. Superindices are also included when necessary to identify the point referred by that magnitude (e.g., the material point whose deformation is being tracked).

In Earth coordinates, all three contributions to the position vector in Equation (6.1) are time-dependent. The instantaneous inertial (total) velocity of point P, expressed in its components in the E frame, $\mathbf{v}_E^P(t)$, is then obtained directly from differentiation with respect to the time variable, that is,

$$\mathbf{v}_E^P(t) = \frac{\mathrm{d}}{\mathrm{d}t}\mathbf{p}_E^P(t) = \frac{\mathrm{d}}{\mathrm{d}t}\left(\mathbf{p}_E(t) + \mathbf{r}_E^P(t) + \mathbf{u}_E^P(t)\right). \tag{6.2}$$

Each term in the right-hand side of this expression needs some discussion. The first one is the inertial velocity of the origin of the body-attached reference frame, B, for which we can use the definition already introduced in Equation (4.5), that is, $\mathbf{v}_E = \frac{\mathrm{d}}{\mathrm{d}t}\mathbf{p}_E(t)$. The second term comes from the relative position of the material point P with respect to the origin of frame B in the undeformed body, which is time-independent if expressed in components in body-attached axes, that is, it can be written as $\mathbf{r}_E^P(t) = \mathbf{R}_{EB}(t)\mathbf{r}_B^P$, where only $\mathbf{R}_{EB}(t)$ is a function of time. Therefore, it is $\dot{\mathbf{r}}_E^P = \dot{\mathbf{R}}_{EB}(t)\mathbf{r}_B^P$. Finally, the third term is the rate of change of the local displacement, which for convenience it is also written in its components in the B frame, that is, $\mathbf{u}_B^P = \mathbf{R}_{BE}\mathbf{u}_E^P$. As a result, Equation (6.2) can be written as

$$\mathbf{v}_E^P(t) = \mathbf{v}_E(t) + \dot{\mathbf{R}}_{EB}(t)\left(\mathbf{r}_B^P + \mathbf{u}_B^P(t)\right) + \mathbf{R}_{EB}(t)\dot{\mathbf{u}}_B^P(t). \tag{6.3}$$

Next, we can write Equation (6.3) in body-attached frame components by premultiplying it by $\mathbf{R}_{BE}$, since $\mathbf{v}_B^P = \mathbf{R}_{BE}\mathbf{v}_E^P$. Introducing the angular velocity ω_B of the body-attached reference frame, defined as in Equation (4.10) as $\tilde{\omega}_B = \mathbf{R}_{BE}\dot{\mathbf{R}}_{EB}$, we have

$$\mathbf{v}_B^P(t) = \mathbf{v}_B(t) + \tilde{\omega}_B(t)\left(\mathbf{r}_B^P + \mathbf{u}_B^P(t)\right) + \dot{\mathbf{u}}_B^P(t). \tag{6.4}$$

Equation (6.4) is the generalization to a flexible body of the rigid-body velocity field derived in Equation (4.11). It states that the inertial velocity of an arbitrary material point in the body can be written in terms of its relative velocity with respect to a non-inertial reference frame with known kinematics, and the translational and angular velocities of that reference frame. Note that the description of the kinematics so far would have been identical should we have considered a floating reference frame, with the only difference that $\mathbf{r}_B^P$ would be a function of time. However, for a floating frame we would also need additional conditions to determine the instantaneous coordinates of the origin and orientation of the reference system B.

General solutions have been proposed in the literature (Meirovitch, 1991) that use the continuous displacement field as primary variable. They result in hybrid state equations coupling ordinary differential equations on the rigid-body DoFs, and partial differential equations on the elastic DoFs. Here, we assume *a priori* a finite-dimensional approximation to the elastic displacement field. If the coordinates of P are (x, y, z) in the B frame of reference, the approximation is

$$\mathbf{u}_B^P(t) = \mathbf{u}_B(x, y, z; t) = \mathbf{N}(x, y, z)\boldsymbol{\xi}(t), \tag{6.5}$$

where $\mathbf{N}$ is a matrix function defining a time-invariant projection basis in the body-attached frame, which is assumed to be known. It is typically obtained from a finite-element discretization of the structure and therefore we refer to it as the matrix of

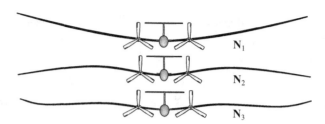

Figure 6.5 Sample projection basis for the displacement field with $N_s = 3$.

shape functions (see Example 6.2 for an illustration of this). In a first approximation, a smaller set of functions defined over the whole structure can also be chosen, as shown in Figure 6.5. Since we are considering the relative displacements to the origin of coordinates, it is always $\mathbf{N}(0,0,0) = \mathbf{0}$. Finally, if N_s elastic DoFs are retained, the matrix $\mathbf{N}$ has dimension $3 \times N_s$, while $\boldsymbol{\xi}(t)$ contains the amplitudes of the N_s elastic DoFs that need to be solved for.

We are also interested in the perturbations of the velocity field defined by Equation (6.4), since they are necessary to construct the EoMs using the variational formulation of Equation (1.1). Small perturbations (or variations) on the three components in frame B of the velocity vector at point P are written as $\delta \mathbf{v}_B^P$. Using this notation, the variation of Equation (6.4) is written as

$$\delta \mathbf{v}_B^P = \delta \mathbf{v}_B - \left(\tilde{\mathbf{r}}_B^P + \tilde{\mathbf{u}}_B^P \right) \delta \boldsymbol{\omega}_B + \tilde{\boldsymbol{\omega}}_B \delta \mathbf{u}_B^P + \delta \dot{\mathbf{u}}_B^P, \qquad (6.6)$$

where we have used the properties in Appendix C.4. It is convenient now to express the variations of the linear and angular velocities at the reference point ($\delta \mathbf{v}_B$ and $\delta \boldsymbol{\omega}_B$, respectively) in terms of small changes in the body-attached frame's position and orientation with respect to the Earth's frame. Noting that the translational velocity of the reference frame is $\mathbf{v}_B = \mathbf{R}_{BE} \mathbf{v}_E$ and that, by definition, $\mathbf{v}_E = \dot{\mathbf{p}}_E$, variations of $\mathbf{v}_B$ can be written as

$$\delta \mathbf{v}_B = \mathbf{R}_{BE} \delta \dot{\mathbf{p}}_E + \delta \mathbf{R}_{BE} \mathbf{v}_E = \delta \dot{\mathbf{p}}_B + \tilde{\boldsymbol{\omega}}_B \delta \mathbf{p}_B + \delta \mathbf{R}_{BE} \mathbf{R}_{EB} \mathbf{v}_B. \qquad (6.7)$$

Since the angular velocity of the body-attached frame of reference is defined as $\tilde{\boldsymbol{\omega}}_B = \mathbf{R}_{BE} \dot{\mathbf{R}}_{EB}$, then its variation yields

$$\widetilde{\delta \boldsymbol{\omega}_B} = \delta \mathbf{R}_{BE} \dot{\mathbf{R}}_{EB} + \mathbf{R}_{BE} \delta \dot{\mathbf{R}}_{EB}. \qquad (6.8)$$

Introducing the relation $\frac{\mathrm{d}}{\mathrm{d}t} \left(\mathbf{R}_{BE} \delta \mathbf{R}_{EB} \right) = \dot{\mathbf{R}}_{EB} \delta \mathbf{R}_{EB} + \mathbf{R}_{BE} \delta \dot{\mathbf{R}}_{EB}$, we can rewrite the previous expression as

$$\begin{aligned} \widetilde{\delta \boldsymbol{\omega}_B} &= \delta \mathbf{R}_{BE} \dot{\mathbf{R}}_{EB} + \tfrac{\mathrm{d}}{\mathrm{d}t} \left(\mathbf{R}_{BE} \delta \mathbf{R}_{EB} \right) - \dot{\mathbf{R}}_{EB} \delta \mathbf{R}_{EB} \\ &= \delta \mathbf{R}_{BE} \mathbf{R}_{EB} \mathbf{R}_{BE} \dot{\mathbf{R}}_{EB} + \tfrac{\mathrm{d}}{\mathrm{d}t} \left(\mathbf{R}_{BE} \delta \mathbf{R}_{EB} \right) - \dot{\mathbf{R}}_{EB} \mathbf{R}_{BE} \mathbf{R}_{EB} \delta \mathbf{R}_{EB}. \end{aligned} \qquad (6.9)$$

If we define now the virtual rotation of the reference frame, $\delta \boldsymbol{\varphi}_B$, as $\widetilde{\delta \boldsymbol{\varphi}_B} = \mathbf{R}_{BE} \delta \mathbf{R}_{EB}$ (see Appendix C.3) and using the properties of the skew-symmetric operator given in Equation (C.23), this expression can be finally simplified as

$$\widetilde{\delta\omega_B} = -\widetilde{\delta\varphi_B}\widetilde{\omega}_B + \widetilde{\delta\dot{\varphi}_B} + \widetilde{\omega}_B\widetilde{\delta\varphi_B}$$
$$= \widetilde{\omega}_B\widetilde{\delta\varphi_B} + \widetilde{\delta\dot{\varphi}_B}.$$
(6.10)

When written in terms of virtual displacements and rotations, Equations (6.7) and (6.10) give the following expressions for the variations of the linear and angular velocities of the reference frame,

$$\delta\mathbf{v}_B = \delta\dot{\mathbf{p}}_B + \tilde{\omega}_B\delta\mathbf{p}_B + \tilde{\mathbf{v}}_B\delta\varphi_B,$$
$$\delta\omega_B = \delta\dot{\varphi}_B + \tilde{\omega}_B\delta\varphi_B.$$
(6.11)

Substituting Equations (6.5) and (6.11) into Equation (6.6), we can finally write the variations of the inertial velocity at the material point P, as

$$\delta\mathbf{v}_B^P = \delta\dot{\mathbf{p}}_B + \tilde{\omega}_B\delta\mathbf{p}_B + \tilde{\mathbf{v}}_B\delta\varphi_B$$
$$- \left(\tilde{\mathbf{r}}_B^P + \widetilde{\mathbf{N}\boldsymbol{\xi}}\right)(\delta\dot{\varphi}_B + \tilde{\omega}_B\delta\varphi_B)$$
(6.12)
$$+ \mathbf{N}\delta\dot{\boldsymbol{\xi}} + \tilde{\omega}_B\mathbf{N}\delta\boldsymbol{\xi}.$$

As it can be seen, perturbations of the translational velocity at point P in the deforming body can appear due to perturbations of the time derivatives of the position of the origin of the body-attached reference frame, the orientation of the body-attached frame, and the rate of change of the elastic deformations of the body. Furthermore, as the time derivatives are evaluated with respect to a moving (non-inertial) frame, gyroscopic terms appear associated with each of those derivatives.

6.3.2 Dynamics

The previous description of the kinematics of the deforming body is used next to evaluate its instantaneous energy state, which can then be directly fed into Hamilton's principle, Equation (1.1). In the following subsections, we evaluate successively the total kinetic and strain energies, and the virtual work of the applied forces.

Kinetic Energy
Let ρ_s be the mass density of the particle P and $\mathcal{V}$ the total volume of the body. As the deformations on the flexible body have been assumed to be small, changes in either the total volume or the local mass density during the elastic deformation are considered to be negligible. Consequently, the instantaneous kinetic energy of the body is defined as

$$\mathcal{T} = \frac{1}{2}\int_{\mathcal{V}} \rho_s \mathbf{v}_E^{P\top} \mathbf{v}_E^P \, d\mathcal{V},$$
(6.13)

and since $\mathbf{R}_{EB}^{\top}\mathbf{R}_{EB} = \mathcal{I}$, it can also be written in terms of the components of the velocity vector in the body-attached reference frame as

$$\mathcal{T} = \frac{1}{2}\int_{\mathcal{V}} \rho_s \mathbf{v}_B^{P\top} \mathbf{v}_B^P \, d\mathcal{V}.$$
(6.14)

Therefore, the kinetic energy displays the well-known property of *frame invariance*, that is, it does not depend on the reference frame that is used to project the inertial velocity (Goldstein et al., 2011, Ch. 2). Substituting Equation (6.4) and the approximation of Equation (6.5) for the displacement field, and after integration over the body, the instantaneous kinetic energy can be written as

$$T = \frac{1}{2} \left\{ \mathbf{v}_B^\top \quad \boldsymbol{\omega}_B^\top \quad \dot{\boldsymbol{\xi}}^\top \right\} \mathbf{M}(\boldsymbol{\xi}) \left\{ \begin{array}{c} \mathbf{v}_B \\ \boldsymbol{\omega}_B \\ \dot{\boldsymbol{\xi}} \end{array} \right\},$$

(6.15)

with the mass matrix defined as

$$\mathbf{M}(\boldsymbol{\xi}) = \begin{bmatrix} m\mathcal{I} & -m\tilde{\mathbf{r}}_B^{CM}(\boldsymbol{\xi}) & \mathbf{M}_{\rho\xi} \\ m\tilde{\mathbf{r}}_B^{CM}(\boldsymbol{\xi}) & \mathbf{I}_B(\boldsymbol{\xi}) & \mathbf{M}_{\varphi\xi}(\boldsymbol{\xi}) \\ \mathbf{M}_{\rho\xi}^\top & \mathbf{M}_{\varphi\xi}^\top(\boldsymbol{\xi}) & \mathbf{M}_{\xi\xi} \end{bmatrix}.$$

(6.16)

This is a symmetric matrix of dimension $N_s + 6$ that depends on the total mass of the aircraft, m, which is assumed to be constant, the instantaneous position of the center of mass (measured from the origin of the body-attached frame), and the moment of inertia of the body. These last two magnitudes change as the flexible body deforms and are defined, respectively, as

$$\mathbf{r}_B^{CM}(\boldsymbol{\xi}) = \frac{1}{m} \int_{\mathcal{V}} \rho_s \left(\mathbf{r}_B^P + \mathbf{N}\boldsymbol{\xi} \right) d\mathcal{V},$$

(6.17a)

$$\mathbf{I}_B(\boldsymbol{\xi}) = -\int_{\mathcal{V}} \rho_s \left(\tilde{\mathbf{r}}_B^P + \widetilde{\mathbf{N}\boldsymbol{\xi}} \right) \left(\tilde{\mathbf{r}}_B^P + \widetilde{\mathbf{N}\boldsymbol{\xi}} \right) d\mathcal{V}.$$

(6.17b)

Note that for the case of a rigid aircraft, these expressions simplify into those of Section 4.3.1, with the difference that there has been no assumption here (so far) about the origin of the body-attached frame being the center of mass. Indeed, Equation (6.17a) shows that the CM of a deforming body does not track the motions of a particular material point.

The second set of coefficients in Equation (6.16) corresponds to the distributed inertia of the deforming body, which is described by the following three matrices,

$$\mathbf{M}_{\rho\xi} = \int_{\mathcal{V}} \rho_s \mathbf{N} d\mathcal{V},$$

$$\mathbf{M}_{\varphi\xi}(\boldsymbol{\xi}) = \int_{\mathcal{V}} \rho_s \left(\tilde{\mathbf{r}}_B^P + \widetilde{\mathbf{N}\boldsymbol{\xi}} \right) \mathbf{N} d\mathcal{V},$$

(6.18)

$$\mathbf{M}_{\xi\xi} = \int_{\mathcal{V}} \rho_s \mathbf{N}^\top \mathbf{N} d\mathcal{V},$$

where $\mathbf{M}_{\rho\xi}$ and $\mathbf{M}_{\varphi\xi}$ are coupling inertia terms between the rigid-body and the elastic DoFs. As it can be seen, the mass matrix is a function of the instantaneous shape of the body, which is determined by the vector of elastic DoFs, $\boldsymbol{\xi}$. Variations of the kinetic energy given in Equation (6.14) yield

$$\delta T = \frac{1}{2} \int_{\mathcal{V}} \rho_s \delta \mathbf{v}_B^{P\top} \mathbf{v}_B^P d\mathcal{V} + \frac{1}{2} \int_{\mathcal{V}} \rho_s \mathbf{v}_B^{P\top} \delta \mathbf{v}_B^P d\mathcal{V} = \int_{\mathcal{V}} \rho_s \delta \mathbf{v}_B^{P\top} \mathbf{v}_B^P d\mathcal{V},$$

(6.19)

where we have used in the last equality the property that the transpose of a scalar returns the scalar itself. Next, we substitute Equations (6.4) and (6.12) into this expression and integrate them by parts to remove the time derivatives on the variations of the DoFs. This naturally results in initial and end conditions on the time integration interval, which are not explicitly included here to simplify the algebra. Note also that, in practice, end conditions are replaced by a full definition of the dynamic state of the system at the initial time.

After some algebraic manipulation (including the properties shown in Equation (C.23)), the variation of the kinetic energy can be written as

$$\delta\mathcal{T} = \left\{ \delta\mathbf{p}_B^\top \quad \delta\boldsymbol{\varphi}_B^\top \quad \delta\boldsymbol{\xi}^\top \right\} \left(-\mathbf{M}(\boldsymbol{\xi}) \left\{ \begin{matrix} \dot{\mathbf{v}}_B \\ \dot{\boldsymbol{\omega}}_B \\ \ddot{\boldsymbol{\xi}} \end{matrix} \right\} - \mathbf{D}(\mathbf{v}_B, \boldsymbol{\omega}_B, \boldsymbol{\xi}) \left\{ \begin{matrix} \mathbf{v}_B \\ \boldsymbol{\omega}_B \\ \dot{\boldsymbol{\xi}} \end{matrix} \right\} \right), \quad (6.20)$$

where the perturbation of the aircraft kinematics on a moving (i.e., non-inertial) reference frame has introduced the following *gyroscopic matrix* function,

$$\mathbf{D}(\mathbf{v}_B, \boldsymbol{\omega}_B, \boldsymbol{\xi}) = \begin{bmatrix} m\tilde{\boldsymbol{\omega}}_B & -m\tilde{\boldsymbol{\omega}}_B\tilde{\mathbf{r}}_B^{\mathrm{CM}}(\boldsymbol{\xi}) & 2\tilde{\boldsymbol{\omega}}_B\mathbf{M}_{\rho\xi} \\ m\tilde{\boldsymbol{\omega}}_B\tilde{\mathbf{r}}_B^{\mathrm{CM}}(\boldsymbol{\xi}) & \tilde{\boldsymbol{\omega}}_B\mathbf{I}_B(\boldsymbol{\xi}) - m\tilde{\mathbf{v}}_B\tilde{\mathbf{r}}_B^{\mathrm{CM}}(\boldsymbol{\xi}) & 2\mathbf{M}_{\varphi\omega\xi}(\boldsymbol{\omega}_B, \boldsymbol{\xi}) \\ \mathbf{M}_{\rho\xi}^\top\tilde{\boldsymbol{\omega}}_B & -\mathbf{M}_{\varphi\omega\xi}^\top(\boldsymbol{\omega}_B, \boldsymbol{\xi}) & 2\mathbf{M}_{\xi\omega\xi}(\boldsymbol{\omega}_B) \end{bmatrix}, \quad (6.21)$$

in which all terms have been previously introduced, except for two new ones,

$$\begin{aligned} \mathbf{M}_{\varphi\omega\xi}(\boldsymbol{\omega}_B, \boldsymbol{\xi}) &= \int_{\mathcal{V}} \rho_s \left(\tilde{\mathbf{r}}_B^P + \widetilde{\mathbf{N}\boldsymbol{\xi}} \right) \tilde{\boldsymbol{\omega}}_B \mathbf{N} \, \mathrm{d}\mathcal{V}, \\ \mathbf{M}_{\xi\omega\xi}(\boldsymbol{\omega}_B) &= \int_{\mathcal{V}} \rho_s \mathbf{N}^\top \tilde{\boldsymbol{\omega}}_B \mathbf{N} \, \mathrm{d}\mathcal{V}. \end{aligned} \quad (6.22)$$

In this general form of the EoMs, the gyroscopic effects introduce up to cubic terms in the DoFs ($\mathbf{v}_B, \boldsymbol{\omega}_B, \boldsymbol{\xi}$). Consequently, the definition of the gyroscopic matrix in Equation (6.21) is not unique, as multiple (but equivalent) combinations of the products of the three variables in the problem could be chosen. Note finally that the gyroscopic matrix, $\mathbf{D}$, as opposed to the mass matrix, $\mathbf{M}$, does not display particular symmetries.

Strain Energy

As it was mentioned earlier, the evaluation of the internal restoring forces on the flexible body is greatly simplified by the assumption that the elastic displacements are small. In this case, the strain energy of the system is defined by a constant (symmetric) *stiffness matrix*, $\mathbf{K}_{\xi\xi}$, of dimension N_s associated with the vector of elastic DoFs defined in Equation (6.5), that is,

$$\mathcal{U} = \frac{1}{2}\boldsymbol{\xi}^\top \mathbf{K}_{\xi\xi}\boldsymbol{\xi}. \quad (6.23)$$

Note that $\boldsymbol{\xi}$ measures the displacement on the deforming body with respect to the origin of the body-attached reference frame and therefore does not include rigid-body motions. Consequently, $\mathbf{K}_{\xi\xi}$ is nonsingular. Note also that the linear elasticity assumption resulting from the small displacements is independent of whether or not those displacements change the inertia of the vehicle, as those are two different physical processes. As a result, the stiffness matrices in Equations (5.5) and (6.23) can be chosen to be the same.

Taking variation of Equation (6.23), we have

$$\delta \mathcal{U} = \frac{1}{2}\delta \boldsymbol{\xi}^{\top} \mathbf{K}_{\xi\xi} \boldsymbol{\xi} + \frac{1}{2}\boldsymbol{\xi}^{\top} \mathbf{K}_{\xi\xi} \delta \boldsymbol{\xi} = \delta \boldsymbol{\xi}^{\top} \mathbf{K}_{\xi\xi} \boldsymbol{\xi}, \tag{6.24}$$

where the last relation results from the symmetry of the stiffness matrix. Note that variation of the strain energy is independent of the rigid-body DoFs.

Virtual Work

The most general external loading is a volumetric force distribution $\mathbf{f}_B^P$ applied through the body. Following on the elastic assumption above, we also assume that the deformations are sufficiently small so that the external forces can be considered as applied on the undeformed structure. We then define the resultant forces and moments at the origin of the body-attached frame, written in terms of their components in that reference frame, as

$$\mathbf{f}_B = \int_{\mathcal{V}} \mathbf{f}_B^P \, d\mathcal{V},$$
$$\mathbf{m}_B = \int_{\mathcal{V}} \tilde{\mathbf{r}}_B^P \mathbf{f}_B^P \, d\mathcal{V}, \tag{6.25}$$

and the *generalized forces* associated with the elastic DoFs, obtained from the projection of the distributed forces on the shape functions as

$$\boldsymbol{\chi} = \int_{\mathcal{V}} \mathbf{N}^{\top} \mathbf{f}_B^P \, d\mathcal{V}. \tag{6.26}$$

Similar expressions are obtained for external forces applied on the surface of the body (which include the aerodynamic forces) and even point forces (typically used to model the propulsion system), but they are not explicitly included here to simplify the description. The force distribution resulting from Equation (6.26) is known as *consistent* loading, to differentiate it from other schemes for mapping external forces onto the structural equations not built on variational principles. Note that Equations (6.25) and (6.26) correspond, respectively, to the rigid-body and distributed loading shown in Figure 5.1. They give the following expression for the virtual work in Equation (1.1),

$$\delta \mathcal{W} = \delta \mathbf{p}_B^{\top} \mathbf{f}_B + \delta \boldsymbol{\varphi}_B^{\top} \mathbf{m}_B + \delta \boldsymbol{\xi}^{\top} \boldsymbol{\chi}. \tag{6.27}$$

For the applications in this book, the external forces include the aerodynamic, propulsive, and gravitational loads on an aircraft. We refer to standard descriptions in flight mechanics for the propulsion system (Section 4.3.4), and we will discuss nonstationary aerodynamic forces in Section 6.5.1. The gravitational forces in body-attached axes are obtained noting that the z axis in the Earth frame is commonly defined to be vertical and positive down, as it was done here for the rigid aircraft in Section 4.2.1. As a result, if the acceleration of gravity is g, then $\mathbf{f}_{gB}^P = \mathbf{R}_{BE}\rho_s g \mathbf{e}_3$, with $\mathbf{e}_3 = (0,0,1)^{\top}$ the unit vector along the z axis, which after integration gives

$$\begin{Bmatrix} \mathbf{f}_{gB} \\ \mathbf{m}_{gB} \\ \boldsymbol{\chi}_g \end{Bmatrix} = g\mathbf{M}(\boldsymbol{\xi}) \begin{Bmatrix} \mathbf{R}_{BE}\mathbf{e}_3 \\ \mathbf{0} \\ \mathbf{0} \end{Bmatrix}, \tag{6.28}$$

that is, in the local frame, the applied forces due to Earth's gravitation scale directly with the mass matrix defined in Equation (6.16). As a result, changes on attitude of the body produce rotations of the gravitational acceleration at the center of mass, but also a redistribution of the gravitational loads among the elastic DoFs. If Euler angles are used to describe the orientation of the body-attached reference frame, as in Figure 4.3, then $\mathbf{R}_{EB}$ is written as in Equation (4.1), which allows introducing the gravity acceleration vector in body-attached frame components, i.e.,

$$\mathbf{g}_B = g\mathbf{R}_{BE}\mathbf{e}_3 = \begin{Bmatrix} -g\sin\theta \\ g\cos\theta\sin\phi \\ g\cos\theta\cos\phi \end{Bmatrix}. \tag{6.29}$$

Equations of Motion

Substituting (6.20), (6.24), and (6.27) into Hamilton's principle, Equation (1.1), and collecting terms associated with each virtual displacement variable, we obtain the flexible-body EoMs as

$$\mathbf{M}(\boldsymbol{\xi}) \begin{Bmatrix} \dot{\mathbf{v}}_B \\ \dot{\boldsymbol{\omega}}_B \\ \ddot{\boldsymbol{\xi}} \end{Bmatrix} + \mathbf{D}(\mathbf{v}_B,\boldsymbol{\omega}_B,\boldsymbol{\xi}) \begin{Bmatrix} \mathbf{v}_B \\ \boldsymbol{\omega}_B \\ \dot{\boldsymbol{\xi}} \end{Bmatrix} + \begin{bmatrix} \mathbf{0} & & \\ & \mathbf{0} & \\ & & \mathbf{K}_{\xi\xi} \end{bmatrix} \boldsymbol{\xi} = \mathbf{M}(\boldsymbol{\xi}) \begin{Bmatrix} \mathbf{g}_B \\ \mathbf{0} \\ \mathbf{0} \end{Bmatrix} + \begin{Bmatrix} \mathbf{f}_B \\ \mathbf{m}_B \\ \boldsymbol{\chi} \end{Bmatrix}, \tag{6.30}$$

with the mass, gyroscopic, and stiffness matrices defined in (6.16), (6.21), and (6.23), respectively. The configuration space (see Section 1.4.1) is, therefore, given by a mix of velocities that describe the dynamics of the body-attached frame, and displacements, which describe the deformed state of the aircraft. This is typical of problems involving both elastic and rigid-body effects. As the gyroscopic terms are associated with the velocities in the equation, they are often referred to as damping terms. This may be however misleading, as the system thus far is conservative (it has no dissipative elements) and damping will only explicitly appear in our models when the aerodynamic forces are included in the description of the aircraft dynamics.

Equation (6.30) needs to be solved together with the initial conditions on the rigid-body velocities, the elastic displacements, and the time derivatives of the elastic displacements. This model for the aircraft dynamics has retained the nonlinear (gyroscopic) terms associated with the rigid-body DoFs and the (also nonlinear) effect of the instantaneous shape on the rigid-body inertia constants while assuming that the amplitude of the elastic displacements is sufficiently small to lead to a linear stiffness matrix. As the gravitational acceleration, Equation (6.29), depends on the orientation of the aircraft, Equation (6.30) needs to be solved together with the propagation equation, Equation (4.16), which is used here to compute the instantaneous attitude of the body-attached reference frame B from the angular velocity $\boldsymbol{\omega}_B = (\omega_x, \omega_y, \omega_z)^\top$, that is,

$$\begin{aligned} \dot{\theta} &= \omega_y\cos\phi - \omega_z\sin\phi, \\ \dot{\phi} &= \omega_x + \omega_y\sin\phi\tan\theta + \omega_z\cos\phi\tan\theta. \end{aligned} \tag{6.31}$$

It is important to remark that Equation (6.31) gives the Euler angles of the body-attached frame B, but that the definition of that frame is not unique. As a result, both angles generally do not coincide with those of a frame at the aircraft CM, with the difference being captured by the coupling terms in the mass matrix.

Therefore, the dynamics of a flexible aircraft under linear elastic assumptions is described by a set of $N_s + 8$ nonlinear equations. The nonlinearities include polynomial terms up to third order in the gyroscopic forces in Equation (6.30), as well as a trigonometric dependency of the aircraft orientation in Equations (6.29) and (6.31). The number of equations can, however, increase if the external forcing terms also have internal dynamics (in particular, unsteady aerodynamics and actuator models as considered in Section 6.5). Similar expressions have been derived by Meirovitch (1991) and Haghighat et al. (2012a), among others. Solution methods will be discussed within the more general fully nonlinear description of Chapter 9, but for simple problems, as in Example 6.2, it is normally sufficient to transform the equations into first-order ODEs, with state vector $\left\{ \mathbf{v}_B^\top \quad \boldsymbol{\omega}_B^\top \quad \boldsymbol{\xi}^\top \quad \dot{\boldsymbol{\xi}}^\top \right\}^\top$ and use Runge–Kutta algorithms or other time-marching schemes that provide the desired accuracy. As the system involves global and local equations on the resultant forces/moments and the local structural response, the solution process typically benefits from matrix preconditioning to ensure that all equations are scaled to have comparable magnitudes (Zwölfer and Gerstmayr, 2019).

Nonlinear Rigid-Linear Elastic Equations of Motion

A major simplification of the solution process is obtained if the elastic displacements in Equation (6.30) are sufficiently small so as not to affect the instantaneous global inertia characteristics of the deforming body. This was already assumed in Section 5.3, but there we also assumed very slow dynamics, which is not the case here. This situation corresponds here to setting $\boldsymbol{\xi} = \mathbf{0}$ in both the mass and the gyroscopic matrices defined earlier. As a result, the mass matrix becomes constant while the gyroscopic terms only retain the couplings associated with the rigid-body velocities. Note, however, that the actual displacements are still used to define the instantaneous shape of the aircraft in the computation of the aerodynamic forces, although this has not yet been considered here. It also becomes convenient in this case to take the origin of coordinates at the center of mass of the body. As result, the aircraft dynamics are approximated as

$$
\mathbf{M}(0) \begin{Bmatrix} \dot{\mathbf{v}}_B \\ \dot{\boldsymbol{\omega}}_B \\ \ddot{\boldsymbol{\xi}} \end{Bmatrix} + \begin{bmatrix} m\tilde{\boldsymbol{\omega}}_B & \mathbf{0} & 2\tilde{\boldsymbol{\omega}}_B \mathbf{M}_{\rho\xi} \\ \mathbf{0} & \tilde{\boldsymbol{\omega}}_B \mathbf{I}_B(0) & 2\mathbf{M}_{\varphi\omega\xi}(\boldsymbol{\omega}_B,0) \\ \mathbf{M}_{\rho\xi}^\top \tilde{\boldsymbol{\omega}}_B & -\mathbf{M}_{\varphi\omega\xi}^\top(\boldsymbol{\omega}_B,0) & 2\mathbf{M}_{\xi\omega\xi}(\boldsymbol{\omega}_B) \end{bmatrix} \begin{Bmatrix} \mathbf{v}_B \\ \boldsymbol{\omega}_B \\ \dot{\boldsymbol{\xi}} \end{Bmatrix}
$$
$$
+ \begin{bmatrix} \mathbf{0} \\ & \mathbf{0} \\ & & \mathbf{K}_{\xi\xi} \end{bmatrix} \begin{Bmatrix} \mathbf{0} \\ \mathbf{0} \\ \boldsymbol{\xi} \end{Bmatrix} = \mathbf{M}(0) \begin{Bmatrix} \mathbf{g}_B \\ \mathbf{0} \\ \mathbf{0} \end{Bmatrix} + \begin{Bmatrix} \mathbf{f}_B \\ \mathbf{m}_B \\ \boldsymbol{\chi} \end{Bmatrix},
$$

$$(6.32)$$

with, as we have chosen origin at the CM,

$$\mathbf{M}(0) = \begin{bmatrix} m\mathcal{I} & \mathbf{0} & \mathbf{M}_{\rho\xi} \\ \mathbf{0} & \mathbf{I}_B(0) & \mathbf{M}_{\varphi\xi}(0) \\ \mathbf{M}_{\rho\xi}^\top & \mathbf{M}_{\varphi\xi}^\top(0) & \mathbf{M}_{\xi\xi} \end{bmatrix}. \tag{6.33}$$

This is a common approximation in the dynamics of flexible vehicles (Waszak and Schmidt, 1988). We refer to them as the *nonlinear rigid-linear elastic* EoMs, which are solved together with the propagation equations, Equation (6.31). Note that it is not a consistent linearization of the elastic part of the problem, as discussed by Hesse and Palacios (2012) and Guimaraes Neto et al. (2016), since we have neglected the structural displacements but have retained the structural velocities in the inertia terms. The linearization of the flexible body dynamic equations is presented in the following section.

Under the previous assumptions, the only remaining nonlinear terms in Equation (6.32) involve the angular velocity ω_B of the reference frame and possibly a second variable, as it was the case in the rigid-body EoMs. Indeed, in the limit case when the elastic displacements are identically zero, only the first six rows and columns are retained in Equation (6.32), which then coincides with the rigid-body equations, given by Equation (4.26).

6.3.3   Linearized Equations of Motion

Linearized EoMs are often required for stability analysis and control design. They are also useful to accelerate the convergence of the nonlinear equations in gradient-based numerical schemes. In this section, they are first derived from Equation (6.30) around an arbitrary reference condition, and then particularized for Equation (6.32) to describe small-perturbation dynamics about straight flight.

General Case

We consider, as we did for rigid aircraft in Section 4.4.2, small perturbations of Equation (6.30) around an equilibrium (trim) condition. Let $(\mathbf{v}_{B0}, \omega_{B0}, \boldsymbol{\xi}_0, \theta_0, \phi_0)$ be the (constant) values of the DoFs at the equilibrium trajectory, including nonzero pitch and roll angles, while $\dot{\boldsymbol{\xi}}_0 = \mathbf{0}$, that is, the aircraft is allowed to have static deformations at the reference trajectory while moving as a rigid body. This equilibrium is necessarily associated with some trim inputs, which are not explicitly discussed here and define the corresponding steady maneuver. Finally, and without loss of generality, the origin for the body-attached reference frame, B, is set at the location of the CM of the deformed vehicle *at the equilibrium condition*, that is, we take $\mathbf{r}_{B0}^{CM} = \mathbf{0}$. The perturbation DoFs are then defined as

$$\begin{aligned} \mathbf{v}_B &= \mathbf{v}_{B0} + \Delta\mathbf{v}_B, & \theta &= \theta_0 + \Delta\theta, \\ \omega_B &= \omega_{B0} + \Delta\omega_B, & \phi &= \phi_0 + \Delta\phi, \\ \boldsymbol{\xi} &= \boldsymbol{\xi}_0 + \Delta\boldsymbol{\xi}. \end{aligned} \tag{6.34}$$

While integration of the angular velocity via Equation (4.16) yields time-varying attitude of the aircraft, those changes are typically considered to be sufficiently slow for the analysis to have local validity at a reference point during the maneuver, as it was done in Section 4.5. After some straightforward, although also a bit tedious, algebra, the linearized dynamics can be described by ordinary differential equations with constant coefficients as

$$\mathbf{M}_0 \left\{ \begin{array}{c} \Delta \dot{\mathbf{v}}_B \\ \Delta \dot{\boldsymbol{\omega}}_B \\ \Delta \ddot{\boldsymbol{\xi}} \end{array} \right\} + \mathbf{D}_0 \left\{ \begin{array}{c} \Delta \mathbf{v}_B \\ \Delta \boldsymbol{\omega}_B \\ \Delta \dot{\boldsymbol{\xi}} \end{array} \right\} + \mathbf{K}_0 \left\{ \begin{array}{c} \mathbf{0} \\ \mathbf{0} \\ \Delta \boldsymbol{\xi} \end{array} \right\} = \mathbf{G}_0 \left\{ \begin{array}{c} \Delta \theta \\ \Delta \phi \end{array} \right\} + \left\{ \begin{array}{c} \Delta \mathbf{f}_B \\ \Delta \mathbf{m}_B \\ \Delta \boldsymbol{\chi} \end{array} \right\}, \tag{6.35}$$

where $\mathbf{M}_0 = \mathbf{M}(\boldsymbol{\xi}_0)$ with $\mathbf{r}_{B0}^{\mathrm{CM}} = \mathbf{0}$, and the tangent gyroscopic and stiffness matrices are given, respectively, by

$$\mathbf{D}_0 = \begin{bmatrix} m\tilde{\boldsymbol{\omega}}_{B0} & -m\tilde{\mathbf{v}}_{B0} & 2\tilde{\boldsymbol{\omega}}_{B0}\mathbf{M}_{\rho\xi} \\ \mathbf{0} & \tilde{\boldsymbol{\omega}}_{B0}\mathbf{I}_{B0} - \tilde{\mathbf{h}}_{B0} & 2\mathbf{M}_{\varphi\omega\xi0} \\ \mathbf{M}_{\rho\xi}^{\top}\tilde{\boldsymbol{\omega}}_{B0} & -\mathbf{M}_{\rho\xi}^{\top}\tilde{\mathbf{v}}_{B0} - \mathbf{M}_{\varphi\xi0}^{\top}\tilde{\boldsymbol{\omega}}_{B0} - 2\mathbf{M}_{\varphi\omega\xi0}^{\top} & 2\mathbf{M}_{\xi\omega\xi0} \end{bmatrix}, \tag{6.36}$$

and

$$\mathbf{K}_0 = \begin{bmatrix} \mathbf{0} & \mathbf{0} & \tilde{\boldsymbol{\omega}}_{B0}\tilde{\boldsymbol{\omega}}_{B0}\mathbf{M}_{\rho\xi} \\ \mathbf{0} & \mathbf{0} & \tilde{\boldsymbol{\omega}}_{B0}\tilde{\boldsymbol{\omega}}_{B0}\mathbf{M}_{\varphi\xi0} + \tilde{\mathbf{g}}_{B0}\mathbf{M}_{\rho\xi} \\ \mathbf{0} & \mathbf{0} & \mathbf{K}_{\xi\xi} + \mathbf{K}_{\omega\omega0} \end{bmatrix}. \tag{6.37}$$

Note that they are all constant matrices. To simplify the notation, we have introduced the zero subindex on all variables that are evaluated at the equilibrium condition. The final term in Equation (6.35) corresponds to the changes in the external forces on the system, which includes both incremental values of the resultant forces and moments at the origin of the body-attached frame, $\Delta \mathbf{f}_B$ and $\Delta \mathbf{m}_B$, respectively, and the (incremental) generalized forces associated with the elastic DoFs, $\Delta \boldsymbol{\chi}$. It is also interesting to note that linearization of the gyroscopic terms has resulted in explicit damping terms on the system dynamics, and that there is a contribution to the tangent stiffness matrix, $\mathbf{K}_0$, from the gravitational forces (the term that includes the gravity acceleration vector, Equation (6.29), at the equilibrium point, $\mathbf{g}_{B0}$). They are associated with the resultant moments produced by the changes on CM position that occur as the aircraft deforms. Although they are typically very small and can be often neglected, they have been retained here for consistency and to keep the generality of the formulation. We have also introduced in Equation (6.35) the constant matrix of *gravitational stiffness*, which is defined as

$$\mathbf{G}_0 = \begin{bmatrix} mg\mathcal{I} \\ \mathbf{0} \\ g\mathbf{M}_{\rho\xi}^{\top} \end{bmatrix} \begin{bmatrix} -\cos\theta_0 & 0 \\ -\sin\theta_0\sin\phi_0 & \cos\theta_0\cos\phi_0 \\ -\sin\theta_0\cos\phi_0 & -\cos\theta_0\sin\phi_0 \end{bmatrix}. \tag{6.38}$$

In the expressions above we have introduced the total angular momentum about the origin of coordinates, $\mathbf{h}_B(t) = \mathbf{I}_B(\boldsymbol{\xi}(t))\boldsymbol{\omega}_B(t)$, as we did in Section 4.3.1, and have also defined the *centrifugal stiffness* as

$$\mathbf{K}_{\omega\omega0} = \int_{\mathcal{V}} \rho_s \mathbf{N}^{\top} \tilde{\boldsymbol{\omega}}_{B0}\tilde{\boldsymbol{\omega}}_{B0}\mathbf{N} \, \mathrm{d}\mathcal{V}. \tag{6.39}$$

Similar to the gravitational stiffness defined in Equation (6.38), additional contributions to the stiffness, damping, and even mass matrices need to be included when

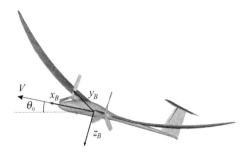

Figure 6.6 Reference conditions: steady climb of a flexible aircraft.

the external forces also depend on the perturbation DoFs. Aerodynamic forces, which vary with the aircraft velocity, are a case in point and are discussed in Section 6.5.

Finally, the gravitational forces in Equation (6.35) depend on the perturbation angles in pitch and roll angles, and, therefore, this system needs to be integrated together with the linearized kinematic relations between Euler angles and angular velocities resulting from Equation (4.16).

Linearization Around a Steady Climb

While the previous section has described the most general linearized dynamics, very often the analysis is restricted to the linear response around straight flight conditions, that is, small perturbations from level, climb, or descent flight. At the equilibrium condition, the aircraft has $\omega_{B0} = \mathbf{0}$ and $\phi_0 = 0$, and although it may display a significant elastic deformation, it will be following a rectilinear path with angle θ_0 with respect to the horizontal. This reference condition is shown in Figure 6.6.

The linearized EoMs around this reference are still of the form of Equation (6.35) but with the constant matrices now defined as

$$
\mathbf{K}_0 = \begin{bmatrix} \mathbf{0} & \mathbf{0} & \mathbf{0} \\ \mathbf{0} & \mathbf{0} & \mathbf{0} \\ \mathbf{0} & \mathbf{0} & \mathbf{K}_{\xi\xi} \end{bmatrix}, \qquad
\mathbf{M}_0 = \begin{bmatrix} m\mathcal{I} & \mathbf{0} & \mathbf{M}_{\rho\xi} \\ \mathbf{0} & \mathbf{I}_B(\boldsymbol{\xi}_0) & \mathbf{M}_{\varphi\xi}(\boldsymbol{\xi}_0) \\ \mathbf{M}_{\rho\xi}^\top & \mathbf{M}_{\varphi\xi}^\top(\boldsymbol{\xi}_0) & \mathbf{M}_{\xi\xi} \end{bmatrix},
$$

$$
\mathbf{D}_0 = \begin{bmatrix} \mathbf{0} & -m\tilde{\mathbf{v}}_{B0} & \mathbf{0} \\ \mathbf{0} & \mathbf{0} & \mathbf{0} \\ \mathbf{0} & -\mathbf{M}_{\rho\xi}^\top\tilde{\mathbf{v}}_{B0} & \mathbf{0} \end{bmatrix}, \qquad
\mathbf{G}_0 = \begin{bmatrix} mg\mathcal{I} \\ \mathbf{0} \\ g\mathbf{M}_{\rho\xi}^\top \end{bmatrix} \begin{bmatrix} -\cos\theta_0 & 0 \\ 0 & \cos\theta_0 \\ -\sin\theta_0 & 0 \end{bmatrix},
$$

(6.40)

while the linearization of the kinematic relations results again in Equation (4.76), which needs to be solved simultaneously with Equation (6.35). It is clear that the gyroscopic matrix, $\mathbf{D}_0$, is much simplified here, but not so the mass matrix, $\mathbf{M}_0$, which still includes coupling terms between the elastic and rigid-body DoFs. We will see in the following section how the mass matrix can be simplified by adopting modal coordinates.

In summary, Equation (6.30) gives the more general description of the dynamics of a flexible aircraft under the assumption of linear elastic structural displacements. A first simplification was obtained in Equation (6.32) when it was further assumed that the elastic deformations are small enough that the resulting changes in geometry

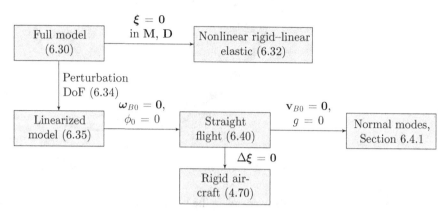

Figure 6.7 Modeling approaches for flexible aircraft dynamics with small-amplitude vibrations.

do not modify the inertia characteristics of the vehicle. It is worth emphasizing this, as it is not always acknowledged in the flight dynamics literature, and it is a more restrictive assumption than the requirement of linear elastic behavior. If the rigid-body velocities also remain small, then any of the previous two descriptions can be linearized around a reference condition – typically the aircraft flying in straight flight (and possibly with deformed wings at that equilibrium), but it could also be of interest to linearize around other conditions for asymptotic stability analysis and/or control design. Equation (6.35) has defined the mass, damping, and stiffness matrices for the more general linear formulation, while Equation (6.40) shows the reduction of that problem to linearizations about straight flight conditions and with inertia constants that are independent of the equilibrium deformation. Finally, if the elastic displacements are identically zero, Equation (6.35) with the coefficients given in Equation (6.40) can be easily seen to reduce to the perturbation equations for the rigid aircraft around steady climb, Equation (4.70). The flow chart in Figure 6.7 summarizes the different modeling assumptions discussed here.

Note that the body-attached reference frame for a flexible aircraft can be defined, without loss of generality, to coincide with the stability axes at a reference condition – typically, the aircraft in level, climb, or descent flight. Stability axes have been defined in Section 4.2.1 and that definition is still valid here. In particular, the reference speed used above in stability axis becomes $\mathbf{v}_{B0} = (V, 0, 0)^\top$. However, the definition of the stability axes changes with the reference flight condition (and more so on a flexible vehicle), while the mass and stiffness of the aircraft are usually known on a single set of axes, often referred to as *structural axes*. This is the axis system used for the generation of the finite-element model of the aircraft. In the discussion above we have assumed, to simplify the discussion, that both set of axes (structural and stability) coincide. In practice, however, the mass and stiffness matrices would need to be rotated to align the structural axes with the forward flight velocity (Baldelli et al., 2006; Goizueta et al., 2021).

6.4 Flexible Body Dynamics in Modal Coordinates

6.4.1 Normal Modes

In the previous section, we have derived the EoMs for a flexible body undergoing small elastic deformations. It has assumed a discretization of the body with known information of both the local inertia and the local stiffness. For complex configurations, such as air vehicles described by a finite-element discretization, this typically results in structural models with hundreds of thousands of DoFs. In practice, the interest is often restricted to dynamics within a relatively narrow frequency bandwidth, either to design a flight control system, or to study the in-flight vibrations of a given subsystem. In those scenarios, it becomes very convenient to reduce the size of the models using the *linear normal modes* (LNMs) (also known as natural vibration modes) of the structure to project the full EoMs.

Perturbations on the equilibrium state of a solid produce elastic waves that propagate within the solid with minimal losses (which come from the structural damping). Those waves are reflected at the boundaries of the solid until they cover the whole solid domain. The multiple reflections lead to interference of the many waves with each other, yielding spatial construction and destruction of the local deformation. For a given structural geometry, material properties (which define the wave propagation speed), and boundary conditions, there are particular wavelengths in which the interference between those traveling waves results in *standing waves*, in a phenomenon known as resonance. In a standing wave, the solid synchronously vibrates with a fixed spatial distribution and periodic amplitude. These periods define the natural or *resonant frequencies* of the structure, while the associated small-amplitude resonant states are the *LNM*. While in theory there is an infinite number of resonant states in a continuum, in practice, internal dissipation in built-up structures allows only the lower-frequency modes to be measurable.

We can define the LNMs around any arbitrary equilibrium state, but for the purpose of vibration analysis, it is customary (and mathematically convenient) to define them on the structure at rest, in vacuum, and in the absence of gravity. Under those assumptions there are neither external nor gyroscopic forces on the system, and the LNMs are the nontrivial solutions of the resulting homogeneous equations. We allow, however, the presence of an initial deformation of the structure, that is, $\boldsymbol{\xi}_0 \neq \mathbf{0}$.[2] This considers the most general case where the mass matrix is $\mathbf{M}_0 = \mathbf{M}(\boldsymbol{\xi}_0)$, although in practice it is often assumed that the inertia is known on the undeformed vehicle, that is, $\mathbf{M}_0 = \mathbf{M}(\mathbf{0})$, Equation (6.33). As a result, the LNMs are identified as the nontrivial solutions of the homogeneous equations derived from Equations (6.35) and (6.40) with $\mathbf{v}_{B0} = \mathbf{0}$ (i.e., the LNMs do not depend on the flight condition) and $g = 0$ (no gravitational effects). Under those assumptions, the damping matrix $\mathbf{D}_0$ is identically zero.

---

[2] Note that there is an inconsistency here, as the deformation can only result from a forcing term. We can, however, mathematically define the mass and, as we will see in Chapter 8, the (tangent) stiffness matrix for any elastic state of the aircraft.

For the modal analysis, we define for convenience a new variable $\Delta \nu$, such that $\Delta \dot{\nu} = \left\{ \Delta \mathbf{v}_B^{\top} \quad \Delta \boldsymbol{\omega}_B^{\top} \right\}^{\top}$, that is, a *fictitious* displacement/rotation vector obtained from integration of the incremental rigid-body velocities. The components of $\Delta \nu$ are known as the rigid-body *quasi-coordinates* (Meirovitch, 1991), which result in a configuration space given by the concatenation of rigid and elastic displacements. With them, the LNMs are obtained from the solutions of the homogeneous system

$$\mathbf{M}_0 \left\{ \begin{matrix} \Delta \ddot{\nu} \\ \Delta \ddot{\xi} \end{matrix} \right\} + \mathbf{K}_0 \left\{ \begin{matrix} \Delta \nu \\ \Delta \xi \end{matrix} \right\} = \mathbf{0}, \tag{6.41}$$

where $\mathbf{M}_0$ and $\mathbf{K}_0$ are the symmetric matrices defined in Equation (6.40). Furthermore, the mass matrix is positive definite while the stiffness is a positive semidefinite matrix (with a null space of dimension 6). It is well-known (Géradin and Rixen, 1997, Ch. 2) that the solutions to Equation (6.41) are harmonic functions in time domain, that is,

$$\left\{ \begin{matrix} \Delta \nu \\ \Delta \xi \end{matrix} \right\} = \sum_{j=1}^{N_s+6} \boldsymbol{\Phi}_{qj} e^{i \omega_j t}, \tag{6.42}$$

where ω_j are the (nonnegative) natural frequencies of the structure and $\boldsymbol{\Phi}_{qj}$ the mode shapes, whose dimension is $N_s + 6$. Substituting Equation (6.42) into Equation (6.41) defines a generalized eigenvalue problem of the form

$$\mathbf{K}_0 \boldsymbol{\Phi}_q = \mathbf{M}_0 \boldsymbol{\Omega}^2 \boldsymbol{\Phi}_q, \tag{6.43}$$

where $\boldsymbol{\Omega} = \mathrm{diag} \left(0,0,0,0,0,0, \omega_1, \omega_2, \ldots, \omega_{N_s} \right)$, that is, the diagonal matrix of the natural frequencies, and the columns of $\boldsymbol{\Phi}_q$, that is, the eigenvectors of the problem, are the corresponding mode shapes known as the *free-vehicle mode shapes*. Due to the structure of the stiffness matrix, this *modal matrix* can be further written as

$$\boldsymbol{\Phi}_q = \left\{ \begin{matrix} \boldsymbol{\Phi}_{rr} & \boldsymbol{\Phi}_{rs} \\ \mathbf{0} & \boldsymbol{\Phi}_{ss} \end{matrix} \right\}. \tag{6.44}$$

Therefore, there are six rigid-body modes that are solely described in terms of the quasi-coordinates in the problem, and N_s *elastic modes*, which, in general have nonzero quasi-coordinate components, $\boldsymbol{\Phi}_{rs}$, due to the rigid-elastic couplings that exist in the mass matrix defined in Equation (6.40).

The natural frequencies are typically ordered such as $\omega_j \leq \omega_k$ for $j < k$. It is also common to normalize the eigenvectors with the mass matrix, in which case they satisfy the following orthogonality conditions,

$$\boldsymbol{\Phi}_q^{\top} \mathbf{M}_0 \boldsymbol{\Phi}_q = \boldsymbol{\mathcal{I}},$$
$$\boldsymbol{\Phi}_q^{\top} \mathbf{K}_0 \boldsymbol{\Phi}_q = \boldsymbol{\Omega}^2. \tag{6.45}$$

This normalization is usually introduced in structural dynamics. However, diagonalization of the mass matrix implies that the rigid-body modes are expressed as translations and rotations about the principal axis of inertia of the undeformed structure, which typically is not a convenient parameterization for the flight dynamics

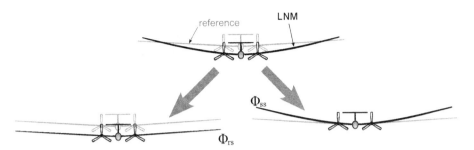

Figure 6.8 Decomposition of a linear normal mode into rigid and elastic contributions.

equations (as we have discussed in Section 4.2.1). A more appropriate normalization of the mode shapes for flexible aircraft dynamic applications is introduced in Section 6.4.3.

The coupling matrix $\mathbf{\Phi}_{rs}$ in Equation (6.44) has dimension $6 \times N_s$ and its columns give the rigid-body contribution to each elastic mode. Enforcing orthonormality between rigid and elastic modes with respect to the mass matrix given in Equation (6.40), one has

$$\mathbf{\Phi}_{rs} = - \begin{bmatrix} m\mathcal{I} & \mathbf{0} \\ \mathbf{0} & \mathbf{I}_B \end{bmatrix}^{-1} \begin{bmatrix} \mathbf{M}_{\rho\xi} \\ \mathbf{M}_{\varphi\xi} \end{bmatrix} \mathbf{\Phi}_{ss}. \tag{6.46}$$

The physical meaning of this equation can be explained as follows. The structural DoFs in each elastic mode shape, given by the columns of $\mathbf{\Phi}_{ss}$, define a deformed state of the aircraft. This elastic deformation would result in a shift of the CM position and/or orientation. The rigid-body component of each elastic mode, defined by $\mathbf{\Phi}_{rs}$, will be such that the CM reference frame remains unaffected by the elastic modes. This is shown schematically in Figure 6.8, where a LNM has been split between its rigid and elastic contributions. Equation (6.46) also implies that, if the inertia distribution is known, only the elastic part, $\mathbf{\Phi}_{ss}$, of the flexible mode shapes needs to be given. The rigid-body contributions for each mode can then be determined by the invariance of the frame at the center of mass.

Normal modes are intrinsic properties of a structure and, therefore, independent of the selection of DoFs. Here, we have used quasi-coordinates to capture the rigid-body motions of the reference frame and elastic displacements relative to that motion. In practice, LNMs of built-up structures are typically obtained from a finite-element discretization (or a ground vibration test) in which the DoFs are the displacements of the structural nodes with respect to an Earth-fixed reference frame. Note that this is effectively only a linear transformation of the problem defined in Equation (6.41), as is exemplified in Example 6.2.

6.4.2 Modal Solution of Linear Problems

The linearized formulation of Equation (6.35) is rarely solved directly, and instead LNMs are first used to project the system onto a few modal coordinates. This is greatly facilitated by ordering the modes by their natural frequencies, which can be used to define a bandwidth of interest beyond which the high-frequency modes can be truncated. More sophisticated strategies that retain static equilibrium (static residualization) or minimize the error for a particular set of sensors and actuators (system balancing) are available but are not discussed here. Further details can be found, for example, in Chapter 6 of Gawronski (2004). In what follows, the modal matrix, $\mathbf{\Phi}_q$, obtained from the solution of Equation (6.43) is used to define the linear transformation

$$\left\{ \begin{matrix} \Delta\nu \\ \Delta\xi \end{matrix} \right\} = \mathbf{\Phi}_q \mathbf{q}, \tag{6.47}$$

where $\mathbf{q}(t)$ is the vector of unknown modal amplitudes that is now used to describe the system dynamics. Therefore, they define a new set of generalized coordinates. In general, a truncated modal expansion is considered, which removes columns of $\mathbf{\Phi}_q$ and results in a narrow rectangular matrix of dimension $(6+N_s) \times N_m$, with $N_m \ll N_s$ the number of retained normal modes. Since the mode shapes are time independent, it is also the case that

$$\left\{ \begin{matrix} \Delta\dot{\nu} \\ \Delta\dot{\xi} \end{matrix} \right\} = \mathbf{\Phi}_q \dot{\mathbf{q}} \quad \text{and} \quad \left\{ \begin{matrix} \Delta\ddot{\nu} \\ \Delta\ddot{\xi} \end{matrix} \right\} = \mathbf{\Phi}_q \ddot{\mathbf{q}}. \tag{6.48}$$

First, note that the quasi-coordinates can be now directly used to obtain the linearized pitch and roll angles since, from Equation (4.76), we have

$$\begin{bmatrix} \Delta\dot{\theta} \\ \Delta\dot{\phi} \end{bmatrix} = \begin{bmatrix} 0 & 0 & 0 & 0 & 1 & 0 \\ 0 & 0 & 0 & 1 & 0 & \tan\theta_0 \end{bmatrix} \Delta\nu, \tag{6.49}$$

which can be directly integrated. It is more convenient to write it as

$$\begin{bmatrix} \Delta\theta \\ \Delta\phi \end{bmatrix} = \mathbf{T}_g \left\{ \begin{matrix} \Delta\nu \\ \Delta\xi \end{matrix} \right\} \tag{6.50}$$

by augmenting the previous relation with rows of zeros.

Substituting now Equations (6.47) and (6.50) into Equation (6.35) and premultiplying by $\mathbf{\Phi}_q^\top$ results in the linear dynamics in modal coordinates being written as

$$\mathbf{M}_q \ddot{\mathbf{q}} + \mathbf{D}_q \dot{\mathbf{q}} + \mathbf{K}_q \mathbf{q} = \mathbf{G}_q \mathbf{q} + \mathbf{f}_q, \tag{6.51}$$

where $\mathbf{M}_q = \mathbf{\Phi}_q^\top \mathbf{M}_0 \mathbf{\Phi}_q = \mathcal{I}$ and $\mathbf{K}_q = \mathbf{\Phi}_q^\top \mathbf{K}_0 \mathbf{\Phi}_q = \mathbf{\Omega}^2$ are the (diagonal) modal mass and stiffness matrices, respectively, $\mathbf{D}_q = \mathbf{\Phi}_q^\top \mathbf{D}_0 \mathbf{\Phi}_q$ is the modal (gyroscopic) damping matrix, and $\mathbf{G}_q = \mathbf{\Phi}_q^\top \mathbf{G}_0 \mathbf{T}_g \mathbf{\Phi}_q$ is the modal gravitational stiffness matrix. Note that the modal damping and gravitational matrices are, in general, not diagonal. Finally, the forcing term in the right-hand side of the equation, $\mathbf{f}_q$, is the vector of modal *generalized forces*, which is obtained from the modal projection of the right-hand side term in Equation (6.35), i.e.,

$$\mathbf{f}_q = \mathbf{\Phi}_q^\top \left\{ \begin{matrix} \Delta\mathbf{f}_B \\ \Delta\mathbf{m}_B \\ \Delta\chi \end{matrix} \right\}. \tag{6.52}$$

As it was mentioned earlier, there are in principle as many mode shapes as DoFs in the original system and, if all modes are retained, the modal matrix $\boldsymbol{\Phi}_q$ is square. For systems with zero gyroscopic damping and gravitational effects, a projection onto modal coordinates results in the dynamics of the structure described by a set of independent single-DoF linear oscillators of the form

$$\ddot{q}_j + \Omega_{jj}^2 q_j = f_{q_j}(t), \tag{6.53}$$

and the frequency content of the external forcing determines which oscillators (which vibration modes) are affected by a particular loading. For example, aeroelastic analysis focuses on the lower frequency spectrum, as higher-frequency vibrations are damped out by the aerodynamic forces. A typical cutoff frequency for a mid-size aircraft would be between 50 and 100 Hz. It is also common to use the modal coordinates to describe the deformed shape of the aircraft in its static (aeroelastic) equilibrium conditions. In that case, a relatively large number of modes may be needed for convergence since the frequency ordering of the modes is no longer relevant (Karpel et al., 1996).

When gyroscopic and gravitational effects are also considered in the linearized dynamics, the additional matrices are in general fully populated and thus introduce couplings between the normal modes. Typically, those couplings are increasingly weaker as the frequency difference between the interacting modes increases, and the modal matrices can still be truncated as part of a convergence analysis (in which enough modes are retained to achieve a certain numerical error). Moreover, even though it has not been considered here, real structures display internal damping whose effect increases with the frequency and dissipates mechanical vibrations beyond a certain frequency.

Example 6.1 Flexible Body with Three Point Masses. This example investigates the linear dynamics of the three-mass system in the x–z plane shown in Figure 6.9a. Two point masses $m/4$ are linked via massless lateral (bending) springs of length l and stiffness $12EI/l^3$ to a central body with mass $m/2$ and moment of inertia $ml^2/2$ about the y axis. This ensemble is restricted to have only planar motion, and initially moves with velocity of magnitude V and direction along a trajectory with a positive angle θ_0 with respect to the horizontal, as shown in Figure 6.9. This could be considered as a simplified model of the longitudinal dynamics of an aircraft with a flexible fuselage, but we do not consider either aerodynamic or propulsion forces in our setup.

The body-fixed reference frame is chosen to be at the middle mass, which is also the CM of the underformed body. This results in a description of the dynamics with the five DoFs shown in Figure 6.9b: two elastic DoFs, ξ_1 and ξ_2, which correspond to normal displacements of each of the edge masses, and three rigid-body DoFs, corresponding to the normal and tangential incremental velocities, v_x and v_z, and the angular velocity, ω_y, at the middle mass. Although they are not explicitly used here, the quasi-coordinates associated with the rigid-body motions are defined, following Equation (6.41), from the integration of $\Delta \dot{\boldsymbol{\nu}} = \left\{ \Delta v_x \quad \Delta v_z \quad \Delta \omega_y \right\}^\top$. This system is

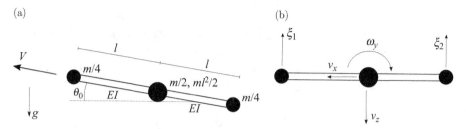

Figure 6.9 Flexible body with three lumped masses. (a) Problem setup and (b) degrees of freedom in body-fixed axes.

subject to gravity, with g the gravitational acceleration. No other external forces are considered.

With this parameterization of the system kinematics, its linearized equations of motion, Equation (6.35), are written as

$$\mathbf{M}_0 \left\{ \begin{matrix} \Delta \dot{v}_x \\ \Delta \dot{v}_z \\ \Delta \dot{\omega}_y \\ \Delta \ddot{\xi}_1 \\ \Delta \ddot{\xi}_2 \end{matrix} \right\} + \mathbf{D}_0 \left\{ \begin{matrix} \Delta v_x \\ \Delta v_z \\ \Delta \omega_y \\ \Delta \dot{\xi}_1 \\ \Delta \dot{\xi}_2 \end{matrix} \right\} + \mathbf{K}_0 \left\{ \begin{matrix} 0 \\ 0 \\ 0 \\ \Delta \xi_1 \\ \Delta \xi_2 \end{matrix} \right\} = \mathbf{G}_0 \Delta \theta, \tag{6.54}$$

which are solved together with the kinematic relation $\Delta\dot\theta = \Delta\omega_y$ from Equation (6.49). The constant matrices in Equation (6.54) are defined as in Equation (6.40). In particular, the nonzero contributions to the stiffness matrix are given by

$$\mathbf{K}_{\xi\xi} = \begin{bmatrix} \frac{12EI}{l^3} & 0 \\ 0 & \frac{12EI}{l^3} \end{bmatrix}, \tag{6.55}$$

and the mass matrix becomes

$$\mathbf{M}_0 = \begin{bmatrix} m\mathcal{I} & \mathbf{0} & \mathbf{M}_{\rho\xi} \\ \mathbf{0} & ml^2 & \mathbf{M}_{\varphi\xi} \\ \mathbf{M}_{\rho\xi}^\top & \mathbf{M}_{\varphi\xi}^\top & \mathbf{M}_{\xi\xi} \end{bmatrix}, \tag{6.56}$$

where

$$\mathbf{M}_{\rho\xi} = -\frac{m}{4} \begin{bmatrix} 0 & 0 \\ 1 & 1 \end{bmatrix}, \qquad \mathbf{M}_{\varphi\xi} = \frac{ml}{4} \begin{bmatrix} 1 & -1 \end{bmatrix}, \qquad \mathbf{M}_{\xi\xi} = \frac{m}{4} \begin{bmatrix} 1 & 0 \\ 0 & 1 \end{bmatrix}. \tag{6.57}$$

Note that the negative terms in the coupling matrix $\mathbf{M}_{\rho\xi}$ are due to the sign convention, as v_z is positive down but the elastic DoFs are positive up. Note also that the existence of negative terms in the $\mathbf{M}_0$ is not inconsistent with it being a positive definite matrix. Submatrices $\mathbf{M}_{\rho\xi}$ and $\mathbf{M}_{\varphi\xi}$ are the coupling terms between the elastic and the rigid-body DoFs, and the inertia of the latter is always obtained after aggregating the distributed inertia on the elastic DoFs.

The gyroscopic (damping) and gravitational matrices defined in Equation (6.40) can be finally written for this problem as, respectively,

$$\mathbf{D}_0 = mV \begin{bmatrix} 0 & 0 & 0 & 0 & 0 \\ 0 & 0 & -1 & 0 & 0 \\ 0 & 0 & 0 & 0 & 0 \\ 0 & 0 & \frac{1}{4} & 0 & 0 \\ 0 & 0 & \frac{1}{4} & 0 & 0 \end{bmatrix}, \qquad \mathbf{G}_0 = mg \begin{bmatrix} -\cos\theta_0 \\ -\sin\theta_0 \\ 0 \\ \frac{1}{4}\sin\theta_0 \\ \frac{1}{4}\sin\theta_0 \end{bmatrix}. \tag{6.58}$$

Note that if the elastic DoFs are removed from those equations, which implies removing the last two rows and columns on all matrices, we are left with the linearized longitudinal equations for the rigid aircraft in climb/descent, Equation (4.77).

The LNM of this structure are defined under still conditions ($V = 0$) and without gravitational effects. That results on a realization of Equation (6.43) of dimension 5, which gives three zero-frequency rigid-body modes and two elastic modes of equal frequency $\omega_1 = \omega_2 = \sqrt{\frac{96EI}{ml^3}}$. The contributions to the modal matrix, Equation (6.44), are

$$\boldsymbol{\Phi}_{rr} = \begin{bmatrix} 1 & 0 & 0 \\ 0 & 1 & 0 \\ 0 & 0 & 1 \end{bmatrix}, \qquad \boldsymbol{\Phi}_{rs} = \frac{1}{4}\begin{bmatrix} 0 & 0 \\ 1 & 1 \\ -1/l & 1/l \end{bmatrix}, \qquad \boldsymbol{\Phi}_{ss} = \begin{bmatrix} 1 & 0 \\ 0 & 1 \end{bmatrix}. \tag{6.59}$$

Note that these mode shapes have not been mass normalized, as this makes for an easier understanding of their physical meaning. As the location of the middle mass coincides with the CM at the reference condition, the rigid-body modes are the same as the rigid-body DoFs, that is, $\boldsymbol{\Phi}_{rr}$ is an identity matrix. Each elastic mode corresponds to a unit displacement of one of the elastic DoFs, that is, $\boldsymbol{\Phi}_{ss}$ is also a unit matrix. However, as the mass matrix, $\mathbf{M}_0$, includes couplings between elastic and rigid DoFs (since both $\mathbf{M}_{\rho\xi}$ and $\mathbf{M}_{\phi\xi}$ are nonzero), the elastic modes also include a nonzero rigid-body component, $\boldsymbol{\Phi}_{rs}$. This enforces the orthogonality (through the mass matrix) between rigid and flexible modes, which for this case means a rigid-body normal displacement and rotation (the quasi-coordinates associated with Δv_z and $\Delta \omega_y$) that keep the center of mass unchanged by the mode shape. It can be easily checked that the matrices above satisfy Equation (6.46).

Combining all the results above, we can write the EoMs in modal coordinates for the three-mass system. They will be of the form of Equation (6.51), with the following matrices

$$\mathbf{M}_q = m\,\text{diag}\left(1,1,l^2,1/8,1/8\right), \qquad \mathbf{K}_q = \frac{12EI}{l^3}\,\text{diag}\left(0,0,0,1,1\right),$$

$$\mathbf{D}_q = -mV \begin{bmatrix} 0 & 0 & 0 & 0 & 0 \\ 0 & 0 & 1 & \frac{1}{4l} & \frac{1}{4l} \\ 0 & 0 & 0 & 0 & 0 \\ 0 & 0 & 0 & \frac{1}{8l} & \frac{1}{8l} \\ 0 & 0 & 0 & \frac{1}{8l} & \frac{1}{8l} \end{bmatrix}, \qquad \mathbf{G}_q = -mg \begin{bmatrix} 0 & 0 & \cos\theta_0 & \frac{1}{4l}\cos\theta_0 & \frac{1}{4l}\cos\theta_0 \\ 0 & 0 & \sin\theta_0 & \frac{1}{4l}\sin\theta_0 & \frac{1}{4l}\sin\theta_0 \\ 0 & 0 & 0 & 0 & 0 \\ 0 & 0 & 0 & \frac{1}{8l}\sin\theta_0 & \frac{1}{8l}\sin\theta_0 \\ 0 & 0 & 0 & \frac{1}{8l}\sin\theta_0 & \frac{1}{8l}\sin\theta_0 \end{bmatrix}.$$

$$\tag{6.60}$$

As it was discussed above, while the modal mass and stiffness are always diagonal, the damping and gravitational matrices are not necessarily so. Indeed, the transformation from physical to modal coordinates typically simplifies the mass matrix at the cost of a more complex gravitational matrix. The main advantage of the modal projection in flexible aircraft dynamics is, however, not a simplification of the algebra, but that it gives the means to achieve a significant reduction in the number of DoFs. While this was not apparent in this example, since it had very few physical DoFs, it will be in Example 6.2.

Example 6.2 Free-flying Beam in Zero Gravity.   Consider the unsupported straight beam shown in Figure 6.10. The aim is to investigate, using modal coordinates, its planar dynamics in the absence of gravitational forces (or moving in a frictionless horizontal surface). An inextensional beam is considered, with (small) elastic deformations in bending about the y axis and negligible shear compliance and rotational inertia. This defines an *Euler–Bernoulli beam*, given by its length L, bending stiffness EI, and mass per unit length $\rho_s A$. The body-fixed reference frame can be chosen to be at any location along the beam, and here it is initially located at one end of the beam, such that the undeformed beam lies along the positive x axis, as shown in Figure 6.10a. Therefore, as the beam deforms, the x–z axes in the figure track the instantaneous position and orientation of one of the free ends of the beam.

The EoMs of the flexible beam can be derived using Hamilton's principle, Equation (1.1). In addition to the rigid-body DoFs, we need to describe the elastic deformations of the beam, for which we use the *relative displacement* along the body-attached z axis, $w(x,t)$, as shown in Figure 6.10a. This results, according to the (linear) Euler–Bernoulli model, in an instantaneous strain energy given as

$$\mathcal{U}(t) = \frac{1}{2} \int_0^L EI w''^2 \, \mathrm{d}x. \tag{6.61}$$

It is important to highlight that the deformed shape depicted in Figure 6.10a does not imply that the left end of the beam is clamped, since that point is moving. A distinctive feature of describing the dynamics of a flexible body in a body-fixed reference frame is that the origin of the frame tracks a point of the structure, and the displacement field is measured with respect to it. The stiffness can be obtained by taking variations

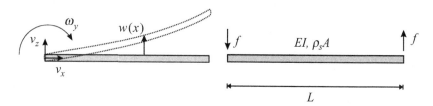

Figure 6.10 Free-flying beam problem setup. (a) Definitions of reference frame and relative elastic displacement and (b) applied forces.

(a)   (b)

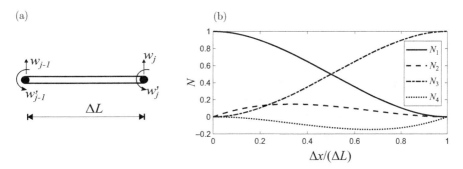

Figure 6.11 Element discretization and shape functions in an Euler–Bernoulli beam. (a) The j-th bending element of the 1-D beam and (b) element shape functions (N_1 and N_3 interpolate nodal displacements and N_2 and N_4 interpolate nodal rotations). [06FlexibleAircraft/freebeam/plotshapes.m]

in Equation (6.61), i.e., $\delta \mathcal{U} = \int_0^L \delta w'' EI w'' \, \mathrm{d}x$, and then seeking numerical integration using a finite-element discretization.

For that purpose, the beam is split into N elements of equal length, $\Delta L = L/N$. For each element j, with $j = 1, \ldots, N$, we take as independent DoFs the nodal values of the displacements, w_j, and rotations, w'_j, as shown in Figure 6.11a, which also shows the sign convention used in the solution. The displacements within each element are now interpolated from the nodal DoFs. For a 2-D Euler–Bernoulli beam, the simplest interpolation uses Hermite polynomials within each pair of nodes (Géradin and Rixen, 1997, Ch. 5), and this is assumed in what follows, with the element shape functions being shown in Figure 6.11b. We can then arrange a vector of discrete elastic variables as

$$\boldsymbol{\xi} = \left(w_1, w'_1, w_2, w'_2, \ldots, w_N, w'_N \right)^\top, \tag{6.62}$$

where, following the definitions above, we have not included the node at the origin of the body-attached reference frame. The finite-element interpolation results in

$$w(x,t) = \mathbf{N}(x)\boldsymbol{\xi}(t), \tag{6.63}$$

where $\mathbf{N}$ denotes the row matrix with $2N$ shape functions that determines the interpolation across the full domain (each term is zero everywhere except for the element where they perform the interpolation).

As a result, the perturbation to the strain energy resulting from the virtual displacements $\delta\boldsymbol{\xi}$ can be written as

$$\delta\mathcal{U} = \int_0^L \delta\boldsymbol{\xi}^\top \mathbf{N}''^\top EI \mathbf{N}'' \boldsymbol{\xi} \, \mathrm{d}x = \delta\boldsymbol{\xi}^\top \left(\int_0^L \mathbf{N}''^\top EI \mathbf{N}'' \, \mathrm{d}x \right) \boldsymbol{\xi}, \tag{6.64}$$

which defines the stiffness matrix as

$$\mathbf{K}_{\xi\xi} = EI \int_0^L \mathbf{N}''^\top \mathbf{N}'' \, \mathrm{d}x, \tag{6.65}$$

for a homogenous beam (EI is constant).

Since we have not included the displacement at the origin in the elastic DoFs in Equation (6.62), this matrix is not singular. This matrix is then augmented with three columns and rows of zeros to include the rigid body DoFs in the planar motions, as in $\mathbf{K}_0$ in Equation (6.40). The mass matrix is $\mathbf{M}_0 = \mathbf{M}(0)$, defined as in Equation (6.33), with $m = \int_0^L \rho_s A \, dx = \rho_s A L$, $I_{zz} = \int_0^L \rho_s x^2 A \, dx = \frac{1}{3} m L^2$, and the distributed inertia constants given by Equation (6.18), with $\boldsymbol{\xi} = \mathbf{0}$ and the shape functions in Equation (6.63).

Next, we compute the LNM of the undeformed structure, which are given by the nontrivial solutions to Equation (6.41). For the numerical results, consider $EI = 2$ Nm^2, $\rho_s A = 1$ kg/m, and $L = 10$ m, with a discretization $N = 25$. This planar structure has three rigid-body modes with zero frequency. The corresponding mass-normalized modal matrix is

$$
\boldsymbol{\Phi}_{rr} = \frac{1}{\sqrt{\rho_s A L}} \begin{bmatrix} 1 & 0 & 0 \\ 0 & 1 & \sqrt{3} \\ 0 & 0 & \frac{2\sqrt{3}}{L} \end{bmatrix} = \begin{bmatrix} 0.316 & 0 & 0 \\ 0 & 0.316 & 0.548 \\ 0 & 0 & 0.109 \end{bmatrix}, \tag{6.66}
$$

that is, translations along the horizontal and vertical directions, as well as rigid-body rotations about the beam midpoint. As the DoFs are written with respect to a free end, this results in both rotations and displacements at the end for that third mode (noting that $0.109 \times L/2 = 0.548$). All elastic mode shapes have nonzero rigid and elastic components, $\boldsymbol{\Phi}_{rs}$ and $\boldsymbol{\Phi}_{ss}$. The rigid component for the first four modes is

$$
\boldsymbol{\Phi}_{rs} = \begin{bmatrix} 0 & 0 & 0 & 0 \\ 0.6325 & 0.6325 & 0.6325 & 0.6325 \\ 0.294 & 0.497 & 0.695 & 0.894 \end{bmatrix}, \tag{6.67}
$$

and the corresponding elastic component, $\boldsymbol{\Phi}_{ss}$, of those modes is shown in Figure 6.12. It can be easily checked that the elastic modes satisfy Equation (6.46). As the beam has constant density, it means that the translational and rotational rigid-body contributions of each mode (i.e., the nonzero rows in Equation (6.67)) are equal to the area integral and the first moment about the origin, respectively, for each of the elastic displacements in Figure 6.12.

As we have chosen a reference at a beam end, splitting the modes in the rigid and elastic contribution makes them hard to visualize. If their rigid-body component is also included, one obtains the mode shapes in Figure 6.13, which are also the free–free modes that would be obtained from a conventional finite-element solver. In the figure, modes 2 and 3 are rigid-body modes associated with vertical displacements and rotations about the axis (mode 1 represents the rigid-body translation along the horizontal axis), and modes 4–7 are the first four elastic modes. Note, however, that this is just a transformation of DoFs from the results shown in Figure 6.12 and the LNMs describe the same beam dynamics in all cases. The natural frequencies of the first four elastic modes are $\omega_1 = 0.50$ rad/s, $\omega_2 = 1.38$ rad/s, $\omega_3 = 2.70$, rad/s, and $\omega_4 = 4.47$ rad/s. These results can be easily verified against the analytical solution for this problem, which gives natural frequencies $\omega_j = \kappa_j^2 \sqrt{\frac{EI}{\rho_s A}}$, with κ_j the positive roots of (Le, 1999, Ch. 4)[3]

---

[3] The analytical solution to this problem is further investigated in Problem 6.3.

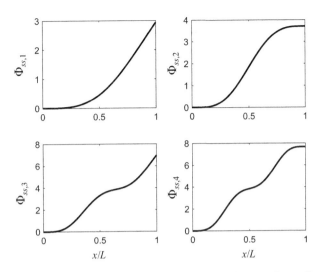

Figure 6.12 Elastic component of the first four flexible modes using quasi-coordinates. [06FlexibleAircraft/freebeam/runLNM.m]

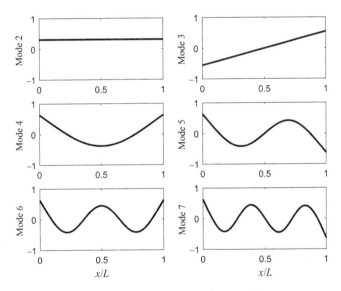

Figure 6.13 Vertical displacemenet of modes 2–7 in absolute coordinates. [06FlexibleAircraft/freebeam/runLNM.m]

$$\cos(\kappa_j L)\cosh(\kappa_j L) = 1. \tag{6.68}$$

A pair of equal forces in opposite direction are now applied on both ends of the unsupported beam, as shown in Figure 6.10b. As before, we select $L = 10$ m and $\rho_s A = 1$ kg/m, and the loading is defined with a linear ramp up to a maximum value $f_{max} = 0.25$ N at time $t_{max} = 5$ s and constant afterwards. Three different stiffness values are considered, $EI = 20\,\mathrm{Nm}^2$, $EI = 5\,\mathrm{Nm}^2$, and $EI = 2\,\mathrm{Nm}^2$. Note that rescaling the stiffness changes the natural frequencies (which grow as $\sqrt{EI}$), but does not affect

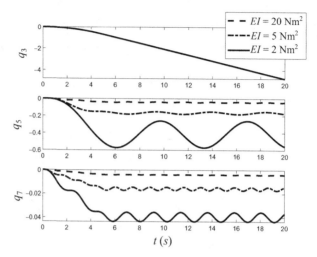

Figure 6.14 Nonzero amplitudes of linear modal analysis with varying stiffness values.
[06FlexibleAircraft/freebeam/run_dynamic.m]

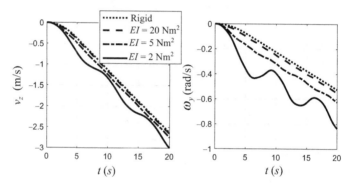

Figure 6.15 Linear and angular velocities of the body-attached reference frame for rigid and linear modal solutions with varying stiffness.

the mode shapes. For the numerical results, we retain the three rigid-body modes and the first four elastic modes. Simulations are all solved with a time step $\Delta t = 0.01$ s. As the EoMs are solved in modal coordinates, the results do not depend on the choice of body-attached reference frame.

Figure 6.14 shows the time history of the nonzero modal amplitudes for the three different stiffness values in this problem. Note first that the response of the rigid-body mode that defines the evolution of the CM is independent of the stiffness. Furthermore, as the loading is antisymmetric, only the antisymmetric elastic modes (modes 5 and 7 in Figure 6.13) are excited, and only those are displayed. As the current model includes no structural damping, the amplitudes of the elastic modes continue to oscillate after the steady load is achieved. This is reflected in the fluctuation of the value of Δw and even in the reference velocities in Figure 6.15.

Premultiplying the modal amplitudes by the modal matrix, Equation (6.44), one recovers the time histories of the linear and angular velocities of the reference frame

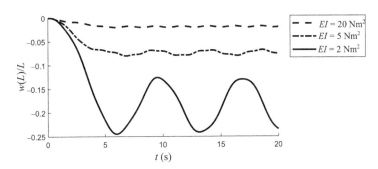

Figure 6.16 Instantaneous relative normal displacement between both beam ends for linear modal solutions, with varying stiffness values.

as well as the elastic displacements relative to that frame. The resulting linear and angular velocities at the reference point are shown in the Figure 6.15, while Figure 6.16 shows the relative normal displacement $w(L)$ between both ends of the beams to quantify the structural deformations. Results are shown again for the three values of stiffness, as well as for a rigid beam. The rigid-body problem (infinite stiffness) has an analytical solution in which the beam rotates with no translation about its center of mass ($x_{CM} = L/2$), with angular velocity growing parabolically up to t_{max} and linearly afterwards. At the body-attached reference frame considered here and located at a beam end, there is also a nonzero normal velocity, v_z, equal to the instantaneous rigid-body angular velocity times the beam semispan $L/2$. Those are shown in the dotted lines in Figure 6.15. For the rigid-body problem, the gyroscopic terms are exactly zero and, therefore, the linear Equations (6.35) provide an excellent approximation of the flexible beam dynamics if displacements are sufficiently small.

Note that the solution of the elastic problem in Figure 6.15 converges to the rigid-body predictions if the beam is sufficiently stiff. As the LNM are orthogonal to each other, the CM in the elastic solutions (whose dynamics are given by the third normal mode shown in Figure 6.13) moves with the same rotational speed that in the rigid problem. However, the velocities of the body-fixed reference frame shown in Figure 6.15 also include the effect of local deformation. Finally, it should be noted that the assumption of linear dynamics for the more flexible problem is not guaranteed, as the elastic displacements are nearly 25% of the span and would result in significant changes to the instantaneous inertia properties of the beam. Such low stiffness effectively needs a geometrically nonlinear beam model, which will be discussed in Chapter 8.

6.4.3 Modal Solution of Nonlinear Rigid-Linear Elastic Problems

The modal transformation defined in Equation (6.47) can also be applied to the general description of Equation (6.30) if $\Delta\dot{\nu}$ and $\Delta\ddot{\nu}$ are replaced in Equation (6.48) by the translational, $\mathbf{v}_B$, and angular, $\boldsymbol{\omega}_B$, velocities of the body-fixed reference frame, as well as their time derivatives. This typically helps in the analysis of the vehicle dynamics, as the truncation in the number of modes reduces system size and removes unnecessary high-frequency dynamics. However, the numerical advantage is still limited, since several inertia terms in the EoMs depend on the integral of the elastic DoFs, ξ, which would need to be constantly updated in any solution process.[4] If, instead, the elastic displacements produce negligible changes in the vehicle inertia, as was assumed in Equation (6.32), then the dynamics of a flexible aircraft can be greatly simplified when written in modal coordinates, as we will see here. We remark, as we did above, that the assumption is that the small elastic displacements do not change the global inertia characteristics of the vehicle, which are obtained from the undeformed geometry, yet we still distinguish between the velocity of the body-attached frame and that of a floating frame attached to the instantaneous CM.

First, the body-fixed reference frame is chosen at the CM of the undeformed vehicle, as it was done in Equation (6.32) for the general nonlinear rigid-linear elastic problem. In the definition of the normal modes, we have enforced Equation (6.45), which implies that the rigid-body modes include rotations about the principal axis of inertia. However, this normalization is not strictly necessary, and the rigid-body modes can be chosen as any set that span the null eigenspace of Equation (6.43). We choose then $\Phi_{rr} = \mathcal{I}$ in Equation (6.47). Since the B frame of reference is at the CM in the undeformed body and, from Equation (6.46), the elastic modes do not define velocities at the instantaneous CM, it results that the rigid-body modes are the translational and angular velocities of the CM. In other words, Equation (6.47) has become

$$\begin{Bmatrix} \mathbf{v}_B \\ \omega_B \end{Bmatrix} = \begin{Bmatrix} \mathbf{v}_B^{CM} \\ \omega_B^{CM} \end{Bmatrix} + \Phi_{rs}\dot{\zeta}, \tag{6.69a}$$

$$\xi = \Phi_{ss}\zeta, \tag{6.69b}$$

where ζ are the amplitudes of the elastic modes and $\mathbf{v}_B^{CM}$ and ω_B^{CM} are effectively the velocities of a floating frame at the CM, while $\mathbf{v}_B$ and ω_B are the velocities at the origin of the body-attached frame (which coincides with the floating frame if $\dot{\zeta} = 0$). Equation (6.69) defines a modal transformation that directly solves for the CM velocities. It can be rewritten as

$$\begin{Bmatrix} \begin{Bmatrix} \mathbf{v}_B \\ \omega_B \end{Bmatrix} \\ \dot{\xi} \end{Bmatrix} = \begin{bmatrix} \mathcal{I} & \Phi_{rs} \\ \hline \mathbf{0} & \Phi_{ss} \end{bmatrix} \begin{Bmatrix} \begin{Bmatrix} \mathbf{v}_B^{CM} \\ \omega_B^{CM} \end{Bmatrix} \\ \dot{\zeta} \end{Bmatrix} = \Psi \begin{Bmatrix} \begin{Bmatrix} \mathbf{v}_B^{CM} \\ \omega_B^{CM} \end{Bmatrix} \\ \dot{\zeta} \end{Bmatrix}. \tag{6.70}$$

Since the stiffness matrix only acts on the elastic DoFs, this transformation can still be normalized to satisfy $\Psi^\top \mathbf{K}_0 \Psi = \Omega^2$. However, due to our choice of Φ_{rr}, the rigid-body modes are no longer mass normalized. Substituting Equation (6.70) on the nonlinear rigid-linear elastic problem, Equation (6.32), results in

---

[4] Solving the problem in intrinsic variables, as is done in Section 8.6, is a way around this problem, as they remove global spatial dependencies on the EoMs of a flexible body.

$$\mathbb{M} \left\{ \begin{matrix} \dot{\mathbf{v}}_B^{CM} \\ \dot{\boldsymbol{\omega}}_B^{CM} \\ \ddot{\boldsymbol{\zeta}} \end{matrix} \right\} + \mathbb{D}(\boldsymbol{\omega}_B) \left\{ \begin{matrix} \mathbf{v}_B^{CM} \\ \boldsymbol{\omega}_B^{CM} \\ \dot{\boldsymbol{\zeta}} \end{matrix} \right\} + \mathbb{K}\boldsymbol{\zeta} = \mathbb{G} \left\{ \begin{matrix} -\sin\theta \\ \cos\theta\sin\phi \\ \cos\theta\cos\phi \end{matrix} \right\} + \left\{ \begin{matrix} \mathbf{f}_B \\ \mathbf{m}_B \\ \boldsymbol{\Phi}_{ss}^\top \boldsymbol{\chi} \end{matrix} \right\}, \qquad (6.71)$$

with the constant matrices

$$\mathbb{M} = \boldsymbol{\Psi}^\top \mathbf{M}_0 \boldsymbol{\Psi}, \quad \mathbb{K} = \begin{bmatrix} \mathbf{0} \\ \mathbf{0} \\ \boldsymbol{\Omega}_s^2 \end{bmatrix}, \quad \text{and} \quad \mathbb{G} = g\boldsymbol{\Psi}^\top \mathbf{M}_0 \begin{bmatrix} \mathcal{I} \\ \mathbf{0} \\ \mathbf{0} \end{bmatrix}, \qquad (6.72)$$

with $\boldsymbol{\Omega}_s$ being the diagonal matrix of the natural frequencies of the elastic modes (obtained by removing the six zero rows and columns of $\boldsymbol{\Omega}$), and with the gyroscopic coupling terms defined as

$$\mathbb{D}(\boldsymbol{\omega}_B) = \boldsymbol{\Psi}^\top \begin{bmatrix} m\tilde{\boldsymbol{\omega}}_B & \mathbf{0} & 2\tilde{\boldsymbol{\omega}}_B \mathbf{M}_{\rho\xi} \\ \mathbf{0} & \tilde{\boldsymbol{\omega}}_B \mathbf{I}_B(0) & 2\mathbf{M}_{\varphi\omega\xi}(\boldsymbol{\omega}_B,0) \\ \mathbf{M}_{\rho\xi}^\top \tilde{\boldsymbol{\omega}}_B & -\mathbf{M}_{\varphi\omega\xi}^\top(\boldsymbol{\omega}_B,0) & 2\mathbf{M}_{\xi\omega\xi}(\boldsymbol{\omega}_B) \end{bmatrix} \boldsymbol{\Psi}, \qquad (6.73)$$

where $\boldsymbol{\omega}_B$ is a function of both the CM angular velocity, $\boldsymbol{\omega}_B^{CM}$, and the rates of the elastic DoFs, $\dot{\boldsymbol{\zeta}}$, following Equation (6.70). Finally, the orthogonality of the LNMs results in a block-diagonal modal mass matrix as

$$\mathbb{M} = \begin{bmatrix} m\mathcal{I} & \mathbf{0} & \mathbf{0} \\ \mathbf{0} & \mathbf{I}_B(0) & \mathbf{0} \\ \mathbf{0} & \mathbf{0} & \mathcal{I} \end{bmatrix}. \qquad (6.74)$$

Partial Modal Projection

As it has been seen, the modal projection does not so much simplify the algebra as it reduces the number of DoFs. This becomes particularly relevant when considering nonlinear gyroscopic couplings in the projected equations. To address this, a modified projection of the nonlinear rigid-linear elastic equations has been proposed by Hesse and Palacios (2012) in which the elastic DoFs are still transformed using Equation (6.69b), while $\mathbf{v}_B$ and $\boldsymbol{\omega}_B$ are retained as the DoFs in the modal description. This results in Equation (6.70) being replaced by

$$\left\{ \begin{matrix} \mathbf{v}_B \\ \boldsymbol{\omega}_B \\ \dot{\boldsymbol{\xi}} \end{matrix} \right\} = \begin{bmatrix} \mathcal{I} & \mathbf{0} & \mathbf{0} \\ \mathbf{0} & \mathcal{I} & \mathbf{0} \\ \mathbf{0} & \mathbf{0} & \boldsymbol{\Phi}_{ss} \end{bmatrix} \left\{ \begin{matrix} \mathbf{v}_B \\ \boldsymbol{\omega}_B \\ \dot{\boldsymbol{\zeta}} \end{matrix} \right\} = \boldsymbol{\Psi} \left\{ \begin{matrix} \mathbf{v}_B \\ \boldsymbol{\omega}_B \\ \dot{\boldsymbol{\zeta}} \end{matrix} \right\}. \qquad (6.75)$$

As before, $\boldsymbol{\Phi}_{ss}$ is the truncated modal matrix in which only a few modes are retained, thus hugely reducing the size of the problem. Introducing this transformation results in a mass matrix with rigid-elastic couplings (i.e., it can no longer be written as in Equation (6.74)). However, it also allows an easy evaluation of the gyroscopic terms through the linear operator defined in Equation (6.73), which now can be further written as

$$\mathbb{D}(\boldsymbol{\omega}_B) = \sum_{i=1}^{3} \frac{\partial \mathbb{D}}{\partial \omega_i} \omega_i, \qquad (6.76)$$

where ω_i refers to the three components of ω_B and $\frac{\partial \mathbb{D}}{\partial \omega_i}$ is a third-order tensor of constant coefficients, which is uniquely determined from the inertia constants. Importantly, the number of those coefficients is only three times the size of problem after projection in modal coordinates. This results in the following equations

$$\mathbb{M} \begin{Bmatrix} \dot{\mathbf{v}}_B \\ \dot{\omega}_B \\ \ddot{\zeta} \end{Bmatrix} + \mathbb{D}(\omega_B) \begin{Bmatrix} \mathbf{v}_B \\ \omega_B \\ \dot{\zeta} \end{Bmatrix} + \mathbb{K}\zeta = \mathbb{G} \begin{Bmatrix} -\sin\theta \\ \cos\theta \sin\phi \\ \cos\theta \cos\phi \end{Bmatrix} + \begin{Bmatrix} \mathbf{f}_B \\ \mathbf{m}_B \\ \Phi_{ss}^\top \chi \end{Bmatrix}, \quad (6.77)$$

with the mass, stiffness, and gravitational matrices defined as in Equation (6.72). These equations are augmented with Equation (6.31), which give the instantaneous attitude of the vehicle, θ and ϕ. The price to be paid by a projection that retains the original rigid-body DoFs is a slower convergence rate with the number of elastic modes (i.e., more modes are needed to achieve a given accuracy). However, since the modal amplitudes do not appear in the nonlinear terms, the overall computational effort can be much smaller than for Equation (6.71). Once Equations (6.31) and (6.77) are solved, the instantaneous velocity of the CM can be obtained. For that, recall the matrix Φ_{ss} in Equation (6.75) still satisfies Equation (6.46) and, as a consequence, the linear transformation defined in Equation (6.69a) is still valid to determine both $\mathbf{v}_B^{CM}$ and ω_B^{CM} from the modal velocities and the velocities of the B reference frame.

Practical Mean Axes

Traditionally, however, the rigid aircraft flight dynamics equations are written with respect to the (instantaneous) center of mass. For flexible aircraft dynamics, that reference has often been maintained in the literature (Milne, 1962; Waszak and Schmidt, 1988). This effectively means retaining the transformation Equation (6.69a), which has decoupled the rigid and elastic DoFs in the mass matrix in Equation (6.71), but that has also resulted in a gyroscopic matrix that depends not only on the velocity at the center of mass of the aircraft but also on the modal velocities, $\dot{\zeta}$. The latter dependencies are often neglected in the literature, which gives a description typically referred to as the EoMs in *practical mean axes* (Schmidt, 2012).

The resulting EoMs are still of the form of Equation (6.71), but with modified damping terms, i.e.,

$$\mathbb{M} \begin{Bmatrix} \dot{\mathbf{v}}_B^{CM} \\ \dot{\omega}_B^{CM} \\ \ddot{\zeta} \end{Bmatrix} + \mathbb{D}_{CM}(\omega_B^{CM}) \begin{Bmatrix} \mathbf{v}_B^{CM} \\ \omega_B^{CM} \\ \dot{\zeta} \end{Bmatrix} + \mathbb{K}\zeta = \mathbb{G} \begin{Bmatrix} -\sin\theta \\ \cos\theta \sin\phi \\ \cos\theta \cos\phi \end{Bmatrix} + \begin{Bmatrix} \mathbf{f}_B \\ \mathbf{m}_B \\ \Phi_{ss}^\top \chi \end{Bmatrix}, \quad (6.78)$$

with

$$\mathbb{D}_{CM}(\omega_B) = \begin{bmatrix} m\tilde{\omega}_B & \mathbf{0} & \mathbf{0} \\ \mathbf{0} & \tilde{\omega}_B I_B(0) & \mathbf{0} \\ \mathbf{0} & \mathbf{0} & \mathbf{0} \end{bmatrix}. \quad (6.79)$$

Here, all gyroscopic effects on the elastic DoFs are neglected, while the mass, stiffness, and gravitational matrices are still those of Equation (6.72), with the mass matrix in particular written again as in Equation (6.74). Note that in practice, the deformations due to the gravitational loads are often also neglected and only the weight at the center of mass is considered.

The mean-axes approximation, as presented above, results in the rigid-body flight dynamic equations on the instantaneous CM, augmented with the dynamics of the elastic modes, which have been decoupled from the rigid-body problem. This makes this formulation very attractive for incorporating elastic effects in flight dynamics analysis framework (Kier, 2011), but it should be remarked that there is an inconsistency in neglecting only the contribution of $\dot{\xi}$ to the gyroscopic terms, and the assumptions need to be justified for each particular application. Indeed, the inertial decoupling between rigid and elastic DoFs implies that the mean-axes approximation is, for frequencies below the elastic natural frequencies, equivalent to the elastified aircraft dynamics of Equations (5.4) and (5.5) with negligible gyroscopic effects on the structural equations.

The assumptions above can be of course avoided with the partial projection of Equation (6.77), although that implies a body-fixed description for the rigid-body velocities. Alternatively, several consistent mean-axes (i.e., floating-frame) formulations have been derived in the literature (Schmidt, 2015; Guimaraes Neto et al., 2016; Saltari et al., 2017) that explicitly enable tracking CM dynamics without the assumptions on the gyroscopic forces. In practice, there is little difference from adopting a body-fixed reference with the CM dynamics obtained as a derived magnitude (i.e., in the output equations), which in our experience provides a more amenable mathematical framework on which to build additional complexity (see Chapters 8 and 9).

6.4.4 Dynamic Loads

In this chapter, we have so far described the kinematic state, that is, the rigid-body velocity and elastic displacements, of a flexible aircraft. Those are of interest to obtain aircraft performance estimates, and for stability analysis or control design, among others. For structural design, it is also necessary to monitor the *internal loads* appearing on the airframe, which determine the instantaneous stress state of the vehicle during a dynamic event (e.g., a gust encounter or a maneuver). This is known as the *dynamic loads* analysis. An overview of the dynamic loads analysis methods and design rules can be found in the books of Lomax (1996) and Wright and Cooper (2007). Our focus will be on the process for evaluating the internal loads after the flexible aircraft dynamics have been determined, but not on how that information enters the design process. Note that the loads calculation occurs in a postprocessing stage, provided appropriate data is gathered during simulations or tests.

By definition, the internal loads, $\chi_s(t)$, on a structure are the strain energy conjugates of the elastic DoFs, $\xi(t)$, namely[5]

---

[5] The transpose appears because the derivative of a scalar by a column vector results in a row vector (such that $\delta \mathcal{U} = \frac{\partial \mathcal{U}}{\partial \xi} \delta \xi(t)$ is also a scalar).

$$\chi_s(t) = \left(\frac{\partial \mathcal{U}}{\partial \boldsymbol{\xi}}\right)^\top. \tag{6.80}$$

For the particular case of a structure with small elastic deformations, the internal energy is given by Equation (6.23). Substituting that expression into Equation (6.80) results in the internal forces given simply by $\chi_s(t) = \mathbf{K}_{\xi\xi}\boldsymbol{\xi}$. As structures are normally discretized using finite elements, the internal loads, $\chi_s(t)$, are a distribution of point forces and moments at the nodes of the structural model, which are then integrated along load paths (e.g., along the span of the wing) to obtain distribution of bending moments, axial and shear forces, and torsional moments (known collectively as *MAST loads*) along all the primary structures of the aircraft. For each component (e.g., the wing root or an engine pylon), the relevant MAST loads determine the local stress state for which that component needs to be sized. MAST loads are typically displayed as univariate distributions along each primary structure (e.g., a moment or shear diagram), or as local bivariate distributions on a particular station (e.g., the phase plot between bending and torsional moment at the wing root) (Niu, 1999). Both representations are used to identify the loads envelope, that is, the worst-case scenarios that size each structural component.

For small-amplitude dynamics, it is common to solve the EoMs in modal coordinates, as it has been done in this section. In such case, the displacement field is obtained from Equation (6.69b), thus resulting in

$$\chi_s(t) = \mathbf{K}_{\xi\xi}\boldsymbol{\Phi}_{ss}\boldsymbol{\zeta}. \tag{6.81}$$

Evaluating the internal loads from the modal amplitudes is known as the *mode-displacement* approach (Karpel et al., 2008). The internal loads are determined through a linear output equation with constant matrix $\mathbf{K}_{\xi\xi}\boldsymbol{\Phi}_{ss}$ (known as *load modes*) from the elastic states. In practice, only a few normal modes are often solved for, which results in poor spatial resolution in the evaluation of the distribution of the internal loads. This is because the spectral decomposition that determines the normal modes brings out dominant global features of the structure, and yet local details primarily determine the actual stress distribution. Consequently, an alternative postprocessing method is often preferred for dynamic loads calculation, known as the *summation-of-forces* (SOF) approach. In the SOF approach, once the solution in modal coordinates is available, the original equations in elastic displacements are first retrieved. This could be either Equation (6.30), (6.32) or (6.35), depending on the approximation used in the analysis, but we only use the linear equations in this example. Then, from Equations (6.35) and (6.70), we solve for the elastic forces as

$$\chi_s(t) = \chi - \boldsymbol{\Psi}_s^\top \mathbf{M}_0 \boldsymbol{\Psi} \begin{Bmatrix} \dot{\mathbf{v}}_B^{CM} \\ \dot{\omega}_B^{CM} \\ \ddot{\boldsymbol{\zeta}} \end{Bmatrix} - \boldsymbol{\Psi}_s^\top \mathbf{D}_0 \boldsymbol{\Psi} \begin{Bmatrix} \mathbf{v}_B^{CM} \\ \omega_B^{CM} \\ \dot{\boldsymbol{\zeta}} \end{Bmatrix} + \boldsymbol{\Psi}_s^\top \mathbf{G}_0 \begin{Bmatrix} \theta \\ \phi \end{Bmatrix} - \mathbf{K}_{\omega\omega 0}\boldsymbol{\Phi}_{ss}\boldsymbol{\zeta},$$

$$\tag{6.82}$$

where we have *not* denoted incremental variables with Δ, as in Equation (6.35), to simplify the notation. We have also only retained the EoMs on the elastic DoFs in Equation (6.35) by defining $\boldsymbol{\Psi}_s = \begin{bmatrix} \mathbf{0} & \boldsymbol{\mathcal{I}} \end{bmatrix}^\top$, where the null matrix has dimension

$N_s \times 6$. Depending on the characteristics of the problem at hand, the last three terms of this equation, associated with gyroscopic, gravitational, and centrifugal forces, respectively, can be removed. Evaluation of dynamic loads using the SOF method is clearly more involved than the direct evaluation through the stiffness matrix using Equation (6.81), but by retrieving the external forces and vehicle inertia at the finite-element nodes, it provides higher-quality estimates of the local strains. Note finally that the external forces χ, which have been defined in Equation (6.26), are determined from the aerodynamic and propulsion loads on the vehicle, which are also a function themselves of the CM velocities and the elastic state. Evaluation of the aerodynamic forces is the focus of the following section.

6.5 Linearized Unsteady Aerodynamics

Now that a full description of the coupled structural and rigid-body dynamics of the aircraft has been completed, we need models to describe the external forces acting on the vehicle. As we did in Section 4.3.5 for the rigid aircraft, we focus on the instantaneous aerodynamic forces appearing on the vehicle, and we limit ourselves to small-amplitude vibrations and, therefore, linearized aerodynamic models. When dealing with flexible aircraft, there are however some major differences. First, in a rigid aircraft, the description was based on aerodynamic derivatives at the aircraft level, as only resultant forces and moments at the center of mass are needed. For a flexible aircraft, however, the detailed distribution of aerodynamic loading over the airframe needs to be known, as it was schematically shown in Figure 5.1. This effect is captured by the forcing term on the elastic DoFs, namely the generalized force $\chi(t)$ in Equation (6.27). Second, the dynamics of rigid aircraft have rather low characteristic frequencies and quasi-steady aerodynamics can usually be assumed; this is no longer necessarily the case for a vibrating airframe, where the period of the dominant vibrations can often be comparable to the convective time scales in the flow, as we have described in Section 3.3.3.

The consequences of the above are significant. A mathematical description of the flight dynamics of a rigid aircraft is based on aerodynamic derivatives, in which only a few nondimensional coefficients need to be estimated or measured. They can then be readily tabulated across all relevant flight conditions and retrieved, including some interpolation, for simulation and control purposes. Flexible aircraft dynamics often needs unsteady aerodynamic models to describe the instantaneous distributed loading on the vehicle's wetted surfaces. The description in this chapter is kept at a relatively high level to introduce the methods currently used for industrial-level applications to compute linearized aerodynamic forces. In particular, the focus is on the doublet-lattice method (DLM) of Albano and Rodden (1969), although in Section 6.5.4 we will also see how the DLM can be replaced by a higher-fidelity aerodynamic model if that was deemed necessary (e.g., in the transonic regime). Chapter 7 will present details of a more general description that is not limited to small deformations of the aerodynamic surfaces, although it is restricted to incompressible flows.

6.5.1   Linear 3-D Potential Flow Aerodynamics

We introduce here a mathematical model for the aerodynamic loads, that is, those that deform the airframe. For attached flows outside the transonic range, they can be very well approximated by potential flow theory and that framework is used here. Thus, we assume in what follows an inviscid flow (Re $\to \infty$) at subsonic speeds (Ma $\lesssim 0.8$) on streamlined aerodynamic surfaces. Under the additional assumption of small perturbations, the linearized potential flow equations can be written in boundary integral form using Green's theorem (Katz and Plotkin, 2001, Ch. 3), resulting in an integral relation between the instantaneous normal velocity on the lifting surfaces (which defines the boundary conditions on the flow) and the resulting pressure distribution. The basic features of the theory have been outlined in Section 3.3.1 for the unsteady aerodynamics of thin airfoils. A single wing is assumed now for simplicity in our derivations, but the solution can be easily extended to multiple interfering lifting surfaces.

As in Section 3.3.1, we assume linearized conditions (small perturbations) and an infinitesimally thin wing of planform area S and reference chord c lying on the x-y plane. This is schematically shown in Figure 6.17. We follow again in this section the usual convention in aerodynamics, where the x axis goes along the freestream incoming flow speed. The instantaneous vertical coordinates of the moving and deforming wing are known and given as $\zeta(x,y;t)$, which includes the angle of attack of the wing. Under the small-perturbation assumption we have $|\zeta(x,y;t)| \ll c$. Next, solutions are sought in the frequency domain for a given Mach number, Ma. As a result, if $w_b(\mathbf{x},t)$ is the normal velocity of the flow (upwash) at point $\mathbf{x} = (x,y)$ on the wing, defined as in Equation (3.36), we will write it as $w_b(\mathbf{x},t) = \overline{w}_b(\mathbf{x},i\omega)e^{i\omega t}$, and

$$\frac{\overline{w}_b(\mathbf{x},i\omega)}{V} = -\int_S \overline{\Delta c_p}(\mathbf{x}_0,i\omega)\mathfrak{K}(\mathbf{x}-\mathbf{x}_0,i\omega,\text{Ma})\,d\mathbf{x}_0, \tag{6.83}$$

where $\overline{\Delta c_p}(\mathbf{x}_0,i\omega)$ is the complex-valued pressure coefficient at a point $\mathbf{x}_0$ on the wing ($\Delta\bullet$ indicates the difference between the upper and lower side of the wing, as

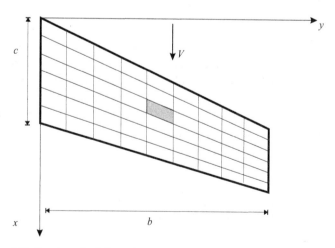

Figure 6.17 Panel-method discretization of a wing initially flat in the $x-y$ plane.

in Equation (3.23)) and $\mathfrak{K}$ is the *aerodynamic kernel function*. The kernel is obtained from fundamental solutions to the potential flow boundary-element problem (Katz and Plotkin, 2001). An example of this is the 1-D kernel of the airfoil equation for $\omega = 0$, Equation (3.35), which can be written as $\mathfrak{K}(x-x_0) = \frac{1}{4\pi(x-x_0)}$. For the case of unsteady aerodynamics on lifting surfaces with oscillatory motion, Albano and Rodden (1969) derived a closed-form expression for the kernel that is assumed to be known here and which, after spatial discretization, results in the DLM.

The solution to Equation (6.83) is generally sought numerically, typically using *collocation methods* (also known as panel methods): the mean surface of the wing is discretized into N_p panels, on which the non-penetrating boundary condition (zero fluid velocity in the direction normal to the wing surface, $\mathbf{n}$) is enforced only at certain *collocation points*. A typical panel discretization of a swept wing is shown in Figure 6.17.

Similar to Equation (3.16) for the 2-D problem, in the linear approximation the components of the normal vector to the planar wing become $\mathbf{n}(\mathbf{x},t) = \left(-\frac{\partial\zeta}{\partial x}, -\frac{\partial\zeta}{\partial y}, 1\right)^T$. Only the chordwise component of the normal is retained on the linearized non-penetrating boundary conditions on the wing surface. We also include a nonzero *background velocity field* in the normal direction, $w_g(\mathbf{x},t)$, to characterize the nonstationary atmospheric conditions discussed in Chapter 2. As a result, the non-penetrating conditions on the wing midsurface are written as

$$w_b(\mathbf{x},t) - w_g(\mathbf{x},t) = -\frac{\partial\zeta}{\partial t} - V\frac{\partial\zeta}{\partial x}. \tag{6.84}$$

As this is a linear theory, the normal component of the background velocity field (the gust velocity) is assumed to be much smaller than the freestream velocity V. This background field is further assumed to be known at all times and to remain undistorted as it encounters the wing. It can, therefore, be described using the models we introduced in Chapter 2.

For convenience, we introduce now the nondimensional time $s = \frac{Vt}{c/2}$ and the corresponding nondimensional frequency (the *reduced frequency*) $k = \frac{\omega c}{2V}$, as we did in Section 3.3.3. The boundary conditions are then written as

$$w_b(\mathbf{x},s) - w_g(\mathbf{x},s) = -\frac{V}{c/2}\frac{\partial\zeta}{\partial s} - V\frac{\partial\zeta}{\partial x}. \tag{6.85}$$

For harmonic oscillations of the wing, the dimensionless time and frequency are such that $w_b = \overline{w}_b e^{i\omega t} = \overline{w}_b e^{iks}$. Assuming $\zeta(\mathbf{x},s) = \overline{\zeta}(\mathbf{x},ik)e^{iks}$ and similarly for the gust velocity, we have

$$\frac{\overline{w}_b - \overline{w}_g}{V} = -2ik\frac{\overline{\zeta}}{c} - \frac{\partial\overline{\zeta}}{\partial x}. \tag{6.86}$$

With these definitions, Equation (6.83) is expressed in discrete form as

$$2ik\frac{\overline{\zeta}(\mathbf{x}_n,ik)}{c} + \overline{\zeta}'(\mathbf{x}_n,ik) + \frac{\overline{w}_g(\mathbf{x}_n,ik)}{V} = \sum_{m=1}^{N_p}\overline{\Delta c_p}(\mathbf{x}_m,ik)\mathfrak{K}(\mathbf{x}_n - \mathbf{x}_m,ik,\mathrm{Ma})S_m, \tag{6.87}$$

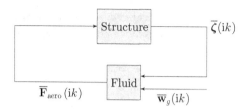

Figure 6.18 Aerodynamic inputs/outputs in the aeroelastic process.

with $n = 1, \ldots, N_p$ and S_m being the area of the m-th panel. Equation (6.87) defines a linear algebraic equation between the (known) upwash at the collocation points and the (unknown) constant pressure on each panel. We can also define the column vector $\mathbf{w}_g(s) = \left\{ w_g(\mathbf{x}_1, s), \ldots, w_g(\mathbf{x}_{N_p}, s) \right\}^\top$ with the instantaneous gust velocities at all the collocation points.

Next, we define $\boldsymbol{\zeta}(s) = \left\{ \zeta(\mathbf{x}_1, s), \zeta(\mathbf{x}_2, s), \ldots, \zeta(\mathbf{x}_{N_p}, s) \right\}^\top$ as the column vector with the out-of-plane coordinates of all the collocation points (one per panel in a zero-order formulation), and equally $\boldsymbol{\zeta}_x(s) = \left\{ \frac{\partial \zeta}{\partial x}(\mathbf{x}_1, s), \ldots, \frac{\partial \zeta}{\partial x}(\mathbf{x}_{N_p}, s) \right\}^\top$ the column vector of chordwise derivatives. Since the total force (which is along the z axis in our linear approximation) at each panel is $\frac{1}{2}\rho V^2 S_n \overline{\Delta c}_p(\mathbf{x}_n, ik)$, solving for the pressure jump in Equation (6.87) results in the column vector of the aerodynamic forces at all panel centers being written as

$$\overline{\mathbf{f}}_{\mathrm{aero}} = \frac{\rho V^2}{2} \mathbf{E}(ik, \mathrm{Ma}) \left(\frac{2ik}{c} \overline{\boldsymbol{\zeta}} + \overline{\boldsymbol{\zeta}}_x + \frac{1}{V} \overline{\mathbf{w}}_g \right). \tag{6.88}$$

The complex matrix $\mathbf{E}(ik, \mathrm{Ma})$ is an aerodynamic influence coefficient (AIC) matrix often referred to as the *unsteady aerodynamic (pressure) influence coefficient* matrix. It only depends on the geometry of the wing, the flight regime (via the Mach number), and the reduced frequency, and determines the distribution of forces on the wing as a function of its instantaneous shape and in the presence of gusts.

We have now completed the description of the lower block in the diagram shown of Figure 6.18. The upper block is determined by the structural descriptions of either Sections 6.3 or 6.4. To close the problem, we are still missing some additional details on the gust model and on the interfacing between the structural and aerodynamic models. They are discussed next.

Gust Modes

We have defined the nonstationary background velocity field in Equation (6.87) by means of an arbitrary distribution of velocities on the lifting surfaces. In practice, this is often simplified by using Taylor's frozen turbulence hypothesis. As done in Section 3.3.4, we consider instead a traveling gust that convects with the freestream velocity. For frequency-domain analysis, a harmonic gust of frequency ω is assumed as in Equation (3.55), although as this is a 3-D problem we can, in general, consider a spanwise distribution of the gust velocities as discussed in Section 2.3.4. For a point

(x, y) on the lifting surface in the coordinate system of Figure 6.17, the local upwash is given as

$$
\begin{aligned}
\overline{w}_g(x, y, ik) &= \overline{w}_g(x_0, y, ik)e^{-i\omega(x-x_0)/V} \\
&= \overline{w}_g(x_0, y, ik)e^{-i2k(x-x_0)/c}
\end{aligned}
\tag{6.89}
$$

where $\overline{w}_g(x_0, y, ik)$ is the spanwise distribution of gust velocities at a chordwise location x_0 upstream of the wing. We can now define the vector of independent gust input velocities, $\overline{\mathbf{w}}_{g0}(ik) = \left\{ \overline{w}_g(x_0, y_1, ik), \ldots, \overline{w}_g(x_0, y_{N_g}, ik) \right\}^{\top}$, with N_g as the number of spanwise points on which the gust is discretized, which does not need to coincide with any other spanwise discretization of the wing. The discrete gust distribution over the wing that appeared in Equation (6.88) is then written as

$$
\overline{\mathbf{w}}_g = \boldsymbol{\Phi}_g(ik)\overline{\mathbf{w}}_{g0}.
\tag{6.90}
$$

The rectangular matrix $\boldsymbol{\Phi}_g(ik)$ has, therefore, dimension $N_p \times M$ and it is obtained directly from evaluating Equation (6.89) at the collocation modes in the aerodynamic discretization. It introduces the delay in the convection of the traveling gust along the chordwise direction, and it is known as the matrix of *gust modes*. When a uniform spanwise inflow is assumed, then $\boldsymbol{\Phi}_g(ik)$ becomes a column vector. However, the more general definition is retained here to accommodate the 2-D models introduced in Section 2.3.4.

A simplified approach to define gust modes was introduced by Ripepi and Mantegazza (2013). Instead of considering the penetration of the gust along the wing using chordwise modes, the instantaneous gust profile is approximated by splitting the wing on multiple patches of several panels, each with an associated amplitude. The delay in the penetration of the gust appears now between the amplitudes of those patches. This results in a matrix relation as in Equation (6.90) without the reduced frequency dependency but with a larger size of $\mathbf{w}_{g0}$ to include the chordwise evolution of the gust.

Generalized Aerodynamic Forces

The instantaneous shape of the (discretized) lifting surfaces, which determines the coordinates of the collocation points, $\boldsymbol{\zeta}(t)$, has been assumed to be known in the evaluation of the aerodynamic forces. It is defined in fact by the kinematics of the underlying wing structure, as well as the deflection of any control surfaces. Using the modal projection (6.47), the vector of vertical displacements at the panel centers can be written as

$$
\boldsymbol{\zeta}(t) = \mathbf{T}_{a\leftarrow s}\begin{bmatrix} \boldsymbol{\Phi}_q & \boldsymbol{\Phi}_\delta \end{bmatrix} \begin{Bmatrix} \mathbf{q} \\ \boldsymbol{\delta} \end{Bmatrix} = \boldsymbol{\Phi}_u \mathbf{u}(t),
\tag{6.91}
$$

where $\boldsymbol{\Phi}_q$ is the (constant) structural modal matrix, which includes both elastic and rigid-body DoFs. It has been augmented with $\boldsymbol{\Phi}_\delta$ to include the control surfaces. The columns of this matrix define the wing displacements after unit deflection of each control surface, as shown schematically in Figure 6.19, and $\boldsymbol{\delta}(t)$ is a column vector of dimension N_δ with the instantaneous deflection of all control surfaces. Note that while

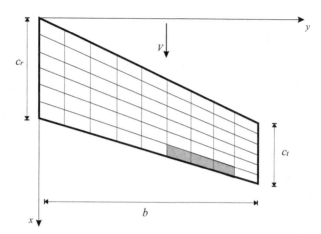

Figure 6.19 Panel-method discretization of a trailing-edge control surface (in grey).

the control surfaces are here defined on the aerodynamic model, in practice details of the hinge mechanisms can have a significant importance on the structure as well.[6]

Finally, the matrix $\mathbf{T}_{a \leftarrow s}$ in Equation (6.91) is an interpolation matrix that generates a mapping between the displacements in the structural and the aerodynamic meshes that, in general, do not coincide. However, to simplify the derivations in what follows, this matrix has been embedded in the definition of the matrix of input modes, $\boldsymbol{\Phi}_u$. Note finally that this description assumes that the control surfaces are defined on the structural model. Substituting Equations (6.90) and (6.91) into Equation (6.88), we have the following expression for the distribution of aerodynamic forces in the discretized wing

$$\bar{\mathbf{f}}_{\text{aero}}(ik, \text{Ma}) = \tfrac{1}{2}\rho V^2 \mathbf{E}(ik, \text{Ma}) \left(\tfrac{2ik}{c} \boldsymbol{\Phi}_u \bar{\mathbf{u}} + \boldsymbol{\Phi}_{u,x} \bar{\mathbf{u}} + \boldsymbol{\Phi}_g(ik) \tfrac{\overline{w}_{g0}}{V} \right), \qquad (6.92)$$

where $\boldsymbol{\Phi}_{u,x}$ refers to the slope of the input modes along the chordwise coordinate. The distributed forces are finally integrated for each mode shape, as in Equation (6.52), resulting in

$$\bar{\mathbf{f}}_q = \boldsymbol{\Phi}_q^\top \bar{\mathbf{f}}_{\text{aero}} = \tfrac{1}{2}\rho V^2 \boldsymbol{\Phi}_q^\top \mathbf{E} \left(\tfrac{2ik}{c} \boldsymbol{\Phi}_u \bar{\mathbf{u}} + \boldsymbol{\Phi}_{u,x} \bar{\mathbf{u}} + \boldsymbol{\Phi}_g(ik) \tfrac{\overline{w}_{g0}}{V} \right). \qquad (6.93)$$

This defines a multiple-input multiple-output transfer function in frequency domain that relates the generalized aerodynamic forces (GAFs) to the modal amplitudes and control/gust inputs. It is usually known as the matrix of GAFs. For a given aircraft or wing, it is a function of the reduced frequency and the Mach number and can be written as the $N_q \times (N_q + N_\delta + N_g)$ rectangular (complex) matrix:

$$\mathcal{A}(ik, \text{Ma}) = \boldsymbol{\Phi}_q^\top \mathbf{E}(ik, \text{Ma}) \left[\tfrac{2ik}{c} \boldsymbol{\Phi}_u + \boldsymbol{\Phi}_{u,x} \quad \boldsymbol{\Phi}_g(ik) \right]. \qquad (6.94)$$

---

[6] The flexibility of the aileron mount, coupled with aerodynamic feedback, indeed resulted in *aileron flutter* on early wing designs (Erickson and Mannes, 1949).

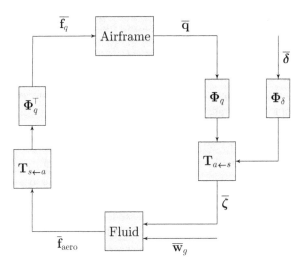

Figure 6.20 Fluid-structure coupling in frequency domain with control inputs and gust disturbances.

We can also partition this equation in terms of the original DoFs in Equation (6.91) as

$$\overline{\mathbf{f}_q} = \tfrac{1}{2}\rho V^2 \mathcal{A}(ik,\mathrm{Ma}) \left\{ \begin{matrix} \overline{\mathbf{u}} \\ \frac{\overline{\mathbf{w}}_{g0}}{V} \end{matrix} \right\} = \tfrac{1}{2}\rho V^2 \left(\mathcal{A}_q \overline{\mathbf{q}} + \mathcal{A}_\delta \overline{\boldsymbol{\delta}} + \mathcal{A}_g \frac{\overline{\mathbf{w}}_{g0}}{V} \right). \qquad (6.95)$$

While the AICs of Equation (6.88) only depend on the geometry, reduced frequency, and Mach number, the GAFs, $\mathcal{A}(ik,\mathrm{Ma})$, also depend on the structural characteristics through its natural vibration modes, $\boldsymbol{\Phi}_q$. Both matrices are in general full (and complex) matrices, but the size of $\mathbf{E}(ik,\mathrm{Ma})$ is the number of aerodynamic panels, which is typically much larger than the (small) number of retained vibration modes and the (even smaller) number of control surfaces that define the size of $\mathcal{A}(ik,\mathrm{Ma})$. Note finally that by construction $\mathcal{A}_q$ is a square matrix. The full block diagram of the frequency-domain coupling between structural (*Airframe*) and aerodynamic (*Fluid*) subsystems, including control surface and gust inputs, is shown in Figure 6.20. The airframe is described in modal coordinates and, therefore, needs the projection of forces and displacements from physical space using the modal matrix to interface with the fluid model. Additional interpolation matrices may be needed to accommodate nonmatching discretization on aerodynamics and structure, but they are not included here for simplicity. The reader is referred to Smith et al. (1995, 2000) and Brown (1997) for further details on typical interpolation algorithms.

The DLM outlined here has been the workhorse for linear aeroelastic simulation in subsonic flows for nearly five decades and it has well-established numerical implementations. See Blair (1992) for a full detail of its mathematical underpinnings and numerical solutions and Rodden (1997) for a historical account of its development.

Steady Aerodynamic Forces

Equation (6.95) is a general expression for the linear unsteady aerodynamics of a flexible aircraft, which in general may include the effect of steady forces. The only restriction in the derivations above is that the motions of the lifting surfaces occur normal to their reference frame. In practice, however, implementations of the DLM outlined earlier have been derived for aeroelastic analysis and do not consider the effect of steady forces. Those were seen in Section 4.3.5 to be essential for the prediction of the flight dynamics, and they are, therefore, relevant when that problem is expanded to include the elastic response of the aircraft. A typical solution in those cases is to add the aerodynamic derivatives associated with the static forces in Equation (4.44) to the frequency-domain GAF of Equation (6.95) (Baldelli et al., 2006; Saltari et al., 2017). A more generic formulation for incompressible flows that uses a linearization of the unsteady vortex-lattice method and considers arbitrary wing kinematics is described in some detail in Section 7.3.3.

The DLM also neglects skin friction, which is necessary to estimate the drag derivatives on the aircraft. Those are also critical for flight dynamics. Usually drag is only necessary at the global aircraft level (drag forces do not deform the airframe) and it is, therefore, rather straightforward (Baldelli et al., 2006; Ouellette and Valdez, 2020) to augment the linear aerodynamic description with measured quasi-steady aerodynamic derivatives from wind-tunnel, flight tests, or high-fidelity simulations. This gives the required fidelity necessary for the prediction of the flight dynamics characteristics that strongly depend on the changes of the drag forces on the vehicle, such as the phugoid frequency and damping.

6.5.2   Flutter Predictions in the Frequency Domain

Frequency-domain aerodynamics have been historically the basis for flutter prediction methods. We have defined flutter in Section 3.5.2 as a dynamic instability of the coupled system resulting from the interconnection of aerodynamics and structure. Importantly, the onset of flutter manifests itself in self-excited neutrally stable oscillations of the aeroelastic system (e.g., a wing or an aircraft). Those are harmonic oscillations at the *flutter frequency*, which is unknown, and can therefore be directly sought in frequency domain. To exemplify this, consider the linear dynamics of an aeroelastic system in modal coordinates as given by Equation (6.51) but without gyroscopic and gravitational terms. Assuming harmonic oscillations, $\mathbf{q}(t) = \overline{\mathbf{q}}e^{i\omega t}$, and aerodynamic forces given by Equation (6.95) under calm air conditions and without control inputs, we have

$$\left(-\omega^2\mathbf{M} + \mathbf{K} - \tfrac{1}{2}\rho V^2 \mathcal{A}_q(ik, \mathrm{Ma})\right)\overline{\mathbf{q}} = \mathbf{0}, \tag{6.96}$$

which is known as the *flutter equation* and has the same structure as Equation (3.101), which was derived for the airfoil problem. The flutter onset conditions are determined by the combination of airspeed, air density, and frequency of oscillations that result in nontrivial solutions to Equation (6.96). Direct "one-shot" solution methods have been developed to solve this problem (Li and Ekici, 2018), but it is more common and robust to use iterative solution methods instead. Iterative methods for flutter pose a modified but easier to solve problem for which Equation (6.96) is a particular solution. One such method is the *p-k* method, which was first proposed by Hassig (1971) and is outlined here. Noting that the frequency can be written in terms of the reduced frequency as $\omega = 2Vk/c$, the p-k method seeks solutions to the following modified equation

$$\left(\frac{4V^2}{c^2} p^2 \mathbf{M} + \mathbf{K} - \tfrac{1}{2}\rho V^2 \mathcal{A}_q(ik, \mathrm{Ma}) \right) \bar{\mathbf{q}} = \mathbf{0}, \qquad (6.97)$$

where $p = \gamma k + ik$ is a complex number that includes a nondimensional rate of decay, γ. Clearly, when $\gamma = 0$ we recover the flutter equation. For given air density (i.e., altitude) and Mach number, pairs (k_j, γ_j) that satisfy Equation (6.97) are found for increasing values of the airspeed, V, using the following iterative method. First, an initial guess k_j is set for the reduced frequency of each aeroelastic mode, with $j = 1, \ldots, N_m$. That fixes the GAF matrix as $\mathcal{A}_q(ik_j, \mathrm{Ma})$ and Equation (6.97) becomes then a generalized eigenvalue problem in p, which is solved. Then, k_j is updated from the imaginary part of the closest eigenvalue, resulting in a fixed-point iteration of the form $k_j = \mathrm{Im}(p_j)$, in which the left-hand side defines the reduced frequency on $\mathcal{A}_q$, and the right-hand side the solution to the resulting eigenvalue problem. This process is repeated for each mode and then the airspeed is updated. The flutter speed is obtained when one of the roots satisfies $\mathrm{Re}(p_j) = 0$. The corresponding Mach number is then obtained and compared to the one for which the GAF matrix was obtained. Therefore, a double matching condition must be satisfied, that is, Mach number and reduced frequency, in this iterative method. Further details on iterative solution methods for the flutter equation can be found in Section 4.4 of Hodges and Pierce (2014). An implementation for the problem in Example 3.3 can be found in the software repository of this book.

6.5.3   Rational-Function Approximations

The evaluation of the matrix of GAFs defined in Equation (6.94) is typically done via sampling on a range of both reduced frequencies and Mach numbers of interest. For flutter analyses, this results in efficient numerical strategies, as the goal is to identify the conditions for neutrally stable oscillations of the airframe and effectively only unsteady aerodynamic information at the flutter frequency (and Mach) is needed. For transient dynamics, sampled frequency data with sufficient resolution would allow frequency-domain solutions that can be combined with an inverse Fourier transform, but this limits the available tools for analysis. Instead, the resulting tabulated data is often used to construct an identified LTI aerodynamic model written in state-space form. This is known as the *realization problem* (Antoulas, 2005), which has been introduced in Section 1.4.2. For applications in unsteady aerodynamics, multiple algorithmic strategies have been proposed (see Ripepi and Mantegazza (2013) and Quero et al. (2019) for some recent contributions to the literature) and here the focus is on rational-function approximations (RFAs), which are by far the most commonly found approach in practice. RFAs have already been introduced for the airfoil problem in Section 3.4 to obtain state-space approximations from the analytic solutions of unsteady airfoil theory. However, we have also seen there that the aerodynamic forces at sufficiently high frequencies introduce effective mass and damping onto the structure, while under stationary conditions produce an effective stiffness. It is numerically convenient to enforce those limit values for very high and very low frequencies in the solution to the realization problem,[7] which approximates the GAFs as (Karpel, 1982, 1990)

$$\mathcal{A}(ik) \approx \mathcal{A}_0 + ik\mathcal{A}_1 + (ik)^2 \mathcal{A}_2 + ik\mathbf{C}_r (ik\mathcal{I} - \mathbf{A}_r)^{-1} \mathbf{B}_r, \qquad (6.98)$$

where all matrices on the right-hand side are real and constant and $\mathbf{A}_r$ in particular has only stable eigenvalues. The real matrices $\mathcal{A}_0$, $\mathcal{A}_1$, and $\mathcal{A}_2$, as well as the strictly proper LTI system defined by $\mathbf{A}_r$, $\mathbf{B}_r$, and $\mathbf{C}_r$, are obtained from a least-squares approximation from a sufficiently large sampling of DLM-generated GAFs over a range of reduced frequencies (Tiffany and Adams, 1988; Hoadley and Karpel, 1991). The approximation in Equation (6.98) was first proposed by Karpel (1982) and it is known as the *minimum-state method*. Note that the identification needs to be done for each Mach number of interest, although this has not been explicitly included in this section for simplicity. Matrices $\mathcal{A}_0$, $\mathcal{A}_1$, and $\mathcal{A}_2$ can be identified, respectively, as aerodynamic stiffness, damping, and inertia (apparent mass) matrices, and should be obtained first, as we mentioned earlier, from the limit values at zero and very high frequencies of $\mathcal{A}(ik)$. The last term in Equation (6.98) is a generic frequency-response function between modal velocities and aerodynamic forces written in the form of Equation (1.19), with $\mathbf{A}_r$, $\mathbf{B}_r$ and $\mathbf{C}_r$ as its state, input, and output matrices, respectively, and no feedthrough. This residual term is called the lag term and is essentially associated with unsteady wake effects, which, as it was seen for the 2-D problem in Chapter 3, modulate the amplitude and introduce delays ("memory" effects) on the instantaneous aerodynamic forces resulting from wall motions. The dimension of $\mathbf{A}_r$ depends on both problem complexity and required accuracy and typically ranges between 10 and 100.

A simpler formulation can be obtained assuming a diagonal state matrix in the aerodynamic lag model in Equation (6.98), namely $\mathbf{A}_r = \mathrm{diag}\left(-\gamma_1, -\gamma_2, \ldots, -\gamma_{n_a}\right)$. This results in the RFA with the same structure as those seen in Section 3.4 for the 2-D problem, which for general 3-D problems results in the approximation that was originally proposed by Roger (1977), that is,

$$\mathcal{A}(ik) \approx \mathcal{A}_0 + ik\mathcal{A}_1 + (ik)^2 \mathcal{A}_2 + \sum_{j=1}^{n_a} \frac{ik}{\gamma_j + ik} \mathcal{A}_{j+2}, \qquad (6.99)$$

where $\mathcal{A}_j$, with $j = 0, \ldots, n_a + 2$, are real constant matrices and γ_j, with $j = 1, \ldots, n_a$, are real positive coefficients. Note that, from the partition of Equation (6.95), all $\mathcal{A}_j$ are rectangular matrices which can be written as the concatenation of submatrices associated with structural, control surface, and gust inputs as $\mathcal{A}_j = \begin{bmatrix} \mathcal{A}_{qj} & \mathcal{A}_{\delta j} & \mathcal{A}_{gj} \end{bmatrix}$ for $j = 0, \ldots, n_a + 2$. While the last term in Equation (6.99) does not have a direct physical meaning, some intuition can be built from Figure 6.21, which shows the step response of a single-input single-output (SISO) system whose FRF is $\frac{ik}{\gamma + ik}$ (or, equivalently, the impulse response of $\frac{1}{\gamma + ik}$). As it can be seen, it results in functions with exponential decay, which is determined by the value of the aerodynamic lags. They

---

[7] We refer to this as *polynomial preconditioning* (Maraniello and Palacios, 2019).

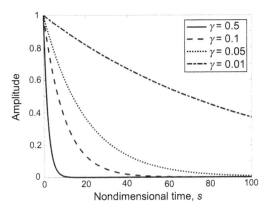

Figure 6.21 Step response of a system with frequency-response function $\frac{ik}{\gamma+ik}$.

are intimately linked to the transient process described by Wagner's function in the rigid airfoil problem of Section 3.4.3.

Compared to the minimum state-method described above, the RFA in Equation (6.99) needs substantially more internal states for a given accuracy, but it also defines an easier optimization problem for the identification of the RFA coefficients (Eversman and Tewari, 1991). Since Roger's approximation links directly to the approximation introduced in Section 3.4 to obtain the EoMs of a 2-DoF airfoil, they will be used for consistency in what follows to construct the time-domain aeroelastic EoMs. Note, however, that a very similar formulation would be obtained if Equation (6.98) were used instead.

Next, we construct time-domain expressions for the aerodynamic forces resulting from structural, control surface or gust inputs. To simplify the algebra, we first define the corresponding vector of inputs in Equation (6.95) as

$$\mathbf{z}(t) = \left\{\mathbf{q}^\top \quad \boldsymbol{\delta}^\top \quad \tfrac{1}{V}\mathbf{w}_{g0}^\top\right\}^\top. \tag{6.100}$$

Noting that for harmonic oscillations we have $\omega t = ks$, with the (nondimensional) reduced time of Equation (3.33), the instantaneous input vector is further written as $\mathbf{z}(t) = \bar{\mathbf{z}}e^{i\omega t} = \bar{\mathbf{z}}e^{iks}$. Substituting Equation (6.99) into Equation (6.95) results in

$$\overline{\mathbf{f}}_q(ik) = \tfrac{1}{2}\rho V^2 \left[\boldsymbol{\mathcal{A}}_0 + ik\boldsymbol{\mathcal{A}}_1 + (ik)^2\boldsymbol{\mathcal{A}}_2 + \sum_{j=1}^{n_a} \frac{ik}{\gamma_j + ik}\boldsymbol{\mathcal{A}}_{j+2}\right]\bar{\mathbf{z}}(ik). \tag{6.101}$$

A new set of variables referred to as the *aerodynamic states* (or aerodynamic lags) is introduced next to rewrite the lag terms, that is,

$$\overline{\boldsymbol{\lambda}}_j(ik) = \frac{ik}{\gamma_j + ik}\bar{\mathbf{z}}(ik) \quad \text{for } j = 1,\ldots,n_a. \tag{6.102}$$

Since the input vector, $\mathbf{z}$, includes structural, control surface, and gust components, there are $N_m + N_\delta + N_g$ aerodynamic states associated with each aerodynamic lag γ_j, which leads to the split as $\boldsymbol{\lambda}_j = \left\{\boldsymbol{\lambda}_{qj}^\top \quad \boldsymbol{\lambda}_{\delta j}^\top \quad \boldsymbol{\lambda}_{gj}^\top\right\}^\top$. This is the approach of Pasinetti and Mantegazza (1999), where the same lags are utilized for all inputs to the aerodynamics. Alternatively, one can independently identify optimal aerodynamic lags for the structure, control, and gust GAF. Collecting all the lag terms into a single column vector, we define $\boldsymbol{\lambda} = \left\{ \boldsymbol{\lambda}_1^\top \;\; \cdots \;\; \boldsymbol{\lambda}_{n_a}^\top \right\}^\top$ and as a result, the size of the identified aerodynamic internal model is $N_a = n_a \left(N_m + N_\delta + N_g \right)$. Equations (6.101) and (6.102) can now be written as two linear algebraic relations as

$$\overline{\mathbf{f}_q} = \tfrac{1}{2} \rho V^2 \left[\mathcal{A}_0 + ik \mathcal{A}_1 + (ik)^2 \mathcal{A}_2 \right] \overline{\mathbf{z}} + \tfrac{1}{2} \rho V^2 \sum_{j=1}^{n_a} \mathcal{A}_{j+2} \overline{\boldsymbol{\lambda}}_j ,$$

$$\left(\gamma_j + ik \right) \overline{\boldsymbol{\lambda}}_j = ik \overline{\mathbf{z}}, \quad \text{for } j = 1, \ldots, n_a. \tag{6.103}$$

The associated state-space equations in (dimensionless) time that define the aerodynamic loading for an arbitrary evolution in time of the aerodynamic inputs, Equation (6.100), can be finally obtained by applying the inverse Fourier transform on Equation (6.103), that is,

$$\mathbf{f}_q(s) = \tfrac{1}{2} \rho V^2 \left[\mathcal{A}_0 \mathbf{z} + \mathcal{A}_1 \mathbf{z}' + \mathcal{A}_2 \mathbf{z}'' + \sum_{j=1}^{n_a} \mathcal{A}_{j+2} \boldsymbol{\lambda}_j \right] , \tag{6.104a}$$

$$\boldsymbol{\lambda}_j' + \gamma_j \boldsymbol{\lambda}_j = \mathbf{z}', \quad \text{for } j = 1, \ldots, n_a, \tag{6.104b}$$

where primes $(\bullet)'$ are used to indicate derivatives with respect to the reduced time, $s = \frac{Vt}{c/2}$. Equation (6.104) is a state-space realization of the identified unsteady aerodynamic model, with Equation (6.104b) and Equation (6.104a) the state and output equations, respectively. It should be remarked that all $\mathcal{A}_k$, for $k = 0, \ldots, n_a + 2$, and the aerodynamic lags, γ_j, for $j = 1, \ldots, n_a$, are constant coefficients that depend only of the aircraft characteristics (geometry, material and inertial distribution, with the last two appearing via the normal modes) and the Mach number. They are therefore not unlike the aerodynamic derivatives in Equations (4.41) and (4.42), noting that for a flexible aircraft on the nonstationary atmosphere we have many more dependencies, and that by restricting our description to potential flow we have neglected Reynolds-number effects. In order to make Equation (6.104) more general, we have normalized time with the convective time scale of the freestream, and this needs to be rescaled to physical time for aeroelastic integration in, for example, Equation (6.51), as we will see in Section 6.5.5.

In summary, for a given value of the dynamic pressure, the time-domain GAFs are directly computed using Equation (6.104a) from the input vector $\mathbf{z}(s)$, which includes the time histories of structural modes, control inputs and gusts, and on its nondimensional-time derivatives. The EoMs for the flexible aircraft, Equation (6.51), provide the link between the displacements, velocities, and accelerations on the fluid boundaries, which are not independent from each other. Equally, a model of the actuator dynamics that links its inputs and its derivatives is also needed and it is briefly discussed in the following section.

Simplified Descriptions

When the time evolution of the aerodynamic forces is sufficiently slow, it may be possible to neglect both the effect of aerodynamic lags and apparent mass in Equation (6.104a), that is, $\mathbf{f}_q(s) = \frac{1}{2}\rho V^2 \left(\mathcal{A}_0 \mathbf{z} + \mathcal{A}_1 \mathbf{z}'\right)$. This *quasi-steady aerodynamic* model is typically used to study the flight dynamics of rigid aircraft. It is sometimes still valid for flexible aircraft, but wing vibrations may introduce faster dynamics and the validity of this approximation for a given problem needs to be checked by investigating the frequency content in the vehicle responses of interest. Unsteady airfoil theory gives an indication of the validity of the quasi-steady approximation (see, for example, Figure 3.10). A quasi-steady approximation is typically sufficient when the reduced frequency k_{max} corresponding to the maximum frequency of interest satisfies $k_{max} \lesssim 0.05$. The main advantage of the quasi-steady aerodynamic assumption is that there is no need to evaluate the internal states on the aerodynamic model, which then simply defines a set of constant feedback gains on the EoMs of the vehicle. Those gains include the *aerodynamic derivatives* that were introduced in Chapter 4 and also aeroelastic effects due to quasi-steady deformation, as in Section 5.7.

Finally, the parameter that determines the relatively importance of the apparent mass effects is still the mass ratio that was defined in Equation (3.109) for a single-DoF system. For large vehicles with heavy wings (e.g., most current commercial airlines), apparent mass effects can be often neglected and, therefore, $\mathcal{A}_2 = \mathbf{0}$ in Equation (6.104a).

Example 6.3 RFA of the GAFs of a Cantilever Wing. As an example, we consider the unsteady aerodynamics on a cantilever rectangular wing for which the structure is a simple box beam built with shells and longitudinal reinforcements. The wing has a 4-ft., structural chord and 6-ft., aerodynamic chord with geometry defined by Goland (1945) and structural properties given in Beran et al. (2004). The span for the current simulations is stretched from the original to 40 ft., which gives an aspect ratio of 13.3. We refer to this wing as the stretched Goland wing, for which the geometry is shown in Figure 6.22. A DLM aerodynamic model is built with 24 panels in the span direction and 20 panels along the chord. The results in this example have been presented in Cea and Palacios (2021).

Three LNMs are retained from the structural model, that is, $N_m = 3$. The first mode can be identified as the first bending of the wing, the second one as the first torsion, and the third one as the second bending mode. However, there is an offset in the chordwise direction between the sectional center of mass and the elastic axis, which results in all those modes having both nonzero bending and torsional components. The corresponding 3×3 matrices of GAFs, $\mathcal{A}_q(\text{Ma}, ik)$ are then calculated using a DLM solver for reduced frequencies in the range $k = (10^{-9}, 0.01, 0.02, ..., 1)$. This is repeated for each Mach number of interest. Figures 6.23 and 6.24 show the imaginary versus real parts of all the components of the GAFs for two Mach numbers. The dots in the figures correspond to the sampled reduced frequencies in the DLM. When $k \to 0$ the imaginary part goes to zero and we are left with the steady forces (normalized

Figure 6.22 Structural and aerodynamic models of the "stretched" Goland wing (reprinted from Cea and Palacios (2021) with permission from Elsevier).

by the dynamic pressure) when the wing deforms with unit amplitude of the mode shape. Note also that it is a full matrix, which implies that deformation on one mode shape generate nonzero aerodynamic forces on the rest of the modes. As frequency increases, the real and imaginary parts of the GAFs take comparable values which indicates a substantial phase shift between the harmonic inputs (modal coordinates) and outputs (modal forces).

Figures 6.23 and 6.24 include the discrete results from the DLM and the RFA results with two and four aerodynamic states. Two choices of aerodynamic lags are considered. First, we consider lags in a geometric progression on the negative real axis as in Castrichini et al. (2020), that is $\gamma_j = \frac{k_{max}}{j}$, with $k_{max} = 1$. After that, we identify the constant aerodynamic matrices, $\mathcal{A}_j$, using least squares. These results are identified as "RFA" in the figures. Second, we also include the aerodynamic lags on the least-squares minimization (enforcing a minimum separation of 0.1 between them to avoid clustering), which results in optimized lags at $[0.265, 0.077]$ and $[0.435, 0.335, 0.165, 0.054]$ for Ma $= 0$, and $[1, 0.067]$ and $[1, 0.9, 0.8, 0.096]$ for Ma $= 0.7$. As it can be seen, the higher the Mach number, the more complex the interpolation of the aerodynamic matrices gets and therefore a higher-order approximation (i.e., a higher number of lags, n_a) is needed for an accurate representation.

6.5.4   System Identification from Unsteady RANS

At transonic speeds, or for problems at moderate Reynolds numbers, potential flow theory is no longer sufficiently accurate and one needs to consider additional physical effects in the aerodynamic models. Moreover, Reimer et al. (2015) have shown that potential flow aerodynamics overestimates gust loads even for subsonic conditions. In such cases, the GAFs are computed instead from solutions to the unsteady Reynolds-Averaged Navier–Stokes (RANS) equations over the whole fluid domain, which is often discretized using either finite-volume or finite-element methods.

It is possible to obtain frequency-domain solutions directly from the unsteady RANS equations using Fourier expansions, which results in the so-called *harmonic-balance method* (Thomas et al., 2002; Li and Ekici, 2019; Simiriotis and Palacios,

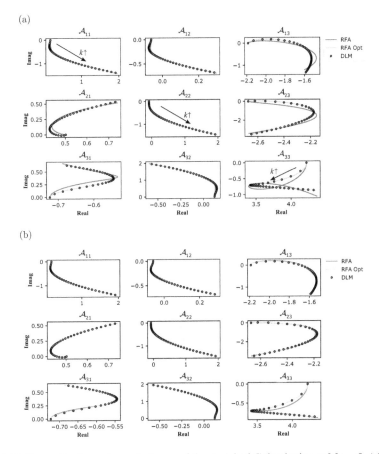

Figure 6.23 Generalized aerodynamic forces of the stretched Goland wing at Ma $= 0$. (a) $n_a = 2$ and (b) $n_a = 4$.

2023). As our interest is in the small-amplitude vibrations of the airframe, it is computationally advantageous to, first, linearize the fluid dynamic equations and, second, to retain only the first harmonic in the resulting system. This approach is known as the *Linear Frequency-Domain* method (LFD), which can give reductions of computational cost of over an order of magnitude in the computation of GAFs from the RANS equations (Thormann and Widhalm, 2013).

It is also common to use system identification techniques on time-marching solutions using the methods discussed in Section 1.4.3, since this allows the direct use of off-the-shelf general-purpose RANS solvers. The application of system identification techniques in computational aeroelasticity was pioneered by Silva and collaborators (Silva, 1997; Silva et al., 2001; Silva and Bartels, 2004). First, the fluid equations are written using an Arbitrary Lagrangian–Eulerian (ALE) formulation, such that the grid used to discretize the RANS equations can be deformed to follow the structural motions. Next, an interpolation scheme is introduced to map the displacements and velocities of the solid walls onto the fluid boundaries, in a process similar to that of

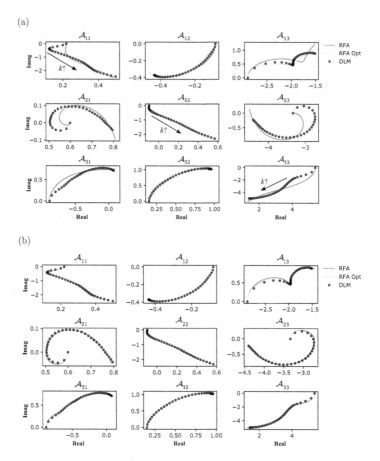

Figure 6.24 Generalized aerodynamic forces of the stretched Goland wing at Ma $= 0.7$ (reprinted from Cea and Palacios (2021) with permission from Elsevier). (a) $n_a = 2$ and (b) $n_a = 4$.

Figure 6.20. Finally, the impulse response[8] applied on a mode shape is computed in the fluid equations around a static aeroelastic equilibrium condition, and the tractions on the walls are recorded and used to compute the modal forces, again as in Figure 6.20. In the early implementation of the approach, the fluid equations were solved with nonzero initial conditions on the mode of interest, in a process known as the *unit sample response*. Then, the Markov parameters of the system response were obtained numerically and fed into an Eigenvalue Realization Algorithm to determine a low-order state-space model of the system, as in Equation (1.31). In that approach, each column of the matrix of GAFs is obtained from an individual time-domain simulation of the fluid equations. Later formulations introduced excitation in multiple structural modes in a single fluid dynamics simulation by introducing orthogonal functions, such

---

[8] In practice, however, the step response is often preferred to impulse response to achieve improved numerical stability of the CFD response (Raveh et al., 2001).

as Walsh functions (Silva, 2008). They enable the extraction of the different modal contributions from a single run in a postprocessing step. Additional enhancements to the basic method accounting for the effects of excitation amplitude and different Mach number regimes can be found in Skujins and Cesnik (2014).

The ERA results in compact state-space models for the unsteady aerodynamics, from which it is straightforward to obtain the frequency-response functions in terms of the reduced frequency using Equation (1.18). In what follows, it will be assumed for simplicity that the structure of the identified aerodynamic model obtained from unsteady RANS simulations is still that of Equation (6.104). In practice, slightly different models are typically obtained, although the discussion that follows would still be relevant to them.

6.5.5 Generalized Aerodynamic Forces in Physical Time

As it has been seen, unsteady aerodynamic loads under linear assumptions are naturally expressed in terms of the reduced frequency $k = \frac{\omega c}{2V}$, obtained from a normalization of the period of oscillations with the convective time scale, and, for a sufficiently high airspeed, also in terms of the Mach number, Ma. This was also the case for the airfoil problem in Section 3.3 and there we wrote the aeroelastic EoMs in dimensionless time, $s = \frac{Vt}{c/2}$ (Section 3.5). Here, the interest is in the analysis of the dynamics of a flexible aircraft, which are better understood in physical time and, therefore, we first rewrite all the unsteady aerodynamic models as dimensional LTI systems.

In Equation (6.104), we have identified three contributions to the aerodynamic loads on the airframe, namely loads from control surface deflections, wind gusts, and the deformations of the airframe itself. They are the three different inputs that feed into the "Fluid" block in Figure 6.20. Under linear assumptions, each of them can be computed independently in what follows.

Control Surface Aerodynamic Loads

Consider first the unsteady aerodynamics generated by time-varying deflections of the control surfaces. Control surfaces are typically driven by servo actuators, whose dynamics depend on its detailed specifications as well as their particular assembly on a wing. Their response is often nonlinear, including *freeplay* of the control surfaces (a gap region of aileron operation where small tolerances result in control surface deflections occurring without a restoring hinge moment (Panchal and Benaroya, 2021)), *deadband* of the electric motors that move the control surfaces (Pankonien et al., 2020), and delays between the commanded control input and the actual control surface deflections. Dynamic models are typically identified through experimental characterization and are valid for a specific frequency bandwidth, which needs to be known in the design of flight control systems.

A common approach is to define the linearized dynamics of each actuator (within a relatively small range of operation) by a linear third-order system, which is a characteristic response of hydraulic servoactuators (Waszak and Fung, 1996; Karpel, 2001).

Assuming N_δ actuators, we have

$$\frac{d^3\delta_i}{dt^3} + \Xi_{2i}\frac{d^2\delta_i}{dt^2} + \Xi_{1i}\frac{d\delta_i}{dt} + \Xi_{0i}\delta_i = \Xi_{0i}\delta_{ci}(t), \tag{6.105}$$

where the coefficients Ξ_{ji} are identified for the i-th actuator. This basic model is often augmented to include actuator delays by replacing $\delta_c(t)$ by $\delta_c(t-\tau)$ for the relevant input signals, with τ representing a fixed delay (Li et al., 2010). While this can have a significant effect in the controller design, they are neglected here to restrict this discussion to linear models. The resulting vector form of the actuator equation is then

$$\dddot{\boldsymbol{\delta}} + \Xi_2\ddot{\boldsymbol{\delta}} + \Xi_1\dot{\boldsymbol{\delta}} + \Xi_0\boldsymbol{\delta} = \Xi_0\boldsymbol{\delta}_c, \tag{6.106}$$

with $\Xi_j = \text{diag}\left(\Xi_{j1},\cdots,\Xi_{jN_\delta}\right)$ and $j = 0$, 1, and 2. This expression determines the actuator inputs in Equation (6.104) as a function of the control surface command $\delta_c(t)$.

From Equation (6.104b), the lag equations associated with the control surfaces in physical time can be written as

$$\dot{\boldsymbol{\lambda}}_{\delta j}(t) + \hat{\gamma}_j\boldsymbol{\lambda}_{\delta j}(t) = \dot{\boldsymbol{\delta}}(t), \quad \text{for } j = 1,\ldots,n_a, \tag{6.107}$$

where we have introduced for convenience the dimensional aerodynamic lags $\hat{\gamma}_j = \frac{V}{c/2}\gamma_j$. For a given control command, $\delta_c(t)$, Equations (6.106) and (6.107) can be used to solve the actual instantaneous control deflections, $\boldsymbol{\delta}(t)$, and its time derivatives, and the associated aerodynamic lags, $\boldsymbol{\lambda}_\delta$. After that, Equation (6.104a) can be directly used to compute the evolution of the resulting GAFs as

$$\mathbf{f}_{qc}(t) = \tfrac{1}{2}\rho V^2\left[\mathcal{A}_{\delta 0}\boldsymbol{\delta} + \frac{c}{2V}\mathcal{A}_{\delta 1}\dot{\boldsymbol{\delta}} + \frac{c^2}{4V^2}\mathcal{A}_{\delta 2}\ddot{\boldsymbol{\delta}} + \sum_{j=1}^{n_a}\mathcal{A}_{\delta,j+2}\boldsymbol{\lambda}_{\delta j}\right]. \tag{6.108}$$

To facilitate integration into the vehicle model, it is convenient to aggregate all the dependent variables into a single *actuation state vector*, that is,

$$\mathbf{x}_c = \left\{\begin{array}{c}\boldsymbol{\delta}\\ \dot{\boldsymbol{\delta}}\\ \ddot{\boldsymbol{\delta}}\\ \boldsymbol{\lambda}_{\delta 1}\\ \vdots\\ \boldsymbol{\lambda}_{\delta n_a}\end{array}\right\}, \tag{6.109}$$

whose time evolution is given in state-space form, with the actuator command, δ_c, as input, and the normalized GAFs, $\mathbf{y}_c = \mathbf{f}_{qc}/\left(\tfrac{1}{2}\rho V^2\right)$, as output. This results in a state-space representation without feedthrough of the form

$$\dot{\mathbf{x}}_c = \mathbf{A}_c\mathbf{x}_c + \mathbf{B}_c\delta_c,$$
$$\mathbf{y}_c = \mathbf{C}_c\mathbf{x}_c. \tag{6.110}$$

The output equation gives the contribution of the control surface deflections onto the aerodynamic forces. Equation (6.110) defines a dynamical system of dimension $N_\delta(3 + n_a)$, where the state and input matrices are given, respectively, as

$$
\mathbf{A}_c =
\begin{bmatrix}
\mathbf{0} & \mathcal{I} & \mathbf{0} & \mathbf{0} & \cdots & \mathbf{0} \\
\mathbf{0} & \mathbf{0} & \mathcal{I} & \mathbf{0} & \cdots & \mathbf{0} \\
-\Xi_0 & -\Xi_1 & -\Xi_2 & \mathbf{0} & \cdots & \mathbf{0} \\
\mathbf{0} & \mathcal{I} & \mathbf{0} & -\hat{\gamma}_1\mathcal{I} & \cdots & \mathbf{0} \\
\cdots & \cdots & \cdots & \cdots & \ddots & \vdots \\
\mathbf{0} & \mathcal{I} & \mathbf{0} & \mathbf{0} & \cdots & -\hat{\gamma}_{n_a}\mathcal{I}
\end{bmatrix}
\quad \text{and} \quad
\mathbf{B}_c =
\begin{bmatrix}
\mathbf{0} \\
\mathbf{0} \\
\Xi_0 \\
\mathbf{0} \\
\vdots \\
\mathbf{0}
\end{bmatrix},
\qquad (6.111)
$$

and where the output matrix is

$$
\mathbf{C}_c = \begin{bmatrix} \mathcal{A}_{\delta 0} & \frac{c}{2V}\mathcal{A}_{\delta 1} & \frac{c^2}{4V^2}\mathcal{A}_{\delta 2} & \mathcal{A}_{\delta 3} & \cdots & \mathcal{A}_{\delta, n_a + 2} \end{bmatrix}. \qquad (6.112)
$$

Wind Gust Aerodynamic Loads

Next, we can obtain similar expressions for the aerodynamic forces generated by a given gust profile. Recall that we have defined the gust profile in general through the vector $\mathbf{w}_{g0}$ in Equation (6.100) such that we can consider either spanwise variations of the background flow and/or the different components of the velocity vector. Equation (6.104b) defines the evolution of the associated aerodynamic lags, which can be aggregated into a *gust state vector* as

$$
\mathbf{x}_g = \begin{Bmatrix} \lambda_{g1} \\ \vdots \\ \lambda_{gn_a} \end{Bmatrix}. \qquad (6.113)
$$

If we also define $\mathbf{y}_g = \mathbf{f}_{qg} / \left(\frac{1}{2}\rho V^2 \right)$ as the normalized GAFs due to the gust, we have that the gust forces are evaluated from

$$
\begin{aligned}
\dot{\mathbf{x}}_g &= \mathbf{A}_g \mathbf{x}_g + \mathbf{B}_g \frac{\mathbf{w}_{g0}}{V}, \\
\mathbf{y}_g &= \mathbf{C}_g \mathbf{x}_g + \mathbf{D} \frac{\mathbf{w}_{g0}}{V},
\end{aligned}
\qquad (6.114)
$$

with the following state and input matrices:

$$
\mathbf{A}_g =
\begin{bmatrix}
-\hat{\gamma}_1\mathcal{I} & \cdots & \mathbf{0} \\
\cdots & \ddots & \vdots \\
\mathbf{0} & \cdots & -\hat{\gamma}_{n_a}\mathcal{I}
\end{bmatrix}
\quad \text{and} \quad
\mathbf{B}_g =
\begin{bmatrix}
\mathcal{I} \\
\vdots \\
\mathcal{I}
\end{bmatrix}. \qquad (6.115)
$$

We have seen in the analytical solution for the 2-D problem, Equation (3.71), that the gust loads on an airfoil are adequately described by a steady-state value and Padé approximants for the transient response. This structure is normally assumed in the realization problem for the 3-D unsteady aerodynamics, that is, we select $\mathcal{A}_{g1} = \mathbf{0}$ and $\mathcal{A}_{g2} = \mathbf{0}$, and only the zero-order term in the polynomial preconditioning of the RFA on the gust terms is included. This results in the output and feedthrough matrices:

$$\mathbf{C}_g = \begin{bmatrix} \mathcal{A}_{g3} & \cdots & \mathcal{A}_{g,n_a+2} \end{bmatrix},$$
$$\mathbf{D}_g = \mathcal{A}_{g0}. \tag{6.116}$$

Modal Aerodynamic Loads

Finally, we can write the state equations for the aerodynamic states associated with the structural DoFs in compact form. For that purpose, we define the *aerodynamic modal state vector* as

$$\mathbf{x}_a = \begin{Bmatrix} \lambda_{q1} \\ \vdots \\ \lambda_{qn_a} \end{Bmatrix} \tag{6.117}$$

of dimension $n_a N_m$, and the corresponding outputs $\mathbf{y}_a = \mathbf{f}_{qa}/\left(\frac{1}{2}\rho V^2\right)$, whose state-space equations are written as

$$\dot{\mathbf{x}}_a = \mathbf{A}_a \mathbf{x}_a + \mathbf{B}_a \dot{\mathbf{q}},$$
$$\mathbf{y}_a = \mathbf{C}_a \mathbf{x}_a, \tag{6.118}$$

with $\mathbf{A}_a = \mathbf{A}_g$, $\mathbf{B}_a = \mathbf{B}_g$ as in Equation (6.115), and

$$\mathbf{C}_a = \begin{bmatrix} \mathcal{A}_{q3} & \cdots & \mathcal{A}_{q,n_a+2} \end{bmatrix}. \tag{6.119}$$

For convenience, we have not included here the forces proportional to the modal amplitudes and their derivatives of Equation (6.104a). They are instead retained explicitly in the aggregated equations below.

Aggregated Aerodynamic Loads

The total instantaneous aerodynamic forces on the airframe are obtained from superposition of the three previous effects. As a result, the GAFs in physical time domain can be written as

$$\mathbf{f}_q = \frac{1}{2}\rho V^2 \mathcal{A}_{q0}\mathbf{q} + \frac{1}{4}c\rho V \mathcal{A}_{q1}\dot{\mathbf{q}} + \frac{1}{8}c^2\rho \mathcal{A}_{q2}\ddot{\mathbf{q}} + \mathbf{f}_{qa} + \mathbf{f}_{qc} + \mathbf{f}_{qg}, \tag{6.120}$$

or, in terms of the gust, control, and modal aerodynamic states defined above, as

$$\mathbf{f}_q = \frac{1}{2}\rho V^2 \left[\mathcal{A}_{q0}\mathbf{q} + \frac{c}{2V}\mathcal{A}_{q1}\dot{\mathbf{q}} + \frac{c^2}{4V^2}\mathcal{A}_{q2}\ddot{\mathbf{q}} + \frac{1}{V}\mathcal{A}_{g0}\mathbf{w}_{g0} + \mathbf{C}_a\mathbf{x}_a + \mathbf{C}_c\mathbf{x}_c + \mathbf{C}_g\mathbf{x}_g \right]. \tag{6.121}$$

In summary, for a given time history of the commanded inputs on the control surfaces, a gust profile, and amplitudes of the structural modes, Equations (6.110), (6.114) and (6.118) determine the time evolution of the corresponding aerodynamic states $\mathbf{x}_c$, $\mathbf{x}_g$, and $\mathbf{x}_a$, respectively. Those are then used to add the effect of actuator dynamics and aerodynamic delays to the instantaneous generalized (modal) forces appearing on the aircraft using Equation (6.121). This process is summarized by the flowchart in Figure 6.25, which also includes the modeling of the nonstationary atmospheric conditions described in Chapter 2, and the structural model of the airframe (which also

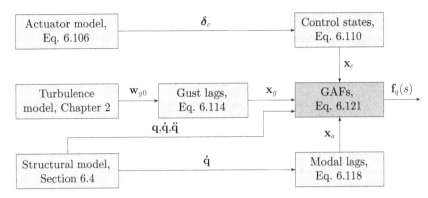

Figure 6.25 Flow diagram for evaluation of the generalized aerodynamic forces in time domain.

includes, in general, rigid-body dynamics). The resulting forces, $\mathbf{f}_q(t)$, finally allow us to fill the lower block of the diagram in Figure 6.20.

As a final remark, it is interesting to note that Equation (6.121) has the same structure to that obtained for the airfoil problem in Equation (3.76), with the difference that here the gust velocity $\mathbf{w}_{g0}$ explicitly appears in the force equations. This is because in this 3-D formulation the relation between gust and forces is obtained by "black box" identification methods that do not exploit known information of the mathematical description of the underlying physical system. This can be prevented with a physics-based linearized model of the aerodynamics, such as the one outlined in Chapter 7. Equation (6.121) can be compared to Equation (7.66), which is obtained using that second approach to represent the same physical process.

6.6 Linear Dynamics of Flexible Aircraft

The external forces on a powered aircraft include the distributed forces due to gravity and aerodynamic loading, as well as localized thrust forces. Since they depend on the instantaneous external geometry of the vehicle, the problem can be represented as the feedback system shown in Figure 6.20. Depending on its kinematics, the aircraft dynamics can be approximated by the rigid-body–linear elastic EoMs, Equation (6.71), the mean-axes approximation, Equation (6.78), or the linearized equations, given by Equation (6.51). Only this latest case is discussed in this section.

The linearization occurs around a reference flight condition, which is assumed to be known – either level, climb, or descent flight, with given payload distribution, airspeed, and altitude. The stability axes are then established for the aircraft around that point, and the linearized variables and the corresponding aircraft mass, damping, and stiffness properties, as well as the aerodynamic derivatives, are all defined with respect to that reference. This typically implies a rotation of both the global moments of inertia and the LNM of the structure, such that x is aligned with the forward flight velocity.

6.6.1   State-Space Equations of Motion

Substituting the GAFs defined by Equation (6.121) into the linearized EoMs for the flexible aircraft, Equation (6.51), and collecting terms in $\mathbf{q}$, $\dot{\mathbf{q}}$ and $\ddot{\mathbf{q}}$, we have the linearized EoMs of flexible aircraft dynamics as

$$\mathbf{M}_{ae}\ddot{\mathbf{q}} + \mathbf{D}_{ae}\dot{\mathbf{q}} + \mathbf{K}_{ae}\mathbf{q} = \tfrac{1}{2}\rho V \mathcal{A}_{g0}\mathbf{w}_{g0} + \mathbf{C}_a\mathbf{x}_a + \mathbf{C}_c\mathbf{x}_c + \mathbf{C}_g\mathbf{x}_g + \mathbf{f}_{qt}, \qquad (6.122)$$

which need to be solved together with the internal equations for the control, gust, and aerodynamic states, Equations (6.110), (6.114) and (6.118), respectively. In this expression we have aggregated rigid-body, structural, and aerodynamic properties into the following matrices

$$\begin{aligned}
\mathbf{M}_{ae} &= \mathbf{M}_q - \tfrac{1}{8}c^2\rho \mathcal{A}_{q2}, \\
\mathbf{D}_{ae} &= \mathbf{D}_q - \tfrac{1}{4}c\rho V \mathcal{A}_{q1}, \\
\mathbf{K}_{ae} &= \mathbf{K}_q - \tfrac{1}{2}\rho V^2 \mathcal{A}_{q0} - \mathbf{G}_q,
\end{aligned} \qquad (6.123)$$

referred to as the *aeroelastic mass, damping, and stiffness* matrices, respectively. Equation (6.122) also incorporates a generic modal forcing term $\mathbf{f}_{qt}$, which may include engine thrust and any other additional external loading that may be applied onto the vehicle. Equation (6.122) is known as the *NDOF model* of a flexible aircraft, where N is typically of the same dimension as $\mathbf{q}$, that is, the sum of both rigid and elastic DoFs.

Finally, we define the following aeroelastic state vector by appending the rigid, structural and aerodynamic states in the problem as

$$\mathbf{x}_{ae} = \begin{Bmatrix} \mathbf{q} \\ \dot{\mathbf{q}} \\ \mathbf{x}_a \\ \mathbf{x}_c \\ \mathbf{x}_g \end{Bmatrix}. \qquad (6.124)$$

The state-space form of the aeroelastic system then becomes

$$\dot{\mathbf{x}}_{ae} = \mathbf{A}_{ae}\mathbf{x}_{ae} + \mathbf{B}_{ae} \begin{Bmatrix} \mathbf{f}_{qt} \\ \delta_c \\ \frac{\mathbf{w}_{g0}}{V} \end{Bmatrix}, \qquad (6.125)$$

with the aeroelastic state and input matrices defined, respectively, as

$$\mathbf{A}_{ae} = \begin{bmatrix} \mathbf{0} & \mathcal{I} & \mathbf{0} & \mathbf{0} & \mathbf{0} \\ -\mathbf{K}_{ae}^* & -\mathbf{D}_{ae}^* & \mathbf{C}_a^* & \mathbf{C}_c^* & \mathbf{C}_g^* \\ \mathbf{0} & \mathbf{B}_a & \mathbf{A}_a & \mathbf{0} & \mathbf{0} \\ \mathbf{0} & \mathbf{0} & \mathbf{0} & \mathbf{A}_c & \mathbf{0} \\ \mathbf{0} & \mathbf{0} & \mathbf{0} & \mathbf{0} & \mathbf{A}_g \end{bmatrix} \quad \text{and} \quad \mathbf{B}_{ae} = \begin{bmatrix} \mathbf{0} & \mathbf{0} & \mathbf{0} \\ \mathbf{M}_{ae}^{-1} & \mathbf{0} & \mathbf{D}_g^* \\ \mathbf{0} & \mathbf{0} & \mathbf{0} \\ \mathbf{0} & \mathbf{B}_c & \mathbf{0} \\ \mathbf{0} & \mathbf{0} & \mathbf{B}_g \end{bmatrix}, \qquad (6.126)$$

where we have introduced $\mathbf{X}^* = \mathbf{M}_{ae}^{-1}\mathbf{X}$ to simplify the notation. Equation (6.125) has resulted from the dynamic coupling between the structure, including rigid-body dynamics, unsteady aerodynamics, and actuator system. Given external forces, $\mathbf{f}_{qt}(t)$, actuator commands, $\delta_c(t)$, and/or gust profiles, $\mathbf{w}_{g0}(t)$, it determines the linearized aircraft response. Further analysis of the results using the methods of Section 6.4.4 also determines the associated internal loads on the airframe. It can be used to establish the response to particular atmospheric conditions if the gust profiles are defined by the nonstationary atmospheric models of Chapter 2.

Note that for the state-space form of the EoMs to exist the aeroelastic mass matrix, $\mathbf{M}_{ae}$, has to be nonsingular. This is typically the case as the mass of the vehicle is much higher than the apparent mass. In practical applications, the size of the system (the dimension of $\mathbf{x}_{ae}$) is of order $\mathcal{O}(10^2)$. As an example, the model for the X-56A aircraft used in Example 6.6 has around 400 states (Burnett et al., 2016). Equation (6.125) allows the direct solution, in a computationally efficient manner, of the EoMs of the wing or aircraft for prescribed deflections of the control surfaces and under known atmospheric velocities, that is, it describes the more general open-loop linear response of a flexible air vehicle. As it has been written in state-space form, all the methods for the analysis of linear systems can also be applied directly herein. Typical analyses include the evaluation of admittances (frequency-response functions) between input and relevant outputs in the aircraft dynamics, or assessment of asymptotic stability. Some of those basic concepts are introduced next to conclude this chapter, together with examples on actual aircraft. In Chapter 10, we investigate the closed-loop dynamics, where the deflections of the control surfaces are not prescribed but rather are calculated based on the current state of the aircraft using automatic control methods.

6.6.2   Stability Analysis

The eigenvalues of the aeroelastic state matrix, $\mathbf{A}_{ae}$, determine the asymptotic stability in the flexible aircraft dynamics. As our description includes both flight dynamics and aeroelastic DoFs, it captures simultaneously the longitudinal and lateral modes of the (flexible) vehicle, and the aeroelastic modes that determine flutter or divergence (as in Example 6.5.2). As it will be seen in the examples below, that distinction gets blurred here when flexible effects may affect the static aerodynamic derivatives or when dynamic coupling between rigid and flexible modes occurs.

From Equation (6.123), but also from the definition of the dimensional aerodynamic lags in Equation (6.107), the state matrix depends on both the airspeed, V, and air density, ρ, at the current flight condition. Therefore, the stability of a flexible aircraft needs to be investigated independently at every point across the flight envelope. For example, for very light aircraft, the effect of apparent mass can be very high at low altitudes, where the air density is high (and, therefore, low mass ratio, Equation (3.109)), while it can be neglected at high altitudes. That may change the effective vibration characteristics in a way that substantially modifies the dynamic stability of the vehicle with altitude, even if the dynamic pressure is kept constant to maintain lift.

For a given altitude (i.e., air density ρ), the onset of instability occurs at the smallest airspeed at which one of the eigenvalues of $\mathbf{A}_{ae}$ crosses the imaginary axis. If the imaginary part of that eigenvalue is positive, we have either an unstable flight dynamics or aeroelastic mode, depending on the frequency (or a mixed of both as it will be seen in Example 6.6). If the imaginary part is zero, that is, the migrating pole is a zero eigenvalue, then we have a static instability.

Moreover, it was seen in Equation (6.94) that the GAFs depend on the Mach number. Since the Mach number in the Earth atmosphere is determined uniquely by altitude (i.e., density) and airspeed, this defines an iterative process to determine the unstable conditions at each altitude: The GAFs are obtained for an initially assumed value on the flight Mach number, which is then updated after the stability analysis has determined the flutter speed for those initial GAFs. This process is repeated until convergence, resulting in the so-called *matched conditions*, at which the GAFs correspond to the actual Mach numbers at the unstable points in the flight envelope. In some situations however, it is convenient to compute stability on *unmatched conditions*, that is, the GAF refer to a nominal flight condition that is used as constant reference to estimate the flutter speed. In that case, the difference between the limit and the nominal airspeed is considered as a safety factor.

Example 6.4 Dynamic Stability of a Cantilever Wing.   The flutter speed of the stretched Goland wing introduced in Example 6.3 is calculated from the eigenvalues of $\mathbf{A}_{ae}$ for a range of Mach values up to Ma $= 0.85$. An *unmatched flutter condition* solution is considered, in which the Mach number used to compute the aerodynamic forces remains fixed and may not correspond to the Mach number at the flutter speed. For matched flutter conditions, an iterative process would need to be considered – however it is not uncommon that the flutter speed is in a different flight regime (e.g., supersonic) where the aerodynamic model is not valid, and therefore unmatched flutter is often used to establish stability margins.

An RFA of the GAFs with $n_a = 4$ is first computed for each Mach number (see Example 6.3). Then, the state-space model is constructed, as in Equation (6.125). The eigenvalues of the state matrix, for varying airspeeds, are used to determine the aeroelastic stability of the wing. The stability boundary is referred to as RFA in Figure 6.26. As a reference, the direct frequency-domain solution is also included, in which neutrally stable solutions of the system are sought using the *p-k* method (see Example

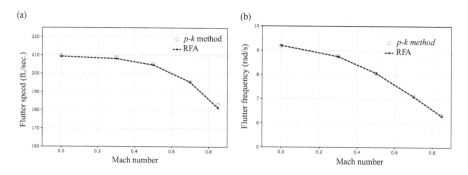

Figure 6.26 Stretched Goland wing flutter at different Mach numbers. (a) Flutter speed and (b) flutter frequency.

6.5.2). Both methods compare very well, with small differences due to the truncation in the RFA, and show the strong dependency of the flutter onset conditions with compressibility effects.

Example 6.5 Dynamic Stability of the Predator UAV. The Predator Unmanned Aerial Vehicle (UAV) is used to exemplify the differences between rigid- and flexible aircraft dynamics on a conventional wind-body-tail configuration with relatively high wing flexibility. The Predator is a medium-altitude long-endurance reconnaissance aircraft first developed by General Atomics in 1994 (see Figure 6.27). The vehicle length and wingspan are 8.23 m and 14.84 m, respectively, and its sea-level cruise speed is around 200 km/h. A coupled aeroelastic/flight dynamics model was built by Baldelli et al. (2006) using DLM for the unsteady aerodynamics corrected with flight data for the rigid-body aerodynamic derivatives, and with a linear flexible-dynamic model using up to 13 "free-free" elastic modes. This corresponds to a cutoff frequency of 10.3 Hz in the vibration response.

The dynamic stability of this aircraft is reported using quasi-steady and unsteady aerodynamic models, which were transformed into state-space form using the minimum-state method. For the quasi-steady model, only $\mathcal{A}_1$ and $\mathcal{A}_0$ are retained in Equation (6.98), as done in flight dynamics. The unsteady model introduces both apparent mass and lag terms into the aerodynamics. The first six dynamic modes of the aircraft as a function of the number of elastic modes, N_s, are shown in Table 6.1 using quasi-steady aerodynamics and in Table 6.2 under full unsteady aerodynamic assumptions. They are referred to as flight-dynamic modes, as they can still be identified to have the main features of a phugoid, short-period, spiral, Dutch roll, and roll subsidence, but most of them include contributions of the elastic DoFs.

The first column in Table 6.1 includes the dynamic modes obtained with a rigid aircraft model, such as those in Chapter 4. It can be seen that a rigid aircraft would be predicted to be stable except for a marginally unstable spiral mode. The first column of Table 6.2 shows that considering unsteady aerodynamics in a rigid aircraft model can already have a nonnegligible effect, and indeed the short-period mode moves from marginally stable to marginally unstable if wake effects are included.

Figure 6.27 Predator UAV flies on a simulated reconnaissance flight off the coast of Southern California on Dec. 5, 1995 (photo by the U.S. Navy).

Table 6.1 Predator UAV flight-dynamic modes with quasi-steady aerodynamics as a function of the number of elastic modes (Baldelli et al., 2006).

Mode	$N_s = 0$	$N_s = 1$	$N_s = 6$
Short-period	$-0.015 \pm 0.118i$	$-0.017 \pm 0.117i$	$0.033 \pm 0.083i$
First phugoid	-2.294	-2.247	-2.471
Second phugoid	-1.159	-1.137	-1.179
Roll subsidence	-8.372	-8.373	-8.732
Dutch roll	$-0.280 \pm 1.227i$	$-0.280 \pm 0.227i$	$-0.277 \pm 1.256i$
Spiral	0.054	0.054	0.059

Table 6.2 Predator UAV flight-dynamic modes with unsteady aerodynamics as a function of the number of elastic modes (Baldelli et al., 2006).

Mode	$N_s = 0$	$N_s = 1$	$N_s = 6$
First short-period	$0.019 + 0.108i$	$-2.253 + 1.060j$	-1.913
Second short-period	$-0.019 - 0.108i$	$-2.253 - 1.060i$	-2.793
First phugoid	-3.037	0.012	0.013
Second phugoid	-1.208	0.778	0.415
Roll subsidence	-10.722	-11.623	-8.648
Dutch roll	$-0.185 \pm 1.137i$	$-0.216 \pm 1.185i$	$-0.272 \pm 1.216i$
Spiral	0.033	0.040	0.057

The first structural mode is the wing bending mode, and if this mode is added to the problem, the longitudinal stability characteristics are rather significantly affected. In particular, bending vibrations stabilize the short period mode and quite dramatically so if the unsteady aerodynamics are considered in the analysis. On the other hand, the phugoid mode becomes marginally unstable when both bending and the unsteady aerodynamics are included.

Typically, the first few structural modes are expected to have an impact in the flight dynamic stability of a flexible aircraft, and here results with $N_s = 6$ are included. They show first a minimal influence of the elastic modes on the lateral dynamics in a converged model, while elastic effects mitigate the phugoid instability that was found when considering unsteady aerodynamics. Overall, very large differences can be observed in the longitudinal modes predicted by the last column of Table 6.2 (the most accurate model) to the rigid aircraft results in the first column Table 6.1.

Example 6.6 Body Freedom Flutter of NASA X-56A UAV. The X-56A Multi-Utility Technology Testbed (MUTT) is an experimental UAV designed to explore the effect of airframe flexibility in aircraft performance (Beranek et al., 2010; Burnett et al., 2016). The vehicle has swept wings with 8.5-m span and a maximum weight of just over 200 kg, with the tailless configuration shown in Figure 6.28. This flying wing

Figure 6.28 X-56A Multi-Utility Technology Testbed (photo by NASA).

has strong rigid-flex coupling and indeed it has been designed to be aeroelastically unstable within its flight envelope. However, the vehicle dynamics can be stabilized using a flight control system and different feedback control strategies have been successfully demonstrated in a flight test campaign (Grauer and Boucher, 2020). This has provided a wealth of data for the subsequent development of lightweight flying wings with superior aerodynamic performance.

Due to their compact geometry, tailless aircraft have a much lower pitch moment of inertia, I_y, than aircraft with a conventional (wing-body-tail) configuration. This has two interrelated effects on the aircraft dynamics. First, the short-period frequency increases as the pitch inertia reduces (it can be seen from analysis of Equation (4.94) that this frequency goes as $V/\sqrt{I_y}$). Second, the flexible wings can no longer be seen as "clamped" about a effectively rigid fuselage, that is, an upward deformation of a swept wing corresponds to a nonnegligible downward translation and pitch of the aircraft to keep the CM constant. Furthermore, changes in pitch imply changes in aerodynamic forces and, since the short-period mode is now closer to the first bending mode, the likelihood of unstable coupling between both modes significantly increases. This results in a characteristic dynamic instability of flying wings known as *body-freedom flutter* (BFF). BFF occurs between the short-period pitch oscillations and the first bending mode of the wings, which interact following a very similar mechanism to those described for pitch-and-plunge systems in Chapter 3. Historically, BFF was first reported in the 1940s in the Horten wings, which were the first viable design for a swept flying wing (Nickel and Wohlfahrt, 1994), and have also been reported on the B-2 bomber (Jacobson et al., 1998). BFF also appear within the flight envelope of other nonconventional aircraft configurations that display an appropriate combination of wing bending stiffness, fuselage pitch inertia and location of the aircraft CM. For example, using a simplified (linear) analytical model, Weisshaar and Lee (2002) identified the potential occurrence of BFF in joined-wing aircraft. Other historical examples have been reviewed by Love et al. (2005). The traditional solution to avoid BFF has been to increase the wing bending frequency by means of higher wing stiffness. The problem is that this increases weight and cancels out most (if not all) the efficiency gains from the aerodynamic design. Active flutter suppression systems

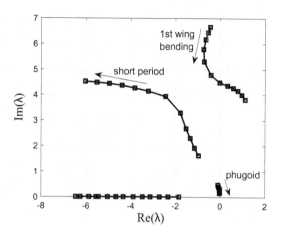

Figure 6.29 Stability diagram of the SC005 configuration for the X-56A UAV.

(Livne, 2018) are, therefore, investigated in the X-56A as an alternative solution with minimal mass penalty.

For an early prototype (configuration SC005) of the X-56A a linear finite-element model was built by Love et al. (2005) with frequency-domain unsteady aerodynamics given by the DLM and rewritten in time domain using RFA. This results in a coupled (linear) flight dynamic-aeroelastic model for which the eigenvalues were computed for different flight speeds. The resulting stability diagram is shown in Figure 6.29. Only the low-frequency longitudinal dynamics are displayed, showing the first elastic mode, the short period and phugoid modes, and the first aerodynamic lag along the negative real axis. The airspeed sweep goes from 120 to 340 ft./sec. in increments of 20 ft./sec. with arrows indicating the evolution of the roots with increased airspeed. As it can be seen, the first frequency of the wing bending decreases with the airspeed, while that of the short period increases. They coalesce around 240 ft./sec., thus resulting in a BFF instability and the elastic mode becoming unstable.

Based on what we have seen here, flutter is expected to be dependent on the mass of the system. It is interesting to note that for the case of the X-56A, payload is only carried in the center body, and the wings are dry. Therefore, considering a cantilever wing in isolation, no change in its flutter behavior is expected with the change in payload. However, when investigating the flutter behavior of the free-flying vehicle for three different payload (fuel and ballast water) configurations, we find that the flutter mode changes due to the changes in the rigid body dynamics of the vehicle. This was studied for the *as-built* X-56A by Jones and Cesnik (2016) and Jones (2017), and it is summarized here. Consider three payload configurations: empty-fuel empty-water (EFEW), 7-lb. fuel empty-water (7FEW), and full-fuel empty-water (FFEW) cases. For each case, the model was trimmed at level flight at every speed point before the aeroelastic modes and frequencies were extracted for the flutter calculation. The results are given in Table 6.3, and the corresponding modes are shown in Figure 6.30.

In the free-flight condition, the flutter speed increases as the payload weight increases. The flutter mode of the EFEW configuration is a symmetric body-freedom

Table 6.3 As-built X-56A flutter onset in free level flight. Speed is normalized based on a reference speed, V_{ref} (Jones, 2017).

Payload configuration	Normalized flutter speed	Flutter frequency (Hz)	Flutter mode (Figure 6.30)
EFEW	0.89	0.79	SBFF
7FEW	0.95	2.70	ABFF
FFEW	1.04	2.87	ABFF

(a)   (b)

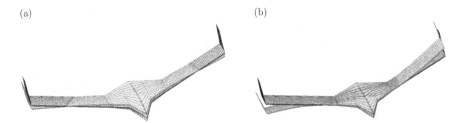

Figure 6.30 X-56A body-freedom flutter modes associated with different payload configurations. (a) Symmetric body-freedom flutter (SBFF) and (b) antisymmetric body-freedom flutter (ABFF).

flutter, as expected. However, the flutter mode of the heavier configurations is an antisymmetric wing-bending–torsion structural mode coupled with rigid-body pitch DoF. These results highlight the importance of the rigid-body characteristics of the aircraft on its flutter onset.

6.7 Summary

This chapter has presented the mathematical models that describe the dynamics of flexible aircraft that display dynamic interactions between rigid and elastic DoFs but whose deformations are small enough to be in the linear elastic regime. The EoMs of such vehicles have been derived from first principles using a body-fixed reference frame to define the aircraft's instantaneous attitude and position with respect to Earth. The assumption of linear elastic behavior leads to the deformation being described by a stiffness matrix, which can be easily obtained from a linear finite-element discretization of the airframe. Deformations of the aircraft may still affect its instantaneous inertia characteristics, and this was therefore first considered in Equation (6.30), which includes explicit (nonlinear) dependencies of the elastic DoFs on the inertia and gyroscopic matrix.

A further simplification can be introduced when flexibility effects are sufficiently small to have a negligible effect on the inertia properties. This yield what we referred to as the nonlinear rigid-linear elastic EoMs, which are better described using a modal projection of the elastic DoFs. The resulting description is computationally efficient as it has few DoFs and its only nonlinear terms are those found in rigid aircraft dynamics, and it integrates very well with standard methods of linear unsteady aerodynamics. Those have been described using the DLM to exemplify the features required on an aerodynamic description. The more general case, in which the aerodynamic forces are obtained from a discretization of the Navier–Stokes equations, has also been outlined. It was seen that frequency-domain aerodynamics can be easily cast into state-space formulations using a RFA.

The last section has integrated all of the above to describe linear dynamics of a flexible aircraft under commanded inputs on its control surfaces and potentially in nonstationary atmospheric conditions. It was seen that the EoMs can be cast in state-space form with state and input matrices that vary for each point in the flight envelope. This has finally enabled the investigation of some unconventional stability problems, such that the BFF characteristic of low-pitch-inertia flexible flying wings.

6.8   Problems

Exercise 6.1. A flexible square plate of side length $2a$, originally in the $x-y$ plane and with its sides parallel to the axes, is free to move in 3-D space. The out-of-plane elastic deformations measured with respect to the origin of the axes can be approximated by two parabolic *shape functions*, $N_1(x,y) = \frac{x^2}{a^2}$ and $N_2(x,y) = \frac{y^2}{a^2}$, leading to $w(x,y,t) = N_1(x,y)\xi_1(t) + N_2(x,y)\xi_2(t)$. Consider a point P at coordinates $x=a$ and $y=a$ on the undeformed plate, as shown in the figure, and assume that the translational and angular velocities at the center of the plate are known for $t > 0$.

(i) Using a body-attached reference frame B at the center of the plate, obtain the matrix of derivatives of the instantaneous velocity at point P with changes in the elastic degrees of freedom, $\frac{\partial v_B^P}{\partial \xi}$, and its rates, $\frac{\partial v_B^P}{\partial \dot{\xi}}$, in terms of the velocities at the center of the plate.

(ii) Determine the instantaneous location of the center of mass as the plate deforms and use it to define a CM-based floating reference frame.

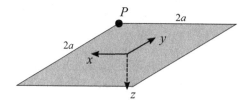

Exercise 6.2. Consider the 2-D free-flying mechanical system made of two massless beams of length l and bending stiffness EI, and three equal point masses of magnitude m. This system is initially aligned with the vertical axis and it is moving in the horizontal direction at a speed V, as shown in the figure. Its subsequent dynamics are driven by gravitational acceleration, g. A body-attached reference frame is considered,

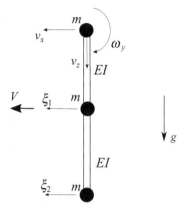

which is chosen to be at the top mass at $t = 0$, as shown in the figure. Assume that the deformations of the beams is sufficiently small for both linear elastic behavior and for a negligible effect in changing the global inertia characteristics.

(i) Derive the linearized EoMs for this 5-DoF system.
(ii) Compute its mode shapes and rewrite the EoMs in modal coordinates.

Exercise 6.3. Consider a free-flying straight beam with bending flexibility and constant properties. The beam has length L, bending stiffness EI, and mass per unit length ρA and it is restricted to move in the $x-z$ plane. An analytical solution for the linear normal modes is available for this problem, with the natural frequencies of the elastic modes given by $\omega_j = \kappa_j^2 \sqrt{\frac{EI}{\rho A}}$, with κ_j being the positive roots of Equation (6.68), that is,

$$\cos\left(\kappa_j L\right)\cosh\left(\kappa_j L\right) = 1. \tag{6.127}$$

The associated eigevectors can be expressed analytically when written in terms of the absolute displacements in an inertia reference frame as

$$r_{z,j} = \sinh\kappa_j x + \sin(\kappa_j x) - \frac{\sin\kappa_j L - \sinh\kappa_j L}{\cos\kappa_j L - \cosh\kappa_j L}\left(\cosh\kappa_j x + \cos(\kappa_j x)\right). \tag{6.128}$$

(i) Demonstrate numerically that the first four roots of Equation (6.127) are
$\kappa L = \{4.730, 7.853, 10.996, 14.137\}$.
(ii) Show that the elastic modes are orthogonal to rigid displacements and rotations about the center of the beam.
(iii) Express the mode shapes with respect to a body-fixed reference frame located at the midpoint of the beam.
(iv) If the beam is attached to a rotating engine at its midpoint such that it has constant rigid-body rotations of rate ω_{y0}, write the linearized equations in modal coordinates for the resulting system and determining its resonant characteristics (the rotating natural frequencies).

Exercise 6.4. Consider a linear aeroelastic system with N structural DoFs, $\mathbf{q}$, and known associated mass, $\mathbf{M}$, damping, $\mathbf{D}$, and stiffness, $\mathbf{K}$, matrices. Its flow conditions are defined for a freestream with speed V and density ρ, and obtained from a frequency-domain fluid solver that estimates the GAFs as $\frac{1}{2}\rho V^2 \mathcal{A}(ik)\bar{\mathbf{q}}$, with $\mathbf{q}(t) = \bar{\mathbf{q}}e^{i\omega t}$ and $k = \frac{\omega c}{2V}$. Assume a rational-function approximation for the aerodynamic influence coefficient matrix given by n lags and negligible apparent mass as

$$\mathcal{A}(ik) = \mathcal{A}_0 + ik\mathcal{A}_1 + \sum_{j=1}^{n} \frac{ik}{\gamma_j + ik}\mathcal{A}_{j+1}.$$

(i) Write the structural dynamics equations in state-space form.
(ii) Write the unsteady aerodynamics equations in state-space form.
(iii) Write the coupled time-domain equations for the resulting feedback system, identify its state vector and state matrix.
(iv) Explain how you would determine the aeroelastic stability of this system.

7 Time-Domain Unsteady Aerodynamics for Low-Speed Flight

7.1 Introduction

Chapter 6 has introduced a unified model of the dynamics of flexible aircraft, which captures all the major inertial and aerodynamic interactions between the flight dynamics and the aeroelastic response of a deforming aircraft. A key assumption has been that the elastic deformations of the airframe are sufficiently small for (1) linear deformation models to be valid (i.e., the elastic state of the structure is defined by a constant stiffness matrix) and (2) the aerodynamic forces to have a linear dependency with the deformations. As nonstationary effects can be important, the DLM has been the main strategy to compute the aerodynamic loads on the vehicle. However, a major drawback from that approach, already discussed in Section 6.5.1, is that the DLM does not capture in-plane kinematics and forces (e.g., induced drag), which are essential in flight dynamics. Therefore, for applications beyond traditional aeroelastic simulations, the DLM needs to be augmented with rigid-body aerodynamic derivatives, obtained through other means (see, e.g., Castrichini et al. (2020)). It is also increasingly possible to build the aerodynamic model using methods of computational fluid dynamics (CFD) (Farhat, 2004), typically considering viscous effects with a turbulence model. However, the routine use of CFD solutions for flexible aircraft simulation is still mostly confined to problems with linear structural dynamics, as we have done in Section 6.5.4, while the focus in this and the next chapters is on vehicles with very flexible wings.

Thus, we introduce here a more general description of the unsteady aerodynamics for low-speed (i.e., incompressible) flows that accounts for arbitrary kinematics of the lifting surfaces, namely the unsteady vortex-lattice method (UVLM). The UVLM will go a long way into ameliorating the intrinsic limitations of the DLM for flexible-aircraft dynamics at low Mach numbers, but it is still based on potential flow theory and, therefore, neglects the viscous drag. Furthermore, as with the DLM, it uses a lifting surface model that may need thickness corrections to predict the steady lift-curve slope. The UVLM is also relevant for some other problems with small-amplitude vibrations; in particular, it captures the dependency of the unsteady loads on the steady forces in the reference configuration (Murua et al., 2012, 2014), and it is essential in the dynamics of very flexible aircraft undergoing large wing displacements, which is the focus of Chapter 9.

The main interest here is in the nonstationary aerodynamic loads that generate airframe deformations, and we will restrict ourselves to points of the flight envelope where wing stall does not occur. For such lift-dominated problems, potential flow theory provides adequate estimates of the aerodynamic forces acting on the airframe with a rather modest computational effort and, consequently, it is the most commonly used approach (Afonso et al., 2017). In this chapter, boundary-element solutions in nonstationary potential flow aerodynamics will be presented in terms of vortex segment discretizations in Section 7.2, for 2-D problems, and vortex rings for 3-D problems, which result in the UVLM (Section 7.3). Contrary to frequency-domain theories of previous chapters, which are limited to linearized problems, this gives numerical solutions for lifting surfaces undergoing arbitrary kinematics (provided no significant flow separation occurs) in the time domain. These general descriptions are then written in Section 7.4, after local linearization around an arbitrary equilibrium point, in a state-space descriptor form, so that aeroelastic models can be obtained from the feedback interconnection between the fluid and the structure. Going back to the distinction between internal and external representations of dynamical systems that were introduced in Section 1.4.3, the DLM results in an external description of the unsteady aerodynamics, while the UVLM results in an internal description in which system states and their dynamics are explicitly defined. However, the resulting models are typically of large order, and Section 7.4 outlines some standard model-order reduction techniques in linear dynamic systems; in particular, the focus is on model reduction based on system balancing, which becomes essential for the routine application of the UVLM in the analysis of flexible aircraft. Balancing-based ROMs are first exemplified in 2-D unsteady aerodynamic problems in Section 7.4.1. For general problems, the computationally efficient balancing algorithm of Maraniello and Palacios (2020) is considered. This method is tailored to both the relevant physics and the mathematical structure of the UVLM, and it is presented in Section 7.4.2. Numerical results using the open-source aeroelastic toolbox SHARPy (del Carre et al., 2019a), which is introduced in some more detail in Section 9.4, are included throughout the chapter to illustrate the methods.

7.2   Discrete-Vortex Methods for Nonstationary Airfoils

Chapter 3 has introduced analytical solutions for some problems in 2-D unsteady aerodynamics. Those solutions exist only for the linearized airfoil problem with infinitesimally small thickness, small-amplitude out-of-plane kinematics, and a flat wake. More general solutions can be numerically obtained within the assumptions of potential flow theory by using time-dependent singularity distributions, for which an extensive discussion can be found in the book of Katz and Plotkin (2001). They give good estimates of lift forces in complex nonstationary conditions, including some separated flows (Gonzalez-Salcedo et al., 2017; Zou et al., 2015). The interest in this section is limited, however, to the generation of state-space descriptions of the unsteady aerodynamics directly from the singularity solutions, as originally proposed by Hall (1994) and further explored in Dowell et al. (2004). To exemplify the problem, we restrict ourselves again here to the assumptions of Section 3.3, that is, a linear theory for vortex dynamics under small-amplitude kinematics and with a flat wake. The results of this section will be later extended to more general 3-D situations and complex kinematics in Section 7.3.

7.2.1 Linearized Vortex Dynamics

Consider again the thin-airfoil problem in the x–z plane that was defined in Section 3.3.1. As we did there, we seek a vortex sheet solution using the singularity distribution of Figure 3.5b, which results in the unsteady thin-airfoil equation, Equation (3.25). While in Section 3.3.1 we could identify analytical solutions of this integral equation using Chebyshev polynomials, here we seek numerical solutions using a *collocation method*. In such an approach, a spatial discretization of the vortex sheet is introduced (lumped vortices in our case), and the nonpenetrating boundary conditions are only enforced on a subset of points, known as collocation points. This results in a set of algebraic equations that determine the circulation strength of the lumped vortices. As it will be seen, this method has a much higher computational cost than the analytical solutions in Section 3.3.1, but it can be easily extended to 3-D geometries with arbitrary kinematics, as we do in Section 7.3.

For our demonstration problem, the airfoil is divided into N_b segments of equal length. Following Pistolesi's rule (Kuchemann, 1952), the (unknown) bound vorticity in each segment is concentrated in a lumped vortex at its 1/4-length point, while the induced velocity is evaluated at a *collocation point* located at the 3/4-length point on the segment. As we have seen in Section 3.3.2, this choice results in an exact solution with a single vortex in quasi-steady problems. The resulting discretization follows a pattern of alternating lumped vortices and collocation points, as shown in Figure 7.1. The figure shows the particular case of a flat plate of chord c, discretized by four lumped vortices, $N_b = 4$. The wake convects vorticity downstream according to Equation (3.27). Although this is not strictly necessary, we discretize the bound and wake circulation with segments of equal length, that is, $\frac{c}{N_b}$, since similar comparable spatial features are expected in both. In practice, it is convenient for computational efficiency to progressively increase the size of the wake panels in the streamwise direction (Muñoz Simón et al., 2022), but this is not done here for simplicity. The time step is then defined as $\Delta t = \frac{1}{N_b}\frac{c}{V}$, and a finite wake is considered and discretized by N_w segments of length $V\Delta t$, such that the shed vorticity (which is also discretized using lumped vortices) convects from one segment to the next downstream in each time step.

We define now the vector of unknown (bound) circulation strengths on all N_b segments on the airfoil, $\mathbf{\Gamma}_b(t)$, the vector of circulation strengths of the N_w segments in the wake, $\mathbf{\Gamma}_w(t)$, and the vector with the (known) upwash at the N_b collocation points on the airfoil, $\mathbf{w}_b(t)$. As we did in Chapter 3, we define the upwash as the normal component of the flow velocity (the vertical component in this case) on the airfoil, measured from an observer moving with the airfoil. All the previous column vectors, as well as the time variable, are now nondimensionalized as

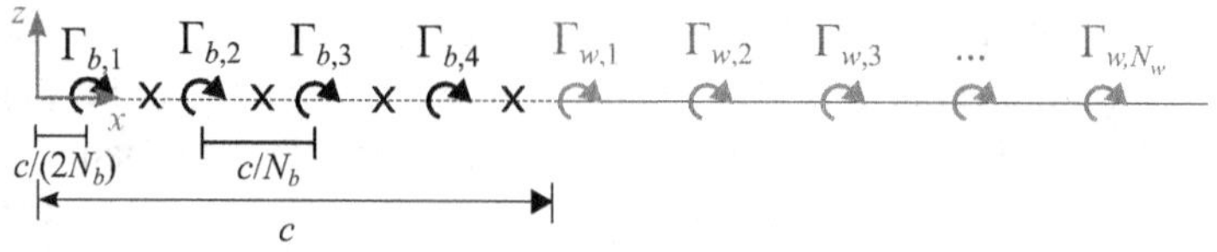

Figure 7.1 Example of discrete-vortex discretization of the problem of Figure 3.5 with $N_b = 4$. Crosses represent collocation points along the airfoil.

$$\hat{\Gamma}_b = \frac{\Gamma_b}{cV}, \quad \hat{\Gamma}_w = \frac{\Gamma_w}{cV}, \quad \hat{w}_b = \frac{w_b}{V}, \quad \text{and} \quad \Delta s = \frac{V\Delta t}{c/2} = \frac{2}{N_b}. \tag{7.1}$$

The discrete form of the nonpenetrating boundary conditions, Equation (3.25), at time t_{n+1} gives N_b-independent equations,

$$\mathfrak{K}_b \hat{\boldsymbol{\Gamma}}_b^{n+1} + \mathfrak{K}_w \hat{\boldsymbol{\Gamma}}_w^{n+1} + \hat{\boldsymbol{w}}_b^{n+1} = \mathbf{0}, \tag{7.2}$$

where if the dimensionless coordinate along the airfoil is defined as $\hat{x} = \frac{x}{c}$, then the discrete kernel of the problem is $\mathfrak{K}_{ij} = \left[2\pi(\hat{x}_j - \hat{x}_i - \frac{1}{2N_b}) \right]^{-1}$. This is the Biot–Savart law in 2-D, which depends on the relative (dimensionless) positions between the lumped vortex on the jth segment, $\hat{x}_j$, and the collocation point on the ith segment, $\hat{x}_i + \frac{1}{2N_b}$. Note that $\frac{c}{2N_b}$ is half the total length of a segment, that is, the distance between its 1/4- and 3/4-length points, where the lumped vortex and collocation point are, respectively, located. Note also that in Equation (7.2), we have separated the contributions to the induced velocity of the bound and wake vorticity, as was done in Equation (3.25). However, the coefficients in both the square matrix $\mathfrak{K}_b$ and the rectangular matrix $\mathfrak{K}_w$ are defined using the same discrete kernel function.

The instantaneous circulation at the first wake segment that has just been shed from the trailing edge, $\hat{\Gamma}_{w,1}$, is determined by the unsteady Kutta–Joukowski condition, Equation (3.28). The discrete form of this expression, between time steps t_n and t_{n+1}, becomes now

$$\sum_{j=1}^{N_b} \hat{\Gamma}_{b,j}^{n+1} + \hat{\Gamma}_{w,1}^{n+1} = \sum_{j=1}^{N_b} \hat{\Gamma}_{b,j}^{n}. \tag{7.3}$$

Finally, due to our choice of time step and wake discretization, the subsequent convection of the shed vorticity in the wake simply becomes the *shift operator*

$$\hat{\Gamma}_{w,j}^{n+1} = \hat{\Gamma}_{w,j-1}^{n}, \quad \text{for} \quad j = 2, ..., N_w. \tag{7.4}$$

This expression implies that the instantaneous circulation at t_{n+1} at a segment $j > 2$ within the wake is the circulation of the panel immediately upstream $(j-1)$ at the previous time step, t_n. In theory, the wake would grow to become infinitely long, but in practice, it is truncated to a fixed number of segments, N_w, as was already assumed in Equation (7.2). The truncation of the wake and the associated loss of circulation in the truncated wake region affect, in particular, the lower frequency aerodynamic response. To avoid having to create models with a extremely large number of wake segments,

Hall (1994) proposed a simple correction to Equation (7.4), in which the vorticity accumulates (with a very small dissipation) at the last segment downstream, that is, $\hat{\Gamma}_{w,N_w}^{n+1} = \hat{\Gamma}_{w,N_w-1}^{n} + (1-\varepsilon)\,\hat{\Gamma}_{w,N_w}^{n}$, with $0 < \varepsilon \ll 1$. Note that $\varepsilon = 0$ would correspond to the conservation of total vorticity (Kelvin's theorem). However, this generates numerical instabilities as the size of that end vortex grows and a small dissipation improves numerical performance. The numerical examples in this section use a wake extending over 30 chords, $N_w/N_b = 30$, and $\varepsilon = 2.5 \times 10^{-4}$.

Once the distribution of circulation at the new time step is obtained, the instantaneous lift and moment are given by the discrete form of Equation (3.32), which becomes

$$L = \rho V \sum_{i=1}^{N_b} \Gamma_{b,i} + \rho V \Delta t \sum_{i=1}^{N_b}\sum_{j=1}^{i} \dot{\Gamma}_{b,j},$$

$$M_{ac} = -\rho V \sum_{i=1}^{N_b} \Gamma_{b,i}\Delta x_i - \rho V \Delta t \sum_{i=1}^{N_b}\sum_{j=1}^{i} \dot{\Gamma}_{b,j}\Delta x_i,$$

(7.5)

with $\Delta x_i = x_i - x_{ac}$ being the coordinate with respect to the point where the resultant forces are computed (the aerodynamic center). Their corresponding expressions in a dimensionless form at time t_n for the lift coefficient are

$$c_l^n = 2\sum_{i=1}^{N_b} \hat{\Gamma}_{b,i}^{n} + \frac{4}{N_b}\sum_{i=1}^{N_b}\sum_{j=1}^{i} \hat{\Gamma}_{b,j}^{\prime n} = \sum_{i=1}^{N_b}\left(2\hat{\Gamma}_{b,i}^{n} + \frac{4(N_b-i+1)}{N_b}\hat{\Gamma}_{b,i}^{\prime n}\right),$$

(7.6)

and

$$c_m^n = -2\sum_{i=1}^{N_b} \hat{\Gamma}_{b,i}^{n}\Delta\hat{x}_i - \frac{4}{N_b}\sum_{i=1}^{N_b}\sum_{j=1}^{i} \hat{\Gamma}_{b,j}^{\prime n}\Delta\hat{x}_i$$

$$= -\sum_{i=1}^{N_b}\left(2\Delta\hat{x}_i\hat{\Gamma}_{b,i}^{n} + \frac{4\sum_{j=i}^{N_b}\Delta\hat{x}_j}{N_b}\hat{\Gamma}_{b,i}^{\prime n}\right)$$

(7.7)

for the moment coefficient about the aerodynamic center. In these equations, we have introduced the derivative of the circulation with the nondimensional time as $\Gamma_b' = \frac{d\Gamma_b}{ds}$, which includes, as we know from Section 3.3.1, the forcing due to the apparent mass effect. Circulation rates are approximated here by finite differences as

$$\hat{\Gamma}_b^{\prime n+1} = \sum_{i=1-a}^{1} \beta_i \hat{\Gamma}_b^{n+i},$$

(7.8)

where a is the order of the approximation. That is, for $a = 1$, one has a (first-order) forward Euler time integration, where $\beta_1 = \frac{1}{\Delta s}$ and $\beta_0 = -\frac{1}{\Delta s}$, which is used in the section to follow. The convergence rate can, however, be improved with a second-order approximation of the time derivatives, that is, $a = 2$, with $\beta_1 = \frac{3}{2\Delta s}$, $\beta_0 = -\frac{2}{\Delta s}$ and $\beta_{-1} = \frac{1}{2\Delta s}$, as shown by Maraniello and Palacios (2019).

Equations (7.2)–(7.4) and (7.6)–(7.8) define a well-posed linear problem in time. For a given initial distribution of circulation (in general, on both airfoil and wake) and

time history of the upwash distribution on the airfoil, it gives the time history of lift
and aerodynamic moment coefficients.

7.2.2   State-Space Unsteady Aerodynamics

Defining now the aerodynamic state vector $\mathbf{x}_a^\top = \left\{ \hat{\boldsymbol{\Gamma}}_b^\top \quad \hat{\boldsymbol{\Gamma}}_w^\top \right\}$ and the aerodynamic
output vector $\mathbf{y}_a^\top = \left\{ c_l \quad c_m \right\}$, Equations (7.2)–(7.4), (7.6), and (7.7), with the first-
order approximation for the time derivatives, Equation (7.8), can be rewritten as
discrete-time linear time-invariant (DLTI) state-space equations (see Section 1.4.2),
with constant (nondimensional) time step Δs. Including the circulation rates, $\hat{\boldsymbol{\Gamma}}_b'$, in
the state vector improves the convergence rate, and this is done in Section 7.3.3. How-
ever, it is not strictly necessary and is not done here for simplicity. The natural form of
the state-space equations is in a descriptor form, but the mass matrix of the resulting
system is nonsingular, and they can be written in the regular form as

$$\begin{aligned} \mathbf{x}_a^{n+1} &= \mathbf{A}\mathbf{x}_a^n + \mathbf{B}\hat{\mathbf{w}}_b^{n+1}, \\ \mathbf{y}_a^{n+1} &= \mathbf{C}\mathbf{x}_a^{n+1} + \mathbf{D}\hat{\mathbf{w}}_b^{n+1}. \end{aligned} \tag{7.9}$$

Note that the apparent mass terms in Equations (7.6) and (7.7) have resulted,
after the finite-difference approximation in Equation (7.8), in a *feedthrough* from
the upwash to the instantaneous lift and moment, that is, a nonzero matrix $\mathbf{D}$ in
Equation (7.9). This feature dominates the high-frequency dynamics of the system,
which, from the initial value theorem, generates nonzero aerodynamic forces at time
zero under any airfoil motions. This has already been observed in the noncirculatory terms in Wagner's analytical solution to a step change of angle of attack,
Equation (3.83).

Equation (7.9) gives the time-domain resultant forces for an arbitrary definition
of the upwash distribution on the airfoil using a state-space description with $N_a = N_b + N_w$ states. Note that the input vector is naturally defined at the next time step, t_{n+1}.
This is because incompressible potential flows are described by elliptic equations, in
which disturbances propagate instantaneously in the whole domain (i.e., at infinite
speed of sound). However, the system can be transformed into a conventional DLTI
state-space description with the change of state variable $\mathbf{x}_a = \mathbf{x}_a + \mathbf{B}\hat{\mathbf{w}}_b$ (Franklin et al.,
1998), thus resulting in

$$\begin{aligned} \mathbf{x}_a^{n+1} &= \mathbf{A}\mathbf{x}_a^n + \mathbf{A}\mathbf{B}\hat{\mathbf{w}}_b^n, \\ \mathbf{y}_a^n &= \mathbf{C}\mathbf{x}_a^n + (\mathbf{C}\mathbf{B} + \mathbf{D})\hat{\mathbf{w}}_b^n. \end{aligned} \tag{7.10}$$

It is often sufficient, however, and this is the approach followed in the numerical
examples below, to approximate the upwash at t_{n+1} by its value at the current time
step, t_n. The DLTI state-space system defined in Equation (7.10) is stable for any
choice of parameterization. This can be seen from the eigenvalues λ_j of the discrete-
time system matrix, which are always within the unit circle in the complex plane.
They are shown in Figure 7.2 for a moderately coarse discretization of the airfoil
($N_b = 20$ and $N_w/N_b = 30$). As has been discussed in Section 1.4.2, the corresponding

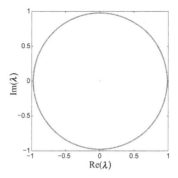

Figure 7.2 Eigenvalues of the discrete-time state matrix $\mathbf{A}$ ($N_b = 20$, $N_w/N_b = 30$).

continuous-time eigenvalues can be obtained directly from them as $\log(\lambda_i)/\Delta s$. A numerical investigation shows that, as the number of segments grows, the eigenvalues get ever closer to the unit circle, but the system remains stable.

Finally, frequency-domain representations of Equation (7.9) can also be obtained by the direct manipulation of the discrete-time system, together with the Z-transform (see Appendix A.3). Assuming zero initial conditions, the discrete-time FRF associated with Equation (7.9) is $\mathbf{G}_a(z) = \mathbf{C}\,(z\mathbf{\mathcal{I}} - \mathbf{A})^{-1}\,\mathbf{B} + \mathbf{D}$. However, $\mathbf{A}$ in Equation (7.9) is a very sparse matrix and, for computational efficiency, it is convenient instead to use the mathematical structure of the original time-domain problem defined in Equations (7.2)–(7.4). Let $(\overline{\bullet})$ indicate the Z-transform of a discrete variable; then Equations (7.2)–(7.4) become

$$\mathfrak{K}_b\overline{\boldsymbol{\Gamma}}_b + \mathfrak{K}_{w,}\overline{\boldsymbol{\Gamma}}_w = \overline{\mathbf{w}}_b,$$

$$z\overline{\Gamma}_{w1} = (1-z)\sum_{k=1}^{N_b}\overline{\Gamma}_{bk}, \qquad (7.11)$$

$$z\overline{\Gamma}_{wj} = \overline{\Gamma}_{w,j-1}, \quad \text{for } j = 2,\ldots,N_w.$$

After recursively applying the third equation, and then computing $\overline{\Gamma}_{w1}$ from the second one, we obtain

$$\overline{\boldsymbol{\Gamma}}_w = \begin{bmatrix} 1 & z^{-1} & z^{-2} & \cdots & z^{1-N_w} \end{bmatrix}^{\top} \overline{\Gamma}_{w1} = \mathbf{Z}\overline{\boldsymbol{\Gamma}}_b, \qquad (7.12)$$

with the $N_w \times N_b$ matrix

$$\mathbf{Z} = (1-z)\begin{bmatrix} z^{-1} & z^{-2} & \cdots & z^{-N_w} \end{bmatrix}^{\top}\begin{bmatrix} 1 & \cdots & 1 \end{bmatrix}, \qquad (7.13)$$

which can then be substituted into the first line of Equation (7.11) to yield

$$(\mathfrak{K}_b + \mathfrak{K}_w\mathbf{Z})\overline{\boldsymbol{\Gamma}}_b = \overline{\mathbf{w}}_b. \qquad (7.14)$$

This expression gives, therefore, a simpler expression for the discrete-time FRF between inputs (upwash) and outputs (lift and moments) of the system described in Equation (7.9). It only involves the inversion of an $N_b \times N_b$ matrix and can be written as

$$\mathbf{G}_a(z) = \mathbf{C}\,(\mathfrak{K}_b + \mathfrak{K}_w\mathbf{Z})^{-1} + \mathbf{D}. \qquad (7.15)$$

Expressions in the frequency domain can finally be generated by means of a bilinear transformation, Equation (A.12), restricted to the unit circle, namely $z = e^{ik\Delta s} \approx \frac{1+ik\Delta s/2}{1-ik\Delta s/2}$, where k is the reduced frequency and Δs is the dimensionless time step, which has been defined in Equation (7.1). As will be seen in Section 7.3.3, the large reduction in computational effort derived from the direct solution of circulation via Equation (7.15) is also useful to obtain frequency-domain expressions in more general 3-D problems.

7.2.3   Vertical Gust

The effect of a vertical gust can be directly included in the relative velocity in Equation (7.9). In general, one could consider any instantaneous distribution of velocities at all collocation points. We consider instead, as in Section 3.3.4, a *frozen* traveling gust that moves with the airspeed V and is, therefore, uniquely defined by its time history, $w_g(t)$, at the airfoil leading edge. Next, we define $\boldsymbol{\nu}^n$ as the vector with the instantaneous distribution of vertical velocities on the N_b collocation points along the airfoil due to the gust. As the gust advances with airspeed V, it penetrates into the airfoil. Due to the time discretization in Equation (7.1), the instantaneous distribution of gust-induced velocities is given by the *lag operator*

$$\nu_1^{n+1} = \hat{w}_g(t_{n+1}),$$
$$\nu_j^{n+1} = \nu_{j-1}^n, \qquad \text{for} \quad j = 2, \ldots, N_b,$$

(7.16)

where we have defined, following the notation convention in this section, the dimensionless gust velocity $\hat{w}_g = \frac{w_g}{V}$. For a given time history of the gust, Equation (7.16) can be solved independently of Equation (7.9) to obtain the instantaneous induced velocity due to the gust, which is then added to the input vector $\hat{\mathbf{w}}_b$. This process is the time-domain equivalent in 2-D of the frequency-domain gust modes introduced in Equation (6.90). If $\mathbf{G}_g(ik)$ is the FRF resulting from Equation (7.16) and $\mathbf{G}_a(ik)$ is the corresponding one for Equation (7.9), then the process to obtain the aerodynamic forces due to gust inputs can be described by the flow diagram of Figure 7.3. The figure also includes the inputs from the airfoil motion, $\mathbf{u}_a$, which are defined in the following section.

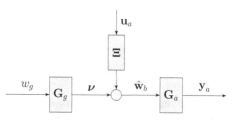

Figure 7.3 Flow diagram to compute lift and moment coefficients due to airfoil velocities, $\mathbf{u}_a$, and a vertical gust with a known time history $w_g(t)$ at the airfoil trailing edge.

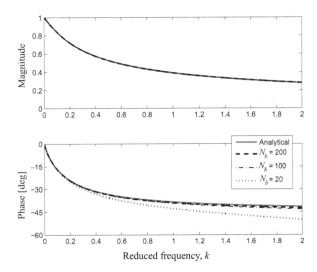

Figure 7.4 Lift coefficient (divided by 2π) due to a harmonic gust. The wake in the vortex solution extends for 30-chord lengths, $N_w/N_b = 30$. [07UnsteadyAero/vortex_gust.m]

Figure 7.4 shows the FRF from leading-edge gust velocity, w_g, to the lift coefficient, that is, the series connection $\mathbf{G}_a(ik)\mathbf{G}_g(ik)$, after transformation to the reduced frequency, $k = \frac{\omega c}{2V}$. Results are obtained using the state-space description of the vortex dynamics on the airfoil, with a wake extending over 30-chord lengths ($N_w/N_b = 30$), and compared with Sears's analytical solution that was presented in Section 3.3.4. As was discussed there, a traveling gust in a potential flow generates lift with zero moment at the aerodynamic center as it sweeps through the airfoil. Numerical results show a very quick convergence for the amplitude with the discretization of the wake. The convergence in phase, however, is much slower, and over 100 segments ($N_b = 100$) are needed to accurately capture the phase at higher frequencies.

7.2.4   2-DoF Airfoil with a Control Surface

We consider next the particular case of the rigid airfoil with pitch and plunge DoFs defined in Figure 3.1. Rotations are assumed about the elastic axis, which is located at a distance x_{ea} from the leading edge. The airfoil also has a trailing-edge flap with a hinge located at the coordinate x_{fh}, as defined in Equation (3.41). Finally, the airfoil may also be subject to a traveling vertical (normal) gust with an instantaneous distribution of velocities $\boldsymbol{\nu}$, defined by Equation (7.16). This will be expressed as

$$\hat{\mathbf{w}}_b = \boldsymbol{\Xi}\mathbf{u}_a + \boldsymbol{\nu}, \tag{7.17}$$

where the input vector corresponding to the airfoil kinematics is defined as

$$\mathbf{u}_a^\top = \left\{ \alpha \quad \alpha' \quad \frac{h'}{c/2} \quad \delta \quad \delta' \right\}. \tag{7.18}$$

With this definition, the discretization matrix, $\Xi \in \mathbb{R}^{N_b \times 5}$, is

$$\Xi_{i1} = 1,$$
$$\Xi_{i2} = 2\left(\hat{x}_i + \frac{1}{2N_b} - \hat{x}_{ea}\right),$$
$$\Xi_{i3} = 1, \qquad\qquad (7.19)$$
$$\Xi_{i4} = 1, \quad \text{if} \quad x_i > x_{fh}, \quad \text{and } 0 \text{ otherwise,}$$
$$\Xi_{i5} = 2\left(\hat{x}_i + \frac{1}{2N_b} - \hat{x}_{fh}\right), \quad \text{if} \quad x_i > x_{fh}, \quad \text{and } 0 \text{ otherwise.}$$

Substituting Equation (7.17) into Equation (7.9) with staggered inputs, and defining $\mathbf{B}_a = \mathbf{B}\Xi$ and $\mathbf{D}_a = \mathbf{D}\Xi$, we finally have

$$\mathbf{x}_a^{n+1} = \mathbf{A}\mathbf{x}_a^n + \mathbf{B}_a \mathbf{u}_a^n + \mathbf{B}\boldsymbol{\nu}^n,$$
$$\mathbf{y}_a^n = \mathbf{C}\mathbf{x}_a^n + \mathbf{D}_a \mathbf{u}_a^n + \mathbf{D}\boldsymbol{\nu}^n. \qquad\qquad (7.20)$$

This is the discrete-time linear state-space equation for the unsteady aerodynamics of an airfoil with 3-DoF under an arbitrary vertical gust profile. It should be remarked that, even though we have explicitly separated the effect of α and δ and their rates in the generation of lift and moment in Equation (7.18), they need to be combined to obtain the actual instantaneous loads. As an example, let $G_{l\alpha}(ik)$ and $G_{l\alpha'}(ik)$ be the FRFs between the pitch and the pitch rate, respectively, and the lift, as obtained from the state-space system defined in Equation (7.20). The instantaneous lift obtained from a given time history of the angle of incidence is obtained by combining them, as shown in the flow diagram of Figure 7.5.

Figure 7.6 shows the magnitude and phase of the FRFs, written in terms of the reduced frequency, between the angle of incidence and flap deflection (as inputs) and lift and moment coefficients (as outputs). The pitch rotation is around the 1/4-chord point ($x_{ea} = 0.25c$) and the flap hinge is at the 3/4-chord point ($x_{fh} = 0.75c$). The numerical results are obtained for varying number of segments on the airfoil ($N_b = 20$ and 100) and with a 30-chord wake length. They are compared with the analytical solutions in the frequency domain obtained in Section 3.3.3, Equation (3.48).

The numerical results have converged with a discretization of the airfoil chord using 100 segments ($N_b = 100$), although an excellent approximation to the analytical solution is already obtained with 20 segments. Results in Figure 7.6 are displayed for reduced frequencies $k \le 2$, but recall that the approximations of thin-airfoil theory are no longer valid for reduced frequencies above $k \approx 0.6$. As for the gust response, it can also be noticed that the error in the numerical approximation grows with the reduced frequency. This is mainly due to the apparent mass terms (or noncirculatory lift, Equation (3.50)) that dominate the solution at higher frequencies. A *polynomial*

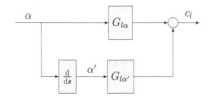

Figure 7.5 Instantaneous airfoil lift due to the angle of attack.

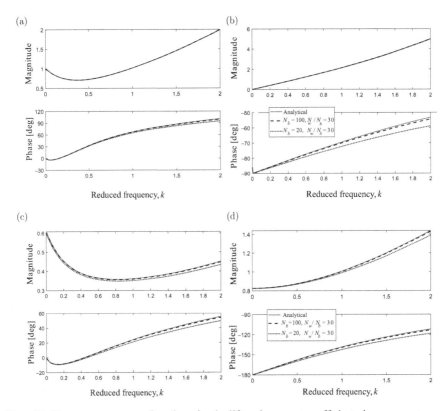

Figure 7.6 Frequency-response functions for the lift and moment coefficients in response to a commanded angle of incidence and flap deflection [$\hat{x}_{ea} = 1/4$ and $\hat{x}_{fh} = 3/4$]. (a) α to $\frac{c_l}{2\pi}$, (b) α to $\frac{c_m}{\pi/4}$, (c) δ to $\frac{c_l}{2\pi}$, and (d) δ to $\frac{c_m}{\pi/4}$. [07UnsteadyAero/vortex_bode.m]

preconditioning of the system, where a structure such as that of Equation (6.98) is identified *a priori*, can be included in the solution process to accelerate convergence. Further details on that approach with numerical examples can be found in Maraniello and Palacios (2019), but it is not pursued further here.

The size of the linear system in Equation (7.20) is $N_a = N_b + N_w$, with $N_w \gg N_a$, since a wake measuring over 20 chord lengths needs to be included to capture aerodynamic lag effects, as was seen in Figure 3.16. The lift and moment resultants, however, have quite smooth responses in the low-frequency range, and, indeed, it was already seen in Section 3.4.1 that its dynamics can be captured using very few aerodynamic lags. In Section 7.4, model-order reduction strategies are applied to Equation (7.20) to reduce the computational cost of the analysis.

7.3 Unsteady Vortex Lattice for Lifting Surfaces

A more general numerical solution to the potential flow unsteady aerodynamics on lifting surfaces is considered now, namely the UVLM. It shares with the DLM, which was outlined in Section 6.5.1, that the 3-D streamlined geometries are idealized as

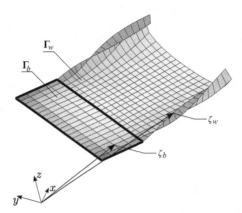

Figure 7.7 Vortex-lattice geometry, given in an inertial reference frame *xyz*, and circulation distribution on a rigid free-flying wing with a free wake (Simpson et al., 2017).

(zero-thickness) lifting surfaces, but here a time-dependent vortex sheet is used as the singularity distribution. In the UVLM, each lifting surface is discretized using a lattice of quadrilateral constant-circulation vortex rings, as schematically shown for a single free-flying wing in Figure 7.7. As it was seen for the 2-D problem, nonstationary conditions result in vorticity shedding from the trailing edge and then forming a wake downstream of each lifting surface. In potential-flow theory, the wake is also idealized as a vortex sheet, whose instantaneous shape needs to be determined. The wake is also discretized in the UVLM using constant-circulation vortex rings that are also included in Figure 7.7. The instantaneous flow velocity is finally calculated by superimposing any background flow (free stream and atmospheric turbulence) and the velocities induced by all vortex rings across the fluid domain. The process to determine the circulation strength on the lifting surfaces and their wakes, as well as the instantaneous shape of the wake, is outlined in the following section. Only the main results are summarized here and further details of the formulation, as well as printouts of a typical implementation, can be found in the book of Katz and Plotkin (2001).

   As in the 2-D problem, Section 7.2, the instantaneous circulation in each vortex ring is determined by two conditions. First, the instantaneous induced velocity field needs to satisfy the nonpenetration boundary condition on the (moving) lifting surfaces. A *collocation method* is used by the UVLM, in which the nonpenetrating condition is imposed at the center of each bound vortex ring (the collocation points). Second, we need a model for the wake dynamics. As shown in Equation (7.4), shed vorticity stems from the Kutta–Joukowski condition at the trailing edge (Morino and Bernardini, 2001) and is convected downstream with the local flow velocity. Linear assumptions in the 2-D problem resulted in the wake convecting at the freestream airspeed, V. Here, a higher order description is considered, in which the wake moves with the actual local flow velocity following *Helmholtz' laws of vortex motions* (Cottet and Koumoutsakos, 2000), namely (1) circulation strength is constant along a vortex filament, (2) a vortex filament must either close on itself or onto the boundaries of the fluid, and (3) in the absence of rotational external forces, an inviscid fluid that is initially irrotational remains irrotational. This results in the so-called *free-wake* model, which is the situation depicted in Figure 7.7. The following section expands on the details of the solution process to obtain the instantaneous distribution of singularities, which is then used to obtain the instantaneous loading on the airframe in Section 7.3.2.

7.3.1   Bound and Wake Circulation

In what follows, the description is done for a single lifting surface, although this can be easily generalized for multiple surfaces. As shown in Figure 7.7, the lifting surface is discretized using a structured grid (the *bound lattice*) of quadrilateral vortex rings, with M rings in the spanwise direction and N_b along the chordwise direction. Its corresponding *wake lattice* has MN_w rings, with $N_w \gg N_b$ being the (typically, very large) number of rings along the wake streamlines. As stated earlier, each vortex ring, whether on the lifting surface or on the wake, has a scalar circulation strength that may vary with time. In a similar manner to Section 7.2, we define $\Gamma_b(t)$ as the vector including the instantaneous values of all circulation strengths on the lifting surfaces (the *bound circulation*) and $\Gamma_w(t)$ as the corresponding vector resulting from the discretization of the wake (the *wake circulation*). Note that the size of $\Gamma_w(t)$ is typically much larger than that of $\Gamma_b(t)$. A collocation point is defined at the center of each vortex ring in the bound lattice and will be used to enforce the nonpenetrating boundary conditions.

The instantaneous lattice geometry is given by the coordinates, in an inertial reference frame, of the vertices (i.e., the corner points of all vortex rings) of both lattices. Let $\zeta_b(t)$ be the vector storing the spatial coordinates (in all three dimensions) of the vertices of the bound lattice and $\zeta_w(t)$ the corresponding vector for the wake lattice. For problems in aerodynamics, the time history of the wing kinematics, which is given by $\zeta_b(t)$, is known,[1] and the evolution of the wake geometry, $\zeta_w(t)$, is computed from the local flow velocity. Additionally, a nonstationary *background flow* field is considered on both the wing surface and the wake. It is defined by the components, in the inertial reference frame, of the flow velocity vector at all vertices of the bound and wake lattices, $\nu_b(t)$ and $\nu_w(t)$, respectively. This background flow may include both a slow-varying freestream velocity, atmospheric turbulence and gusts, as well as any other flow features discussed in Chapter 2. It is assumed that it is known *a priori* and, importantly, that it is not distorted by the presence of the lifting surfaces. The problem can be now nondimensionalized following Equation (7.1), with c being a reference chord, but, for simplicity in notation, we keep a dimensional description in what follows.

The instantaneous distribution of circulation in the bound and wake lattices produces a 3-D flow field, *induced velocity*. Let $\mathbf{w}_{\mathrm{ind}}(\mathbf{x})$ be that induced velocity at a point of coordinates $\mathbf{x}$. The induced velocity at that point is evaluated from the

---

[1] In aeroelastic problems, it is, of course, given by the solution of the structural problem. The resulting coupled problem is studied in Chapter 9.

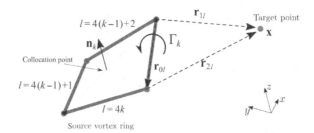

Source vortex ring

Figure 7.8 Geometry in the Biot–Savart law between the kth vortex ring in a lattice and a target point. Dashed lines indicate vectors corresponding to the third segment, $l = 4(k-1) + 3$ in Equation (7.21).

Biot–Savart law in three dimensions using the instantaneous geometry and distribution of circulation on the bound and wake lattices. The Biot–Savart law determines a rotational field with linearly decaying intensity around a vortex line. For the case of a quadrilateral vortex ring, this results in a closed-form expression obtained from the sum of the contribution of each of its four constituent linear segment as (Katz and Plotkin, 2001, Ch. 2)

$$\mathbf{w}_{ind}(\mathbf{x}) = \frac{1}{4\pi} \sum_{k} \sum_{l=4(k-1)+1}^{4k} \mathbf{r}_{0l}^{T} \left(\frac{\mathbf{r}_{1l}}{||\mathbf{r}_{1l}||} - \frac{\mathbf{r}_{2l}}{||\mathbf{r}_{2l}||} \right) \frac{\widetilde{\mathbf{r}}_{1l}\mathbf{r}_{2l}}{||\widetilde{\mathbf{r}}_{1l}\mathbf{r}_{2l}||^2} \Gamma_{k}, \qquad (7.21)$$

where the index k runs through all vortex rings and the definition of the segment vectors $\mathbf{r}_{0l}, \mathbf{r}_{1l}$, and $\mathbf{r}_{2l}$ depends on the geometry of the kth vortex ring and the relative position of the target point $\mathbf{x}$, which is illustrated in Figure 7.8. The figure also includes the unit normal vector $\mathbf{n}_k$ at the vortex ring at its center point (the collocation point). This normal vector is defined to point toward the suction side of the lifting surface, which corresponds to the left-hand-side convention for the given ordering of the four segments. Note that all terms in Equation (7.21) are, in general, time dependent, even though this has not been explicitly included in the expression.

Nonpenetrating Boundary Conditions
The nonpenetrating boundary conditions on the lifting surface imply that the normal component of the fluid velocity, which results from the addition of the induced and background velocities, is equal to the local normal velocity of the structure (or, equivalently, that the relative normal velocity is zero). This is enforced using a collocation method, as we did in the 2-D problem in Equation (7.2). The collocation points are now at the center of each vortex ring. As the background velocity, $v_b(t)$, and the structural velocity, $\dot{\zeta}_b(t)$, are known at the vertices of the bound lattice, an interpolation from corners to the center of each quadrilateral is needed. As a result, the nonpenetrating boundary conditions at time t_{n+1} can be written as

$$\Re_c(\zeta_b^{n+1}, \zeta_b^{n+1})\Gamma_b^{n+1} + \Re_c(\zeta_b^{n+1}, \zeta_w^{n+1})\Gamma_w^{n+1} + \mathcal{W}_c(\zeta_b^{n+1})\left(v_b^{n+1} - \dot{\zeta}_b^{n+1}\right) = \mathbf{0},$$

$$(7.22)$$

which is the generalization of Equation (7.1) to a lifting surface with potentially large geometry changes and in a nonstationary background velocity field, $\boldsymbol{\nu}_b(t)$. The discrete kernel matrix $\mathfrak{K}_c$ is obtained from the *upwash* resulting from Equation (7.21) at the collocation points, noting that, if $\mathbf{x}_k$ are the coordinates of the collocation point at the kth vortex ring, with local normal $\mathbf{n}_k$, the local upwash is simply $-\mathbf{w}_{\text{ind}}(\mathbf{x}_k) \cdot \mathbf{n}_k$. For convenience, we distinguish between the contributions to the induced velocity of the circulations on the bound and wake lattices. The first dependency of $\mathfrak{K}_c$ in Equation (7.22) refers to the target points (the collocation points, defined through the bilinear interpolation of the corner points of each vortex ring) and the second to the coordinates of the source points. Finally, matrix $\mathcal{W}_c$ is a sparse matrix that includes (1) a bilinear interpolation from the vertices (where the grid velocity is known) to the center (where the boundary conditions are applied) of each vortex ring, and (2) a projection onto the normal vector at each collocation point.

Wake Dynamics

The instantaneous wake is obtained by a two-step process. First, as the lifting surface moves (or the flow moves with respect to the lift surface), circulation is shed from the trailing edge on to the first row of wake vortex rings, as shown in Figure 3.6. The new circulation is then convected downstream, in a process analogous to the one followed by the 2-D problem, but along streamlines. Second, each vortex ring evolves following the local flow velocity, which is obtained by adding the velocity induced by the instantaneous lattice and the background velocity field. The (inertial) velocity in the corner points of the wake, $\dot{\boldsymbol{\zeta}}_w(t)$, is therefore given at the time instant t_n as

$$\dot{\boldsymbol{\zeta}}_w^n = \mathfrak{K}_w(\boldsymbol{\zeta}_w^n, \boldsymbol{\zeta}_b^n)\boldsymbol{\Gamma}_b^n + \mathfrak{K}_w(\boldsymbol{\zeta}_w^n, \boldsymbol{\zeta}_w^n)\boldsymbol{\Gamma}_w^n + \boldsymbol{\nu}_w^n, \tag{7.23}$$

where the discrete kernel, $\mathfrak{K}_w$, is obtained again from the Biot–Savart law, Equation (7.21). The subindex w on the kernel matrix is used here to denote that the target points are now the vertices of the wake lattice. Finally, $\boldsymbol{\nu}_w(t)$ is the vector with the instantaneous atmospheric flow velocities (the background field) at all wake lattice vertices. Note that there is no need for an interpolation matrix as shown in Equation (7.22).

The wake shape is updated at each time step from the integration of those velocities. A first-order forward Euler integration is typically used here. As a result, the wake geometry at time t_{n+1} is given by (Murua et al., 2012)

$$\boldsymbol{\zeta}_w^{n+1} = \mathbf{C}_{\zeta b}\boldsymbol{\zeta}_b^{n+1} + \mathbf{C}_{\zeta w}\boldsymbol{\zeta}_w^n + \Delta t \dot{\boldsymbol{\zeta}}_w^n, \tag{7.24}$$

where $\mathbf{C}_{\zeta b}$ is a sparse matrix with constant coefficients that relates the coordinates at the trailing edge with those on its downstream wake, and $\mathbf{C}_{\zeta w}$ shifts the wake lattice downstream on streamwise row at every time step. This process results in a characteristic *wake roll-up* downstream from the wingtips, which have been sketched in Figure 7.7. If we further assume that $\dot{\boldsymbol{\zeta}}_w^{n+1} = \boldsymbol{\nu}_w^{n+1}$, Equation (7.24) results in a prescribed wake model, in which the wake follows the local background velocity.

Finally, the convection of the wake circulation is written as

$$\mathbf{\Gamma}_w^{n+1} = \mathbf{C}_{\Gamma b}\mathbf{\Gamma}_b^n + \mathbf{C}_{\Gamma w}\mathbf{\Gamma}_w^n, \tag{7.25}$$

in which $\mathbf{C}_{\Gamma w}$ is again a sparse constant matrix that, in general, is obtained from a discretization of the convection equation on an irregular mesh (Muñoz Simón et al., 2022). For simplicity, we assume here again a constant discretization such that the wake circulation moves downstream by one chordwise row at each time step, as shown in Equation (7.4), in the 2-D problem. Finally, matrix $\mathbf{C}_{\Gamma b}$ introduces the same operation between the trailing-edge vortex rings on the lifting surface and the first vortex rings on the wake, that is, in a sectional enforcement of the Kutta–Joukowski condition, as shown in Equation (7.3). As in the 2-D problem, this links the nondimensional time step, Δs, to the discretization of the wake along the streamwise direction, yielding $V\Delta t = \frac{c}{2}\Delta s$.

Equation (7.22), (7.24), and (7.25) define a first-order, explicit, time-stepping scheme for the UVLM that has been widely used for low-speed air vehicles (Katz and Plotkin, 2001; Murua et al., 2012; Wang et al., 2010).

7.3.2 Aerodynamic Forces on the Lifting Surfaces

So far, we have obtained expressions for the instantaneous distribution of circulation that generates a flow field that satisfies the boundary conditions. Next, we need to obtain the distribution of forces on the lifting surfaces resulting from that flow state. As the interest is on wings with complex kinematics, we consider the general case in which all components of the local force vector are retrieved. In lifting-surface methods, which assume zero thickness, the main problem is to compute the forces due to *leading-edge suction*, as was already discussed for the 2-D problem in Section 3.3.5. From classic aerodynamic theory (von Kármán and Burgers, 1968), we know that the leading-edge suction corresponds to an integrable singularity associated with infinite vorticity over an infinitesimal domain near the leading edge. Fortunately, a vortex ring discretization provides a straightforward solution to that problem, as the finite size of the rings near the leading-edge segments seamlessly removes the singularity. This is a quasi-steady effect, and the associated force on each vortex ring segment can be obtained by the application of the *Joukowski's theorem*, which states that the forces per unit span length on a stationary vortex filament in a steady flow are obtained from the cross product of the circulation and the local flow velocity, with the fluid density being a scaling factor. Consider then the four segments $l = 4(k-1)+j$ corresponding to the kth vortex ring, with $j = 1, \ldots, 4$, as defined in Figure 7.8. The instantaneous resultant force generated by each segment at its center is

$$\mathbf{f}_{q_l} = \rho \Gamma_{bk} \tilde{\mathbf{v}}_l \mathbf{r}_{0l}, \tag{7.26}$$

where $\mathbf{v}_l$ is the relative velocity of the fluid with respect to the ring at the center of the lth segment. Recall that the vortex rings move with the local structural velocity, while the fluid velocity is obtained by adding the background flow velocity field and the induced velocity, at the center of the segment, $\mathbf{x}_l$, given by Equation (7.21). As a result, one has

$$\mathbf{v}_l = \mathbf{w}_{\mathrm{ind}}(\mathbf{x}_l) + \mathcal{W}_l(\boldsymbol{\zeta}_b)\left(\boldsymbol{\nu}_b - \dot{\boldsymbol{\zeta}}_b\right), \tag{7.27}$$

where $\mathcal{W}_l$ denotes a sparse matrix that extracts velocities at both vertices of the lth segment and linearly interpolates into its center. Note that the velocities induced by a segment on itself would go to infinite, and this is easily avoided by including a cutoff (single word as typically used in the community and similarly used in Ch. 6) distance in the Biot–Savart kernel, Equation (7.21), below which it is either truncated or made identically zero. Further details of this desingularization process, with relevance to flexible aircraft dynamic applications, can be found in del Carre and Palacios (2020).

Equation (7.26) assumes quasi-steady conditions and does not account for the inertial forces resulting from the acceleration of the vortex filament, that is, it does not consider noncirculatory effects. These forces can be more directly computed from the apparent mass terms in the Bernoulli's principle, that is, the $\frac{\partial \phi}{\partial t}$ term in Equation (3.22) (Pesmajoglou and Graham, 2000; Simpson et al., 2013). Integration of $\frac{\partial \phi}{\partial t}$ along the chordwise direction results in a nonstationary component to the pressure jump on the surface of the kth bound vortex ring equal to $\Delta p_k = -\rho \dot{\Gamma}_k$, as shown in Equation (3.31).[2] The resulting force vector is obtained by multiplying that pressure by the instantaneous area of the vortex ring, $S_k^n = S_k(t_n)$, and the instantaneous normal vector $\mathbf{n}_k^n = \mathbf{n}_k(t_n)$ at the collocation point, which is defined positive toward the suction side, as shown in Figure 7.8. At time t_n, this results in

$$\mathbf{f}_{\mathbf{u}_k}^n = \rho \dot{\Gamma}_k^n S_k^n \mathbf{n}_k^n. \tag{7.28}$$

Note that both the instantaneous area of the ring and the direction of the normal vector are considered, as they may also change with time. Moreover, the time derivative of the bound circulation is needed for these apparent mass terms, and it is obtained by numerical differentiation using Equation (7.8). The distributed aerodynamic forces across the lifting surface are then obtained by adding the quasi-steady forces given by Equation (7.26) at the center of each segment on the bound vortex rings and the nonstationary forces of Equation (7.28) at the center point of those vortex rings. We finally obtain the vector of aerodynamic forces at all vertices of the bound lattice, written in terms of their components in an inertial reference frame, as

$$\mathbf{f}_{\mathrm{aero}} = \mathcal{W}_q \mathbf{f}_q + \mathcal{W}_u \mathbf{f}_u, \tag{7.29}$$

where the sparse constant matrices $\mathcal{W}_q$ and $\mathcal{W}_u$ split the forces at the segments' midpoints and the collocation points, respectively, into the vertices of the bound lattice.

Example 7.1 Nonstationary Lift on a Rigid Prismatic Cantilever Wing. As an example of unsteady aerodynamic modeling using the methods introduced in this chapter, consider a prismatic uncambered wing of chord c, aspect ratio 16, and sweep angle Λ. A UVLM model of the wing is built with a 20×5 bound lattice ($M = 20$, $N_b = 5$) and

---

[2] Note that in the evaluation of the second integral in Equation (3.31) along the chordwise direction, the vortex ring discretization results in pairs of segments of opposite sign, thus canceling its downstream effect on the noncirculatory pressure.

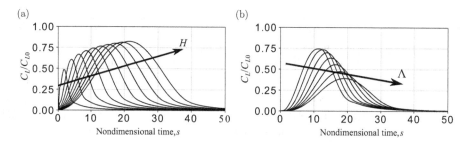

Figure 7.9 Normalized lift coefficient in the response to a "1-cosine" gust of intensity w_0 and gradient H. (a) $\Lambda = 0$, $H = c, 2c, \ldots, 10c$, and (b) $H = 5c$, $\Lambda = 0, 5, \ldots, 30°$. [07UnsteadyAero/dynamic_gust.ipynb]

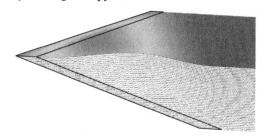

Figure 7.10 Snapshot at time instant $s = 10$ of the lattice for $\Lambda = 30°$, $w_0 = 0.1V$, and $H = 5c$. The wing geometry is the thick line and the wireframe shows the location of the vortex rings at the current instant. The reflected wing and its wake show the instantaneous circulation distribution.

a wake lattice that stretches 25 chords downstream ($N_w = 500$). A prescribed wake is considered. The wing is at a zero angle of attack in an airstream with horizontal speed V and subject to a traveling "1-cosine" gust of intensity w_0 and length H, as defined in Section 2.4.2. Figure 7.9 shows the time history of the normalized lift coefficient on the wing under small-amplitude gusts. The normalization constant is the steady-state lift predicted by linear theory on a thin airfoil after a step gust w_0 (Section 3.3.4), that is, $C_{L0} = 2\pi \frac{w_0}{V}$.

Figure 7.9(a) shows the lift on a straight wing ($\Lambda = 0$) for increasing values of length of the discrete gust. These results are strikingly similar to those obtained for the airfoil problem in Figure 3.25, with the key difference being that the maximum value of the lift is tapered by the wingtip losses, a quasi-steady effect. Therefore, the increase in the maximum lift that occurs with increased value of H, as well as the smaller tail on the curves as the gust vanishes, is related to the wake-induced delays in lift generation, that is, a 3-D manifestation of Wagner's problem (see Section 3.4.3).

For a given gust length, $H = 5c$, Figure 7.9(b) shows the effect of wing sweep in the lift time history. For the zero sweep angle, the gust reaches all wing airfoils simultaneously, and this results in a higher peak value and smaller period of interactions. As the sweep angle increases, the peak value of lift reduces (and rather substantially for a short gust such as the one in this example), as the discrete gust reaches the outer wing sections with a time delay. There is also a longer period of interactions associated with the time for the gust front reaching the leading edge at the wing root to the time it leaves the trailing edge at the wingtip. This can be visually seen in Figure 7.10 that shows a snapshot (at $s = \frac{2Vt}{c} = 10$) of the wing–gust interaction for $\Lambda = 30°$ and a gust profile with gust length $H = 5c$ and 10% intensity. As can be seen, the wake convection is limited to the background field $\boldsymbol{\nu}_w$, and it does not include self-induced velocities.

7.3.3   Discrete-Time Linear Time-Invariant UVLM

A linearization of the unsteady aerodynamic model described by Equations (7.22), (7.25), and (7.29) is now sought under the assumption of a fixed (frozen) wake geometry. For that purpose, we perturb a reference state in each DoF, while assuming that small perturbations on the wake geometry have a negligible effect on the instantaneous circulation on the wing. Contrarily to the DLM (Section 6.5.1), the frozen wake can take an arbitrary shape ζ_{w0}, which may be either prescribed, but not necessarily planar, or the steady-state geometry of a free-wake model in an aeroelastic equilibrium condition. The approach is, therefore, very general and results in a DLTI state-space formulation around nonzero reference conditions. Moreover, the (linearized) aerodynamic forces include induced drag effects, thus generalizing the solution of Section 3.3.5. As always in this chapter, however, its application is restricted to low-speed (incompressible) flows. The description that we present here is independent of the choice of the structural model in applications to flexible aircraft dynamics.

It is obtained from small perturbations about an arbitrary reference configuration. For the circulation states, this can be written as

$$\boldsymbol{\Gamma}_b = \boldsymbol{\Gamma}_b^0 + \Delta\boldsymbol{\Gamma}_b, \quad \boldsymbol{\Gamma}_w = \boldsymbol{\Gamma}_w^0 + \Delta\boldsymbol{\Gamma}_w, \quad \text{and} \quad \dot{\boldsymbol{\Gamma}}_b = \dot{\boldsymbol{\Gamma}}_b^0 + \Delta\dot{\boldsymbol{\Gamma}}_b, \tag{7.30}$$

along with the geometric and kinematic variables, which become

$$\boldsymbol{\zeta}_b = \boldsymbol{\zeta}_b^0 + \Delta\boldsymbol{\zeta}_b, \quad \dot{\boldsymbol{\zeta}}_b = \dot{\boldsymbol{\zeta}}_b^0 + \Delta\dot{\boldsymbol{\zeta}}_b, \quad \text{and} \quad \boldsymbol{\nu}_b = \boldsymbol{\nu}_b^0 + \Delta\boldsymbol{\nu}_b. \tag{7.31}$$

Note that this implies that the reference state may include a nonzero but constant distribution of velocities of the body, as well as a constant (but possibly spatially varying) background velocity field. With these definitions, the linearization of the nonpenetrating boundary conditions at time t_{n+1}, which were given in Equation (7.22), results in

$$\frac{\partial}{\partial \zeta_b}(\mathfrak{R}_{bb}\boldsymbol{\Gamma}_b^0)\Delta\zeta_b^{n+1} + \mathfrak{R}_{bb}^0\Delta\boldsymbol{\Gamma}_b^{n+1} + \frac{\partial}{\partial \zeta_b}(\mathfrak{R}_{bw}\boldsymbol{\Gamma}_w^0)\Delta\zeta_b^{n+1} + \mathfrak{R}_{bw}^0\Delta\boldsymbol{\Gamma}_w^{n+1}$$
$$+ \frac{\partial}{\partial \zeta_b}\left(\mathcal{W}_c(\boldsymbol{\nu}_b^0 - \dot{\boldsymbol{\zeta}}_b^0)\right)\Delta\zeta_b^{n+1} + \mathcal{W}_c^0\left(\Delta\boldsymbol{\nu}_b^{n+1} - \Delta\dot{\boldsymbol{\zeta}}_b^{n+1}\right) = 0, \tag{7.32}$$

where $\mathfrak{R}_{bb}^0 = \mathfrak{R}_c(\zeta_b^0, \zeta_b^0)$, $\mathfrak{R}_{bw}^0 = \mathfrak{R}_c(\zeta_b^0, \zeta_w^0)$, $\mathcal{W}_c^0 = \mathcal{W}_c(\zeta_b^0)$, and the terms $\frac{\partial}{\partial \zeta_b}(\mathfrak{R}_{b\bullet}\boldsymbol{\Gamma}_\bullet^0)$ are obtained by the analytical linearization of the Biot–Savart kernels about the reference geometry. Finally, $\frac{\partial \mathcal{W}_c}{\partial \zeta_b}$ gives the effect of changes of the geometry of the bound lattice on the upwash. It includes, in particular, the effect of the change of orientation of the normal vectors to the lattice. Note that Equation (7.32) gives third-order tensors that are costly to evaluate. This is avoided here by differentiating instead the corresponding matrix–vector product. Equation (7.32) can be conveniently rewritten as

$$\mathcal{R}^0_{bb}\Delta\mathbf{\Gamma}^{n+1}_b + \mathcal{R}^0_{bw}\Delta\mathbf{\Gamma}^{n+1}_w + \mathcal{W}^0_c\Delta\mathbf{\nu}^{n+1}_b - \mathcal{W}^0_c\Delta\dot{\mathbf{\zeta}}^{n+1}_b = \mathcal{B}_0\Delta\mathbf{\zeta}^{n+1}_b. \tag{7.33}$$

The left-hand side of this expression corresponds to the incremental nonpenetrating boundary conditions on the reference geometry. The right-hand-side term has been simplified with the constant matrix $\mathcal{B}_0 = -\frac{\partial\mathcal{W}_c}{\partial\zeta_b}(\mathbf{\nu}^0_b - \dot{\mathbf{\zeta}}^0_b) - \frac{\partial\mathcal{R}_{bb}}{\partial\zeta_b}\mathbf{\Gamma}^0_b - \frac{\partial\mathcal{R}_{bw}}{\partial\zeta_b}\mathbf{\Gamma}^0_w$ and accounts for the effect of the incremental changes in the wing geometry. The first term comes from changes in the direction of the reference upwash, while the last two are due to the reference (steady) circulation being out of balance with the reference upwash in the modified geometry.

The propagation of circulation through the wake, Equation (7.25), is already a linear relation, which simply becomes

$$\Delta\mathbf{\Gamma}^{n+1}_w = \mathbf{C}_{\Gamma b}\Delta\mathbf{\Gamma}^n_b + \mathbf{C}_{\Gamma w}\Delta\mathbf{\Gamma}^n_w, \tag{7.34}$$

and, equally, the linearized circulation rates are obtained from Equation (7.8) as

$$\Delta\dot{\mathbf{\Gamma}}^{n+1} = \sum_i \beta_i \Delta\mathbf{\Gamma}^{n+i}. \tag{7.35}$$

So far, we have obtained the linearized governing equations that determine the incremental distribution of circulation on the bound and wake lattices. Next, the incremental aerodynamic forces at the lattice vertices can be computed. Taking perturbations on Equation (7.29) results in

$$\Delta\mathbf{f}^n_{\text{aero}} = \mathcal{W}_q\Delta\mathbf{f}^n_q + \mathcal{W}_u\Delta\mathbf{f}^n_u, \tag{7.36}$$

where the linearized quasi-steady force on the lth segment, part of the kth vortex ring, at time t_n is

$$\Delta\mathbf{f}^n_{q_l} = \rho\tilde{\mathbf{v}}^0_l\mathbf{r}^0_{0l}\Delta\mathbf{\Gamma}^n_{bk} + \rho\mathbf{\Gamma}^0_{bk}\tilde{\mathbf{v}}^0_{0l}\frac{\partial\mathbf{r}_{0l}}{\partial\zeta_b}\Delta\zeta_b - \rho\mathbf{\Gamma}^0_{bk}\tilde{\mathbf{r}}^0_{0l}\Delta\mathbf{v}^n_l, \tag{7.37}$$

where the perturbation of the velocities at the segment midpoints, which was defined in Equation (7.27), needs to be also expressed in terms of the incremental values to the circulation and bound lattice coordinates and velocities (Maraniello and Palacios, 2019). As with the linearized nonpenetrating boundary condition, the linear expression for the force recovery includes the effect of the rotation of the reference (steady) forces with changes of geometry. This term has been shown by Parenteau and Laurendeau (2020) to have a large effect on the flutter of T-tails (see Example 7.3).

Finally, the noncirculatory (apparent mass) forces at the collocation point of the kth vortex ring are given by the linearization of Equation (7.28):

$$\Delta\mathbf{f}_{u_k} = \rho S^0_k\mathbf{n}^0_k\Delta\dot{\mathbf{\Gamma}}^n_{bk} + \rho\dot{\mathbf{\Gamma}}^0_{bk}\frac{\partial S^n_k\mathbf{n}^n_k}{\partial\zeta_b}\Delta\zeta_b, \tag{7.38}$$

where the product of the instantaneous area enclosed by the vortex ring, S_k^n, and corresponding normal vector, $\mathbf{n}_k^n$, is a quadratic function of the coordinates of vertices of the reference lattice.

After introducing the normalization of Equation (7.1), we can now define the vector of aerodynamic states by the concatenation of the bound and wake circulations (in general, for multiple lifting surfaces), as well as the bound circulation rates, as

$$\mathbf{x}_a = \left[\Delta\hat{\boldsymbol{\Gamma}}_b^\top \quad \tfrac{d}{ds}\Delta\hat{\boldsymbol{\Gamma}}_b^\top \quad \Delta\hat{\boldsymbol{\Gamma}}_w^\top \right]^\top, \tag{7.39}$$

and the input and output vectors from the values at the lattice vertices of the displacement/velocities and forces, respectively, as

$$\mathbf{u}_a = \left[\Delta\hat{\boldsymbol{\zeta}}_b^\top \quad \tfrac{d}{ds}\Delta\hat{\boldsymbol{\zeta}}_b^\top \quad \Delta\hat{\boldsymbol{\nu}}_b^\top \right]^\top, \tag{7.40}$$

$$\mathbf{y}_a = \hat{\mathbf{f}}_{\text{aero}},$$

with the dimensionless variables $\hat{\boldsymbol{\zeta}}_b = \boldsymbol{\zeta}_b/c$, $\hat{\boldsymbol{\nu}}_b = \boldsymbol{\nu}/V$, and $\hat{\mathbf{f}}_{\text{aero}} = \mathbf{f}_{\text{aero}}/\left(\tfrac{1}{2}\rho V^2 S\right)$. Using these definitions, Equations (7.32) and (7.34)–(7.38) can be written in the dimensionless DLTI state-space form as

$$\mathbf{x}_a^{n+1} = \mathbf{A}_a\mathbf{x}_a^n + \mathbf{B}_a\mathbf{u}_a^{n+1},$$
$$\mathbf{y}_a^n = \mathbf{C}_a\mathbf{x}_a^n + \mathbf{D}_a\mathbf{u}_a^n. \tag{7.41}$$

This is the generalization of Equation (7.9) to 3-D problems with arbitrary kinematics, although here we have explicitly introduced the bound circulation rates as states on the system. As in the 2-D problem, the problem is defined as a system with delay because the input appears at time t_{n+1}, and the transformation of Equation (7.10) can also be used here. Alternatively, for coupled fluid–structure interaction problems with a staggered solution process, the structure of Equation (7.41) may be beneficial, as in each iteration between the structure and the fluid, the current update of displacements on the walls would naturally appear as the input to the aerodynamic system. Finally, note that, even though the system is written in the form of Equation (7.41) for convenience in the analysis, an efficient time-marching algorithm can be obtained if, at each time step, the wake circulations are solved first using Equation (7.34), followed by an evaluation of the bound circulation using Equation (7.33).

Equation (7.41) describes the linearized unsteady aerodynamics on the lifting surfaces around an arbitrary reference state. The size of this system is $M(2N_b + N_w)$, which is typically in the tens or even hundreds of thousands. This results in a physics-based linear formulation for incompressible potential flow aerodynamics in the time domain, which has been written as a DLTI system. Note finally that the nondimensionalization of Equation (7.1) has transformed time derivatives to the reduced time $s = \frac{2Vt}{c}$, with c being a reference chord.

The FRF between the inputs and outputs of the difference equation (7.41) is obtained using the Z-transform (see Appendix A.3), as we did in Section 7.2.2.

Assuming zero initial conditions (which is given in a linearization around an equilibrium point), and introducing the reduced frequency through the relation $z = e^{ik\Delta s}$, we have

$$\mathbf{G}_a(ik) = \mathbf{C}_a \left(e^{ik\Delta s}\mathcal{I} - \mathbf{A}_a \right)^{-1} \mathbf{B}_a + \mathbf{D}_a, \tag{7.42}$$

where the cost of evaluating the inverse can be hugely reduced by following the approach of Equation (7.14) along the wake streamlines (Maraniello and Palacios, 2020). This expression generalizes, for all components of forces and displacements, the matrix of frequency-domain unsteady aerodynamic influence coefficients defined in Equation (6.88). It is only valid, however, for incompressible flows (Ma $\ll$ 1). Note also that Equation (6.88) is an external description of the aerodynamics (see Section 1.4.3) and rational interpolation was needed to construct state-space models, while Equation (7.42) has been defined directly from a state-space model. In other words, the UVLM and DLM move in opposite directions between frequency and time, as schematically shown in Figure 7.11. This figure associates each of the three (linearized) unsteady aerodynamic models in this book with one of the representations of the underlying dynamical system that were introduced in Figure 1.10. The DLM aerodynamics naturally provides transfer functions in the frequency domain and data-driven system identification on CFD is built on impulse responses. Both are external descriptions between inputs and outputs. Finally, the DLTI UVLM introduced in this section has led to state-space models.

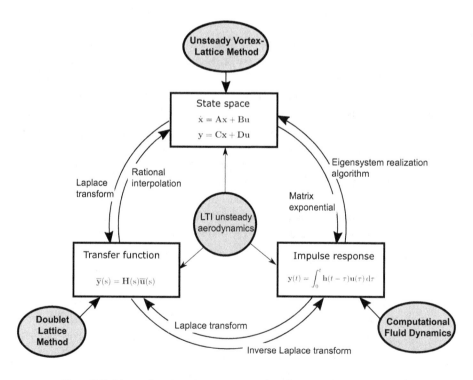

Figure 7.11 Linear time-invariant representations in unsteady aerodynamics.

The main drawback of the UVLM is the system dimensionality, which will be addressed in the following section. However it is first necessary to demonstrate discretization-independent results, that is, the convergence of the results on both spatial and temporal discretizations. The nondimensional formulation of the DLTI UVLM gives a single model across a range of dynamic pressures of interest, which will later facilitate the computation, in particular, of flutter boundaries. The dimensionless sampling rate Δs is associated with a dimensionless Nyquist frequency $k_N = \frac{\pi}{\Delta s}$, which defines the maximum reduced frequency that can be accurately captured by the model.

As Δs is proportional to $V \Delta t$, which is also, in the first approximation, the chordwise length of the wake vortex rings, the Nyquist frequency is directly linked to the spatial resolution of the wake geometry. The physical problem is typically defined by two parameters: first, by the maximum physical (angular) frequency of interest, $\omega_{\lim}$, which is normally associated with the highest relevant structural model in the aeroelastic system; and second, by the smallest true airspeed for the problem, $V_{\lim}$. While one would need, in principle, a prohibitively large number of wake elements at very low speeds, the associated reduced frequency is also very low and those conditions are often well approximated by the quasi-steady aerodynamics. In that case, the discrete wake can be substituted by a *horseshoe wake* (Drela, 2014).

Finally, in applications to aeroelastic analysis in which the UVLM is coupled with a structural model, it is important to assess the physical validity of the potential flow assumptions, as has been already outlined in Section 5.2. First, the amplitude of the deformations must be such that there is negligible separation in the fluid. Second, the maximum frequency of interest in the structural analysis (the highest vibration frequency of interest) needs to correspond to a reduced frequency no higher than $k_{\max} \lesssim 1$ (often even smaller values, around 0.6, are quoted). If important structural frequencies for the aeroelastic response are beyond this range, then higher fidelity aerodynamic models are necessary. In some situations, a modified potential flow model with a leading-edge separation approximation (Ramesh et al., 2014) may be sufficient.

7.4 Reduced-Order Models Using Balanced Realizations

As it was already mentioned, unsteady vortex-lattice models of complex configurations typically require tens or even hundreds of thousands of states for convergence (Murua et al., 2012; Roesler and Epps, 2018). For time-marching solutions of the general problem defined by Equations (7.22), (7.25), and (7.29), fast multiple methods give an easily parallelizable strategy that drastically reduces the computational burden (Willis et al., 2007). As we will see later in this book, for applications in the aeroelasticity of very flexible aircraft, the DLTI UVLM still offers an excellent approximation. As it has been written as the internal description of a large but linear dynamical system, standard methods of model-order reduction are directly applicable here and result in very efficient implementations. For a detailed exposition of methods of model-order reduction of dynamical systems, the readers are referred to the books by Dowell and Tang (2003) and Antoulas (2005), while the focus here remains on projection methods that directly manipulate the system matrices defined in Section 7.3.3. In particular, model-order reductions in this chapter are based on balanced realizations, which are arguably the most commonly used projection method for linear systems, and were first proposed for the UVLM by Baker et al. (1996). Alternative approaches, based on moment-matching and Krylov-subspace projections (Goizueta et al., 2021), are equally attractive but will not be discussed here.

We first outline the main features of the conventional (direct) balancing approach, which will be exemplified in the linear 2-D problem of Section 7.2. We then present an approximated solution process based on an efficient singular-value decomposition that significantly reduces the computational cost for the general 3-D DLTI UVLM. As we have seen that potential flow unsteady aerodynamics always results in stable linear (aerodynamic) systems, this is assumed in what follows.

7.4.1   Balanced Realizations

Consider a stable dynamic system described by a DLTI state-space model G of (large) dimension N_x, with N_u inputs and N_y outputs, written as

$$\mathbf{x}^{n+1} = \mathbf{A}\mathbf{x}^n + \mathbf{B}\mathbf{u}^n,$$
$$\mathbf{y}^n = \mathbf{C}\mathbf{x}^n + \mathbf{D}\mathbf{u}^n. \tag{7.43}$$

The presentation in this first section is for a general DLTI system with real matrices, although the ultimate goal is to apply the balanced realization on the DLTI UVLM, written as in Equation (7.10). Projection-based methods for model reduction seek a linear transformation into a new and smaller set of states, $\tilde{\mathbf{x}}$, of the form $\mathbf{x} = \mathbf{T}\tilde{\mathbf{x}}$, where $\mathbf{T}$ is a narrow rectangular matrix that defines the system dynamics $\tilde{G}$ that best approximates the original one under a certain error metric. Typically, one seeks that: (1) the error in the approximation is sufficiently small, (2) the stability characteristics of the full system are preserved, and (3) the generation of the reduced-order system from the full-order one is obtained by a computationally efficient process. System-balancing methods (Moore, 1981) address all three requirements by means of a linear transformation that identifies and preserves the most controllable and observable states. System controllability has been introduced in Section 5.4.2 to assess whether the available inputs can steer an aircraft into a desired state. *Observability* is its dual metric and assesses whether the current state of the system can be inferred from past output measurements. It is discussed next for DLTI systems.

Consider the solution to Equation (7.43) with known initial conditions $\mathbf{x}(0) = \mathbf{x}_0$ at $t = 0$ and no inputs. The time history of the outputs, up to a certain time t_n, can be obtained as the recursive sequence

$$\mathbf{y}^n = \mathbf{C}\mathbf{x}^n = \mathbf{C}\mathbf{A}\mathbf{x}^{n-1} = \cdots = \mathbf{C}\mathbf{A}^n\mathbf{x}_0. \tag{7.44}$$

We can now compute the square norm of the time history of all current and future values of the output variable from the given initial state as

$$\sum_{i=0}^{\infty} \mathbf{y}^{i\top} \mathbf{y}^i = \sum_{i=0}^{\infty} \mathbf{x}_0^\top \mathbf{A}^{i\top} \mathbf{C}^\top \mathbf{C} \mathbf{A}^i \mathbf{x}_0 = \mathbf{x}_0^\top \mathbf{W}_o \mathbf{x}_0, \tag{7.45}$$

where we have introduced the real symmetric matrix $\mathbf{W}_o$ of dimension $N_x \times N_x$, known as the (discrete-time) *observability Gramian*, as

$$\mathbf{W}_o = \sum_{i=0}^{\infty} \mathbf{A}^{i\top} \mathbf{C}^\top \mathbf{C} \mathbf{A}^i. \tag{7.46}$$

The observability Gramian, thus, measures how much the initial conditions, $\mathbf{x}_0$, affect all future outputs of the system. Conversely, it can be used to assess which system outputs are excited from all possible nonzero initial conditions on the state vector and, as a result, whether there are *unobservable* states in the system (if some are not). Furthermore, as the system is time invariant, we can arbitrarily shift the origin of times and effectively the observability Gramian measures the effect of a state at a given time in all future observations. As $\mathbf{W}_o$ is a real symmetric matrix, its eigenvalues can be used to identify dominant directions in the state vector that have the larger impact on the outputs (or, conversely, the choice of outputs that give the most information about the state of the system).

The dual problem is the evaluation of a future (nominally, at $t \to \infty$) state of the system associated with past histories of inputs, which results in the *controllability Gramian* for the discrete-time system of Equation (7.43). It is defined as the real symmetric matrix

$$\mathbf{W}_c = \sum_{i=0}^{\infty} \mathbf{A}^i \mathbf{B} \mathbf{B}^\top \mathbf{A}^{i\top}, \tag{7.47}$$

which is also of dimension $N_x \times N_x$ and is inversely correlated with the effort required from the actuation to reach a certain state of the system. Moreover, this matrix is singular if the system is uncontrollable, and it can be shown (Dullerud and Paganini, 2013, Ch. 5) that the condition of controllability of the system, which was defined in Equation (5.20), is equivalent to having a full-rank controllability Gramian. That property will be important for control design in Chapter 10, but, as with the observability Gramian, the interest here is in the identification of the dominant directions in the state vector (via eigenvalue analysis) that are most affected by the system inputs.

Therefore, the eigenvalues of the discrete observability and controllability Gramians play an essential role in system balancing. However, before we can discuss them, we need a practical way to compute both matrices that do not involve an infinite series. For that purpose, note first that if we pre- and postmultiply the controllability Gramian by $\mathbf{A}$ and $\mathbf{A}^\top$, respectively, then we have

$$\mathbf{A} \left(\sum_{i=0}^{\infty} \mathbf{A}^i \mathbf{B} \mathbf{B}^\top \mathbf{A}^{i\top} \right) \mathbf{A}^\top = \sum_{i=0}^{\infty} \mathbf{A}^{i+1} \mathbf{B} \mathbf{B}^\top \mathbf{A}^{i+1\top} = \sum_{i=1}^{\infty} \mathbf{A}^i \mathbf{B} \mathbf{B}^\top \mathbf{A}^{i\top}$$

$$= \sum_{i=0}^{\infty} \mathbf{A}^i \mathbf{B} \mathbf{B}^\top \mathbf{A}^{i\top} - \mathbf{B} \mathbf{B}^\top, \tag{7.48}$$

since $\mathbf{A}^0 = \mathcal{I}$. A similar manipulation can be done in the observability Gramian and, as a result, the following two relations can be obtained:

$$\begin{aligned}
\mathbf{A}\mathbf{W}_c\mathbf{A}^\top - \mathbf{W}_c &= -\mathbf{B}\mathbf{B}^\top, \\
\mathbf{A}\mathbf{W}_o\mathbf{A}^\top - \mathbf{W}_o &= -\mathbf{C}\mathbf{C}^\top.
\end{aligned} \tag{7.49}$$

These relations are *discrete-time Lyapunov equations*, which are the counterpart in DTLI systems of the (continuous-time) Lyapunov equations that have been introduced in Equation (2.27). They give a direct evaluation of the Gramians that avoid the infinite series in Equations (7.46) and (7.47). As for the continuous Lyapunov equation, several numerical algorithms are readily available for the solution of Equation (7.49). A description of the most common ones can be found in Chapter 6 of Antoulas (2005).

Consider now the transformation $\mathbf{x} = \mathbf{T}\tilde{\mathbf{x}}$, with $\mathbf{T}$ being a square invertible matrix. Equation (7.43) then becomes

$$\begin{aligned}
\tilde{\mathbf{x}}^{n+1} &= \tilde{\mathbf{A}}\tilde{\mathbf{x}}^n + \tilde{\mathbf{B}}\mathbf{u}^n, \\
\mathbf{y}^n &= \tilde{\mathbf{C}}\tilde{\mathbf{x}}^n + \mathbf{D}\mathbf{u}^n,
\end{aligned} \tag{7.50}$$

where $\tilde{\mathbf{A}} = \mathbf{T}^{-1}\mathbf{A}\mathbf{T}$, $\tilde{\mathbf{B}} = \mathbf{T}^{-1}\mathbf{B}$, and $\tilde{\mathbf{C}} = \mathbf{C}\mathbf{T}$. The objective is to define such transformation in a way that allows the identification of the most controllable and observable states in the original system. If we now construct the controllability and observability Gramians of Equation (7.50), using Equations (7.46) and (7.47), it is easy to see that the transformed Gramians are given by the following congruence transformations:

$$\begin{aligned}
\tilde{\mathbf{W}}_c &= \mathbf{T}^{-1}\mathbf{W}_c\mathbf{T}^{-\top}, \\
\tilde{\mathbf{W}}_o &= \mathbf{T}^\top\mathbf{W}_o\mathbf{T},
\end{aligned} \tag{7.51}$$

and from this, the product of both Gramians satisfies $\tilde{\mathbf{W}}_c\tilde{\mathbf{W}}_o = \mathbf{T}^{-1}\mathbf{W}_c\mathbf{W}_o\mathbf{T}$. This is a similarity transformation (Horn and Johnson, 2012, Ch. 2) and, therefore, the eigenvalues of $\tilde{\mathbf{W}}_c\tilde{\mathbf{W}}_o$ are the same[3] as the eigenvalues of $\mathbf{W}_c\mathbf{W}_o$. Note that this is not the case for the eigenvalues of each individual Gramian.

The balancing problem then becomes the search for a transformation $\mathbf{T}$ such that states with poor controllability in the transformed equations are also difficult to observe. This would be achieved if we could identify a transformation that results in $\tilde{\mathbf{W}}_c = \tilde{\mathbf{W}}_o$, so all states after transformation would have equally good (or poor) observability and controllability. Furthermore, the resulting states would be easy to classify if that matrix were diagonal. Those become the objectives of system balancing, which seeks the simultaneous diagonalization of $\mathbf{W}_c$ and $\mathbf{W}_o$. Fortunately, as we will see in what follows, this operation is possible because they are both (symmetric) positive definite matrices by construction. The standard solution to this problem, known as the *direct method* (Moore, 1981), is summarized next.

---

[3] If λ is an eigenvalue of $\mathbf{A}$, then $\mathbf{A}\mathbf{x} = \lambda\mathbf{x}$ for some $\mathbf{x}$. Defining $\mathbf{x} = \mathbf{T}\tilde{\mathbf{x}}$ gives $\mathbf{A}\mathbf{T}\tilde{\mathbf{x}} = \lambda\mathbf{T}\tilde{\mathbf{x}}$. Premultiplying by $\mathbf{T}^{-1}$ results finally in $\tilde{\mathbf{A}}\tilde{\mathbf{x}} = \lambda\tilde{\mathbf{x}}$, with $\tilde{\mathbf{A}} = \mathbf{T}^{-1}\mathbf{A}\mathbf{T}$, and λ is also eigenvalue of $\tilde{\mathbf{A}}$.

First, we obtain the *Cholesky factorization* of $\mathbf{W}_c$, that is, $\mathbf{W}_c = \mathbf{Z}\mathbf{Z}^\top$, with $\mathbf{Z}$ being an upper diagonal matrix with positive diagonal entries (since $\mathbf{W}_c$ was positive definite). Second, we carry out a singular-value decomposition of the symmetric matrix $\mathbf{Z}^\top \mathbf{W}_o \mathbf{Z} = \mathbf{U}\boldsymbol{\Sigma}^2 \mathbf{U}^\top$, with $\boldsymbol{\Sigma}$ being the diagonal matrix of positive singular values and $\mathbf{U}$ a unitary matrix (i.e., $\mathbf{U}^\top \mathbf{U} = \boldsymbol{\mathcal{I}}$). The balancing transformation is then obtained with

$$\mathbf{T} = \mathbf{Z}\mathbf{U}\boldsymbol{\Sigma}^{-1/2},$$
$$\mathbf{T}^{-1} = \boldsymbol{\Sigma}^{1/2}\mathbf{U}^\top \mathbf{Z}^{-1}. \tag{7.52}$$

It can be easily verified that substituting both expressions into Equation (7.51) results in $\tilde{\mathbf{W}}_c = \tilde{\mathbf{W}}_o = \boldsymbol{\Sigma} = \mathrm{diag}\left(\sigma_1, \ldots, \sigma_{N_x}\right)$, which are ordered such that $\sigma_1 \geq \sigma_2 \geq \cdots \geq \sigma_{N_x} > 0$. The coefficients σ_j are known as the *Hankel singular values* (HSVs) of the system described by Equation (7.43). Due to the similarity transformation defined earlier, the HSVs are also the square roots of the eigenvalues of the matrix product $\mathbf{W}_c \mathbf{W}_o$.

The HSVs provide a metric for the combined effect in the input and output of the system of the balanced states. States with sufficiently small HSV can be disregarded from the reduced-order system. For that purpose, the balanced equations are partitioned as

$$\tilde{\mathbf{x}}_1^{n+1} = \tilde{\mathbf{A}}_{11}\tilde{\mathbf{x}}_1^n + \tilde{\mathbf{A}}_{12}\tilde{\mathbf{x}}_2^n + \tilde{\mathbf{B}}_1\mathbf{u}^n,$$
$$\tilde{\mathbf{x}}_2^{n+1} = \tilde{\mathbf{A}}_{21}\tilde{\mathbf{x}}_1^n + \tilde{\mathbf{A}}_{22}\tilde{\mathbf{x}}_2^n + \tilde{\mathbf{B}}_2\mathbf{u}^n, \tag{7.53}$$
$$\mathbf{y}_2^n = \tilde{\mathbf{C}}_1\tilde{\mathbf{x}}_1^n + \tilde{\mathbf{C}}_2\tilde{\mathbf{x}}_2^n + \tilde{\mathbf{D}}_2\mathbf{u}^n,$$

where $\tilde{\mathbf{x}}_1 \in \mathbb{R}^{N_r}$ are the states corresponding to the N_r HSVs that we wish to preserve. We finally residualize the remaining states, in a similar manner to Equation (5.9), by assuming $\tilde{\mathbf{x}}_2^{n+1} = \tilde{\mathbf{x}}_2^n$, solving $\tilde{\mathbf{x}}_2^n = \left(\boldsymbol{\mathcal{I}} - \tilde{\mathbf{A}}_{22}\right)^{-1}\left(\tilde{\mathbf{A}}_{21}\tilde{\mathbf{x}}_1^n + \tilde{\mathbf{B}}_2\mathbf{u}^n\right)$, and substituting this expression in the first and last equations in Equation (7.53). Despite its higher computational cost, this *residualization* step is a preferred simple truncation method of $\tilde{\mathbf{x}}_2$ in the equations for unsteady aerodynamics problems, as it preserves the steady-state gain of the system (the quasi-steady aerodynamics solution). The inverse of $\boldsymbol{\mathcal{I}} - \tilde{\mathbf{A}}_{22}$ can also be approximated by a truncated Neumann series with no major penalty in accuracy.

The application of balanced transformations to unsteady aerodynamics was first proposed by Baker et al. (1996). Rule et al. (2004) later compared balanced realizations with direct projection on the eigenvectors of the aerodynamic subsystem, as proposed by Dowell and Hall (1997), and showed a much better performance for a given system size. This is because the eigenvalues of $\mathbf{A}$ describe the system internal dynamics, but not its dependency on any specific inputs and outputs and, therefore, typically provide a less optimal basis for model-order reduction. The computational effort of finding balanced realizations using the direct approach described earlier goes as $\mathcal{O}(N_x^3)$ and therefore becomes impractical for system sizes $N_x > 10^4$. For larger systems, numerical approximations may be necessary, and we discuss a suitable one for the DLTI UVLM in Section 7.4.2.

Example 7.2 Balanced Residualization of a Discrete-Vortex Airfoil Model. The balanced residualization described in this section is applied on the DLTI state-space equations for the flat plate defined in Equation (7.20). Results have been obtained for an airfoil pitching about the quarter chord from the leading edge, $\hat{x}_{ea} = 1/4$, a flap hinge located at a quarter-chord distance from the trailing edge, $\hat{x}_{fh} = 3/4$ and no gust inputs. As the normalized plunge rate is interchangeable with the pitch angle in the system, only four independent inputs are considered, namely $\mathbf{u}_a^\top = \left\{ \alpha \quad \alpha' \quad \delta \quad \delta' \right\}$. Finally, from the numerical exploration of Section 7.2.4, a (converged) full-order model with $N_b = 100$ and a wake extending for 30 chord lengths, $N_w/N_b = 30$, are considered. This results in a DLTI system described by Equation (7.20) with 4 inputs, 2 outputs (lift and moment about the aerodynamic center), and 3,100 states.

Figure 7.12 shows the first 50 HSVs of this system. They are normalized such that $\sigma_1 = 1$ and presented in logarithmic scale, as the interest is in their relative importance. As can be seen, the values of the HSVs quickly drop in with the ROM size, and therefore only a few balanced states are necessary to represent the system dynamics. Indeed, the error in a truncated balanced system goes as the sum of the neglected HSVs (Antoulas, 2005, Ch. 7).

The original system is projected on its balanced form and truncated as shown in Equation (7.53). Next, reduced systems that only retain the first N_r HSVs are obtained using residualization. Finally, the resulting reduced DLTI state-space systems are transformed into continuous-time systems using the Tustin transformation, Equation (A.12). The magnitude and phase of the associated FRFs are shown in Figures 7.13 and 7.14 for $N_r = 2$, 3, and 4 (lift and moment coefficients are normalized by 2π). Each figure, therefore, shows four FRFs, corresponding to combinations of the four inputs and the relevant output. Note finally that the FRFs in Figure 7.6 are simply retrieved by the addition of the angles and their derivatives (i.e. $(1+ik)\bar{\alpha}$), which have been left as independent inputs in this example.

These results show that the low-frequency response of the full system (of dimension 3,100) can be accurately described with a reduced system with as little as four internal states. In fact, a very good approximation can be already obtained at the low-frequency

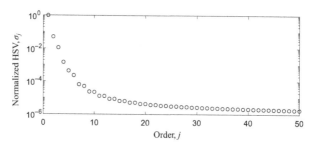

Figure 7.12 Hankel singular values for a thin airfoil with pitch, plunge, and flap DoFs ($\hat{x}_{ea} = 1/4$, $\hat{x}_{fh} = 3/4$, $N_b = 100$, $N_w/N_b = 30$). [07UnsteadyAero/vortex_rom.m]

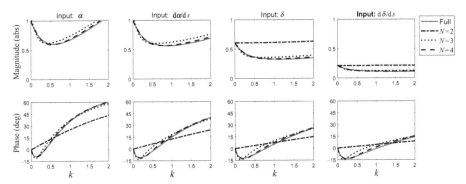

Figure 7.13 Frequency-response functions for a flat plate with pitch, $\hat{x}_{ea} = 1/4$, and trailing-edge flap, $\hat{x}_{fh} = 3/4$, inputs and normalized lift coefficient, $\frac{c_L}{2\pi}$, as output. Full-order model with $N_b = 100$, $N_w/N_b = 30$. [07UnsteadyAero/vortex_rom.m]

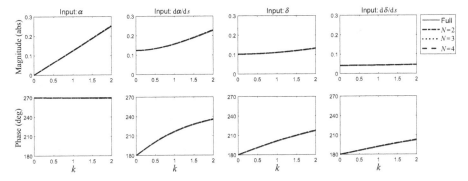

Figure 7.14 Frequency-response functions for a flat plate with pitch, $\hat{x}_{ea} = 1/4$, and trailing-edge flap, $\hat{x}_{fh} = 3/4$, as inputs and normalized moment coefficient, $\frac{c_m}{2\pi}$, as output. Full order is $N_b = 100$, $N_w/N_b = 30$. [07UnsteadyAero/vortex_rom.m]

end with $N_r = 3$. In other words, we have numerically obtained compact expressions that are comparable in complexity to the analytical solution of Equation (3.48).

Note finally that the aerodynamic forces predicted by potential flow theory do not tend to zero at high frequencies, as it is known from Theodorsen's theory. It would be possible to include polynomial preconditioning as shown in Equation (6.98) on the full system and perform the reduction on the resulting residual, but this has not been pursued here.

7.4.2 Frequency-Limited Balancing

Direct methods for balancing, such as those considered in Section 7.4.1, have a computational cost that scales as $\mathcal{O}(N_x^3)$ and, therefore, become impractical as a model-reduction strategy for the DLTI UVLM on complex configurations. In such a case, one may turn instead to *low-rank approximations* to the solutions of the Lyapunov equations (Benner et al., 2016; Gugercin and Antoulas, 2004), which provide a computationally efficient strategy. In particular, both the mathematical structure of the DLTI UVLM and the interest in the lowest end of the (reduced) frequency spectrum facilitate computational-efficient direct integration of approximated system Gramians. Such an approach is described next.

First, it is convenient to write the controllability and observability Gramians, defined in Equations (7.46) and (7.47), in the frequency domain. This is obtained using Plancherel's theorem for discrete-time functions, Equation (A.14). Using the discrete FRF of the system, Equation (7.42), the Gramians can be rewritten as

$$
\begin{aligned}
\mathbf{W}_c &= \frac{1}{2\pi} \int_{-\pi}^{\pi} \left(e^{i\theta}\mathbf{\mathcal{I}} - \mathbf{A} \right)^{-1} \mathbf{B}\mathbf{B}^{\top} \left(e^{-i\theta}\mathbf{\mathcal{I}} - \mathbf{A}^{\top} \right)^{-1} d\theta, \\
\mathbf{W}_o &= \frac{1}{2\pi} \int_{-\pi}^{\pi} \left(e^{-i\theta}\mathbf{\mathcal{I}} - \mathbf{A}^{\top} \right)^{-1} \mathbf{C}^{\top}\mathbf{C} \left(e^{i\theta}\mathbf{\mathcal{I}} - \mathbf{A} \right)^{-1} d\theta,
\end{aligned}
\tag{7.54}
$$

where $\theta = k\Delta s$, with k being the reduced frequency and Δs the nondimensional time step in the DLTI UVLM. This integral is defined for the whole reduced frequency range up to the Nyquist (reduced) frequency, Equation (7.43). For unsteady aerodynamics problems, our interest is only on the lower end of the reduced frequency spectrum and, therefore, it is worth considering modified Gramians that are only evaluated there. This results in the *frequency-limited Gramians*, $\mathbf{W}_{lc}$ and $\mathbf{W}_{lo}$, associated with the discrete-time state-space model of Equation (7.41), which are defined as

$$
\begin{aligned}
\mathbf{W}_{lc} &= \frac{\Delta s}{2\pi} \int_{-k_c}^{k_c} \mathbf{\Phi}_c(ik)\mathbf{\Phi}_c^{\star}(ik)\,dk, \quad \text{with } 0 < k_c < \pi/\Delta s, \\
\mathbf{W}_{lo} &= \frac{\Delta s}{2\pi} \int_{-k_o}^{k_o} \mathbf{\Phi}_o(ik)\mathbf{\Phi}_o^{\star}(ik)\,dk, \quad \text{with } 0 < k_o < \pi/\Delta s,
\end{aligned}
\tag{7.55}
$$

where k_c and k_o define the controllability and observability frequencies of interest, respectively, and where we have also defined the matrices

$$
\begin{aligned}
\mathbf{\Phi}_c(ik) &= \left(e^{ik\Delta s}\mathbf{\mathcal{I}} - \mathbf{A} \right)^{-1} \mathbf{B}, \\
\mathbf{\Phi}_o(ik) &= \left(e^{-ik\Delta s}\mathbf{\mathcal{I}} - \mathbf{A}^{\top} \right)^{-1} \mathbf{C}^{\top}.
\end{aligned}
\tag{7.56}
$$

As will be seen shortly, choosing $k_c \neq k_o$ has numerical advantages, and this is assumed here. Importantly, integration in the range $[-k_c, k_c]$ and $[-k_o, k_o]$ guarantees that both $\mathbf{W}_{lc}$ and $\mathbf{W}_{lo}$ are real matrices. Note that if $k_c = k_o = \pi/\Delta s$, that is, the Gramians are resolved over the full Nyquist range, then internal balancing of Section 7.4.1 is retrieved. A smaller frequency range results in *frequency-limited balancing* (FLB) of the equations, which is discussed next. First, due to the reduced range of integration, the integrals in Equation (7.55) are directly evaluated, that is, instead of solving the associated Lyapunov Equations (7.49), the integrals are numerically approximated using finite sums. For the frequency-limited controllability Gramian, this results in

$$
\mathbf{W}_{lc} \approx \frac{\Delta s}{2\pi} \sum_{i=1}^{N_c} \kappa_i \left[\mathbf{\Phi}_c(ik_i)\mathbf{\Phi}_c^{\star}(ik_i) + \mathbf{\Phi}_c(-ik_i)\mathbf{\Phi}_c^{\star}(-ik_i) \right],
\tag{7.57}
$$

where κ_i are the quadrature weights associated with N_c integration points $k_i \in [0, k_c]$. A similar expression is obtained for the frequency-limited observability Gramian, $\mathbf{W}_{lo}$, using N_o discrete frequencies. Either uniformly spaced trapezoidal or Gauss–Lobatto quadratures are good choices, and they are both used in Example 7.4. Expanding the sums and collecting terms, Equation (7.57) results in

$$\mathbf{W}_{lc} \approx \mathbf{Z}_c \mathbf{Z}_c^\top, \qquad (7.58)$$

with $\mathbf{Z}_c$ being the narrow real matrix of dimension $N_x \times (2N_u N_c)$ obtained by concatenation of the real and imaginary parts of the $\mathbf{\Phi}_c(ik)$ at the sample frequencies, that is,

$$\mathbf{Z}_c = \begin{bmatrix} \mathcal{R}_1 & \mathcal{I}_1 & \mathcal{R}_2 & \mathcal{I}_2 & \cdots & \mathcal{R}_{N_c} & \mathcal{I}_{N_c} \end{bmatrix}, \qquad (7.59)$$

where $\mathcal{R}_i = \sqrt{\frac{\kappa_i \Delta s}{\pi}} \mathrm{Re}\left[\mathbf{\Phi}_c(ik_i)\right]$ and $\mathcal{I}_i = \sqrt{\frac{\kappa_i \Delta s}{\pi}} \mathrm{Im}\left[\mathbf{\Phi}_c(ik_i)\right]$. The approximated observability Gramian can equally be written as $\mathbf{W}_{lo} \approx \mathbf{Z}_o \mathbf{Z}_o^\top$, with $\mathbf{Z}_o$ being a matrix of dimension $N_x \times (2N_y N_o)$ defined in a similar manner to Equation (7.59). In practice, only a few points are necessary to approximate Equation (7.57), that is, $N_c, N_o \ll N_x$, which results in $\mathbf{Z}_c$ and $\mathbf{Z}_o$ being very narrow (i.e., low-rank) matrices. Equation (7.58) is, therefore, a discrete version to the much more expensive Cholesky factorization of Section 7.4.1.

The transformation matrix used to obtain the (frequency-limited) balanced states (i.e., the matrix in the projection $\mathbf{x} = \mathbf{T}\tilde{\mathbf{x}}$) can now be obtained using the square-root method (Penzl, 2006) as

$$\mathbf{T} = \mathbf{Z}_c \mathbf{V} \mathbf{\Sigma}^{-1/2},$$
$$\mathbf{T}^\dagger = \mathbf{\Sigma}^{-1/2} \mathbf{U}^\top \mathbf{Z}_o^\top, \qquad (7.60)$$

where $\mathbf{U}$, $\mathbf{V}$, and $\mathbf{\Sigma} = \mathrm{diag}(\boldsymbol{\sigma})$ are obtained from the singular-value decomposition $\mathbf{Z}_o^\top \mathbf{Z}_c = \mathbf{U} \mathbf{\Sigma} \mathbf{V}^\top$. Note that since, in general, $N_c \neq N_o$, then $\mathbf{Z}_o^\top \mathbf{Z}_c$ is a rectangular matrix for which only the nonzero singular values are retained (Brunton and Kutz, 2019). It is also interesting to compare Equation (7.60) to the transformation matrices obtained by the direct method, Equation (7.52). As the manipulation is now only on the approximation of the Gramian given by Equation (7.58), we have different left and right projection matrices, $\mathbf{U}$ and $\mathbf{V}$. A major advantage is that the SVD is carried out now over a small matrix and results in a narrow rectangular transformation matrix $\mathbf{T}$ with pseudo-inverse[4] $\mathbf{T}^\dagger$. The resulting balanced model, $\tilde{G}$, has system matrices

$$\tilde{\mathbf{A}} = \mathbf{T}^\dagger \mathbf{A} \mathbf{T},$$
$$\tilde{\mathbf{B}} = \mathbf{T}^\dagger \mathbf{B},$$
$$\tilde{\mathbf{C}} = \mathbf{C} \mathbf{T}, \qquad (7.61)$$
$$\tilde{\mathbf{D}} = \mathbf{D}.$$

As we have narrowed the frequency bandwidth in the generation of the *empirical Gramians*, FLB truncation displays much faster convergence rates than internal balancing to capture the physics in the low-frequency range. As a result, residualization,

---

[4] Note that it is easy to see from Equation (7.60) that $\mathbf{T}^\dagger \mathbf{T} = \mathcal{I}$.

which needs a large-matrix inversion, can often be avoided and ROMs are simply obtained from the truncation of $\tilde{G}$. This results in a small system $\tilde{G}_r$, whose dimension N_r can be determined by a convergence analysis. Its accuracy can be easily determined against the FRF of the full system G at a minimal computational cost.

While internal balancing is known to preserve the stability of the system, the approximation of the Gramians on a restricted spectrum does not (Gawronski and Juang, 1990), that is, even though the original equations define a stable system, the transformed ones may not do so. Stability-preservation algorithms for ROM generation using FLB truncation are available in the literature (Benner et al., 2016; Gugercin and Antoulas, 2004), but they are based on the manipulation of the Lyapunov equations associated with the system Gramians, and hence they are not directly applicable to our solution method, based on empirical Gramians. Maraniello and Palacios (2020) have found a solution to this problem inspired by the strategies used in frequency-weighted balancing (FWB) (Enns, 1984). In FWB, the balanced realizations are obtained by connecting two filters to the inputs and outputs of G, and the FLB can be seen as a limit case in which those filters are infinite dimensional (Antoulas, 2005). In other words, Equation (7.55) is equivalent to an FWB with highly-resolved low-pass filters in the inputs and outputs of G with cutoff reduced frequencies k_c and k_o, respectively. It can be further shown (Enns, 1984; Gugercin and Antoulas, 2004) that FWB preserves stability if one of both filters is removed. The analogous process in the FLB is to integrate either the controllability or the observability Gramians in the full Nyquist range. For convenience, we perform this operation on the controllability Gramian, which has proved to be less costly for unsteady aerodynamic systems.

Note finally that the large-matrix inversions can be avoided using the wake structure of the DLTI UVLM, which results in solutions like Equation (7.14) along the wake streamlines (i.e., "columns" of the wake lattice) (Maraniello and Palacios, 2020). The resulting low-rank FLB procedure proceeds as follows:

1. Choose the reduced frequency k_{max} that determines the integration bounds in Equation (7.55), where we choose $k_o = k_{max}$ and $k_c = k_N$. Since the range of validity of potential flow aerodynamics is for frequencies up to $k_{max} = \mathcal{O}(1)$, it is $k_{max} \ll k_N$.

2. Choose a numerical integration scheme for the low-frequency range $(0 \leq k \leq k_{max})$, where a dense sampling is used, and the high-frequency range $(k_{max} < k \leq k_N)$, where very few frequencies are sampled.

3. Evaluate Equation (7.57) for both the observability and controllability Gramians. Only the latter is evaluated in the high-frequency range. In the low-frequency range, both Gramians are computed at the same sampling points, which reduces the number of operations. As the frequency evaluations are done independently, this step can easily be parallelized.

4. Obtain the balanced transformation using Equations (7.60) and (7.61). The full-system matrices are first transformed using Equation (7.10) to obtain system realizations with a suitable definition of the input signal.

For unsteady aerodynamic applications, both $\Phi_c(ik)$ and $\Phi_o(ik)$ are normally smooth functions, and very few sampling points are needed for accurate and stable realizations. It is also worth remarking that finally this approach does not require manipulating the full $\mathbf{A}_a$ matrix in Equation (7.41), and it is sufficient to operate on the linearized kernels in Equation (7.22), which further reduces its memory requirements.

7.5   Linear Aeroelastic Integration

The DLTI UVLM introduced in this chapter can be readily coupled to the linear flexible-body dynamic models in modal coordinates introduced in Section 6.4. It becomes an alternative aerodynamic description for low Mach number problems to the linear unsteady models considered in Section 6.5. The resulting aeroelastic description is the objective of this section, while Chapter 9 considers the more general geometrically nonlinear problems and their linearization about an arbitrary aeroelastic equilibrium.

Therefore, we introduce the vector $\mathbf{q}(t)$ of instantaneous modal amplitudes for the flexible-body dynamics obtained from the solution of Equation (6.51). It includes, in general, both rigid and elastic modes. If $\boldsymbol{\delta}(t)$ is again the vector with the instantaneous deflection in all control surfaces, the instantaneous bound aerodynamic lattice geometry is given by Equation (6.91), which is written now as

$$\Delta\boldsymbol{\zeta}_b = \mathbf{T}_{as} \begin{bmatrix} \boldsymbol{\Phi}_q & \boldsymbol{\Phi}_\delta \end{bmatrix} \begin{Bmatrix} \mathbf{q} \\ \boldsymbol{\delta} \end{Bmatrix}, \tag{7.62}$$

where $\mathbf{T}_{as}$ is the interpolation matrix between the structural and aerodynamic discretized geometries. The background velocity field can be written in terms of (frequency-domain) gust modes $\boldsymbol{\Phi}_g(ik)$, as shown in Equation (6.90), which can be written now as

$$\frac{\overline{\Delta\boldsymbol{\nu}_b}}{V} = \boldsymbol{\Phi}_g(ik)\frac{\overline{w_{g0}}}{V}. \tag{7.63}$$

As the current formulation is in the time domain, the corresponding state-space realization is needed, which results in the lag operator of Equation (7.16) at each spanwise wing location (with a constant offset to account for wing sweep). Let $\mathbf{A}_g$, $\mathbf{B}_g$, and $\mathbf{C}_g$ be the resulting state, input, and output matrices for the gust system, respectively. This results in the following DLTI equations for the propagation of a spanwise-varying gust velocity into the lifting surface:

$$\mathbf{x}_g^{n+1} = \mathbf{A}_g\mathbf{x}_g^n + \mathbf{B}_g\mathbf{w}_{g0}^{n+1}, \tag{7.64}$$
$$\Delta\boldsymbol{\nu}_b^n = \mathbf{C}_g\mathbf{x}_g^n.$$

We can now simultaneously solve Equations (7.41) and (7.64) for given time histories of the modal amplitudes and their derivatives, $\mathbf{q}$ and $\dot{\mathbf{q}}$; the control inputs and their derivatives, $\boldsymbol{\delta}$ and $\dot{\boldsymbol{\delta}}$; and the amplitudes of the gust modes, $\mathbf{w}_{g0}$. The output of Equation (7.41) has been defined in Equation (7.40) as the dimensionless aerodynamic forces at the lattice vertices. The corresponding instantaneous *generalized aerodynamic forces* associated with the structural modes are obtained then by, first, projecting on the structural grid and, then, onto the mode shapes as

$$\mathbf{f}_q = \tfrac{1}{2}\rho V^2 S \Phi_q^\top \mathbf{T}_{as}^\top \mathbf{y}_a, \tag{7.65}$$

which, in terms of the inputs defined above, can be finally written as

$$\mathbf{f}_q(t) = \tfrac{1}{2}\rho V^2 \left[\mathcal{A}_{q0}\mathbf{q} + \mathcal{A}_{\delta 0}\delta + \tfrac{c}{2V}\left(\mathcal{A}_{q1}\dot{\mathbf{q}} + \mathcal{A}_{\delta 1}\dot{\delta} \right) + \mathcal{C}_a \mathbf{x}_a + \tfrac{1}{V}\mathcal{C}_g \mathbf{x}_g \right]. \tag{7.66}$$

Equation (7.66) determines the time-domain generalized aerodynamics forces of the DLTI UVLM. It needs to be solved together with the state equations in Equations (7.41) and (7.64), which determine the aerodynamic and gust states, $\mathbf{x}_a$ and $\mathbf{x}_g$, respectively. All matrices in Equation (7.66) are directly obtained from the output equations in Equations (7.41) and (7.64) and the transformations defined by Equations (7.62) and (7.65). They have been written using the notation introduced for the RFA of frequency-domain aerodynamics, Equation (6.121), although some differences can be observed. First, there are no terms associated with the acceleration of the inputs, $\ddot{\mathbf{q}}$ and $\ddot{\delta}$ in Equation (7.66). Instead, the aerodynamic state vector, which now has a well-defined physical meaning, includes the time derivative of the circulation intensity on the vortex rings. Second, the gust input, $\mathbf{w}_{g0}$, does not explicitly appear and the dependency is on $\mathbf{x}_g$, which gives the instantaneous gust distribution at the corner points of the lifting surface. For simplicity, we have not included here the actuator model, Equation (6.106), although this is straightforward.

Finally, it is important to remark that Equation (7.41) is in practice replaced by a reduced-order approximation, such as the one obtained from the frequency-limited balanced models of Section 7.4.2. The state equations of the FBL ROM have also been defined in discrete time, but the system dimension is much smaller than that in the original DLTI UVLM and can therefore be easily transformed to continuous time using the methods of Appendix A.3.

The time-domain modal forces given in Equation (7.66) can now be substituted into the linear EOMs in modal coordinates for the flexible aircraft, Equation (6.51). The resulting coupled fluid–structure description still follows the flow diagram of Figure 6.20 and can be written in a very similar form to Equation (6.122), with the new definition of the generalized aerodynamic forces. The key difference is that the rational-function approximation needed to bring DLM aerodynamics into the time domain has been replaced by a balanced truncation of the DLTI UVLM. Both processes result in aerodynamic models of similar size, but (1) the UVLM solves a more general problem and (2) FLB defines an (algorithmically) more robust process than the realization problem on large datasets (Antoulas, 2005). The main limitation of the UVLM is the incompressible assumption, which restricts its use to problems with $\mathrm{M_a} < 0.3$.

Example 7.3 Flutter of a T-tail Configuration. To exemplify the DLTI UVLM in a representative problem, we consider the (dynamic) aeroelastic stability of the T-shaped aircraft tail shown in Figure 7.15a. T-tails are a classical design solution

(a)

(b)

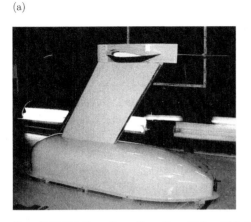
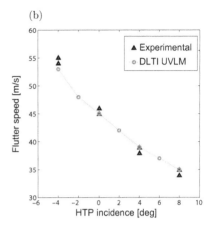

Figure 7.15 T-tail configuration of van Zyl and Matthews (2011). (a) Wind-tunnel model and (b) flutter speed change with HTP incidence angle (copyright of AIAA, published with permission).

for aircraft with wings mounted over the fuselage, such as the Airbus A400M, or with rear-mounted engines, as in the Gulfstream G700. However, the static forces on the horizontal tailplane (HTP) affect the torsional dynamics of the vertical tailplane (VTP), which may result in a destabilizing effect and bending-torsion flutter of the VTP (Murua et al., 2014). This results in a strong dependency of the flutter velocity of a T-tail on the HTP angle of incidence, as this one determines the aerodynamic forces on the HTP.

This can be clearly seen in Figure 7.15b that also includes the wind-tunnel results of van Zyl and Matthews (2011). They tested a T-tail and rear fuselage model with a sweptback fin and a trimmable HTP, and measured the flutter speed as a function of the angle of incidence of the HTP. The wind-tunnel model is shown in Figure 7.15a. Relative changes of the flutter speed of nearly 50% were obtained for a rather modest range of angles of incidence of the HTP ($2° \pm 6°$).

Numerical prediction of this flutter mechanism needs a linear structural dynamics model of the rear vehicle (although typically only the flexibility of the VTP is important), and a linearized unsteady aerodynamic model that includes the static forces on the HTP. This means that a different realization of the DLTI UVLM needs to be evaluated at each angle of incidence. A linear aerodynamic model without the effect of steady forces was also assembled and predicted a constant flutter speed of 49 m/s for all angles of incidence.

The comparison between the flutter speed obtained experimentally and the converged numerical results obtained with the DLTI UVLM is shown in Figure 7.15b. For the simulations, an aeroelastic state-space system is assembled based on Equation (6.125) but using the DLTI UVLM aerodynamics of Equation (7.66), which needs to be obtained for each angle of incidence. The eigenvalues of the system matrix are then computed as a function of the freestream velocity, and the flutter onset is determined as the point at which one of the complex roots migrates to the positive real plane. As can be seen, this approach gives excellent prediction of the experimental results (for which the error bars are unknown). Additional details of this study and a detailed exploration of the flutter mechanisms in T-tails can be found in the work of Murua et al. (2014).

Example 7.4 Model Reduction of a Simplified T-tail Aeroelastic Model. A second and much simpler T-tail configuration is considered here. It has a VTP with 2-m chord and 6-m span, and an HTP also with a chord of 2 m and tip-to-tip length of 8 m, which is mounted at 90° with respect to the VTP. Both VTP and HTP have a zero sweep angle. The bending and torsional stiffness for the VTP are both 10^7 N·m^2, while those values for the HTP are 10 times larger. The rest of the properties are the same for both surfaces, with the elastic axis at 0.25c, the inertial axis at 0.35c, mass per unit length being 35 kg/m, and torsional inertia being 8 kg·m. This configuration was introduced by Murua et al. (2014), and the present numerical investigation is reported in Maraniello and Palacios (2020).

Both VTP and HTP are discretized with 24 vortex rings in both the chordwise and spanwise directions. A 25-chord wake is added for both lifting surfaces, thus resulting in a full-order model with 32,256 states, for which results are considered to be sufficiently converged. The structure is described by the first 8 natural modes, which define the size of the input/output of the DTLI UVLM model. The first two natural modes of the structure are shown in Figure 7.16.

Frequency-limited balanced models are then generated with a maximum reduced frequency $k_{max} = 0.5$. The maximum relative error between the FRF of the full model, $\mathbf{G}_a(ik)$, and the reduced models, $\mathbf{G}_r(ik)$, for $0 \le k \le 0.5$, is then evaluated, and results are shown in Figure 7.17(a) as a function of the number of states in the ROM. Two different integration algorithms (trapezoidal and Gauss–Lobatto quadratures) were considered, as well as different numbers of integration points in Equation (7.57). As an example, a 1% error in the ROM can be obtained with 20 aerodynamic states and as little as two integration points. As expected, the accuracy of the ROM increases with the number of integration points, while the choice of the integration algorithm has little influence on the results for this case.

(a)   (b)

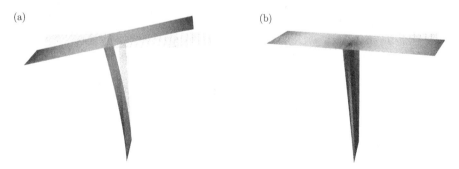

Figure 7.16 First two structural modes of the simple T-tail model displayed on the bound aerodynamic lattice (Maraniello and Palacios, 2020). (a) Mode 1 (VTP bending, 2.86 Hz) and (b) mode 2 (VTP torsion, 5.21 Hz).

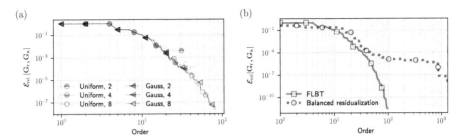

Figure 7.17 Maximum error between full and reduced-order models of the DLTI UVLM of the simple T-tail for $0 \le k \le 0.5$ (Maraniello and Palacios, 2020). (a) Frequency-limited balancing with different integration points and (b) FLB truncation vs. internal balancing with residualization.

Figure 7.17b shows the convergence rate with model size of ROMs obtained from truncation after FLBT. Results are compared against internal balancing with residualization, which are obtained here by integrating Equation (7.55) up to the Nyquist frequency. As was mentioned in Section 7.4.2, integrating the Gramians only over the low frequencies results in the FLBT ROM displaying much faster convergence than the models obtained from internal balancing, which also retain unnecessary higher frequency dynamics.

Finally, the reduced-order aerodynamic model generated by FLB truncation is employed to investigate the flutter characteristics of the T-tail for sea-level conditions and values of the angle of attack, α_0, ranging between $-10°$ and $20°$. For each value of α_0, the aerodynamic ROM is calculated for $k \le 0.5$ and 12 integration points. A Tustin transformation, Equation (A.12), generates then a continuous-time model, which is finally coupled to a structural model of the tail. As before, the first 8 natural models are considered. The flutter speed and the associated reduced frequency are finally sought by investigating the evolution of the eigenvalues of the coupled system. They are shown in Figure 7.18. As can be seen, there is a maximum of flutter speed around $\alpha_0 = -5°$, which then shows a very large drop as the angle of attack is increased. This is the result to large changes in the modal interactions as steady lift builds up, as was already observed in Example 7.3. To demonstrate this here, we consider the flutter mode (the eigenvector of the unstable eigenvalue) at $\alpha_0 = -8°$ and $-1.5°$, for which the flutter speed is nearly identical (around 250 m/s). For $\alpha_0 = -8°$, the contribution to the flutter mode of the first torsional mode of the vertical tail is around 34% of that of the first bending mode of the tail. For $\alpha_0 = -1.5°$, this becomes 83.6%.

7.6 Summary

This chapter has introduced vortex-lattice models for lifting surfaces that move and deform following complex kinematics. Such methods are grounded on the potential theory of incompressible fluids and are, therefore, valid only for attached flows at sufficiently low speeds. Each lifting surface and its wake are approximated by a lattice of vortex rings (or discrete vortex in two dimensions), with the wake evolution being

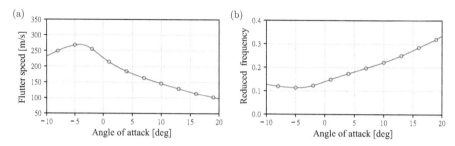

Figure 7.18 Flutter speed and frequency of the simple T-tail as a function of the angle of attack. (a) Flutter speed and (b) flutter reduced frequency.

described by Helmholtz's laws. Furthermore, while UVLM aerodynamics neglects viscous drag the effects, the induced drag is captured. Viscous effects are mainly quasi-steady effects in attached flows and can be added as corrections to match local airfoil polars. Further extensions of the theory exist to capture thickness effects and aerodynamic interference due to nonlifting bodies, although they have not been described here. Instead, the focus has been on approaches to reduce the computational cost of time marching the UVLM equations on complex configurations, which has been historically their main hindrance, since the general UVLM equations need the inversion of a large dense matrix at each step. In most cases, however, linear unsteady aerodynamic models – built in general around nonlinear aeroelastic equilibrium conditions – are sufficient, and those can be obtained by analytical linearization of the UVLM equations around an arbitrary reference. In that case, methods for model-order reduction of linear dynamic systems are readily available, and we have focused in particular on balanced realizations, which define a linear transformation of the system states that result in compact descriptions that give the best controllability and observability for given inputs and outputs. It has been seen that balancing can reduce the dimensionality of the aerodynamic model by several orders of magnitude. Direct balancing methods of large systems are, however, rather expensive themselves, and a highly computational efficient low-rank alternative has been presented in Section 7.4.2 that exploits both the physical meaning and the mathematical structure of the UVLM equations.

The linear aerodynamic methods in this chapter do naturally fit into the general framework of Chapter 6, and this has been illustrated with numerical examples in Section 7.5. There, flutter analysis of T-tails, which has required bespoke corrections in the past for DLM-based aerodynamic models, is investigated using the DLTI UVLM.

7.7   Problems

Exercise 7.1. Using the analytical solutions of thin-airfoil theory from Section 3.3.2, show that the quasi-steady lift on a rigid airfoil is exactly retrieved by a single discrete vortex at its quarter-chord point and a collocation point at the $3c/4$ point.

Exercise 7.2. Using the properties of the Z-transform listed in Appendix A.3, show that a traveling gust defined in discrete time by Equation (7.16) (the lag operator) is equivalent to a gust mode defined by Equation (6.89) in the frequency domain.

Exercise 7.3. Write a discrete-vortex solution for the lift on a pitching airfoil using a single bound vortex at the quarter-chord point and a single wake vortex at one chord downstream from the first one. Write the continuous-time FRF between the pitch angle and the lift using first the Z-transform and then bilinear transformation to express your results in terms of the reduced frequency.

Exercise 7.4. Find the Hankel singular values and use them to propose a suitable balanced realization of the following systems:

(i) $\quad \mathbf{A} = \begin{bmatrix} -1 & 0 & 0 \\ 0 & -2 & 0 \\ 0 & 0 & -3 \end{bmatrix}, \quad \mathbf{B} = \begin{bmatrix} 1 \\ 0.1 \\ 0.1 \end{bmatrix}, \quad \mathbf{C} = \begin{bmatrix} 1 & 1 & 0.1 \end{bmatrix}.$

(ii) $\quad \mathbf{A} = \begin{bmatrix} -1 & 0 & 0 \\ 0 & -0.1 & 0 \\ 0 & 0 & -0.099 \end{bmatrix}, \quad \mathbf{B} = \begin{bmatrix} 1 \\ 1 \\ -1 \end{bmatrix}, \quad \mathbf{C} = \begin{bmatrix} 1 & 1 & 1 \end{bmatrix}.$

Exercise 7.5. Consider again the problem of Example 7.2 but also including atmospheric gust as input, that is, a discrete-vortex airfoil model with input vector $\mathbf{u}^\top = \left\{ \alpha \quad \alpha' \quad \delta \quad \delta' \quad w_g \right\}$. The airfoil aerodynamics are, therefore, given by a model such as the one shown in Figure 7.3. Compute a balanced realization of this augmented system for increasing number of states and compare it with the result of adding a balanced model to the gust input to the results of Example 7.2.

8   Geometrically Nonlinear Composite Beams

8.1   Introduction

We continue in this chapter our presentation of some modeling strategies for flexible aircraft displaying large-amplitude elastic displacements in flight. As we have discussed in Chapter 1, this mainly occurs on air vehicles with wings of a very high aspect ratio, and therefore we restrict ourselves to structural descriptions using beam models. A beam model approximates the structural dynamics of a 3-D deforming slender body by means of a 1-D description along its dominant direction (e.g., the wingspan). Classical beam models do not capture local features, such as the local skin buckling that may result from high wing bending, but give an excellent approximation of the deformed vehicle geometry and its associated distribution of inertia for aerodynamic and flight dynamics purposes, respectively. Conventional linear beam theory (Balachandran and Magrab, 2018; Hodges and Pierce, 2014) is, however, not sufficient for this problem, and one needs to consider a more elaborated version that includes (1) geometrically nonlinear effects required for the accurate kinematic description of the large displacements and (2) potential local elastic couplings that appear between DoFs due to anisotropic material distributions (e.g., torsion–bending coupling with unbalanced composite laminates). This results in *geometrically nonlinear composite beam models*, which are the focus of this chapter.

Starting from the early work of Cesnik et al. (1996), Patil et al. (1998, 2001), and Drela (1999), structural models built using geometrically nonlinear beams have become an indispensable feature of coupled flight dynamics–aeroelastic formulations with large elastic displacements, and their main features are outlined in this chapter. Several formulations have been developed over more than 20 years (Afonso et al., 2017) for application to very flexible aircraft dynamics, including displacement-based, corotational, strain-based, and intrinsic beam models, and the most relevant of them are discussed here and derived from a single set of basic kinematic assumptions. For further discussion on the theory and available models of geometrically nonlinear composite beams, the reader is referred to the monograph by Hodges (2006). We should, however, note that composite beam models have also long been used in the helicopter community for the study of blade dynamics and are also now routinely used in the aeroelastic analysis of wind turbine rotors (Wang et al., 2017). Consequently, many of the methods discussed in this chapter have also found application in rotor blade design.

This chapter presents an overview of geometrically nonlinear composite beam theory. Section 8.2 first outlines some of the key features of geometrically nonlinear structural analysis. Next, we introduce in Section 8.3 a general framework for the description of the dynamics of a beam assembly on a moving reference frame. Within that framework, Sections 8.4–8.6 present three different formulations that have been proposed to solve this problem, including some representative examples. Key issues in the numerical solution of the equations are discussed, including parameterizations of the finite rotations, spatial discretization of the resulting PDEs, and linearization around arbitrary nonlinear equilibrium points.

8.2 Geometric Nonlinearities in Structural Dynamics

Geometrically nonlinear effects result from considering large deflections on either the inertia, the stiffness, or the external forces in Hamilton's principle, Equation (1.1). In Section 6.3, we have considered the effect of shape changes on the inertia of a deforming body, which have resulted in a mass matrix that depends on the deformation, Equation (6.16). As a result, the position of the CM within the body and/or its moment of inertia change with the elastic deflections. When the deflections are sufficiently large, they also result in changes on stiffness, and may even impact the direction in which the external forces are applied, as we discuss now.

Geometric Stiffening

Equilibrium of forces in an elastic body ensures that the internal forces within the structure (the internal stresses) balance themselves with the applied loads. Forces (and moments) are vector quantities and, therefore, their lines of action are as important as their magnitudes. Furthermore, the lines of action for the internal forces are defined by the instantaneous geometry of the structure. If the structural displacements are sufficiently large, then the geometry of the structure changes, and so do the lines of action of the internal forces, thus resulting in a reorientation of the internal stresses that modify the effective stiffness of the structure. This situation is known as geometric stiffening.

As a example, consider the vibrations of a cantilever isotropic beam under a harmonic vertical forcing in its first bending mode (a normal force per unit length, $q(x)$, with the shape of the first bending mode, as shown in the inset of Figure 8.1). Figure 8.1 shows the amplitude of the beam vibrations (measured by the tip displacement, u_z) as a function of the frequency of the force excitation. The frequencies are chosen near the first natural frequency, ω_1, and each line in the figure corresponds to increasing amplitude of the force. The resonant behavior results in much larger amplitudes when the excitation is near ω_1, but as the amplitude of the force increases, the maximum response occurs at slightly larger frequencies. This is due to the changes in both stiffness and inertia (the natural frequencies goes as the square root of their ratio) as the geometry changes. The curve linking the maximum amplitudes is commonly known as the *backbone curve*, and it is also displayed in Figure 8.1.

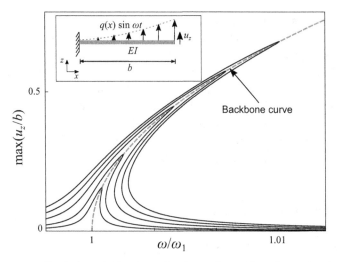

Figure 8.1 Large-amplitude forced vibrations of an isotropic cantilever beam showing hardening behavior (after Vizzaccaro (2021)).

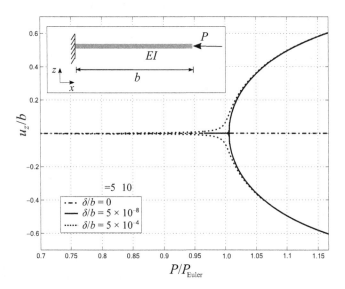

Figure 8.2 Buckling of a cantilever beam under a tip axial force.

A softening behavior, in which frequencies drop with amplitude, also occurs in some situations. When the drop of stiffness under loading reaches a critical threshold, it results in an elastic (static) instability known as *buckling*. This is easily seen with the example of Figure 8.2 that shows the vertical displacement, u_z, at the free end of an isotropic cantilever beam with bending stiffness EI as a function of a static force P. For sufficiently small values of P, the beam compresses without transversal displacements, but a critical load exists beyond which the structure buckles. For an isotropic cantilever beam with no axial compliance, the critical load is $P_{\text{Euler}} = \frac{\pi^2 EI}{4b^2}$,

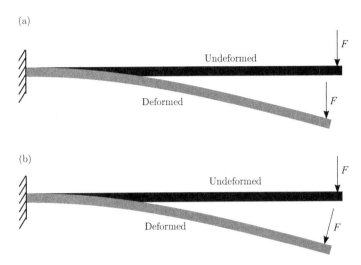

Figure 8.3 Cantilever beam under a tip static force. (a) Dead force and (b) follower force.

known as the Euler buckling load. The different lines in the graph correspond to tiny imperfections (δ is the initial lateral tip displacement) in the model. The amplitude of the displacement due to the loading displays a symmetric bifurcation, with two stable solutions (branches) at postbuckling loads and one unstable branch along the axis. In stability theory, this is known as a *pitchfork bifurcation*.

Follower and Dead Forces

Geometrically nonlinear effects also affect the definition of the external forces, as the displacements at the point where they are applied need to be considered. We distinguish, in general, between follower and dead forces. *Follower forces* keep the relative orientation with the structure (e.g., pressure forces) and, therefore, rotate as the structure deforms, while dead forces maintain their direction regardless of deformation. Aerodynamic loads on lifting surfaces are follower forces, while weight is a dead force. This is schematically shown for a tip force on a cantilever beam in Figure 8.3. Further discussion can be found in Example 8.1.

8.3 Geometrically Nonlinear Composite Beams

As mentioned above, the primary structures of the aircraft are modeled here as a collection of geometrically nonlinear composite beams. We refer to them as *beam assemblies*, although they are often referred to as *stick models* by practicing engineers. The typical situation is illustrated in Figure 8.4, where the dynamics of the structure is described by tracking the instantaneous shape of a *reference line* (a space curve) along the vehicle main load paths. A *Cosserat's rod model* is then postulated in which the instantaneous 3-D shape of the deformed airframe follows the deforming space curve, which in general has nonzero initial twist and curvature, with rigid cross

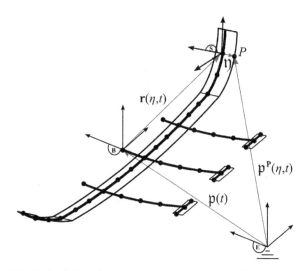

Figure 8.4 Flexible airframe structure modeled using beam elements.

sections attached to it. Inertia and compliance properties are assigned to each cross section to reproduce the dynamics of the original 3-D structure in some average sense (Cesnik and Hodges, 1997). There are no restrictions to the sectional location of the reference line, although some definitions (e.g., elastic axis, if it exists) may simplify the interpretation of results. Note that sectional inertia and compliance are linked to the selection of this reference line. If the location of the reference were to be modified, so would be the sectional properties of the structure. To describe the kinematics of the resulting beam assembly, a set of reference frames are necessary and those are defined next. The kinematic description in this chapter follows the framework introduced by Hodges (1990).

Reference frames. As we are concerned about flying vehicles, which are subject to both rigid body and elastic motions, we seek a kinematic description in which those effects can be easily identified. In the same manner as has been done in Section 6.2, we achieve this by defining a body-attached frame of reference that serves to track the vehicle position and orientation with respect to an Earth (inertial) frame, E. This frame is again identified as the B frame of reference, as shown in Figure 8.4, and is referred to here as the *vehicle axes*. Its definition is not unique, although a good choice is to make it coincide with the stability axes in a reference condition. The instantaneous coordinates of the center of mass after the deformation of the aircraft can then be easily obtained, should they be needed, in postprocessing.

A curvilinear coordinate η is now defined along the undeformed beam assembly such as that arclength of the undeformed curve is $\int_\eta d\eta$. Without loss of generality, it is chosen $\eta = 0$ corresponding to the origin of the vehicle axes (the B frame of reference). To define the local orientation of the deforming beam, we need to define a local structural frame of reference, S, at each beam point with coordinates $\mathbf{r}_B(\eta, t)$ measured from the origin of the frame of reference B, as shown in Figure 8.4. It is

referred to as the *material reference frame*, as the material laws (or constitutive relations) are naturally expressed in this reference. Therefore, $\mathbf{R}_{SB}(\eta, t)$ is the coordinate
transformation matrix from the body-fixed reference frame to the local cross section
at a coordinate η. It is chosen such that the local x axis is tangent to the beam refer-
ence line in the undeformed structure, that is, $\mathbf{R}_{SB}(\eta, 0)\mathbf{r}'_B(\eta, 0) = \mathbf{e}_1$, where $(\bullet') = \frac{\mathrm{d}}{\mathrm{d}\eta}$
indicates differentiation with respect to the curvilinear coordinate η.

As everywhere in this book, subindexes are used to indicate the coordinate system
in which each vector magnitude is projected. Accordingly, $\mathbf{r}_B(\eta, t)$ is the components
of the relative position vector in the vehicle axes. Note also that the variables used to
describe the dynamics of beams are functions of both time t and the position in the
structure, defined by the coordinate η, and these are often made explicit in the text.

Finally, and as we did in Chapters 4 and 6, we need to determine the instantaneous
position of an arbitrary material point on the aircraft, P, not necessarily along the
reference line defined by the beam model of the structure, and with respect to an
inertial observer in the Earth frame of reference. For that purpose, we use Cosserat's
model for the beam kinematics, which assumes rigid cross sections along the reference
line. Let η be the curvilinear coordinate that determines the normal cross section on
the undeformed structure where P lies. The instantaneous coordinates of the material
point P with respect to the Earth frame are

$$p_E^P(t) = p_E(t) + \mathbf{R}_{EB}(t)\mathbf{r}_B(\eta, t) + \mathbf{R}_{EB}(t)\mathbf{R}_{BS}(\eta, t)\mathfrak{y}_S, \qquad (8.1)$$

where, as in Equation (6.1), $p_E(t)$ are the coordinates of the origin of vehicle axes from
the Earth frame. We have now also introduced $\mathfrak{y}_S = (0, y, z)^\top$ to represent the cross-
sectional coordinates of point P in the local material frame S, as shown in Figure 8.4,
which are constant with time under the Cosserat's rigid crosssection assumption. Note
that, in the last term, the transformation of the sectional coordinates of P is done in
two successive rotations. First, they are written in vehicle axes, and then those are
projected onto the Earth frame.

8.3.1 General Formulation

Once the reference frames are defined, the instantaneous state of the deforming solid
under known external forces is determined by the three sets of interrelated equations shown in Figure 8.5, namely the *kinematic relations* that define its deformation,
the *constitutive relations* that introduce the material properties, and the *equations of
motion* resulting from enforcing Newton's laws. Some details for each of them are
discussed in the following sections.

Kinematic Relations

We compute first the distribution of velocities along the reference line with respect to
an inertial observer. This choice of the observer simplifies later the evaluation of the
instantaneous kinetic energy. As the beam cross sections can have both translations
and rotations in all three axes, we need to define both their (inertial) linear and angular velocities. When written in their components on the local material frame, S, they

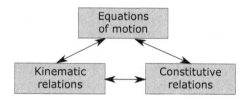

Figure 8.5 The three sets of equations needed in solid mechanics.

will be written as $\bar{\mathbf{v}}_S(\eta,t)$ and $\bar{\boldsymbol{\omega}}_S(\eta,t)$, respectively, where the macron $\bar{\bullet}$ is included to indicate a dependency on the position along the beam reference line, η. This distinguishes them from the linear and angular velocity of the vehicle axes, that is, at $\eta = 0$, which are $\mathbf{v}_B(t) = \bar{\mathbf{v}}_S(0,t)$ and $\boldsymbol{\omega}_B(t) = \bar{\boldsymbol{\omega}}_S(0,t)$. The inertial velocities at the origin $\eta = 0$ will be referred to as rigid-body velocities, even though our choice of body-fixed reference frame B is somehow arbitrary, as we have discussed in Section 6.2.

With this intermediate reference frame, we can obtain the inertial velocities at points along the beam reference line (sectional velocities) through the addition of the global velocity of the vehicle axes (i.e., the origin of coordinates) and the relative velocity with respect to the vehicle axes, due to elastic deformations. Expressed in their components in the local material frame S, this results in the following geometrically nonlinear *velocity-displacement kinematic relations*:

$$\bar{\mathbf{v}}_S(\eta,t) = \mathbf{R}_{SB}(\eta,t)\dot{\mathbf{r}}_B(\eta,t) + \mathbf{R}_{SB}(\eta,t)\left(\tilde{\boldsymbol{\omega}}_B(t)\mathbf{r}_B(\eta,t) + \mathbf{v}_B(t)\right),$$
$$\bar{\boldsymbol{\omega}}_S(\eta,t) = \mathbf{R}_{SB}(\eta,t)\dot{\mathbf{R}}_{BS}(\eta,t) + \mathbf{R}_{SB}(\eta,t)\tilde{\boldsymbol{\omega}}_B(t)\mathbf{R}_{BS}(\eta,t),$$
(8.2)

where the relative angular velocity between S and B has been written in terms of the derivatives of the coordinate transformation matrix using Equation (4.10). As shown elsewhere in this book, $(\bullet) = \frac{\mathrm{d}}{\mathrm{d}t}$ indicates derivatives with physical time, t, and $(\tilde{\bullet})$ the cross-product operator. In Equation (8.2), angular velocities have been written as skew-symmetric matrices for convenience. The first term in both equations corresponds to the relative velocities of the deforming structure with respect to the B frame, which are then projected onto the local material frame, S. This final manipulation is often referred to as a *push forward* operation (Marsden and Hughes, 1994, Ch. 1). The second term includes the rigid-body velocities of the vehicle axes with respect to the Earth frame and expressed in their components in the local material axes (using Equation (C.23) for the angular velocity). Note that, by definition, the velocities at the origin of frame B correspond to the origin, $\eta = 0$, of the curvilinear coordinates and that[1] $\mathbf{R}_{SB}(0,t) = \mathcal{I}$ and $\mathbf{r}_B(0,t) = \mathbf{0}$.

Similarly, we have the geometrically nonlinear *strain–displacement kinematic relations* that determine the current deformation of the structure. For that purpose, we define the local force strain, $\boldsymbol{\gamma}(\eta,t)$, and moment strain, $\boldsymbol{\kappa}(\eta,t)$, along the reference line as

---

[1] In practice, it is often convenient to introduce an additional constant rotation between the body-fixed reference frame B, which describes the global dynamics of the vehicle, and the material frame S at the origin $\eta = 0$, but this is not explicitly included here to simplify the description.

$$\gamma(\eta, t) = \mathbf{R}_{SB}(\eta, t)\mathbf{r}'_B(\eta, t) - \mathbf{R}_{SB}(\eta, 0)\mathbf{r}'_B(\eta, 0),$$
$$\tilde{\kappa}(\eta, t) = \mathbf{R}_{SB}(\eta, t)\mathbf{R}'_{BS}(\eta, t) - \mathbf{R}_{SB}(\eta, 0)\mathbf{R}'_{BS}(\eta, 0). \tag{8.3}$$

Here, $\mathbf{r}'_B(\eta, t)$ measures changes in the position vector with the curvilinear coordinate, written in its components on the instantaneous vehicle axes. In geometric terms, $\mathbf{r}'_B(\eta, t)$ are the components of the instantaneous tangent vector of the deformed curve, which are first pushed forward to the local reference frame by the rotation defined by $\mathbf{R}_{SB}(\eta, t)$, and then compared with the same metric at time $t = 0$. Since, by definition, the tangent vector of the undeformed curve is the unit vector along the axial coordinate, we have $\mathbf{R}_{SB}(\eta, 0)\mathbf{r}'_B(\eta, 0) = \mathbf{e}_1$. In physical terms, the first component of the force strain is the extensional (or axial) strain, that is, it measures the local stretching of the reference line, while the last two correspond to beam shear strains.

The moment strain performs a similar operation. First, note that the local curvature vector, written in the material frame of reference, can be written in a similar manner to the local angular velocity (see Appendix C for details),

$$\tilde{k}_S = \mathbf{R}_{SB}\mathbf{R}'_{BS}. \tag{8.4}$$

Therefore, the moment strain κ in Equation (8.3) effectively compares the values of each component of the curvature vector in the material frame of reference before and after deformation. Its first component (which goes along the reference line on the undeformed structure) is the torsional strain, while the last two are bending strains.

Both force and moment strains are defined by comparing the components of two vectors in two different bases (the local material frame before and after deformation). This is why, contrary to all other variables defined in this section, we do not associate them with a unique reference frame – that is, they do not have subindexes in their definitions. The initial geometric curvature of the reference line, which may come, for example, from initial pretwist of a wing, can be written as $\tilde{k}_0(\eta) = \mathbf{R}_{SB}(\eta, 0)\mathbf{R}'_{BS}(\eta, 0)$. For an initially straight beam, it means $\mathbf{k}_0 = \mathbf{0}$. Moreover, if the structure is clamped at some point, it is convenient to define the origin of coordinates, $\eta = 0$, at that point, such that $\mathbf{v}_B(t) = \mathbf{0}$ and $\omega_B(t) = \mathbf{0}$.

Note the close similarity in the definitions of beam strains and beam velocities. In Equation (8.2), the instantaneous values of the time derivatives of positions and rotations are measured at two different spatial locations (origin and local coordinates); the definition of beam strains, on the other hand, compares their spatial derivatives at two different instants of time (initial and current time) on a fixed spatial location. In fact, for a clamped and initially straight beam, Equations (8.2) and (8.3), when written in terms of displacements, are identical if one replaces time by spatial derivatives. This symmetry between time and space is known as the *Kirchhoff's kinetic analogy* (Love, 1927, Ch. 19) and will be seen in Section 8.6 than it can be used to derive a Hamiltonian formulation of the beam dynamics. Note finally that the displacement vector is defined as the instantaneous relative difference between the initial (reference) and final (current) positions of a material point. It is the natural parameterization in linear theories, and it was used in Section 6.3.1, but it is less useful in a total Lagrangian nonlinear formulation such as this one, where the magnitude of the displacements is comparable to that of the original position vector.

Constitutive Relations

Linear constitutive relations are assumed along the reference line. Constitutive relations are naturally written in material coordinates, and the previous definitions of Equations (8.2) and (8.3) make for rather straightforward expressions for the total kinetic, $\mathcal{T}$, and strain, $\mathcal{U}$, energies of the beam assembly. When written in terms of the cross-sectional mass, $\mathcal{M}(\eta)$, and compliance, $\mathcal{C}(\eta)$, matrices, they are obtained as the following line integrals along the entire structure:

$$\mathcal{T}(t) = \frac{1}{2} \int_{\eta} \left\{ \bar{\mathbf{v}}_S^\top \ \bar{\boldsymbol{\omega}}_S^\top \right\} \mathcal{M} \left\{ \begin{matrix} \bar{\mathbf{v}}_S \\ \bar{\boldsymbol{\omega}}_S \end{matrix} \right\} d\eta,$$

$$\mathcal{U}(t) = \frac{1}{2} \int_{\eta} \left\{ \boldsymbol{\gamma}^\top \ \boldsymbol{\kappa}^\top \right\} \mathcal{C}^{-1} \left\{ \begin{matrix} \boldsymbol{\gamma} \\ \boldsymbol{\kappa} \end{matrix} \right\} d\eta. \tag{8.5}$$

The assumption of linear constitutive relations implies that both $\mathcal{M}(\eta)$ and $\mathcal{C}(\eta)$ do not depend on the deformation. Furthermore, while they can vary along the reference line, they are assumed here to be constant with time. Both matrices are symmetric and, in particular, positive definite matrices, which ensures that the total energy of the structure is always nonnegative. The cross-sectional mass matrix is derived as was done in Section 6.3.2, with all integrals now restricted to the local cross section, which results in

$$\mathcal{M}(\eta) = \begin{bmatrix} m(\eta)\mathcal{I} & -m(\eta)\tilde{\mathfrak{h}}_{cg}(\eta) \\ m(\eta)\tilde{\mathfrak{h}}_{cg}(\eta) & \mathbf{I}_S(\eta) \end{bmatrix}, \tag{8.6}$$

with $m(\eta)$ being the mass per unit beam length, $\tilde{\mathfrak{h}}_{cg}(\eta) = (0, y_{cg}(\eta), z_{cg}(\eta))$ are the coordinates in the local S frame of the center of gravity of the cross section, and $\mathbf{I}_S(\eta)$ is the sectional inertia tensor, defined as in Equation (4.21) but now with (area) integrals over the cross section (therefore, inertias per unit beam length). As mentioned above, they are all constant with time when evaluated in the local material frame, S.

The sectional compliance matrix is, in general, a full (symmetric) matrix. Its inverse, which is the one that appears in the strain energy, is known as the sectional stiffness matrix. Off-diagonal terms in the compliance matrix account for the elastic coupling that would occur with either a complex internal geometry (e.g., a misaligned spar) or with anisotropic materials, such as composite laminates While in general, the sectional compliance need to be obtained by a process of homogenization of the original 3-D structure (Cesnik and Hodges, 1997; Cesnik et al., 1996; Dizy et al., 2013; Palacios and Cesnik, 2005; Riso et al., 2020) it is, however, assumed to be known here. For simple geometries and material distributions, the beam reference line can often be chosen such that the inverse of the compliance matrix is diagonal, whose coefficients are usually defined as

$$\mathcal{C}^{-1}(\eta) = \mathrm{diag}(EA, GA_y, GA_z, GJ, EI_{yy}, EI_{zz}), \tag{8.7}$$

with $EA(\eta)$ being the local axial stiffness, $GA_\bullet(\eta)$ the local shear stiffness (in both sectional axes), $GJ(\eta)$ the local torsional stiffness, and $EI_\bullet(\eta)$ the local bending stiffness (also in both axes). In those cases, the beam reference line is typically referred to as the *elastic axis*, and the beam model is referred to as the *Timoshenko model*. If we further assume zero shear flexibility (i.e., equivalent in the isotropic case to $GA_y \to \infty$, $GA_z \to \infty$) and flexural rotational inertia, we have a geometrically nonlinear model of the well-known Euler–Bernoulli beam model (see Example 8.3). Commonly, Timoshenko beam models also include rotary inertia, that is, the nonzero inertia tensor in the local bending axes (Tessler and Dong, 1981), while those terms are assumed to be zero in the Euler–Bernoulli model.

It is also convenient to introduce the cross-sectional linear and angular momenta (per unit spanwise length) as the energy conjugates of the local inertial velocities, that is, $\mathbf{p}_S^\top = \frac{\partial \mathcal{T}}{\partial \bar{\mathbf{v}}_S}$ and $\mathbf{h}_S^\top = \frac{\partial \mathcal{T}}{\partial \bar{\omega}_S}$, respectively. Analogously, the internal forces and moments (per unit spanwise length) are defined as $\mathfrak{f}_S^\top = \frac{\partial \mathcal{U}}{\partial \gamma}$ and $\mathfrak{m}_S^\top = \frac{\partial \mathcal{U}}{\partial \kappa}$, respectively. This results in the following set of constitutive relations:

$$\begin{cases} \mathbf{p}_S(\eta,t) \\ \mathbf{h}_S(\eta,t) \end{cases} = \mathcal{M}(\eta) \begin{cases} \bar{\mathbf{v}}_S(\eta,t) \\ \bar{\omega}_S(\eta,t) \end{cases},$$

$$\begin{cases} \gamma(\eta,t) \\ \kappa(\eta,t) \end{cases} = \mathcal{C}(\eta) \begin{cases} \mathfrak{f}_S(\eta,t) \\ \mathfrak{m}_S(\eta,t) \end{cases}. \tag{8.8}$$

Physically, the internal forces and moments correspond to the cross-sectional stress resultants – that is, the force and moment resultant of the distributed stresses that appear in the cross section when the beam deforms.

Equations of Motion

The dynamics of the (initially curved and twisted) composite beam in the time interval $[0, t_f]$ needs to satisfy Hamilton's principle, Equation (1.1). For the kinetic and potential energies defined in Equation (8.5), which also imply a parameterization of the beam dynamics, this results in

$$\int_0^{t_f} \int_\eta \left(\frac{\partial \mathcal{T}}{\partial \bar{\mathbf{v}}_S} \delta \bar{\mathbf{v}}_S + \frac{\partial \mathcal{T}}{\partial \bar{\omega}_S} \delta \bar{\omega}_S - \frac{\partial \mathcal{U}}{\partial \gamma} \delta \gamma - \frac{\partial \mathcal{U}}{\partial \kappa} \delta \kappa + \delta \mathcal{W}_{cs} \right) d\eta \, dt = 0, \tag{8.9}$$

where $\delta \mathcal{W}_{cs}$ is the virtual work per unit span associated with the external forces on the cross section, which is considered next.

Let $\bar{\mathbf{f}}_S^P(t)$ be the volumetric force at point P, whose position is given by Equation (8.1), and consider, without loss of generality, that this is the only type of force acting on the solid. The virtual work $\delta \mathcal{W}_{cs}$ is defined as the integral over the cross section of the work of the external forces along virtual displacements, which need to be given with respect to an inertial observer. If A is the area of the local cross section, this is written as

$$\delta \mathcal{W}_{cs} = \int_A \left(\mathbf{R}_{ES} \bar{\mathbf{f}}_S^P \right)^\top \delta \mathbf{p}_E^P \, dA = \int_A \left(\mathbf{R}_{BS} \bar{\mathbf{f}}_S^P \right)^\top \mathbf{R}_{BE} \delta \mathbf{p}_E^P \, dA$$

$$= \int_A \left(\mathbf{R}_{BS} \bar{\mathbf{f}}_S^P \right)^\top \left(\delta \mathfrak{p}_B - \left(\widetilde{\mathbf{r}}_B + \widetilde{\mathbf{R}_{BS} \eta_S} \right) \delta \varphi_B + \mathbf{R}_{SB} \delta \mathbf{r}_B - \widetilde{\eta}_S \delta \phi_S \right) dA, \tag{8.10}$$

where the last expression is obtained after taking variations of Equation (8.1). We have introduced $\delta \mathfrak{p}_B(t)$ and $\delta \varphi_B(t)$, which have already appeared in Equation (6.27) to denote the virtual displacements and rotations of the body-fixed reference frame,

B. We have also introduced the local material virtual rotation as $\widetilde{\delta\phi}_S(\eta,t) = \mathbf{R}_{SB}\delta\mathbf{R}_{BS}$ (see Appendix C.3). With this, we can now define the applied force and moment per unit span as

$$\bar{\mathbf{f}}_S(\eta,t) = \int_A \bar{\mathbf{f}}_S^P \, dA,$$
$$\bar{\mathbf{m}}_S(\eta,t) = \int_A \tilde{\eta}_S \bar{\mathbf{f}}_S^P \, dA. \tag{8.11}$$

Note that we use different fonts for, on the one side, the applied forces and moments per unit length, ($\bar{\mathbf{f}}_S$ and $\bar{\mathbf{m}}_S$, respectively), which result from the external loading on the structure, and, on the other side, the internal forces and moments (also per unit span $\mathfrak{f}_S$ and $\mathfrak{m}_S$), which are the cross-sectional resultants from the instantaneous stress state of the structure. Using these definitions and the linear constitutive relations given in Equation (8.8), Hamilton's principle results in the following *weak form* of the EoMs of a beam assembly as it deforms and moves with respect to an inertial observer,

$$\int_0^{t_f} \int_\eta \left(\delta\bar{\mathbf{v}}_S^\top \mathbf{p}_S + \delta\bar{\omega}_S^\top \mathbf{h}_S - \delta\gamma^\top \mathfrak{f}_S - \delta\kappa^\top \mathfrak{m}_S + \delta\mathbf{r}_S^\top \bar{\mathbf{f}}_S + \delta\phi_s^\top \bar{\mathbf{m}}_S \right) d\eta \, dt$$
$$+ \int_0^{t_f} \left[\delta\mathbf{p}_B^\top \int_\eta \bar{\mathbf{f}}_B \, d\eta + \delta\varphi_B^\top \int_\eta \left(\bar{\mathbf{m}}_B + \tilde{\mathbf{r}}_B\bar{\mathbf{f}}_B \right) d\eta \right] dt = 0. \tag{8.12}$$

Equation (8.12), together with the kinematic relations, Equations (8.2) and (8.3), and the constitutive relations, Equation (8.8), uniquely define the dynamics of a beam assembly under known (distributed) external forces which are given here by the sectional resultants $\bar{\mathbf{f}}_S(\eta,t)$ and $\bar{\mathbf{m}}_S(\eta,t)$, and by the initial conditions on beam coordinates, rotations, and linear and angular velocities. Several solution procedures can be devised for this problem and have been detailed in Hodges (2006). In this work, we consider three of them, which are discussed in the following sections. They differ on the selection of independent DoFs in the solution process, that is, whether one considers displacements, strains, or simultaneously velocities and internal forces. Each of them offers some advantages with respect to the others, and presenting them together allows switching between them depending on the question that needs answering at any given point in the analysis of a very flexible structure. First, we need to derive expressions for the kinematic relations under small perturbations of the displacement field.

8.3.2   Infinitesimal Beam Kinematics

Hamilton's principle results in the weak (or variational) formulation of the (infinite-dimensional) structural dynamics problem. Galerkin projection methods (Reddy, 2002, Ch. 6) can then be used to obtain finite-dimensional approximations to that problem, suitable for computer simulation. This implies the selection of a suitable set of test functions that approximate the spatial dependency of both the variables and their variations on Equation (8.12). However, since the displacement, velocities, and strains are linked through the kinematic relations, the corresponding variations are not independent of each other. The variations of the kinematic variables are computed here by introducing small perturbations in Equations (8.2) and (8.3), as shown in this section.

Perturbation of the first of the strain–displacement kinematic relations, Equation (8.3), results in

$$
\begin{aligned}
\delta\gamma &= \mathbf{R}_{SB}\delta\mathbf{r}'_B + \delta\mathbf{R}_{SB}\mathbf{r}'_B = \mathbf{R}_{SB}\delta\mathbf{r}'_B - \widetilde{\delta\phi}_S\mathbf{R}_{SB}\mathbf{r}'_B \\
&= \mathbf{R}_{SB}\delta\mathbf{r}'_B + \widetilde{\mathbf{R}_{SB}\mathbf{r}'_B}\delta\phi_S,
\end{aligned}
\tag{8.13}
$$

where $\delta\phi_S$ has been introduced in Equation (8.10). Equally, perturbations of the rotation relation in Equation (8.3) yield

$$
\begin{aligned}
\delta\kappa &= \mathbf{R}_{SB}\delta\mathbf{R}'_{BS} + \delta\mathbf{R}_{SB}\mathbf{R}'_{BS} \\
&= \widetilde{\delta\phi}'_S - \mathbf{R}'_{SB}\delta\mathbf{R}_{BS} + \delta\mathbf{R}_{SB}\mathbf{R}'_{BS} \\
&= \widetilde{\delta\phi}'_S - \mathbf{R}'_{SB}\mathbf{R}_{BS}\mathbf{R}_{SB}\delta\mathbf{R}_{BS} + \delta\mathbf{R}_{SB}\mathbf{R}_{BS}\mathbf{R}_{SB}\mathbf{R}'_{BS} \\
&= \widetilde{\delta\phi}'_S + \widetilde{\mathbf{k}}_S\widetilde{\delta\phi}_S - \widetilde{\delta\phi}_S\widetilde{\mathbf{k}}_S,
\end{aligned}
\tag{8.14}
$$

where we have introduced the local curvature in material coordinates that was defined in Equation (8.4). Using the properties of the skew-symmetric matrices, Equation (C.23), we finally obtain closed-form relations between the variations of the force and moment strains and those of the positions and rotations along the beam, which can be written in a rather compact form as

$$
\begin{aligned}
\delta\gamma &= \mathbf{R}_{SB}\delta\mathbf{r}'_B + \mathbf{R}_{SB}\tilde{\mathbf{r}}'_B\mathbf{R}_{BS}\delta\phi_S, \\
\delta\kappa &= \delta\phi'_S + \tilde{\mathbf{k}}_S\delta\phi_S.
\end{aligned}
\tag{8.15}
$$

In a similar manner, we can obtain the relation between the variations of the inertial velocities along the beam and those of the local coordinates and rotations. Equation (8.2) also included the linear and angular velocities at the origin, for which we have obtained linearized relations in Equation (6.11). After some algebraic manipulation, the infinitesimal velocity–displacement kinematic relations become

$$
\begin{aligned}
\delta\bar{\mathbf{v}}_S(\eta,t) &= \mathbf{R}_{SB}(\eta,t)\delta\dot{\mathbf{r}}_B(\eta,t) \\
&\quad + \mathbf{R}_{SB}(\eta,t)\left(\tilde{\omega}_B(t)\delta\mathbf{r}_B(\eta,t) - \tilde{\mathbf{r}}_B(\eta,t)\delta\omega_B(t) + \delta\mathbf{v}_B(t)\right) \\
&\quad + \tilde{\bar{\mathbf{v}}}_S(\eta,t)\delta\phi_S(\eta,t), \\
\delta\bar{\omega}_S(\eta,t) &= \delta\dot{\phi}_S(\eta,t) + \tilde{\bar{\omega}}_S(\eta,t)\delta\phi_S(\eta,t) + \mathbf{R}_{SB}(\eta,t)\delta\omega_B(t),
\end{aligned}
\tag{8.16}
$$

where we have indicated when the velocities at a generic point have explicit dependency on the curvilinear coordinate η and time t so to differentiate from those at the reference $\eta = 0$. Those reference velocities in Equation (8.16) include additional dependencies compared with those in Equation (8.15). Both expressions can now be substituted in the weak form of the EoMs, Equation (8.12), which, after integration by parts in time, results in

$$
\int_0^{t_f}\int_\eta \delta\mathbf{r}_B^T\left[\mathbf{R}_{BS}\left(\dot{\mathbf{p}}_S + \tilde{\omega}_S\mathbf{p}_S - \mathbf{f}_S\right) + \delta\mathbf{r}_B'^\top\mathbf{R}_{BS}\mathfrak{f}_S\right]d\eta\,dt
$$

$$
+ \int_0^{t_f}\int_\eta \delta\phi_S^T\left[\left(\dot{\mathbf{h}}_S + \tilde{\omega}_S\mathbf{h}_S + \tilde{\bar{\mathbf{v}}}_S\mathbf{p}_S - \tilde{\mathbf{k}}_S\mathbf{m}_S - \mathbf{R}_{SB}\tilde{\mathbf{r}}'_B\mathbf{R}_{BS}\mathfrak{f}_S - \bar{\mathbf{m}}_S\right) + \delta\phi'^\top_S\mathbf{m}_S\right]d\eta\,dt
$$

$$+ \int_0^{t_f} \delta \mathbf{p}_B^\top \int_\eta \left[\dot{\mathbf{p}}_B + \tilde{\omega}_B \mathbf{p}_B - \bar{\mathbf{f}}_B \right] \mathrm{d}\eta$$

$$+ \int_0^{t_f} \delta \varphi_B^\top \int_\eta \left[\left(\tfrac{\mathrm{d}}{\mathrm{d}t} + \tilde{\omega}_B \right) (\mathbf{h}_B + \tilde{\mathbf{r}}_B \mathbf{p}_B) + \tilde{\mathbf{v}}_B \mathbf{p}_B + \bar{\mathbf{m}}_B + \tilde{\mathbf{r}}_B \bar{\mathbf{f}}_B \right] \mathrm{d}\eta \, \mathrm{d}t$$

$$= \left\{ \int_\eta \left[\delta \mathbf{r}_B^\top \mathbf{R}_{BS} \mathbf{p}_S + \delta \phi_S^\top \mathbf{h}_S \right] \mathrm{d}\eta + \delta \mathbf{p}_B^\top \int_\eta \mathbf{p}_B \, \mathrm{d}\eta + \delta \varphi_B^\top \int_\eta (\mathbf{h}_B + \tilde{\mathbf{r}}_B \mathbf{p}_B) \, \mathrm{d}\eta \right\} \Bigg|_0^{t_f}.$$

$$(8.17)$$

The first two lines in Equation (8.17) determine the local instantaneous force and moment balance on the beam, the third and fourth lines define the rigid-body dynamics of the reference frame B, while the right-hand side of the equations gives the initial and terminal conditions to the problem. After integration by parts in the spatial coordinate, the first two lines in Equation (8.17) result in the *strong form* of the beam equations in the material frame, written as

$$\dot{\mathbf{p}}_S + \tilde{\omega}_S \mathbf{p}_S = \mathbf{f}_S' + \tilde{\mathbf{k}}_S \mathfrak{f}_S + \bar{\mathbf{f}}_S,$$
$$\dot{\mathbf{h}}_S + \tilde{\omega}_S \mathbf{h}_S + \tilde{\mathbf{v}}_S \mathbf{p}_S = \mathbf{m}_S' + \tilde{\mathbf{k}}_S \mathbf{m}_S + (\tilde{\mathbf{e}}_1 + \tilde{\gamma}) \mathfrak{f}_S + \bar{\mathbf{m}}_S.$$

$$(8.18)$$

The integration by parts also results in the natural boundary conditions on the internal forces and/or displacements of the structure, which are not explicitly included here. The equations for geometrically nonlinear beam dynamics in this form were first derived by Simó and Vu-Quoc (1988). For static problems, where only the right-hand side of both equations needs to be retained, the original derivation is due to Reissner (1973).

Equation (8.18) is valid independently of whether the B frame of reference is stationary or not, since it defines local conditions and it is written in material coordinates (i.e., frame B does not explicitly appear). In the general case of nonzero velocities of the vehicle axes (i.e. a free-flying aircraft), the third and fourth lines in Equation (8.17) determine the rigid-body dynamics of the structure, which are written in a strong form as

$$\left(\tfrac{\mathrm{d}}{\mathrm{d}t} + \tilde{\omega}_B \right) \int_\eta \mathbf{p}_B \, \mathrm{d}\eta = \int_\eta \bar{\mathbf{f}}_B \, \mathrm{d}\eta,$$
$$\left(\tfrac{\mathrm{d}}{\mathrm{d}t} + \tilde{\omega}_B \right) \int_\eta (\mathbf{h}_B + \tilde{\mathbf{r}}_B \mathbf{p}_B) \, \mathrm{d}\eta = \int_\eta \left(\bar{\mathbf{m}}_B + \tilde{\mathbf{r}}_B \bar{\mathbf{f}}_B \right) \mathrm{d}\eta - \tilde{\mathbf{v}}_B \int_\eta \mathbf{p}_B \, \mathrm{d}\eta.$$

$$(8.19)$$

As one should expect, these expressions are actually the rigid-body equations introduced in Chapter 4, Equation (4.24), with the integral expressions in Equation (8.19) being the evaluation of the instantaneous total momenta and applied forces/moments on the vehicle. Note that we track here the vehicle axes, B, which is a body-attached frame of reference that does not necessarily coincide with the CM of the vehicle. Consequently, the last term, $\tilde{\mathbf{v}}_B(t) \int_\eta \mathbf{p}_B \, \mathrm{d}\eta$, has been retained, as we also did in Equation (6.30).

The instantaneous orientation of the vehicle axes with respect to the Earth frame is obtained by integrating its angular velocities using, for example, Equation (4.16). Subsequent integration in time of the translational velocity, $\mathbf{v}_B(t)$, defines the trajectory followed by the origin of the B frame. Finally, the instantaneous coordinates of the CM of the aircraft (measured from the origin of the vehicle axes), $\mathbf{r}_B^{CM}(t)$, can be easily determined in postprocess from the first moment of inertia, that is,

$$\mathbf{r}_B^{CM}(t) = \frac{1}{\int_\eta m(\eta)\,d\eta} \int_\eta \left(\mathbf{r}_B(\eta,t) + \mathbf{R}_{BS}(\eta,t)\mathfrak{y}_{cg}(\eta) \right) m(\eta)\,d\eta. \qquad (8.20)$$

In this section, we have expressed the geometrically nonlinear dynamics of a generic beam assembly in terms of PDEs in the curvilinear coordinate along the reference line and the physical time, Equation (8.18), and also in its equivalent variational form, Equation (8.17). To obtain a numerical solution of this problem, we need next a spatial discretization into a finite-dimensional set of DoFs. The following three sections discuss three alternative approaches that have been proposed for application to very flexible aircraft dynamics, as well as their relative merit.

8.4 Finite-Element Solution Using Displacements and Rotations

The finite-element method is the spatial discretization strategy most commonly used in structural mechanics. For geometrically nonlinear beams, an excellent description, with an extensive literature survey, can be found in Géradin and Cardona (2001) and that reference is followed here. The starting point is the weak form of the EoMs, Equation (8.17), from which a two-step solution strategy follows: first, the selection of a set of independent DoFs along the beam; and second, a Galerkin projection of Equation (8.17) that results in a discrete form of the equations. They are briefly summarized here.

Independent DoFs. First, the instantaneous translational and angular velocities of the body-fixed reference frame (the vehicle axes), $\mathbf{v}_B(t)$ and $\boldsymbol{\omega}_B(t)$, respectively, are used to describe the rigid-body dynamics. This choice is consistent with the parameterization used in the description of both the rigid-body dynamics of aircraft, Equation (4.26), and the dynamics of flexible aircraft with small deformations, Equation (6.30). As was noted in Section 6.2, the reference point on the aircraft does not need to coincide with the velocities at the instantaneous CM. Note however that should the kinematics of the CM of the vehicle be necessary (e.g., by the flight control system), it can still be easily obtained as part of the solution using Equation (8.20).

Second, the instantaneous position vector along the reference line, $\mathbf{r}_B(\eta,t)$, is used to define the current shape of the deforming structure, while a parameterization of rotations is also needed to determine the orientation of the local S reference frame with respect to the vehicle axis, B. Multiple alternative descriptions of finite rotations have been proposed (Euler angles, which we introduced in Section 4.2.2, Bryant angles, quaternions, Rodrigues parameters, etc.), and our description here is based on the *Cartesian rotation vector*, $\boldsymbol{\psi}(\eta,t)$. This parameterization only uses three DoFs and, for sufficiently small values, coincides with the usual linear rotations (see Appendix C.2 for further details). The Cartesian rotation vector writes a coordinate transformation in terms of the *exponential map*, Equation (1.8), as

$$\mathbf{R}_{BS} = \mathbf{R}(\psi) = \exp \tilde{\psi} = \sum_{k=0}^{\infty} \frac{1}{k!} \tilde{\psi}^k. \tag{8.21}$$

This expression defines a unique transformation between the Cartesian rotation vector and its corresponding coordinate transformation matrix for $-\pi < \|\psi\| \leq \pi$. This map is periodic outside this range in multiples of 2π. While this may be an issue in certain applications, it is rarely a concern in problems with nonlinear beam dynamics, as most built-up structures (not to mention aircraft wings!) very rarely fold upon themselves. As with all parameterizations of finite rotations, the Cartesian rotation vector is not additive (i.e., one cannot directly sum rotation vectors to define a compound rotation).

We finally need to express all the rotational variables in the EoMs, namely angular velocity, curvature and virtual rotations, in terms of the Cartesian rotation vector and its derivatives. Noting that it is, by definition, $\widetilde{\delta\phi}_S = \mathbf{R}_{SB}\delta\mathbf{R}_{BS}$, and using the parameterization of the coordinate rotation matrix that has just been introduced in Equation (8.21), one can obtain a closed-form relation between the virtual rotations and the variations of the Cartesian rotation vector. This is done by means of the *tangential rotation operator*, whose derivation can be found in Appendix C.3, and which is written as

$$\mathbf{T}(\psi) = \sum_{k=0}^{\infty} \frac{(-1)^k}{(k+1)!} \tilde{\psi}^k, \tag{8.22}$$

and satisfies $\delta\phi_S = \mathbf{T}(\psi)\delta\psi$. Similar relations are obtained for the local angular velocity, Equation (8.2), which can now be written as

$$\bar{\omega}_S(\eta,t) = \mathbf{T}(\psi(\eta,t))\dot{\psi}(\eta,t) + \mathbf{R}^\top(\psi(\eta,t))\omega_B(t), \tag{8.23}$$

and the local curvature, Equation (8.4), which becomes $\mathbf{k}_S = \mathbf{T}(\psi)\psi'$. These expressions can now be substituted into all the rotation-dependent terms in Equation (8.17) to write all sectional forces, moments, and momenta solely in terms of the positions and rotations along the reference line, and their time and spatial derivatives.

Galerkin projection. Finite elements approximate the spatial dependency of the independent DoFs using test functions with local support commonly known as *shape functions*. This means that the structure (here, the beam reference line) is split into elements (segments) where the unknown variables are approximated between the values at discrete points, known either as *nodes* or *grid points*. If $\boldsymbol{\xi}_d$ is the column vector with all the nodal DoFs (i.e., the deformed coordinates and the components of the Cartesian rotation vector at each node), the following spatial interpolation can be defined:

$$\begin{Bmatrix} \mathbf{r}_B(\eta,t) \\ \psi(\eta,t) \end{Bmatrix} \approx \begin{bmatrix} \mathbf{N}_r(\eta) \\ \mathbf{N}_\psi(\eta) \end{bmatrix} \boldsymbol{\xi}_d(t), \tag{8.24}$$

where $\mathbf{N}_r$ and $\mathbf{N}_\psi$ are the matrices of shape functions that define the local interpolation within each element from a global array of DoFs. Here, three-noded beam elements are chosen, which are then used to approximate all the integrals in space in Equation (8.17). The same interpolation is also used for the virtual displacements and rotations in Equation (8.17). As a result, we have $\delta\mathbf{r}_B(\eta,t) = \mathbf{N}_r(\eta)\delta\boldsymbol{\xi}_d(t)$ for the virtual displacements, while, with $\delta\phi_S = \mathbf{T}(\psi)\delta\psi$ from Equation (C.18), we have $\delta\phi_S(\eta,t) = \mathbf{T}(\mathbf{N}_\psi(\eta)\boldsymbol{\xi}_d(t))\mathbf{N}_\psi(\eta)\delta\boldsymbol{\xi}_d(t)$ for the virtual rotations, which results in symmetric expressions in the discretized equations.

There are known issues with the interpolation of rotations introduced here. In particular, this interpolation does not guarantee the *objectivity* of the formulation, that is, the invariance of the strain field under rigid-body rotations. While more complex interpolation schemes have been proposed by, for example, Crisfield and Jelenic (1999) and Bauchau et al. (2008), quadratic elements have been shown to provide sufficiently good performance for applications in flexible aircraft dynamics, as shown in Example 8.1.

Using this interpolation strategy, all spatial integrals in Equation (8.17) can be computed element by element and finally assembled into a set of nonlinear algebraic equations in terms of the nodal coordinates and rotations and their time derivatives. Details of that process can be found in Géradin and Cardona (2001) or in Hesse (2013) for the particular implementation used in the examples in this book; they are not repeated here. It results in the following discrete form of the EoMs:

$$\mathbf{M}_d(\boldsymbol{\xi}_d)\begin{Bmatrix}\dot{\mathbf{v}}_B\\\dot{\boldsymbol{\omega}}_B\\\ddot{\boldsymbol{\xi}}_d\end{Bmatrix} + \begin{Bmatrix}\mathbf{f}_{\mathrm{gyr},B}\\\mathbf{m}_{\mathrm{gyr},B}\\\boldsymbol{\chi}_{\mathrm{gyr}}\end{Bmatrix} + \begin{Bmatrix}\mathbf{0}\\\mathbf{0}\\\boldsymbol{\chi}_{\mathrm{stif}}(\boldsymbol{\xi}_d)\end{Bmatrix} = \mathbf{M}_d(\boldsymbol{\xi}_d)\begin{Bmatrix}\mathbf{g}_B\\\mathbf{0}\\\mathbf{0}\end{Bmatrix} + \begin{Bmatrix}\mathbf{f}_B\\\mathbf{m}_B\\\boldsymbol{\chi}_d\end{Bmatrix}, \quad (8.25)$$

where $\mathbf{M}_d$ is the discrete mass matrix associated with the choice of DoFs, the discrete gyroscopic forcing is given by $\mathbf{f}_{\mathrm{gyr},B}(\mathbf{v}_B,\boldsymbol{\omega}_B,\dot{\boldsymbol{\xi}}_d,\boldsymbol{\xi}_d)$ and $\mathbf{m}_{\mathrm{gyr},B}(\mathbf{v}_B,\boldsymbol{\omega}_B,\dot{\boldsymbol{\xi}}_d,\boldsymbol{\xi}_d)$ on the rigid-body DoFs and $\boldsymbol{\chi}_{\mathrm{gyr}}(\mathbf{v}_B,\boldsymbol{\omega}_B,\dot{\boldsymbol{\xi}}_d,\boldsymbol{\xi}_d)$ on the elastic ones, and $\boldsymbol{\chi}_{\mathrm{stif}}$ is the discrete stiffness forcing. Note that the stiffness terms only depend on the elastic DoFs, while the gyroscopic terms depend on both the rigid-body velocities and the instantaneous shape. As we did in Equation (6.30), we have split the forcing terms into gravitational loading, with $\mathbf{g}_B$ being defined as in Equation (6.29), and the remaining external loads on the airframe defined as

$$\mathbf{f}_B = \int_\eta \bar{\mathbf{f}}_B\,\mathrm{d}\eta,$$

$$\mathbf{m}_B = \int_\eta \left(\bar{\mathbf{m}}_B + \tilde{\mathbf{r}}_B\bar{\mathbf{f}}_B\right)\mathrm{d}\eta, \quad (8.26)$$

$$\boldsymbol{\chi}_d = \int_\eta \mathbf{N}_r^\top\bar{\mathbf{f}}_B\,\mathrm{d}\eta + \int_\eta \mathbf{N}_\psi^\top\mathbf{T}^\top(\mathbf{N}_\psi\boldsymbol{\xi}_d)\bar{\mathbf{m}}_S\,\mathrm{d}\eta,$$

where the tangent operator appears in the nodal moments to transform the virtual work associated with the virtual rotations $\delta\phi_S(\eta,t)$ in Equation (8.17) to that of the virtual Cartesian rotation vector. Note also that $\mathbf{f}_B(t)$ has units of force, while $\bar{\mathbf{f}}_B(\eta,t)$ is a force per unit length along the beam reference line, and similarly for the moment $\mathbf{m}_B$ and the corresponding moment per unit lengh, $\bar{\mathbf{m}}_B$. For the aeroelastic problems that are considered later in this chapter, these forcing terms are defined in part by the aerodynamic forces which are also dependent on the deformed shape of the structure.

The solution of Equation (8.25) needs to be performed in conjunction with the propagation equations, Equation (6.31), that determine the instantaneous orientation of the vehicle axes (here, the body-attached reference frame B) with respect to the Earth frame. This process is the most common used in the numerical solution of structures that can be modeled as geometrically nonlinear beams. It has a number of advantages. First, it solves the problem of DoFs that are directly measurable, while dealing with dead and follower forces with a similar computational effort. Second, it easily enables additional kinematic constraints to construct flexible multibody systems (Géradin and Cardona, 2001). Third, its linearized form has the same structure as the equations presented in Chapter 6. Finally, there exist a number of efficient time-domain integration algorithms (Newmark, HHT, Generalized-α, etc.) that are readily applicable to this problem. The main disadvantage of this approach is that having nodal rotations as independent DoFs results in numerically inefficient algorithms. This is for two reasons: First, because it needs a relatively large number of operations to build all the functionals in Equation (8.25); and second, because the convergence rate of Newton–Raphson-type solution algorithms slows down as rotations increase.

Nonlinear Rigid-Linear Elastic Dynamics about Nonlinear Equilibrium

One often needs to investigate the dynamics of a preloaded structure around a reference (in general, nonlinear) static equilibrium. Here, we consider the particular case, which is typical of flexible aircraft dynamics, in which the elastic DoFs show a small-amplitude (dynamic) response, while the rigid-body ones can be arbitrarily large. This is, therefore, a generalization of the situation considered in Section 6.3 to the case when the reference state is a deformed structure with large deformations. Considering therefore $\boldsymbol{\xi}_d = \boldsymbol{\xi}_0 + \Delta \boldsymbol{\xi}_d$ in Equation (8.25), we have

$$
\mathbf{M}_d(\boldsymbol{\xi}_0) \begin{Bmatrix} \dot{\mathbf{v}}_B \\ \dot{\boldsymbol{\omega}}_B \\ \Delta \ddot{\boldsymbol{\xi}}_d \end{Bmatrix} + \mathbf{D}(\mathbf{v}_B, \boldsymbol{\omega}_B, \boldsymbol{\xi}_0) \begin{Bmatrix} \mathbf{v}_B \\ \boldsymbol{\omega}_B \\ \Delta \dot{\boldsymbol{\xi}}_d \end{Bmatrix} + \mathbf{K}(\mathbf{v}_B, \boldsymbol{\omega}_B, \boldsymbol{\xi}_0) \begin{Bmatrix} \mathbf{0} \\ \mathbf{0} \\ \Delta \boldsymbol{\xi}_d \end{Bmatrix}
$$
$$
= \mathbf{M}_d(\boldsymbol{\xi}_0) \begin{Bmatrix} \mathbf{g}_B \\ \mathbf{0} \\ \mathbf{0} \end{Bmatrix} + \begin{Bmatrix} \Delta \mathbf{f}_B \\ \Delta \mathbf{m}_B \\ \Delta \boldsymbol{\chi}_d \end{Bmatrix}.
\tag{8.27}
$$

This is the linearization of the elastic DoFs when the external forces are dead loads. If they also include follower forces, one also needs to include the perturbations of the rotation matrix that determine the direction of the applied forces in the nonlinear equilibrium. As in Equation (6.21), the definition of the gyroscopic matrix is not unique. Full details of the construction of the mass, damping, and stiffness terms in the equation can be found in Hesse and Palacios (2012).

The structure of Equation (8.27) is very similar to that obtained for the nonlinear rigid-linear elastic description of the dynamics of a flexible aircraft around an undeformed reference condition, Equation (6.32). The main differences are that the equilibrium is obtained here around a deformed geometry and that the stiffness matrix (which is no longer constant) depends on the instantaneous rigid-body velocity, from evaluating $\frac{\partial \chi_{\text{gyr}}}{\partial \boldsymbol{\xi}_d}$. It can be easily seen that the gyroscopic matrix, $\mathbf{D}$, is linear with the rigid-body velocities (as in Equation (6.30)) and that the stiffness matrix function, $\mathbf{K}$, is quadratic on those velocities (as a result of the centrifugal loads).

As before, Equation (8.27) needs to be solved together with the propagation equations, Equation (6.31).

Linearized Dynamics about Constant Reference

The fully linearized form of the equations around an arbitrary but constant condition is obtained by considering small pertubations of the rigid-body velocities, as we did in Section 6.3.3. We consider a reference condition with, in general, nonzero velocities and deformations. This results in

$$
\mathbf{M}_d(\boldsymbol{\xi}_0) \begin{Bmatrix} \Delta\dot{\mathbf{v}}_B \\ \Delta\dot{\boldsymbol{\omega}}_B \\ \Delta\ddot{\boldsymbol{\xi}}_d \end{Bmatrix} + \mathbf{D}_0(\mathbf{v}_{B0}, \boldsymbol{\omega}_{B0}, \boldsymbol{\xi}_0) \begin{Bmatrix} \Delta\mathbf{v}_B \\ \Delta\boldsymbol{\omega}_B \\ \Delta\dot{\boldsymbol{\xi}}_d \end{Bmatrix} + \mathbf{K}_0(\mathbf{v}_{B0}, \boldsymbol{\omega}_{B0}, \boldsymbol{\xi}_0) \begin{Bmatrix} \mathbf{0} \\ \mathbf{0} \\ \Delta\boldsymbol{\xi}_d \end{Bmatrix}
$$
$$
= \mathbf{G}_0 \begin{Bmatrix} \Delta\theta \\ \Delta\phi \end{Bmatrix} + \begin{Bmatrix} \Delta\mathbf{f}_B \\ \Delta\mathbf{m}_B \\ \Delta\boldsymbol{\chi}_d \end{Bmatrix},
$$
(8.28)

which has an identical structure to Equation (6.35), with the only difference being that the stiffness is now the tangent stiffness matrix obtained from the discrete generalized forces as $\frac{\partial \boldsymbol{\chi}_{\text{stif}}}{\partial \boldsymbol{\xi}_d}$. This is because the equations describing the linearized dynamics of a more flexible aircraft about a nonzero reference are indeed those considered in Chapter 6 for less flexible aircraft – if the reference inertial properties are defined around the deformed shape instead of the undeformed shape, as we did there.

It is finally convenient to consider a projection in modal coordinates of the dynamics under small-amplitude elastic deformations (still under nonlinear static equilibrium conditions). This was presented in Section 6.4, and the analysis and the discussion are again valid here, with the projection now being on either Equation (8.27) or Equation (8.28). As the mass and stiffness change with the equilibrium point, the LNMs of the structure used in the projection are different for each equilibrium point (Hesse and Palacios, 2012).

Example 8.1 Static Equilibrium of a Cantilever under Prescribed Forces. This first example was introduced by Géradin and Cardona (2001). It is a massless prismatic cantilever beam of constant properties and length $b = 5$ m. A diagonal sectional inverse compliance matrix is assumed, as in Equation (8.7), with stiffness constants $EA = 4.8 \times 10^8$ N, $GA = 3.231 \times 10^8$ N (in both axes), $GJ = 1.0 \times 10^6$ N·m^2, and $EI = 9.346 \times 10^6$ N·m^2 (also in both axes). We consider first the static equilibrium when the beam is initially in a horizontal position along the x axis and a heavy mass is attached to its free end, as shown in Figure 8.6. This results in a vertical *dead load* P as the only external force applied on the beam.

Table 8.1 shows the displacements at the free end of the beam for $P = 600$ kN. Results are shown for different discretizations, including two- and three-noded elements, and are compared with converged results from the literature. It can be seen that about 10 three-noded elements, or equivalently 20 two-noded elements, give sufficiently accurate results, but as few as 2 three-noded elements already approximate the

Table 8.1 Displacements and rotations at the free end under a dead force $P = 600$ kN.

Model	Δx (m)	Δz (m)	ψ_2 (rad)
5 two-noded elements	0.586	−2.147	−0.6745
10 two-noded elements	0.594	−2.156	−0.6726
20 two-noded elements	0.596	−2.159	−0.6722
2 three-noded elements	0.550	−2.070	−0.6576
5 three-noded elements	0.589	−2.144	−0.6700
10 three-noded elements	0.596	−2.159	−0.6719
(Géradin and Cardona, 2001)		−2.159	−0.6720

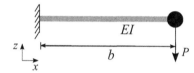

Figure 8.6 Horizontal prismatic cantilever beam under a tip vertical dead load P.

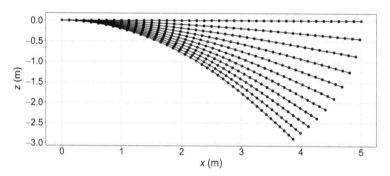

Figure 8.7 Deformed beam under tip dead load with its magnitude ranging from 0 to 1,000 kN, in increments of 100 kN. [08NLBeams/StaticCantilever.ipynb]

solution well. The deformed shapes for increasing tip dead loads up to $P = 1,000$ kN are shown in Figure 8.7. Those results are obtained with a 20 three-noded elements, and all 41 nodes are included in the results shown in the figure. It can be observed that, as the load increases, the resulting changes on beam deformations are increasingly smaller. It is even more apparent that the tip of the beam traces an arc in the x–z plane. This is because if it were to move only on the vertical direction, as linear theory predicts, the beam would substantially increase its length under such large forces.

This is made more apparent in Figure 8.8, where the displacements of the beam tip at its free end ($\eta = b$) are plotted against the tip force P. The two plots in the figure correspond to the horizontal and vertical components of the displacement at $\eta = b$, as seen from an observer at $\eta = 0$. For the situation considered above, identified as *dead load* in the figure, the vertical displacement increases almost linearly until it reaches around 20% of the span (where $P \approx 250$ kN), after which the beam deformation increases less

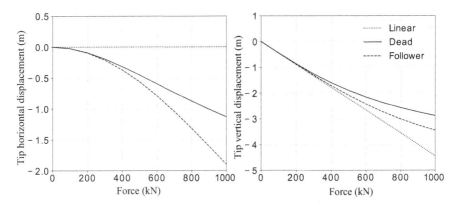

Figure 8.8 Displacements at the beam tip as a function of the tip force.

with the loading. The figure also includes the linear results that do not predict horizontal displacements and are only valid for a small range of forces. A final set of results have been included, corresponding to the tip load P acting as a *follower force*. Follower forces are such that the force acts always in the same direction in the material frame, S. In this example, this corresponds to the force P remaining normal to the beam axis as the beam deforms. This is in contrast to *dead forces* that always have the same direction in an inertial frame (e.g., gravitational forces). Follower forces generate larger displacements than dead ones on the beam, as an increasingly large proportion of the dead force is sustained by the axial stiffness of the beam as the beam deforms. As a result, the vertical displacement under a follower force is very close to the linear one for amplitudes up to around 30% of the span – although for such high loads, the horizontal displacements are already nonnegligible and need a nonlinear model. Note finally that under follower forces, the beam eventually curves backward, that is, the free end starts to point toward the root (although not included in the figures, it occurs at $P \approx 1600$ kN in this example). In aircraft problems, *gravity produces a dead force, while propulsion and aerodynamic forces are always follower ones.*

In both test cases in this example, the beam is subjected to planar deformations in the x–z plane. This has effectively simplified the problem, as the Cartesian rotation vector only has one nonzero component at every node. Indeed, the results in Table 8.1 may lead us to think that the choice between two- and three-noded elements has little impact on the results, and even that the two-noded element slightly outperforms the three-noded one.[2] However, it is rarely the case that complex aircraft-type structures display planar deformations. Wings simultaneously bend and twist, which results in nonlinear couplings for sufficiently large deformations. This is numerically investigated in Table 8.2, in which the cantilever beam considered above is first rotated within the x–y plane, so as to define a nonzero value of the initial Cartesian rotation

---

[2] Remember that a model with 100 two-noded elements has the same DoFs as the one with 50 three-noded elements.

Table 8.2 Tip rotations in the plane of the undeformed beam for varying discretizations and azimuth orientations under a follower force $P = 3{,}000$ kN.

Azimuth	Discretization	ψ_1 (rad)	ψ_2 (rad)	ψ_3 (rad)
0°	10 three-noded elements	0	-2.7553	0
90°	10 three-noded elements	2.339×10^{-5}	-2.7553	-1.621×10^{-4}
180°	10 three-noded elements	4.858×10^{-6}	-2.7553	-3.010×10^{-4}
0°	50 three-noded elements	0	-2.7614	0
90°	50 three-noded elements	5.577×10^{-8}	-2.7614	-4.085×10^{-7}
180°	50 three-noded elements	1.408×10^{-8}	-2.7614	-6.938×10^{-7}
0°	100 two-noded elements	0	-2.7613	0
90°	100 two-noded elements	9.625×10^{-5}	-2.7613	-1.024×10^{-3}
180°	100 two-noded elements	-3.812×10^{-5}	-2.7612	-1.336×10^{-3}

vector, and then it is subjected to a follower force ($P = 3{,}000$ kN) at its free end. The results are then written with respect to a reference frame in which the x–z axes are in the plane of the deformation (a *push-back operation*). This allows us to assess the lack of objectivity in the interpolation of rotations that we have described in the beginning of this section.

Both two- and three-noded elements are used in the study. When the azimuth angle is zero, the Cartesian rotation vector has only one nonzero component, and small but nonnegligible errors can be seen to appear when the problem is solved on a rotated beam. For the two-noded elements, this error remains around 0.1% even for a rather fine discretization. The convergence of the quadratic interpolation scheme with the number of elements is much faster that of the two-noded elements and the error is soon within the tolerance used for the convergence of the solver. It is clear that the quadratic interpolation on the three-noded element gives a much better approximation than the linear interpolation on the two-noded element for problems involving large 3-D rotations.

Example 8.2 "Free-Flying" Very Flexible Beam. As an example of problems with both elastic and rigid-body motions, we consider a free-flying flexible beam in the absence of gravity (a runaway spaghetti from the lunchbox of an astronaut!). This problem was first investigated by Simó and Vu-Quoc (1988), and the results here have been computed by Hesse and Palacios (2012) using SHARPy (see Section 9.4). The unsupported, initially inclined, prismatic isotropic beam of Figure 8.9 is considered. It has the same dimensionless bending and transverse shear stiffness in both axes, $EI = 500$ and $GA = 10{,}000$, respectively. This beam is subjected to two different sets of dead forces and moments on one end, which are also included in the figure.

The first load case defines a 2-D problem in the plane of the undeformed beam. Snapshots of the instantaneous shape of the beam on increments of $\Delta t = 0.5$ are shown

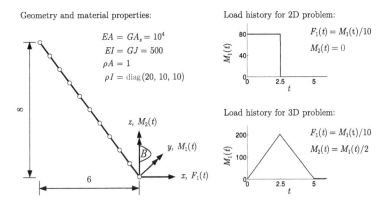

Geometry and material properties:

$EA = GA_s = 10^4$
$EI = GJ = 500$
$\rho A = 1$
$\rho I = \text{diag}(20, 10, 10)$

Load history for 2D problem:

$F_1(t) = M_1(t)/10$
$M_2(t) = 0$

Load history for 3D problem:

$F_1(t) = M_1(t)/10$
$M_2(t) = M_1(t)/2$

Figure 8.9 Geometry, stiffness and inertial properties, and load histories for the flying beam problem (reprinted from Hesse and Palacios (2012), with permission of Elsevier).

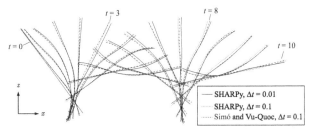

Figure 8.10 Snapshots of the flying beam under the 2-D loading, in time increments of 0.5 (reprinted from Hesse and Palacios (2012), with permission of Elsevier).

in Figure 8.10. Results were obtained with 10 two-noded elements and for two different time steps. First, $\Delta t = 0.1$ is chosen as in the original simulation parameters of Simó and Vu-Quoc (1986), and an excellent agreement between both solutions has been found. Further investigation has shown, however, that a smaller time, $\Delta t = 0.01$, is necessary to achieve a converged solution, and the corresponding results are also included in the figure. This smaller time step is necessary to capture the fourth-bending mode of the beam, which is excited by these impulsive loads in the absence of structural damping in the model.

The second loading in Figure 8.9 also includes a moment $M_2(t)$ acting along the inertial z axis. This introduces 3-D kinematics in the beam response, for which snapshots of the instantaneous shape are shown in Figure 8.11. Results are obtained using SHARPy with 10 and 20 two-noded equal elements along the beam, and a time step $\Delta t = 0.01$. They are also compared with those of Hsiao et al. (1999), who used a corotational formulation with 10 elements and the same time step. An excellent comparison can be observed. Finally, as the external force on the beam always points along the x direction, the instantaneous position of the CM of the beam will follow a simple path that is shown in Figure 8.12 together with the instantaneous beam shape.

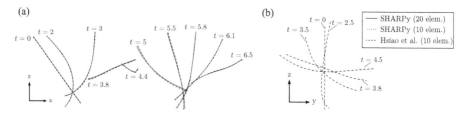

Figure 8.11 Snapshots of the flying beam under 3-D loading: (a) $x-z$ plane and (b) $y-z$ plane (reprinted from Hesse and Palacios (2012), with permission of Elsevier).

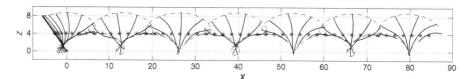

Figure 8.12 Snapshots of the flying beam in the $x-z$ plane under the 3-D loading. Dash lines are the trajectories of the beam ends and stars of its CM. Snapshots with $\Delta t = 0.4$ up to $t = 20$ (reprinted from Hesse and Palacios (2012), with permission of Elsevier).

8.5   Finite-Element Solution Using Element Strains

Equation (8.25) gives a numerical solution for generic structural dynamics problem that can be described by geometrically nonlinear composite beams. However, as mentioned in Section 8.4, the solution in rotations often results in slow convergence, which is particularly difficult for static problems with large deformations. Moreover, the resulting discrete equations can be hard to manipulate, for instance, to investigate approximated solutions. In such instances, alternate solution approaches offer some distinct advantages. We consider two such methods in this book, namely a strain-based formulation and a mixed formulation known as the intrinsic theory of beams. The first one is outlined in this section, while the second one is the focus of Section 8.6.

Computational advantages can be obtained if the geometrically nonlinear beam equations are solved using the force and moment strains, which have been defined in Equation (8.3), as the independent variables. This procedure was first introduced by Cesnik and Brown (2002) using geometrical arguments on the basis of the "elastica" theory – a nice history of it can be found in Levien (2008). While not necessary, the initial formulation does not include transverse shear DoFs in the beam constitutive equation due to the intended application of the formulation: aircraft wing and full aircraft configurations. An alternative derivation, also including shear strains, was presented in Palacios et al. (2010). Since it was based on the analytical integration of the strain–displacement relations, as described in Equation (8.3), we start with it here for consistency. We further assume for simplicity in the derivations that the beam has no initial curvature in its undeformed configuration, although that condition can be easily included as shown in Cesnik and Brown (2002). Under that assumption, and introducing the rotation operator of Equation (8.21), Equation (8.3) can be rewritten as

$$\mathbf{r}'_B(\eta,t) = \mathbf{R}_{BS}(\eta,t)\left(\boldsymbol{\gamma}(\eta,t)+\mathbf{e}_1\right),$$
$$\tfrac{\mathrm{d}}{\mathrm{d}s}\mathbf{R}_{BS}(\eta,t) = \mathbf{R}_{BS}(\eta,t)\tilde{\boldsymbol{\kappa}}(\eta,t). \tag{8.29}$$

We assume a piecewise constant discretization of the strain along the reference line, such that $\boldsymbol{\gamma}(\eta,t) = \boldsymbol{\gamma}_n^*(t)$ and $\boldsymbol{\kappa}(\eta,t) = \boldsymbol{\kappa}_n^*(t)$ in $\eta_{n-1} < \eta \le \eta_n$, with η_{n-1} representing the arclength coordinate at the beginning of the nth element. Within each segment, bending displacements are, therefore, approximated as quadratic functions, while torsional rotations are approximated as linear ones. The result is a scheme that allows us to represent complex deformations with fewer number of states. The local rotations can then be obtained by integrating the second set of equations, Equation (8.29), within an element, that is,

$$\mathbf{R}_{BS}(\eta,t) = \mathbf{R}_{BS}(\eta_n,t)e^{\tilde{\boldsymbol{\kappa}}_n^*(t)\Delta\eta}, \tag{8.30}$$

where, for convenience, we have defined $\Delta\eta = \eta - \eta_{n-1}$ as the incremental arclength within the nth element. Using Equation (C.10), we can define the following operator

$$\mathcal{H}_n^0(\eta,t) = e^{\tilde{\boldsymbol{\kappa}}_n^*(t)\Delta\eta}$$
$$= \mathbf{I} + \frac{\sin\left(\kappa_n^*\Delta\eta\right)}{\kappa_n^*}\tilde{\boldsymbol{\kappa}}_n^*(t) + \frac{1-\cos\left(\kappa_n^*\Delta\eta\right)}{\left(\kappa_n^*\right)^2}\tilde{\boldsymbol{\kappa}}_n^*(t)\tilde{\boldsymbol{\kappa}}_n^*(t), \tag{8.31}$$

with $\kappa_n^* = \|\boldsymbol{\kappa}_n^*\|$, the squared norm of the (constant) moment strain within the element. Similarly, for the displacements, integrating the first set of equations in Equation (8.29) and using the results from Equation (8.30) yields

$$\mathbf{r}_B(\eta,t) - \mathbf{r}_B(\eta_{n-1},t) = \left(\int_{\eta_{n-1}}^{\eta} \mathbf{R}_{BS}(\eta,t)\,\mathrm{d}\eta\right)\left(\boldsymbol{\gamma}_n^*(t)+\mathbf{e}_1\right), \tag{8.32}$$

where the strain has been brought out of the integral as it is constant within the element. The solution for the resulting integral term is now directly obtained using Equation (8.30), that is,

$$\int_{\eta_{n-1}}^{\eta} \mathbf{R}_{BS}(\eta,t)\,\mathrm{d}\eta = \mathbf{R}_{BS}(\eta_{n-1},t)\int_{\eta_{n-1}}^{\eta}\mathcal{H}_n^0(\eta,t)\,\mathrm{d}\eta = \mathbf{R}_{BS}(\eta_{n-1},t)\mathcal{H}_n^1(\eta,t), \tag{8.33}$$

with

$$\mathcal{H}_n^1(\eta,t) = \mathbf{I}\Delta\eta + \frac{1-\cos\left(\kappa_n^*\Delta\eta\right)}{\left(\kappa_n^*\right)^2}\tilde{\boldsymbol{\kappa}}_n(t) + \frac{\kappa_n^*\Delta\eta - \sin\left(\kappa_n^*\Delta\eta\right)}{\left(\kappa_n^*\right)^3}\tilde{\boldsymbol{\kappa}}_n^*(t)\tilde{\boldsymbol{\kappa}}_n^*(t). \tag{8.34}$$

Therefore, integrating Equation (8.29) yields

$$\mathbf{r}_B(\eta,t) = \mathbf{r}_B(\eta_{n-1},t) + \mathbf{R}_{BS}(\eta_{n-1},t)\mathcal{H}_n^1(\eta,t)\left(\boldsymbol{\gamma}_n^*(t)+\mathbf{e}_1\right),$$
$$\mathbf{R}_{BS}(\eta,t) = \mathbf{R}_{BS}(\eta_{n-1},t)\mathcal{H}_n^0(\eta,t), \tag{8.35}$$

for $\eta_{n-1} < \eta \le \eta_n$ and $1 \le n \le N$. Equation (8.35) describes the kinematics of the reference line using constant-strain elements. For given values of the position and orientation at the origin $\eta = 0$ and an instantaneous distribution of force and moment strains along the beam(s), it gives a recursive procedure to obtain the position and orientation in the nth element as

$$\mathbf{r}_B(\eta,t) = \mathbf{r}_B(0,t) + \sum_{i=1}^{n-1} \mathbf{R}_{BS}(\eta_{i-1},t)\mathcal{H}_i^1(\eta_i,t)\left(\boldsymbol{\gamma}_i^*(t) + \mathbf{e}_1\right)$$

$$+ \mathbf{R}_{BS}(\eta_{n-1},t)\mathcal{H}_n^1(\eta,t)\left(\boldsymbol{\gamma}_n^*(t) + \mathbf{e}_1\right), \tag{8.36}$$

$$\mathbf{R}_{BS}(\eta,t) = \mathbf{R}_{BS}(0,t)\mathcal{H}_1^0(\eta_1,t)\mathcal{H}_2^0(\eta_2,t)\cdots\mathcal{H}_{n-1}^0(\eta_{n-1},t),$$

for, again, $\eta_{n-1} < \eta \le \eta_n$ and $1 \le n \le N$. Any discontinuities that may be present in the curvature of the initial reference line can be included as constant rotations at the corresponding nodes (Su and Cesnik, 2010), but they have not been explicitly included above. Taking variations on Equation (8.36) in terms of the discrete strain variables gives the virtual displacements and rotations along the beam. They can be written as

$$\delta\mathbf{r}_B(\eta,t) = \mathcal{J}_{r\gamma}(\boldsymbol{\kappa}^*,\eta)\delta\boldsymbol{\gamma}^* + \mathcal{J}_{r\kappa}(\boldsymbol{\gamma}^*,\boldsymbol{\kappa}^*,\eta)\delta\boldsymbol{\kappa}^*,$$
$$\delta\mathbf{R}_{BS}(\eta,t) = \mathcal{J}_{R\kappa}(\boldsymbol{\kappa}^*,\eta)\delta\boldsymbol{\kappa}^*, \tag{8.37}$$

where we have introduced $\boldsymbol{\gamma}^{*\top} = \left\{\boldsymbol{\gamma}_1^{*\top}\,\boldsymbol{\gamma}_2^{*\top}\cdots\boldsymbol{\gamma}_N^{*\top}\right\}$ as the vector of the discrete force strains and $\boldsymbol{\kappa}^{*\top} = \left\{\boldsymbol{\kappa}_1^{*\top}\,\boldsymbol{\kappa}_2^{*\top}\cdots\boldsymbol{\kappa}_N^{*\top}\right\}$ for the discrete moment strains. The various Jacobian operators, $\mathcal{J}$, can be obtained analytically as a function of $\mathcal{H}^0$, $\mathcal{H}^1$, and the geometry of the problem, as shown in detail in Appendix E of Shearer (2006). Similarly, we can obtain expressions for the time derivative of components of the position vector in the aircraft system and the local rotation matrix as

$$\dot{\mathbf{r}}_B(\eta,t) = \mathcal{J}_{r\gamma}(\boldsymbol{\kappa}^*,\eta)\dot{\boldsymbol{\gamma}}^* + \mathcal{J}_{r\kappa}(\boldsymbol{\gamma}^*,\boldsymbol{\kappa}^*,\eta)\dot{\boldsymbol{\kappa}}^*,$$
$$\ddot{\mathbf{r}}_B(\eta,t) = \mathcal{J}_{r\gamma}(\boldsymbol{\kappa}^*,\eta)\ddot{\boldsymbol{\gamma}}^* + \dot{\mathcal{J}}_{r\gamma}(\boldsymbol{\kappa}^*,\eta)\dot{\boldsymbol{\gamma}}^* + \mathcal{J}_{r\kappa}(\boldsymbol{\gamma}^*,\boldsymbol{\kappa}^*,\eta)\ddot{\boldsymbol{\kappa}}^* + \dot{\mathcal{J}}_{r\kappa}(\boldsymbol{\gamma}^*,\boldsymbol{\kappa}^*,\eta)\dot{\boldsymbol{\kappa}}^*,$$
$$\dot{\mathbf{R}}_{BS}(\eta,t) = \mathcal{J}_{R\kappa}(\boldsymbol{\kappa}^*,\eta)\dot{\boldsymbol{\kappa}}^*,$$
$$\ddot{\mathbf{R}}_{BS}(\eta,t) = \mathcal{J}_{R\kappa}(\boldsymbol{\kappa}^*,\eta)\ddot{\boldsymbol{\kappa}}^* + \dot{\mathcal{J}}_{R\kappa}(\boldsymbol{\kappa}^*,\eta)\dot{\boldsymbol{\kappa}}^*. \tag{8.38}$$

If now we consider again Hamilton's principle, Equation (8.17), but without expanding the terms associated with the internal (strain) energy, we obtain

$$\int_0^{t_f}\int_\eta \delta\mathbf{r}_B^\top\left[\mathbf{R}_{BS}\left(\dot{\mathbf{p}}_S + \tilde{\boldsymbol{\omega}}_S\mathbf{p}_S\right)\right]\mathrm{d}\eta\,\mathrm{d}t$$

$$+ \int_0^{t_f}\int_\eta \delta\boldsymbol{\phi}_S^\top\left[\left(\dot{\mathbf{h}}_S + \tilde{\boldsymbol{\omega}}_S\mathbf{h}_S + \tilde{\mathbf{v}}_S\mathbf{p}_S\right)\right]\mathrm{d}\eta\,\mathrm{d}t$$

$$+ \int_0^{t_f}\delta\mathbf{p}_B^\top\int_\eta\left[\dot{\mathbf{p}}_B + \tilde{\boldsymbol{\omega}}_B\mathbf{p}_B - \bar{\mathbf{f}}_B\right]\mathrm{d}\eta\,\mathrm{d}t \tag{8.39}$$

$$+ \int_0^{t_f}\delta\boldsymbol{\varphi}_B^\top\int_\eta\left[\left(\frac{\mathrm{d}}{\mathrm{d}t} + \tilde{\boldsymbol{\omega}}_B\right)(\mathbf{h}_B + \tilde{\mathbf{r}}_B\mathbf{p}_B) + \tilde{\mathbf{v}}_B\mathbf{p}_B + \bar{\mathbf{m}}_B + \tilde{\mathbf{r}}_B\bar{\mathbf{f}}_B\right]\mathrm{d}\eta\,\mathrm{d}t$$

$$+ \int_0^{t_f}\int_\eta\left[\delta\boldsymbol{\gamma}^\top\mathfrak{f}_S + \delta\boldsymbol{\kappa}^\top\mathfrak{m}_S\right]\mathrm{d}\eta\,\mathrm{d}t = \int_0^{t_f}\int_\eta\left[\delta\mathbf{r}_B^\top\mathbf{R}_{BS}\bar{\mathfrak{f}}_S + \delta\boldsymbol{\phi}_S^\top\bar{\mathfrak{m}}_S\right]\mathrm{d}\eta\,\mathrm{d}t,$$

where the terms in the first two lines correspond to the elastic kinetic energy and the terms in the third and fourth lines to the kinetic energy due to the motion of the body frame B, both already integrated by parts in time. The terms in the fifth line correspond to the original potential energy terms and are in the desired form for this formulation. The right-hand side terms are associated with the work done on the elastic DoFs by the external forces.

With the relations obtained in Equations (8.36)–(8.38), we can express all the terms in Equation (8.39) as a function of the strains and curvatures of the beam. The only term that still needs attention is $\delta\phi_S$. However, from Equation (C.14), we have established that

$$\widetilde{\delta\phi_S}^\top = \delta\mathbf{R}_{BS}\mathbf{R}_{BS}^\top.\tag{8.40}$$

Taking advantage of the inner product equality $\boldsymbol{\alpha}^\top\boldsymbol{\beta} = -\frac{1}{2}\mathrm{tr}(\widetilde{\boldsymbol{\alpha}}\widetilde{\boldsymbol{\beta}})$, Equation (C.23), the term in the integral in the second line of Equation (8.39) becomes

$$\delta\phi_S^\top\left[\left(\dot{\mathbf{h}}_S + \widetilde{\boldsymbol{\omega}}_S\mathbf{h}_S + \widetilde{\mathbf{v}}_S\mathbf{p}_S\right)\right] = -\frac{1}{2}\mathrm{tr}\left[\delta\mathbf{R}_{BS}\mathbf{R}_{BS}^\top\left(\widetilde{\dot{\mathbf{h}}_S} + \widetilde{\widetilde{\boldsymbol{\omega}}_S\mathbf{h}_S} + \widetilde{\widetilde{\mathbf{v}}_S\mathbf{p}_S}\right)\right],$$

which can be finally substituted back, such that Equation (8.39) is now entirely a function of the strains and body-frame motion. Therefore, defining $\boldsymbol{\xi}_s^\top = \{\boldsymbol{\gamma}^{*\top}\ \boldsymbol{\kappa}^{*\top}\}$ as the column vector with all the element strain-independent variables, the corresponding strain-based equilibrium equations can be written as

$$\mathbf{M}_s(\boldsymbol{\xi}_s)\begin{Bmatrix}\dot{\mathbf{v}}_B\\\dot{\boldsymbol{\omega}}_B\\\ddot{\boldsymbol{\xi}}_s\end{Bmatrix} + \mathbf{D}_s(\mathbf{v}_B,\boldsymbol{\omega}_B,\boldsymbol{\xi}_s,\dot{\boldsymbol{\xi}}_s)\begin{Bmatrix}\mathbf{v}_B\\\boldsymbol{\omega}_B\\\dot{\boldsymbol{\xi}}_s\end{Bmatrix} + \begin{Bmatrix}\mathbf{0}\\\mathbf{0}\\\mathbf{K}_s\boldsymbol{\xi}_s\end{Bmatrix} = \mathbf{M}_s(\boldsymbol{\xi}_s)\begin{Bmatrix}\mathbf{g}_B\\\mathbf{0}\\\mathbf{0}\end{Bmatrix} + \begin{Bmatrix}\mathbf{f}_B\\\mathbf{m}_B\\\boldsymbol{\chi}_s\end{Bmatrix}.$$
$$(8.41)$$

Note that in Section 8.5, the generalized mass, $\mathbf{M}_s$, and damping, $\mathbf{D}_s$, matrices, as well as the generalized forces, $\boldsymbol{\chi}_s$, now depend, in general, on the DoFs and the Jacobians introduced in Equation (8.38). However, the generalized stiffness matrix, $\mathbf{K}_s$, is a constant matrix composed of sectional compliance matrices, $\mathcal{C}(\eta)$, along the discrete elements of the beam reference line. This is a direct result of the expression for the variation of internal strain energy, Equation (8.9), when strains are the independent variables (instead of displacements). This is one of the greatest advantages of the strain-based formulation: For nonlinear static solutions, only one generalized stiffness matrix inversion (actually decomposition, in a practical numerical sense) is needed to solve any number of static problems. This is particularly useful when identifying multiple trim solutions to be used in advance of a numerical simulation or flutter analysis (e.g., Section 9.4.1). It is also advantageous for low-frequency dynamic situations, dominated by the stiffness terms. Another advantage of the strain-based formulation is its ability to model complex deformations with fewer number of states. This is particularly useful when using it in connection to control design and simulation. Finally, in this approach, local displacements and rotations are obtained as a postprocessing step. For additional details on the strain-based formulation, the reader is referred to Su and Cesnik (2011b). There are also several applications of the method that can be found in Cesnik and Brown (2002), Cesnik and Su (2005), Shearer and Cesnik (2007), Lupp and Cesnik (2019), Sanghi et al. (2022), and Riso and Cesnik (2023).

8.6   Nonlinear Modal Solution Using Intrinsic Variables

As we have just seen, we can solve the geometrically nonlinear composite beam equations in terms of the instantaneous strain field, with rotations and displacements, if needed, later computed from integration along the reference line. This can be taken further if we consider not only spatial derivatives but also time derivatives of displacements as primary variables, which results in what is known as an *intrinsic formulation*. Here, the solution is sought simultaneously in terms of the internal forces and moments along the beam (i.e., the spatial derivatives of the position and rotation) and the inertial linear and angular velocities (their time derivatives). This results in a solution process that uses twice as many independent DoFs as Section 8.5, but which, as we show in what follows, results in a very elegant formulation with dramatically reduced algorithmic complexity. Such two-field descriptions are often referred to as a mixed or *hybrid formulation* in the solid mechanics literature, and the one presented here was initially proposed by Hodges (2003).

Consider first Equation (8.18), namely the strong form of the equations of motion of a geometrically nonlinear beam. In that expression, we can first write the instantaneous local curvature as $\mathbf{k}_s(\eta,t) = \boldsymbol{\kappa}(\eta,t) + \mathbf{k}_0(\eta)$ using Equation (8.3), with $\mathbf{k}_0 = \mathbf{R}_{SB}(\eta,0)\mathbf{R}'_{BS}(\eta,0)$ the initial curvature at $t=0$. Introducing then the constitutive relations, Equation (8.8), and grouping variables, both force and moment equations can be written in a compact form as (Palacios, 2011)

$$\mathcal{M}\dot{\mathbf{x}}_1 - \mathbf{x}'_2 - \mathsf{E}\mathbf{x}_2 + \mathcal{L}_1(\mathbf{x}_1)\mathcal{M}\mathbf{x}_1 + \mathcal{L}_2(\mathbf{x}_2)\mathcal{C}\mathbf{x}_2 = \mathbf{f}_{\text{ext}}, \qquad (8.42)$$

where we have defined two state vectors $\mathbf{x}_1(\eta,t)$ and $\mathbf{x}_2(\eta,t)$ will be referred to as the velocity and the force states, respectively, and a forcing term as

$$\mathbf{x}_1 = \begin{Bmatrix} \bar{\mathbf{v}}_S \\ \bar{\boldsymbol{\omega}}_S \end{Bmatrix}, \qquad \mathbf{x}_2 = \begin{Bmatrix} \mathsf{f}_S \\ \mathsf{m}_S \end{Bmatrix} \qquad \text{and} \qquad \mathbf{f}_{\text{ext}} = \begin{Bmatrix} \bar{\mathbf{f}}_S \\ \bar{\mathbf{m}}_S \end{Bmatrix}, \qquad (8.43)$$

as well as the linear operators

$$\mathcal{L}_1\left(\begin{Bmatrix} \mathbf{a} \\ \mathbf{b} \end{Bmatrix}\right) = \begin{bmatrix} \tilde{\mathbf{b}} & \mathbf{0} \\ \tilde{\mathbf{a}} & \tilde{\mathbf{b}} \end{bmatrix} \qquad \text{and} \qquad \mathcal{L}_2\left(\begin{Bmatrix} \mathbf{a} \\ \mathbf{b} \end{Bmatrix}\right) = \begin{bmatrix} \mathbf{0} & \tilde{\mathbf{a}} \\ \tilde{\mathbf{a}} & \tilde{\mathbf{b}} \end{bmatrix}, \qquad (8.44)$$

with $\mathbf{a},\mathbf{b} \in \mathbb{R}^3$. The dependency on the initial curvature appears in the coefficient matrix $\mathsf{E} = \mathcal{L}_1\left(\begin{Bmatrix} \mathbf{e}_1 \\ \mathbf{k}_0 \end{Bmatrix}\right)$.

The second set of equations are obtained from the compatibility conditions between the spatial and time derivatives of the position and rotation along the beam. First, note that the angular velocity can be written as $\tilde{\bar{\boldsymbol{\omega}}}_S = \mathbf{R}_{SE}\dot{\mathbf{R}}_{ES}$, and the local curvature as

$\tilde{\mathbf{k}}_S = \mathbf{R}_{SE}\mathbf{R}'_{ES}$, with E being the Earth frame. Differentiating in space the first relation and in time the second one, they become

$$\tilde{\omega}'_S = \mathbf{R}'_{SE}\dot{\mathbf{R}}_{ES} + \mathbf{R}_{SE}\dot{\mathbf{R}}'_{ES},$$
$$\dot{\tilde{\mathbf{k}}}_S = \dot{\mathbf{R}}_{SE}\mathbf{R}'_{ES} + \mathbf{R}_{SE}\dot{\mathbf{R}}'_{ES}. \tag{8.45}$$

Subtracting the first equation from the second results in a relation between curvatures and angular velocities as

$$\dot{\tilde{\mathbf{k}}}_S = \tilde{\omega}'_S + \tilde{\mathbf{k}}_S\tilde{\omega}_S - \tilde{\omega}_S\tilde{\mathbf{k}}_S = \tilde{\omega}'_S + \widetilde{\tilde{\mathbf{k}}_S\tilde{\omega}_S}, \tag{8.46}$$

where we have used the properties given in Appendix C.4. This is a closed-form expression between the instantaneous curvature and angular velocity, and their derivatives in time and space, which, importantly, do not explicitly depend on the coordinate transformation matrix. In terms of the moment strains, this expression can also be written as

$$\dot{\kappa} = \bar{\omega}'_S + \tilde{\kappa}\bar{\omega}_S + \tilde{\mathbf{k}}_0\bar{\omega}_S. \tag{8.47}$$

Similarly, the translational velocity is $\bar{\mathbf{v}}_S = \mathbf{R}_{SE}\dot{\mathbf{r}}_E$, and the force strains is $\gamma = \mathbf{R}_{SE}\mathbf{r}'_E - \mathbf{e}_1$. Differentiating in space the velocity and in time the strain, we have

$$\bar{\mathbf{v}}'_S = \mathbf{R}'_{SE}\dot{\mathbf{r}}_E + \mathbf{R}_{SE}\dot{\mathbf{r}}'_E = \mathbf{R}'_{SE}\mathbf{R}_{ES}\bar{\mathbf{v}}_S + \mathbf{R}_{SE}\dot{\mathbf{r}}'_E = \tilde{\mathbf{k}}_S\bar{\mathbf{v}}_S + \mathbf{R}_{SE}\dot{\mathbf{r}}'_E,$$
$$\dot{\gamma} = \dot{\mathbf{R}}_{SE}\mathbf{r}'_E + \mathbf{R}_{SE}\dot{\mathbf{r}}'_E = \dot{\mathbf{R}}_{SE}\mathbf{R}_{ES}(\gamma + \mathbf{e}_1) + \mathbf{R}_{SE}\dot{\mathbf{r}}'_E = \tilde{\omega}_S(\gamma + \mathbf{e}_1) + \mathbf{R}_{SE}\dot{\mathbf{r}}'_E. \tag{8.48}$$

Substracting the first equation from the second, we have again a closed-form relation between strains and velocities, which, after introducing the moment strains, can be written as

$$\dot{\gamma} = \bar{\mathbf{v}}'_S + \tilde{\kappa}\bar{\mathbf{v}}_S + (\tilde{\gamma} + \tilde{\mathbf{e}}_1)\bar{\omega}_S + \tilde{\mathbf{k}}_0\bar{\mathbf{v}}_S. \tag{8.49}$$

Using the notation introduced for Equation (8.42), Equations (8.47) and (8.49) can be written in a compact form as

$$\mathcal{C}\dot{\mathbf{x}}_2 - \mathbf{x}'_1 + \mathbf{E}^\top\mathbf{x}_1 - \mathcal{L}_1^\top(\mathbf{x}_1)\mathcal{C}\mathbf{x}_2 = \mathbf{0}. \tag{8.50}$$

Equations (8.42) and (8.50) are 12 PDEs that uniquely define the evolution of the intrinsic DoFs, $\mathbf{x}_1(\eta,t)$ and $\mathbf{x}_2(\eta,t)$, given external forces, $\mathbf{f}_{ext}(\eta,t)$, initial conditions, $\mathbf{x}_1(\eta,0) = \mathbf{x}_{10}(\eta)$ and $\mathbf{x}_2(\eta,0) = \mathbf{x}_{20}(\eta)$, and boundary conditions. The boundary conditions for this problem are $\mathbf{x}_1(t,\eta_0) = \mathbf{0}$ if the structure is clamped at η_0, or $\mathbf{x}_2(t,\eta_0) = \mathbf{0}$ if the structure has a free end at η_0. As an example, for a cantilever beam of length b, the boundary conditions are $\mathbf{x}_1(0,t) = \mathbf{0}$ and $\mathbf{x}_2(b,t) = \mathbf{0}$.

The intrinsic equations can be seen as the natural extension to flexible beams of the Newton–Euler equations for a rigid body, Equation (4.26). As with the rigid-body equations, when written in a material (local) reference frame, they have Hamiltonian structure with only quadratic nonlinearities and no explicit dependency on the local coordinate transformation matrix. Therefore, the discretization of the equations guarantees the objectivity of finite strains under a change of reference frame. A final important consideration is that the problem is written in terms of the inertial velocities expressed in a material frame, and it does not explicitly use a global frame to track the vehicle rigid-body dynamics, such as the one shown in Figure 8.4. This means that there is no equivalent to Equation (8.19) in this formulation, and the dynamics are fully described using Equations (8.42) and (8.50) together with the appropriate initial and boundary conditions. The trajectory with respect to an inertial observer at any point of the flexible body can be independently obtained by integrating its velocity. As will be seen in the following section, the projection of the equations in modal coordinates reintroduces the rigid-body DoFs and brings back a similar structure to that of Sections 8.4 and 8.5.

For additional information about the intrinsic EoMs, solution procedures and their properties, the reader is referred to Hodges (2003), Palacios (2011), and Wynn et al. (2013). An extension of the intrinsic model for geometrically nonlinear composite beams that includes a generalized Kelvin–Voigt structural damping model has been proposed by Artola et al. (2021c). The focus in what follows is on the solution in modal coordinates, which will enable nonlinear control strategies in Chapter 10.

8.6.1   Nonlinear Beam Equations in Modal Coordinates

The intrinsic equations are particularly useful to relate the geometrically nonlinear dynamics of a composite beam to their linear vibration characteristics. This is achieved through the projection of the beam equations onto the LNMs of the structure, which have been introduced in Section 6.4. As a first step, we need to obtain the LNMs of a generic beam assembly using the intrinsic description. We note that the LNMs are a physical feature of the structure and do not vary with changes of parameterization in its mathematical description. In other words, one obtains the same LNMs, defined by their natural frequencies and the associated mode shapes, using either the displacement-based, the strain-based, or the intrinsic formulations.

Linearization of Equations (8.42) and (8.50) around the undeformed configuration in the absence of external forces (free-vibration conditions) gives

$$\begin{aligned}\mathcal{M}\Delta\dot{\mathbf{x}}_1 - \Delta\mathbf{x}_2' - \mathsf{E}\Delta\mathbf{x}_2 &= 0,\\ \mathcal{C}\Delta\dot{\mathbf{x}}_2 - \Delta\mathbf{x}_1' + \mathsf{E}^\top\Delta\mathbf{x}_1 &= 0,\end{aligned} \tag{8.51}$$

which, as before, need to be solved with initial and boundary conditions and for, in general, spatially varying matrices $\mathcal{M}$, $\mathcal{C}$, and E. While the spatial boundary conditions are fixed for a given structure, varying the initial conditions determines all possible trajectories of the free-vibrating structure in its configuration space. Define now the state vector $\Delta\mathbf{x} = \left\{\Delta\mathbf{x}_1^\top \quad \Delta\mathbf{x}_2^\top\right\}^\top$. As Equation (8.51) is linear and homogeneous, its solutions are linear combinations of solutions in its eigenspaces, that is,

$$\Delta\mathbf{x}(\eta,t) = \sum_k \boldsymbol{\phi}_k(\eta)e^{i\omega_k t}q_k, \tag{8.52}$$

where $\boldsymbol{\phi}_k(\eta)$ and ω_k are the complex eigenvectors (the mode shapes) and the real eigenvalues (the natural frequencies) of the linear system, respectively, and q_k are the (constant) complex modal amplitudes, whose value depends on the initial conditions.

Substituting Equation (8.52) into Equation (8.51) defines the generalized eigenvalue problem

$$\phi'_k = \mathbf{A}(i\omega_k)\phi_k, \tag{8.53}$$

with

$$\mathbf{A}(i\omega) = \begin{bmatrix} \mathsf{E}^T & i\omega\mathcal{C} \\ i\omega\mathcal{M} & -\mathsf{E} \end{bmatrix} \tag{8.54}$$

and the spatial boundary conditions of the problem. From the inspection of matrix $\mathbf{A}(i\omega)$, solutions of Equation (8.53) are of the form $\phi_k^\top(\eta) = \left\{ \phi_{1k}^\top \quad i\phi_{2k}^\top \right\}$, with $\phi_{1k}(\eta) \in \mathbb{R}^6$ being the velocity component of the modes and $\phi_{2k}(\eta) \in \mathbb{R}^6$ the corresponding force component. Equation (8.53) can be integrated by assuming that, at the origin of coordinates $\eta = 0$, it is $\phi_{1k}(0) = \alpha_k$ and $\phi_{2k}(0) = \beta_k$. Its solutions are then uniquely determined, except for a normalization factor, by 13 constants $(\alpha_k, \beta_k, \omega_k)$, which are determined by enforcing the boundary conditions of the problem. The convention is to normalize the velocity mode shapes with the mass matrix and force, which after substituting in to Equation (8.53) also results in the force modes being normalized with the compliance matrix, that is,

$$\int_\eta \phi_{1j}^\top \mathcal{M}\phi_{1k}\,\mathrm{d}\eta = \delta_{jk},$$
$$\int_\eta \phi_{2j}^\top \mathcal{C}\phi_{2k}\,\mathrm{d}\eta = \delta_{jk}, \tag{8.55}$$

with δ_{jk} being the Kronecker delta. Note that in the projections of Equation (8.51) as presented here, we have made no assumptions regarding the initial geometry of the beam assembly, including initial twist or curvature, or its cross-sectional properties (e.g., isotropic or anisotropic). Typical beam assemblies are, however, only piecewise differentiable, as they may include corners and junctions in the interconnection of different structural elements. This can be addressed in the computation of the LNMs by the piecewise integration of the previous equations, although in practice, the LNMs are more easily obtained from the finite-element solution of Section 8.4 and then written in terms of intrinsic variables through differentiation (Wang et al., 2015). Direct solutions of Equation (8.53) need, in general, to be sought numerically, although under simplified conditions, such as those of Example 8.3, analytical solutions can be obtained. Finally, while there are nominally an infinite number of vibration modes for a structure, in practice, the summation in Equation (8.52) is defined over a finite number of modes, N_m.

The LNMs defined by Equation (8.53) provide an excellent basis to build a finite-dimensional approximation of Equations (8.42) and (8.50). The state vector can be then approximated as $\mathbf{x}(\eta,t) = \mathrm{Re}(\sum_{k=1}^{N_m} \phi_k(\eta)q_k(t))$. To simplify the notation, we write $q_k = q_{1k} - iq_{2k}$, which gives real-number representations of the expansions as

$$\mathbf{x}_1(\eta,t) = \sum_{k=1}^{N_m} \phi_{1k}(\eta)q_{1k}(t) \quad \text{and} \quad \mathbf{x}_2(\eta,t) = \sum_{k=1}^{N_m} \phi_{2k}(\eta)q_{2k}(t). \tag{8.56}$$

As a result, the beam equations in the intrinsic modal coordinates become

$$\dot{\mathbf{q}}_1 = \mathbf{\Omega}\mathbf{q}_2 - \sum_\ell q_{1\ell}\mathbf{\Gamma}_1^\ell\mathbf{q}_1 - \sum_\ell q_{2\ell}\mathbf{\Gamma}_2^\ell\mathbf{q}_2 + \mathbf{f}_q,$$
$$\dot{\mathbf{q}}_2 = -\mathbf{\Omega}\mathbf{q}_1 + \sum_\ell q_{2\ell}\mathbf{\Gamma}_2^{\ell\top}\mathbf{q}_1, \tag{8.57}$$

where $f_{qk} = \int_\eta \boldsymbol{\phi}_{1k}^\top \mathbf{f}_{\text{ext}}\,d\eta$ are the generalized forces on the kth LNM, and $\mathbf{\Omega} =$ diag $(\omega_1,...,\omega_N)$ is the matrix of ordered natural angular frequencies, which may include zeros for a structure with rigid-body DoFs, as in Equation (6.43). The nonlinear coupling terms are given by generalized Christoffel symbols associated with the local velocity (time derivatives) and strain (spatial derivatives) of the reference line, which are defined, respectively, as

$$\Gamma_{1jk}^\ell = \int_\eta \boldsymbol{\phi}_{1j}^\top \mathcal{L}_1(\boldsymbol{\phi}_{1k})\mathcal{M}\boldsymbol{\phi}_{1\ell}\,d\eta,$$
$$\Gamma_{2jk}^\ell = \int_\eta \boldsymbol{\phi}_{1j}^\top \mathcal{L}_2(\boldsymbol{\phi}_{2k})\mathcal{C}\boldsymbol{\phi}_{2\ell}\,d\eta. \tag{8.58}$$

They are third-order tensors, thus defining quadratic nonlinearities between the modal coordinates. It can be further seen that each $\mathbf{\Gamma}_1^\ell$ is a skew-symmetric matrix (Wynn et al., 2013), and, as a result, the EoMs in intrinsic coordinates, Equation (8.57), describe the dynamics of geometrically nonlinear composite beams as a forced Hamiltonian system. This is an important property that can be used in the derivation of nonlinear control laws (van der Schaft and Jeltsema, 2014). Equation (8.57) can now be integrated in time using standard methods for first-order differential equations (e.g., Runge–Kutta). If a large number of modes is required to achieve spatial convergence, higher frequency modes can be easily residualized through time averaging of the corresponding equations (Wang et al., 2015), which become algebraic constraints. As in the strain-based solution of the previous section, Equation (8.57) does not explicitly solve for the displacement and rotation fields, which need to be obtained in postprocessing by either the integration of Equation (8.2) in time or Equation (8.3) in space. An efficient modal projection of the rotation has also been proposed by Artola et al. (2022) to embed the evaluation of the rotation field in the solution process of problems in which the external forces depend on the absolute orientation (e.g., gravity).

The expansion defined in Equation (8.56) has resulted in a finite-dimensional approximation of Equations (8.42) and (8.50). The same process would have been followed if $\boldsymbol{\phi}_{1k}$ and $\boldsymbol{\phi}_{2k}$ were chosen as another set of admissible test functions instead of the LNMs of the structure, with the only difference being that the matrix $\mathbf{\Omega}$ in Equation (8.56) would no longer be diagonal. In particular, a discretization with piecewise constant strains, as shown in Section 8.5, and the linear interpolation of the velocity field between nodes have been proposed by Hodges (2003) and Patil and Hodges (2006), which result in an efficient numerical solution for very flexible aircraft dynamics. However, the modal basis is preferred here as it gives a natural link for mildly geometrically nonlinear problems (Palacios and Cea, 2019) to the standard linear finite-element solutions in aeroelasticity, which have been discussed in Section 6.6.

Finally, it is interesting to compare the mathematical structure of Equation (8.57) and those obtained for the displacement- and the strain-based formulations, Equation (8.25) and section 8.5, respectively. First, by construction, this is a first-order formulation and, therefore, it solves for twice as many DoFs as the previous two solutions. The resulting additional cost is somehow offset by simpler nonlinear terms that can be precomputed using Equation (8.58). If rigid-body motions are present, then the first six LNMs have zero frequency, and the first six equations in Equation (8.57) have a similar structure as the rigid-body equations in both Equation (8.25) and section 8.5. In particular, if the equations are applied to a rigid-body with zero compliance, then the compatibility equations in Equation (8.57) degenerate into $\dot{\mathbf{q}}_2 = \mathbf{0}$, which from the initial conditions implies that $\mathbf{q}_2(t) = \mathbf{0}$ for all times. As a result, the first line in Equation (8.57) simplifies to the Newton–Euler equations for a rigid body, Equation (4.26), written in the principal axes of inertia.

8.6.2   Energy Analysis

The intrinsic description also gives a natural representation to carry out energy balance studies of geometrically nonlinear beams. The instantaneous total energy of a beam assembly, $\mathcal{E}_0$, is obtained by adding the kinetic and strain energies, defined in Equation (8.5). In an intrinsic formulation, it is written as

$$\mathcal{E}_0(\mathbf{x}_1, \mathbf{x}_2) = \frac{1}{2} \int_\eta \mathbf{x}_1^\top \mathcal{M} \mathbf{x}_1 \, d\eta + \frac{1}{2} \int_\eta \mathbf{x}_2^\top \mathcal{C} \mathbf{x}_2 \, d\eta. \tag{8.59}$$

Using Equations (8.42) and (8.50), the energy dissipation rate is given by

$$\frac{d\mathcal{E}_0}{dt} = \int_\eta \mathbf{x}_1^\top \mathbf{f}_{\text{ext}} \, d\eta. \tag{8.60}$$

As one should expect, this equation implies that the total energy remains invariant if there are no external forces, that is, in the free vibrations of the structure, regardless of whether their amplitude is small or large (i.e., linear and nonlinear dynamics, respectively). It also shows that the total energy decays with time if the external force is proportional and opposite to the local velocity (a viscous damping term), and it can also be used to identify other dissipation mechanisms in the external forces. After projection to modal coordinates, Equation (8.56), the total energy of the truncated system is

$$\mathcal{E}_{N_m}(\mathbf{q}_1, \mathbf{q}_2) = \frac{1}{2} \left(\mathbf{q}_1^\top \mathbf{q}_1 + \mathbf{q}_2^\top \mathbf{q}_2 \right), \tag{8.61}$$

that is, the instantaneous total energy in a modal projection of the beam dynamics (with a finite number of modes) is equal to the square norm of its modal amplitudes. Note that since a truncation in modal coordinates is considered, it is always $\mathcal{E}_{N_m} \leq \mathcal{E}_0$. The corresponding expression for the energy dissipation rate is

$$\frac{d\mathcal{E}_{N_m}}{dt} = \mathbf{q}_1^\top \mathbf{f}_q. \tag{8.62}$$

It is important to remark that the previous expressions are valid for beam assemblies that display, in general, rigid-body dynamics and large amplitude (geometrically nonlinear) vibrations. These conservation laws will facilitate the design of nonlinear controllers in Chapter 10. Additional conservation laws can be demonstrated with LNMs defined around nonlinear equilibrium conditions, as shown in Wynn et al. (2013).

Example 8.3 Large-amplitude Vibrations of a Cantilever Beam. Equation (8.57) provides a very convenient framework to investigate nonlinear beam dynamics. As an example, consider a cantilever beam of total arclength b (whose boundary conditions are $\mathbf{x}_1(0,t) = \mathbf{0}$ and $\mathbf{x}_2(b,t) = \mathbf{0}$) and with constant cross-sectional mass, $\mathcal{M}$, and compliance, $\mathcal{C}$, matrices. No further assumptions are made yet regarding initial curvature, $\mathbf{k}_0$, or the sectional mass and stiffness, which may be full (symmetric) matrices. In such a case, Equation (8.53) can be analytically integrated as a function of the values of the (still unknown) eigenvectors at $\eta = 0$ and the associated natural angular frequencies, ω_k. If we define, as before, $\phi_{1k}(0) = \alpha_k$ and $\phi_{2k}(0) = \beta_k$, the integration of Equation (8.53) along the beam from $\eta = 0$ gives the kth LNM as

$$\begin{Bmatrix} \phi_{1k}(\eta) \\ \phi_{2k}(\eta) \end{Bmatrix} = e^{\mathbf{A}(i\omega_k)\eta} \begin{Bmatrix} \alpha_k \\ \beta_k \end{Bmatrix}, \tag{8.63}$$

where the exponential matrix was defined in Equation (1.8). Enforcing now the boundary conditions and mass normalization gives sufficient algebraic equations to determine the LNMs. In particular, the clamped condition at $\eta = 0$ results in $\alpha_k = 0$ for all k, and the free-end condition at $\eta = b$ becomes

$$\begin{Bmatrix} \phi_{1k}(b) \\ \mathbf{0} \end{Bmatrix} - e^{\mathbf{A}(i\omega_k)b} \begin{Bmatrix} \mathbf{0} \\ \beta_k \end{Bmatrix} = \begin{Bmatrix} \mathbf{0} \\ \mathbf{0} \end{Bmatrix}, \tag{8.64}$$

which defines 12 equations in $\phi_{1k}(b)$, β_k, and ω_k. Including one of the normalization conditions of Equation (8.55) closes the problem, but note that only the bottom six equations in Equation (8.64) need to be solved to determine the LNMs. For beams with initial curvature, or elastic or inertial couplings (nondiagonal $\mathcal{C}$ or $\mathcal{M}$, respectively), analytical solutions can be sought using symbolic algebra packages. They simplify to the classical solutions for Euler–Bernoulli beams when $\mathbf{k}_0 = 0$ and the mass and compliance matrices are given by

$$\mathcal{M} = \text{diag}\left(\rho A, \rho A, \rho A, \rho I_1, 0, 0\right), \tag{8.65a}$$

$$\mathcal{C} = \text{diag}\left(\tfrac{1}{EA}, 0, 0, \tfrac{1}{GJ}, \tfrac{1}{EI_2}, \tfrac{1}{EI_3}\right). \tag{8.65b}$$

The solutions to Equations (8.64) and (8.65) can be identified as axial (a), torsional (t), and out-of-plane (o) and in-plane (p) bending modes. After the normalization of Equation (8.55), they can be written as

$$\omega_k^a = \sqrt{\frac{E}{\rho}} \frac{\nu_k}{b}, \qquad\qquad \beta_k^a = \sqrt{\frac{2EA}{b}}(1,0,0,0,0,0)^\top, \qquad (8.66a)$$

$$\omega_k^t = \sqrt{\frac{GJ}{\rho I_1}} \frac{\nu_k}{b}, \qquad\qquad \beta_k^t = \sqrt{\frac{2GJ}{b}}(0,0,0,1,0,0)^\top, \qquad (8.66b)$$

$$\omega_k^o = \sqrt{\frac{EI_2}{\rho A}} \frac{\lambda_k^2}{b^2}, \qquad\qquad \beta_k^o = \sqrt{\frac{4EI_2}{b}}\left(0,0,-\frac{\lambda_k \Lambda_k}{b},0,1,0\right)^\top, \qquad (8.66c)$$

$$\omega_k^p = \sqrt{\frac{EI_3}{\rho A}} \frac{\lambda_k^2}{b^2}, \qquad\qquad \beta_k^p = \sqrt{\frac{4EI_3}{b}}\left(0,\frac{\lambda_k \Lambda_k}{b},0,0,0,1\right)^\top, \qquad (8.66d)$$

with $\nu_k = \frac{2k-1}{2}\pi$, λ_k being the positive roots of $\cos\lambda_k \cosh\lambda_k + 1 = 0$, and $\Lambda_k = \frac{\cos\lambda_k + \cosh\lambda_k}{\sin\lambda_k + \sinh\lambda_k}$. It is, in particular, $\lambda_1 = 0.597\pi$, $\lambda_2 = 1.494\pi$, and $\lambda_k \approx \nu_k$ for $k > 2$. Note that the natural frequencies go, as expected, as the square root of the ratio between sectional stiffness and inertia, but that as the beam length b is increased, they decrease quadratically for the bending modes and linearly for the axial and torsional modes.

Substituting this solution into Equation (8.63) with $\alpha_k = 0$, we can retrieve the mass-normalized mode shapes in intrinsic coordinates. For example, the axial modes, after manipulation, can be written as

$$\phi_{1k}^a = -\sqrt{\frac{\rho}{E}}\beta_k^a \sin(\nu_k \tfrac{\eta}{b}) \quad \text{and} \quad \phi_{2k}^a = \beta_k^a \cos(\nu_k \tfrac{\eta}{b}). \qquad (8.67)$$

As a numerical example, consider a prismatic thin-walled cantilever beam with $E = 10^6$, $\nu = 0.3$, and $\rho = 1$, and a rectangular cross section. This problem was first studied by Wang et al. (2015). The nondimensional length of the beam is $b = 20$, and the cross sections have width $b/20$, height $b/200$, and walls of thickness $b/2,000$. Its low-frequency free-vibration characteristics are therefore well approximated by an Euler–Bernoulli beam model, and the analytical solution above is used to compute its LNMs. Using those modes, Equation (8.57) can be built to describe its geometrically nonlinear response. In particular, we will consider free vibrations for a parabolic initial velocity distribution, given as $x_1(\eta,0) = x_{10}(\eta/b)^2$, where x_{10} is a constant that will define the different test cases. An explicit fourth-order Runge–Kutta is used to solve Equation (8.57) for 25 units of time, with a time step $\Delta t = 0.02$ and no structural damping. Sectional velocities are then obtained from the modal amplitudes using Equation (8.56), and they are finally integrated only at the points of interest using Equation (8.2). Figure 8.13 shows the velocities and displacements at the free end ($\eta = b$) for very small initial velocities, $x_{10} = (0; 0.002; 0.002; 0; 0; 0)$. The solution uses 10 modes and is in perfect agreement with a conventional (linear) finite-element solution, which is also included in the figure. Only out-of-plane displacements and velocities are nonzero.

Geometrically nonlinear effects become clearly visible as we increase the amplitude x_0 of the initial velocities. This can be seen in Figure 8.14 that shows both the tip displacements and velocities for $x_{10} = (0; 2; 2; 0; 0; 0)$. The maximum value of the displacements is around 25% of the beam length, b. A key difference with the linear problem is that a larger modal basis is needed for convergence. This is shown in Figure 8.14, which compares three different cases: (1) a nonlinear simulation with the 10 modes previously used in the linear problem; (2) a second simulation with 20

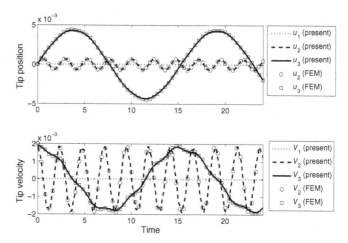

Figure 8.13 Three components of the displacements and velocities at the free end. Small initial conditions with parabollic shape and amplitude $\mathbf{x}_{10} = (0; 0.002; 0.002; 0; 0; 0)$ (Wang et al., 2015).

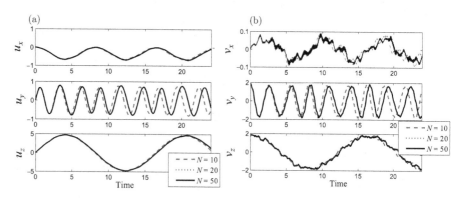

Figure 8.14 Displacements and velocities at $\eta = b$ for large initial conditions, $\mathbf{x}_{10} = (0; 2; 2; 0; 0; 0)$. N is the number of LNMs in the intrinsic model (Wang et al., 2015). (a) Displacements (in global coordinates) and (b) velocities (in material coordinates).

modes, which critically include now the first two axial modes; and (3) a third case with 50 modes, which can be considered converged. Results show a shift in frequency as nonlinear terms are captured, which is the characteristic of large-amplitude dynamics (see Figure 8.1). Importantly, the larger basis is not needed to capture a larger frequency content in the response (as will be the case in a linear setup), but it is needed to approximate the large-amplitude deformations with a modal basis. In fact, it can also be shown analytically (Palacios, 2011) that the axial modes, which are irrelevant in the linear problem, are necessary to capture the nonlinear modal couplings of a 2-D prismatic isotropic beam.

We have seen in Figure 8.14 that converged results are obtained for 20 modes. Figure 8.15 shows the time histories of all the modal amplitudes for the problem. In particular, the force component of the modes, $\mathbf{q}_2$, is shown. Some modes are identified in the figure: Modes 1, 2, and 4 are the first three out-of-plane bending modes; modes

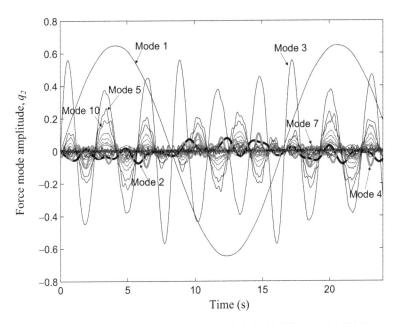

Figure 8.15 First 20 modal amplitudes for $\mathbf{x}_{10} = (0; 2; 2; 0; 0; 0)$ (Wang et al., 2015).

3 and 7 and the first two in-plane bending modes; and modes 5 and 10 are the first two torsional modes. The first few axial modes are also excited at high frequencies and small amplitudes and not easily appreciable in this plot. Significant nonlinear effects can be observed in the response. First, recall that this is a free-vibration problem (the only forcing is the nonzero initial conditions). In a linear system, this would have resulted in a set of independent harmonic oscillators at the natural frequencies of the system. Those can still be observed in the modes with largest amplitude in Figure 8.15, that is, the first out-of-plane and in-plane bending modes (modes 1 and 3 in the graph, respectively). However, multiple additional modes vibrate in sync with those dominant ones, as a result of the nonlinear modal couplings. Note, in particular, that the torsional modes display large-amplitude dynamics, even though they would not be excited in a linear problem. It can be clearly seen in the figure that their amplitude is proportional to the product of the amplitudes of modes 1 and 3. This is a finite-rotation effect that appears when large rotations in two axes generate a rotation on the third axis. Finally, it is possibe to see two modes showing much faster vibrations near the horizontal axis. These are axial modes that weakly enforce the nonextensionality conditions on the beam.

8.7   Summary

In this chapter, we have discussed the basic features of geometrically nonlinear composite beam theories. Using a general variational framework, we have introduced three alternative solution methods to the beam equations written in a moving frame of reference. First, the dynamics of a beam assembly has been described using a finite-element discretization with nodal displacement and finite rotations as DoFs. This was seen to result in a very generic formulation that naturally links to the linear elastic models of Chapter 6 and can be easily expanded to general flexible multibody systems with arbitrary constraints and internal mechanisms. While this is by far the most commonly found solution in the literature, it has a relatively high computational cost that can be reduced with a strain-based formulation. Such a solution process has been discussed next, and it has resulted in a formulation with a very similar mathematical structure to the displacement-based formulation but written in terms of the spatial derivatives of the beam displacements and rotations. This greatly simplifies the numerical solution process for problems with slow dynamics (slow changing in the inertia) and large (geometrically nonlinear) deformations. Finally, to further simplify the characterization of the inertia of the large deformations, we have introduced the intrinsic formulation, in which the geometrically nonlinear beam equations are solved in terms of both strains (internal forces) and inertial velocities along the reference line. This results in twice as many DoFs, but a much simplified mathematical description, which has been used to write the airframe dynamics in terms of the LNMs of the undeformed beam structure.

The geometrically nonlinear structural models introduced here will be used in the next two chapters to investigate the coupled nonlinear aeroelasticity–flight dynamics of very flexible aircraft and to the proposed suitable nonlinear control strategies.

8.8   Problems

Exercise 8.1. Show that the mass matrix defined in Equation (8.6) is positive definite.

Exercise 8.2. Write the strong form of the geometrically nonlinear beam equations, Equation (8.18), in the spatial frame of reference, that is, in terms of the components of the internal forces/moments and momenta in the body-fixed reference frame, B. How would those equations change if they were to be written in an inertial frame of reference, E?

Exercise 8.3. A basic solution of linear algebra (Cayley–Hamilton theorem) states that for any square matrix $\mathbf{A}$ of dimension n, powers of the matrix of order higher than n can be obtained from linear combinations of powers of $\mathbf{A}$ up to order $n - 1$. Using that property and the Taylor series of the harmonic functions, show that the exponential map of the Cartesian rotation vector satisfies

$$\exp \tilde{\psi} = \mathcal{I} + \frac{\sin \psi}{\psi} \tilde{\psi} + \frac{1 - \cos \psi}{\psi^2} \tilde{\psi}\tilde{\psi}.$$

Exercise 8.4. Consider two reference frames A and B, with B defined by a rotation θ about the x axis of A. This results in a Cartesian rotation vector $\psi = (1, 0, 0)$. Using the definitions in Equation (C.10) and (C.21), show that the corresponding coordinate transformation matrix and curvature correspond, respectively, to the usual expressions

$$\mathbf{R}_{BA} = \begin{bmatrix} 1 & 0 & 0 \\ 0 & \cos\theta & -\sin\theta \\ 0 & \sin\theta & \cos\theta \end{bmatrix} \quad \text{and} \quad \mathbf{k}_B = \begin{Bmatrix} \theta' \\ 0 \\ 0 \end{Bmatrix}.$$

Exercise 8.5. Consider an undeformed beam in the shape of a circular helix of radius a and pitch $2\pi b$. Its position vector with respect to a reference frame A at its center is given by

$$\mathbf{r}_A = \begin{Bmatrix} a\cos\lambda\eta \\ a\sin\lambda\eta \\ b\lambda\eta \end{Bmatrix},$$

with η being the arclength coordinate and $\lambda = \dfrac{1}{\sqrt{a^2+b^2}}$. The local reference frame S is defined by the coordinate transformation matrix

$$\mathbf{R}_{AS} = \begin{bmatrix} -\lambda a\sin\lambda\eta & -\cos\lambda\eta & \lambda b\sin\lambda\eta \\ \lambda a\cos\lambda\eta & -\sin\lambda\eta & -\lambda b\cos\lambda\eta \\ \lambda b & 0 & \lambda a \end{bmatrix}.$$

(i) Show that $\mathbf{R}_{SA}(\eta)\mathbf{r}'_A(\eta)$ defines the unit vector along the tangent direction.

(ii) Compute the local curvature vector $\mathbf{k}_S(\eta)$ and show that it is constant along the beam.

(iii) Using this local curvature vector, show that in the limit $b \to 0$, the constant-strain integration formula, Equation (8.35), gives the exact coordinate rotation matrix between the local reference frames of any two points along the curve.

Exercise 8.6. Derive analytical expressions of the LNMs of an unsupported prismatic Euler–Bernoulli beam, from the integration of Equation (8.64). The boundary conditions in this case are $\mathbf{x}_2(0,t) = \mathbf{0}$ and $\mathbf{x}_2(b,t) = \mathbf{0}$.

9   Dynamics of Very Flexible Aircraft

9.1   Introduction

The last two chapters have introduced the building blocks for the study of the coupled aeroelasticity-flight dynamics of aircraft with large and very flexible wings, which are now explored here in some detail. This completes the hierarchy of models that was initiated in Chapter 4, where we have considered aircraft that can be analyzed as perfectly rigid, and it has later been continued in Chapters 5 and 6 to include quasi-steady and dynamic aeroelastic effects, respectively, but still within the assumption of linear elastic behavior of the airframe. Here, we consider problems in which structural deflections may be arbitrarily large, that is, possibly within the geometrically nonlinear regime, although still with linear constitutive relations at the material level (e.g., the airframe does not have elastically nonlinear or plastic deformations). Geometrically nonlinear effects may be a design feature across the flight envelope to achieve extreme efficiency, as in the AeroVironment's Helios solar-power prototype, as shown in Figure 9.1a, or, most commonly, the result of a high load event for a vehicle that otherwise is designed to operate with only moderate wing deflections. The latter situation is exemplified by the very large wing flexing in the ultimate load stress test of modern airliners (see Figure 9.1b for an example). Building on the geometrically nonlinear composite beam models of Chapter 8 and a suitable unsteady aerodynamics model, such as the UVLM introduced in Chapter 7, this chapter discusses simulation and analysis strategies to characterize the dynamics of such vehicles. They are referred to here as *very flexible aircraft*, although the literature also often refers to them as *highly flexible aircraft* (van Schoor and von Flotow, 1990), and both terms are indistinctively used here to identify vehicles displaying geometrically nonlinear response of their primary structures.

With reference to Figure 5.3, the main issues that need to be addressed in the dynamic analysis and design of very flexible aircraft can be summarized as follows: (1) from a *structural* point of view, the description of the flexible airframe needs to consider the geometrically nonlinear effects associated with large-displacement kinematics of the primary structures; (2) from an *aerodynamic* viewpoint, one needs to consider the associated large excursions of the wetted surfaces relative to the mean flow speed and that pressure forces remain normal to those surfaces (i.e., they are *follower forces*, using the definition of Section 8.2); and (3) from a *flight dynamics* point of view, global moments of inertia of the aircraft and the coordinates of the CM

(a)   (b)

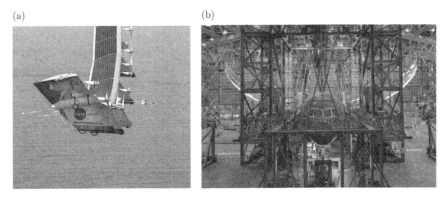

Figure 9.1 Examples of aircraft wings undergoing large deformations: (a) Helios prototype in high dihedral mode (photo by NASA). (b) Ultimate loading of the Boeing 787 wing. Its wingtip displacement is nearly 8 m or 28% of its semispan (photo by The Boeing Company).

are modified by its instantaneous deformation. As all these problems are intrinsically nonlinear, prediction of both the static and dynamic characteristics of very flexible aircraft becomes a multidisciplinary analysis problem involving the vehicle nonlinear structural, aerodynamics, and flight dynamics. Operating vehicles exhibiting nonlinear dynamics may also require nonlinear feedback control strategies, and some of them will be discussed in Chapter 10. Finally, in Section 6.3, we have already seen that the slow vibrations in an airframe with very low stiffness are likely to occur at flight-dynamics timescales – therefore, bringing dynamic couplings between the two. This is even more relevant with very flexible aircraft, with the added complexity that the natural vibration frequencies change with the nonlinear equilibrium, as already shown in the simple example in Section 8.4.

In Chapter 6, we have already explored the full range of couplings that may occur in the inertial characteristics of a flexible air vehicle and provided suitable descriptions to represent the resulting nonlinear dynamics. Chapter 7 has considered the prediction of aerodynamic forces under attached-flow assumptions on deformable wings undergoing arbitrary kinematics. Following on that, Chapter 8 has discussed other geometrically nonlinear effects occurring on wing structures with large displacements (follower forces, geometric stiffening). As they are typically associated with wings of a very high aspect ratio, we have seen that geometrically nonlinear composite beam models provide an efficient, and often also sufficiently accurate, description of the low-frequency dynamics of the primary structures.

This chapter brings together those modeling approaches into a unified description of the coupled rigid-elastic response of the aircraft described by an assembly of geometrically nonlinear beams and with distributed aerodynamic forces on all lifting surfaces. The aerodynamic models that have been more often considered in the literature of very flexible aircraft dynamics are the unsteady vortex-lattice method and *thin-strip aerodynamics* which build directly on the methods of Section 3.3 to compute sectional forces. The main focus here is on the former, since the DLTI UVLM of Section 7.3.3 gives a better approximation at a similar (or even smaller) computational cost than a thin-strip approximation with sectional lags to include unsteady effects. However, a substantial amount of work can be found in the early literature on very flexible aircraft that relies on 2-D aerodynamics (Cesnik and Brown, 2002; Nguyen et al., 2012; Patil and Hodges, 2006; Su and Cesnik, 2011a), and an example of such an approach is included in Section 9.4.2. Section 9.2 discusses the main features of a coupled geometrically nonlinear flight dynamics–aeroelastic environment, which are then exemplified through relevant numerical examples in Section 9.4. Section 9.3 also outlines an alternative formulation for flexible aircraft dynamics with moderately large wing deformations, that builds on the intrinsic modal solution of Section 8.6.1 and the DLTI UVLM of Section 7.3.3. While the aerodynamic approximation limits its applicability for numerical simulation, the formulation is computationally efficient and will form the basis for nonlinear control methods in Chapter 10.

Finally, Section 9.5 reviews the applicability of some conventional structural design practices to the analysis and design of very flexible aircraft. It discusses the effect of large structural deflections and of couplings between aeroelasticity and flight dynamics, in different aspects of the aircraft structural design process. In particular, aeroelastic stability, dynamic loads, and flight dynamics and control are all discussed and are finally illustrated with a numerical example of a representative long-endurance aircraft.

9.2   Fully Coupled Nonlinear Simulation

We can now define a generic environment for the simulation of the open-loop dynamics (i.e., response to input commands or atmospheric disturbance) of very flexible aircraft. Its building blocks are the geometrically nonlinear composite beam models defined in Chapter 8, and the general form of the unsteady vortex-lattice aerodynamics introduced in Section 7.3. However, as was mentioned earlier, a thin-strip approximation built on unsteady thin-airfoil theory (Section 3.3) with semiempirical corrections for dynamic stall, viscous drag, and tip effects can be more suitable in some cases (e.g., Tang and Dowell, 2002). Lahooti et al. (2021) have expanded on that idea to consider local high-fidelity fluid simulations of finite thickness, which have been dubbed *thick-strip* aerodynamics. This is illustrated in Figure 9.2, in which a finite number of thick strips of spanwise length $b_x \ll b$ are considered on a wing of semispan b. Each strip is modeled using a highly resolved large-eddy simulation with spanwise periodic boundary conditions, and b_x is chosen through a convergence exercise on the dominant spanwise local physics of the separated flow. For static aeroelastic applications, the thick strips can be solved offline to compute polars, but for dynamic problems with separated flows, they need to be solved simultaneously with the structural solver. Global 3-D aerodynamic effects, such as tip losses, are included as corrections.

Regarding the beam models introduced in Chapter 8, we restrict ourselves here to the displacement- and strain-based formulations, which are more suitable for nonlinear dynamic simulation, although we will also revisit the intrinsic formulation as an internal model for nonlinear control in Section 9.3. A key feature of the UVLM is

(a)

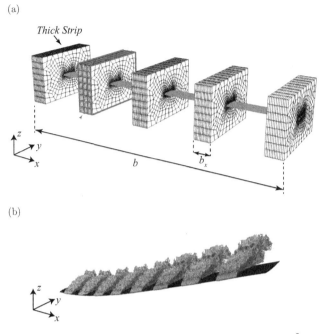

Thick Strip

(b)

Figure 9.2 Very flexible wing at $\alpha = 16°$ in a airstream with Re $= 1.56 \times 10^5$, modeled with *thick-strip* aerodynamics (Lahooti et al., 2021). (a) Fluid mesh; each strip is modeled independently and (b) flow structures within the strips on the deformed wing.

that it is naturally written in discrete time. Consequently, we introduce first a time discretization for the beam equations in Section 9.2.1. The aeroelastic coupling is then described in Section 9.2.2, while some methods for determining static (aeroelastic) trim conditions and the subsequent time marching of the coupled system are discussed in Section 9.2.3.

9.2.1 Time Integration of the Structural Equations

The spatial discretization using finite elements in either Equation (8.25) or Equation (8.41) has resulted in nonlinear second-order semidiscrete equations in time. They can be further reduced to a linear description if only small perturbations from a reference equilibrium need to be considered. Such discretizations need to capture simultaneously the zero-frequency rigid-body dynamics and structural vibrations at relatively high frequencies. This results in a (numerically) stiff system of equations (Géradin and Rixen, 1997), and implicit time-marching methods then become necessary to ensure numerical stability without having to resort to very small time steps. One such numerical scheme, the implicit Newmark integration, is considered here, given both its simplicity and excellent numerical performance.

The starting point is either Equation (8.25) or Equation (8.41), but in what follows, we could also consider the corresponding linearized expressions or the small-deformation model of Equation (6.30). In those equations, we need to solve the elastic

DoFs, their first and second time derivatives, and the rigid-body velocities and their derivatives. Discretization in time implies that the solution to all those variables is sought at discrete points in time, and intermediate values are interpolated.

Consider first the elastic DoFs, that is, either the nodal displacements and rotations or the element strains, which are denoted in general as $\boldsymbol{\xi}(t)$ in what follows. Let $\boldsymbol{\xi}^n$ be its (known) value at a certain time, t_n. We also assume that its derivatives, $\dot{\boldsymbol{\xi}}^n$ and $\ddot{\boldsymbol{\xi}}^n$, are known. Newmark interpolation schemes assume a given evolution of the acceleration between the current, t_n, and the next time step, $t_{n+1} = t_n + \Delta t$. Thus, from the integration of the acceleration within the step, one obtains

$$\boldsymbol{\xi}^{n+1} = \boldsymbol{\xi}^n + \Delta t \dot{\boldsymbol{\xi}}^n + \left(\tfrac{1}{2} - \vartheta_2\right)\Delta t^2 \ddot{\boldsymbol{\xi}}^n + \vartheta_2 \Delta t^2 \ddot{\boldsymbol{\xi}}^{n+1}, \tag{9.1a}$$

$$\dot{\boldsymbol{\xi}}^{n+1} = \dot{\boldsymbol{\xi}}^n + (1 - \vartheta_1)\Delta t \ddot{\boldsymbol{\xi}}^n + \vartheta_2 \Delta t \ddot{\boldsymbol{\xi}}^{n+1}, \tag{9.1b}$$

where the Newmark parameters are typically chosen as (Géradin and Rixen, 1997)

$$\vartheta_1 = \frac{1}{2} + \varepsilon \quad \text{and} \quad \vartheta_2 = \frac{1}{4}\left(\vartheta_1 + \frac{1}{2}\right)^2, \tag{9.2}$$

with $\varepsilon > 0$ being a very small positive number and included to add numerical dissipation. Only one differentiation with time is needed on the rigid-body velocities, $\mathbf{v}_B(t)$ and $\boldsymbol{\omega}_B(t)$, which is carried out using Equation (9.1b). Equation (9.1) can now be substituted into the EoMs evaluated at t_{n+1}, which results in closed-form expressions for the unknown acceleration field $\ddot{\boldsymbol{\xi}}^{n+1}$. Once accelerations are solved for, Equation (9.1) determines the current state of the structure.

In flexible aircraft dynamics applications, we need to consider gravitational forces that depend on the instantaneous orientation of the vehicle with respect to the Earth frame. Those are given by the propagation relations given by Equation (6.31), or more commonly with the equivalent relation in terms of quaternions (Artola et al., 2021a), as they have better numerical properties, although quaternion algebra has not been discussed here for simplicity. An excellent introduction can be found in the book of Bauchau (2011).

Augmenting the EoMs with Equations (6.31) and (9.1), as well as the corresponding approximation for time derivatives of the rigid-body velocities, results in a set of nonlinear algebraic equations, which can be cast in a residual form as

$$\mathcal{F}_s(\mathbf{x}_s^{n+1}, \dot{\mathbf{x}}_s^{n+1}, \mathbf{x}_s^n, \dot{\mathbf{x}}_s^n, \mathbf{u}_s^{n+1}) = 0, \tag{9.3}$$

where the structural states, which also include the rigid-body DoFs, and the force inputs on the structure are defined, respectively, as

$$\mathbf{x}_s = \left\{ \mathbf{v}_B^\top \quad \boldsymbol{\omega}_B^\top \quad \boldsymbol{\xi}^\top \quad \dot{\boldsymbol{\xi}}^\top \quad \theta \quad \phi \right\}^\top,$$
$$\mathbf{u}_s = \left\{ \mathbf{f}_B^\top \quad \mathbf{m}_B^\top \quad \Xi^\top \right\}^\top. \tag{9.4}$$

As it was mentioned earlier, $\boldsymbol{\xi}$ can denote either the nodal displacement and rotations of Equation (8.25) or the element strains of Equation (8.41). For known forces at the current time step, t_{n+1}, and solution of the problem at a previous time step,

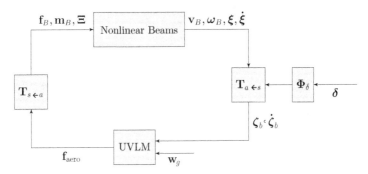

Figure 9.3 Fluid–structure coupling in the time domain with control inputs and gust disturbances.

t_n, Equation (9.3) is solved iteratively until convergence on both $\mathbf{x}_s^{n+1}$ and $\dot{\mathbf{x}}_s^{n+1}$. This is typically done using a Newton–Raphson iteration scheme with (typically approximated) analytical Jacobians. Furthermore, by using the fact that Equations (9.4) are linear equations between the (unknown) displacement/velocities and the accelerations, the first two sets of variables can be implicitly solved to define simplified algebraic equations that solely depend on the instantaneous acceleration.

9.2.2 Fluid–Structure Interface

We need to define the interface conditions from fluid to structure and from structure to fluid. This requires the definition of the instantaneous shape of the airframe at the corner points of the UVLM lattice, and of the resultant aerodynamic forces and moments per unit span length along the reference line of the beam assembly. This process is fundamentally equivalent to that outlined for the linear problem in Figure 6.20, and it can be summarized by the block diagram of Figure 9.3.

First, we consider the mapping from the structure to the aerodynamics. The beam model of the structure gives both the instantaneous rigid-body velocities, $\mathbf{v}_B(t)$ and $\omega_B(t)$, and the relative position and orientation, $\mathbf{r}_B(\eta,t)$ and $\mathbf{R}_{SB}(\eta,t)$, and their time derivatives, along the reference line. This information is available regardless of the choice of DoFs (displacement-based, strain-based, or intrinsic) in the beam equations. All three can be used for this purpose: For the displacement-based formulation, the instantaneous shape of the reference line is obtained from the vector of nodal positions and rotations, $\boldsymbol{\xi}_d(t)$, and the interpolation defined in Equation (8.24), with the coordinate transformation matrix defined from the Cartesian rotation using Equation (C.10). For the strain-based formulation, it is obtained from the vector of element strains, $\boldsymbol{\xi}_s(t)$, with the position and rotation along the reference line given by the spatial integration defined in Equation (8.35). A similar process is followed for the intrinsic formulation, but the local values of the strains need to be obtained first from the modal amplitudes $\mathbf{q}_2$ using Equation (8.56).

As we consider, in general, problems with relatively large changes of the rigid-body velocities of the aircraft, $\mathbf{v}_B$ and ω_B, then we need to consider, as we have

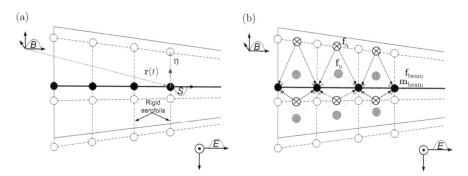

Figure 9.4 Spatial mapping between the aerodynamic lattice and beam discretization. Discontinuous lines show an aerodynamic lattice with two chordwise vortex rings. (a) Mapping of nodal displacements along the beam (black) to the aerodynamic grid (hollow). (b) Discrete aerodynamic forces are integrated into forces and moments at beam nodes. The figure only shows the forces at the leading-edge segments.

done in Chapter 6, the noncirculatory terms associated with the accelerations of the reference frame in the evaluation of the aerodynamic forces. In practice, this implies that we need to compute the coordinates of the UVLM lattice with respect to an inertial observer, which results in evaluations of Equation (8.1) (i.e., the instantaneous geometry given by Cosserat's rod model) at the cornerpoints of the bound lattice. For that purpose, the instantaneous position, $p_E(t)$, and the orientation of the B frame of reference, $\mathbf{R}_{EB}(t)$, are first obtained from integration of the rigid-body velocities in time. Then the instantaneous shape of the reference line, $\mathbf{r}_B(\eta, t)$ and $\mathbf{R}_{SB}(\eta, t)$, is determined from the elastic DoFs. Finally, Cosserat's assumption is invoked to compute the instantaneous position and inertial velocities of the bound lattice. This situation is sketched in Figure 9.4a, where the solid circles represent the discrete points (e.g., finite-element nodes) along the beam reference line, where displacements and rotations (and velocities) are known, and the empty circles represent the vertices of the bound aerodynamic lattice, which can now be obtained.

The result of this process is, therefore, a nonlinear transformation, $\mathbf{T}_{a \leftarrow s}$, from the structural DoFs and the geometry of the bound UVLM lattice, which can be written as $\zeta_b^\top = \left\{ p_E^{(1)^\top} \quad p_E^{(2)^\top} \quad \dots \right\}$, with the superindex going through all lattice vertices. It is important to emphasize that the instantaneous aerodynamic shape needs to be given with respect to an inertial frame (e.g., the Earth reference frame), so that the correct evaluation of the time derivatives $\dot{\zeta}_b$ in Equation (7.27) takes place.

Second, the sectional resultants of the forces on the UVLM bound lattice need to be computed at the beam reference axis. This is schematically shown in Figure 9.4b. The instantaneous circulatory forces on each of the vortex segments, given by Equation (7.26), and the nonciculatory forces at the center of each vortex ring, given by Equation (7.28), are both obtained. In the example in the figure, circles with crosses are used to identify the circulatory forces at the leading-edge segments, and solid gray circles is used for the noncirculatory forces. Next, their force and moment resultant over each cross section is evaluated using again the rigid-section assumption. Those resultant forces, which are defined for each chordwise row of bound vortex rings, are then mapped onto nodal forces/moments at the beam nodes. The (linear) operator that performs this mapping will be referred to as $\mathbf{T}_{s \leftarrow a}$. There is often little computational gains to be made from having different spanwise discretizations on the beam model and the aerodynamic lattice, and it is then convenient to have matching discretizations on both of them. This is the situation shown in Figure 9.4b, where the forces of the chordwise row of vortex rings are equally split between the end nodes of the corresponding beam element.

9.2.3   Nonlinear Trim and Time-Marching Algorithms

The coupled aeroelastic system described in Figure 9.3 is solved using a partitioned procedure. Each domain has been independently discretized in time using appropriate schemes for the underlying physics, but the resulting equations are interdependent and need to be solved together. We have already introduced the nonlinear structural functional, $\mathcal{F}_s$, in Equation (9.3). We can analogously define the discrete-time nonlinear fluid operator, $\mathcal{F}_a$, resulting from writing Equations (7.22), (7.24), and (7.25), together with the first-order approximation in Equation (7.8), which are all written in a residual form. This results in the following coupled aeroelastic system:

$$\mathcal{F}_a(\mathbf{x}_a^{n+1}, \mathbf{x}_a^n, \mathbf{u}_a^{n+1}) = 0, \tag{9.5a}$$

$$\mathcal{F}_s(\mathbf{x}_s^{n+1}, \dot{\mathbf{x}}_s^{n+1}, \mathbf{x}_s^n, \dot{\mathbf{x}}_s^n, \mathbf{u}_s^{n+1}) = 0, \tag{9.5b}$$

where the structural states and inputs have been defined in Equation (9.4), and the aerodynamic states and inputs are, respectively,

$$\mathbf{x}_a = \left\{ \boldsymbol{\Gamma}_b^\top \quad \boldsymbol{\Gamma}_w^\top \quad \dot{\boldsymbol{\Gamma}}_b^\top \quad \boldsymbol{\zeta}_w^\top \right\}^\top,$$
$$\mathbf{u}_a = \left\{ \boldsymbol{\zeta}_b^\top \quad \dot{\boldsymbol{\zeta}}_b^\top \quad \mathbf{w}_g^\top \quad \boldsymbol{\delta}_c^\top \right\}^\top. \tag{9.6}$$

Note the different physical origin that time derivatives have in the structural and aerodynamics states in Equation (9.5). The discrete-time structural functional $\mathcal{F}_s$ is obtained from the approximation of second-order ordinary differential equations in time, and therefore we have explicitly retained the instantaneous value of the structural state, $\mathbf{x}_s$, and its time derivative. The fluid functional comes from the instantaneous solution of Laplace's equation with the time derivatives $\dot{\boldsymbol{\Gamma}}_b$ introduced to capture the apparent mass effects.

The interconnection between both domains is determined by the interpolation operators at the interface that we have introduced in Section 9.2.2. They can be written in a symbolic form as

$$(\boldsymbol{\zeta}_b^n, \dot{\boldsymbol{\zeta}}_b^n) = \mathbf{T}_{a \leftarrow s} \left(\boldsymbol{\xi}^n, \dot{\boldsymbol{\xi}}^n, \mathbf{v}_B^n, \boldsymbol{\omega}_B^n \right), \tag{9.7a}$$

$$(\mathbf{f}_B^n, \mathbf{m}_B^n, \boldsymbol{\Xi}^n) = \mathbf{T}_{s \leftarrow a}(\mathbf{f}_{\text{aero}}^n), \tag{9.7b}$$

where $\mathbf{f}_{aero}$ has been defined in Equation (7.29). This is a generalization to geometrically nonlinear problems of the mapping defined in Equation (6.122) for linear aeroelastic problems.

Equilibrium conditions on a flexible aircraft can be obtained by seeking the fixed point $\mathbf{x}_a^{n+1} = \mathbf{x}_a^n$, $\mathbf{x}_s^{n+1} = \mathbf{x}_s^n$, and $\dot{\mathbf{x}}_s^n = 0$ for constant commands and the background velocity field. This results in generalizations to coupled nonlinear aeroelastic–flight-dynamic problems of the *steady maneuvers* identified for the rigid aircraft in Section 4.5.1 and of the *aeroelastic trim* conditions defined on Section 5.6. As we did there, the problem is identifying the (constant) command input, δ_{c0}, for which the aircraft is in equilibrium for a given velocity and altitude/load factor.

Regarding the time integration of Equation (9.5), note that the fluid operator depends on the instantaneous (and unknown) geometry of the wetted surfaces at t_{n+1}, while the structural operator depends on the instantaneous (and also unknown) aerodynamic loading at t_{n+1}. As the system has been partitioned, we can only advance in time the structure with a known fluid state or the aerodynamics with a known structural state. The simplest coupling approach consist on time marching both domains independently with exchange of information at predetermined time steps, which do not necessarily need to be the same as the internal time steps used to solve the fluid and structural equations. This is known as a *loosely-coupled* integration scheme. While this approach avoids the cost of subiterations between domains, it typically needs small time steps for convergence and may lose numerical stability (Farhat et al., 2006; Smith et al., 1995).

Therefore, coupled time integration schemes that subiterate between structural and aerodynamic solvers to ensure convergence at given time intervals are typically preferred. They are often referred to as *strongly-coupled* integration schemes, although some researchers restrict the use of this term to monolithic schemes in which the solution to both domains is orchestrated by a single solver (Bhardwaj et al., 1998). The most common strongly-coupled solution algorithm is the *conventional serial staggered* solution algorithm that consists of four steps (Strganac and Mook, 1990):

1. *Transfer the instantaneous shape* and wall velocities from the solid to the fluid domains, Equation (9.7a). This is typically accelerated by a predictor of the state of the structure at time t_{n+1}, which can be obtained from Equation (9.1) with constant acceleration.
2. *Advance the aerodynamic equations*, Equation (9.5a), with the current estimate of the geometry and local gust velocity at t_{n+1}. Note that the background velocity field is normally known in the inertial frame, and the instantaneous effect on the wing depends on the evolution of the vehicle.
3. *Transfer the instantaneous aerodynamic loading* into the equivalent forces and moments on the structure, Equation (9.7b).
4. *Advance the structural equations*, Equation (9.5b), with the current estimate of the aerodynamic forces.
5. Go back to Step 1 until a desired level of convergence is achieved.

Finally, the characteristic timescales of the physical processes in the solid and fluid domains may differ. In such a case, it is computationally more efficient to have internal subiterations within the domain with the fastest dynamics, while the time step for the coupled iterations is determined by the subsystem with the slowest dynamics.

Example 9.1 Gust Response of a Flexible Prismatic Cantilever Wing. Consider again the wing of Example 7.1, namely a prismatic cantilever wing of aspect ratio 16. Only the unswept wing is considered here, for which a UVLM model is built with a 20×5 bound lattice with ($M = 20$, $N_b = 5$) and a wake lattice that extends 25 chords downstream from the wing's trailing edge ($N_w = 125$). The wing is at zero angle of attack in an airstream with a spanwise-uniform traveling "1-cosine" gust of intensity w_0 and length H, as defined in Section 2.4.2.

The wing is now allowed to deform under aerodynamic loading. Recall that, for aeroelastic analysis, we need dimensional magnitudes (e.g., the actual lift as opposed to lift coefficient), unlike in Example 7.1. In particular, the ratio between the dynamic pressure of the fluid and the characteristic stiffness of the structure, as in Equation (5.1), determines the importance of aeroelastic effects. This is a fundamental difference between the effect of aerodynamic loads on rigid and flexible aircraft, and can be clearly seen in the different scaling of the apparent mass, damping, and stiffness with the flight conditions in Equation (6.123). Indeed, the complexity of *aeroelastic scaling* often makes it impossible for all relevant nondimensional coefficients to be matched between a wing-tunnel model and the full-scale aircraft. As we discuss in Section 11.3, the implication of this is that one may obtain rather different observations in aeroelastic wind-tunnel experiments when compared to actual flight tests of a flexible aircraft, and the former cannot be used to replace the latter for airworthiness certification.

Here, we consider first a problem with freestream velocity $V = 5$ m/s, air density $\rho = 0.1$ kg/m^3 (corresponding to an altitude of 19,200 m), and chord $c = 1$ m. The beam reference axis goes along the the midchord of the wing, and the relevant stiffness constants are its out-of-plane bending stiffness, $EI = 2 \times 10^4$ N.m^2, and its torsional stiffness $GJ = 10^4$ N.m^2. The CM of the sections is also at the midchord, and the only nonzero constants are the mass per unit length, $m = 0.75$ kg/m, and the torsional inertia per unit length, $\rho I = 0.1$ kg·m/m.

Figure 9.5 shows the time history of the normalized lift coefficient on the wing for "1-cosine" gusts of intensity $w_0 = 0.1V$. The time variable has been normalized again by the convective timescale to highlight unsteady aerodynamics features, as we did in Example 7.1. Results are presented for gusts of varying lengths, with H being between c and $10c$. To facilitate the comparison with the results on a rigid wing, Figure 7.9a, results are nondimensionalized and the lift coefficient is again normalized with $C_{L0} = 2\pi \frac{w_0}{V}$. The interest in a gust encounter is typically in the internal loads originated in the airframe, which are assessed by monitoring a set of the so-called *quantities of interesting*. The choice of quantity of interest in this example is the vertical displacement at the wing tip, expressed in terms of the wing semispan b. It is included in Figure 7.9b for the nominal air density and varying gust lengths. Results

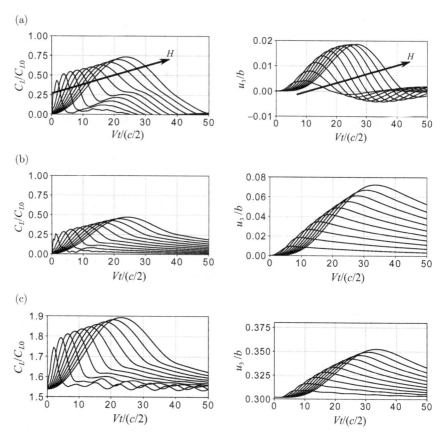

Figure 9.5 Instantaneous lift coefficient C_L and wingtip deflection u_3 for a "1-cosine" gust of intensity $w_0 = 0.1V$ and gradient $H = c, 2c, \ldots 10c$. (a) $\rho = 0.1$ kg/m^3, $\alpha_0 = 0°$. (b) $\rho = 1$ kg/m^3, $\alpha_0 = 0°$. (c) $\rho = 1$ kg/m^3, $\alpha_0 = 10°$.
[07UnsteadyAero/dynamic_gust.ipynb]

show that the maximum wing displacements are below 2% of the semispan, which is well within the linear domain. Note also that the higher excitation frequency of the shorter gusts results in stronger wing vibrations and a double peak in the lift signals in Figure 9.5a. The response for the longer gusts is much smoother than the one for the shorter gusts (although never quasi-steady), and the vertical displacement has much smaller excursions along the negative axis after the gust disappears.

The effect of changes in the aeroelastic scaling is assessed here by increasing the air density in the example to $\rho = 1$ kg/m^3. This situation corresponds to the aircraft flying at the same speed as before, but at an altitude of around 2,000 m. On a rigid wing, the time histories of the total lift in the gust encounter would remain unchanged between both altitudes. However, multiplying the dynamic pressure by a factor of 10 has a significant effect on the lift coefficient of the vehicle. Figure 9.5c shows the lift coefficient on the aeroelastic wing, which drops by around 50% with respect to the results in Figure 9.5b. This is mainly due to the wing having the elastic axis downstream of the aerodynamic axis, such that the lift on its cross sections produces a negative twisting, with a rather marked effect. However, since the maximum lift (in newtons) is almost seven times larger, the resulting wingtip deflections, as shown in Figure 9.5d, are much higher than those in Figure 9.5b (although far from seven times larger!). A second major difference can be found in the transient dynamics of the wing after the gust has disappeared. With higher air density, the aerodynamic damping becomes more important, and in this case results in wing dynamics typical of an *overdamped* system (Shabana, 1991a, Ch. 1).

Finally, this last simulation is repeated in Figure 9.5e and Figure 9.5f but with a wing positioned with a nonzero angle of incidence at the root. A large angle of attack, $\alpha_0 = 10°$, is considered, which results in an initial lift on the aeroelastic equilibrium equal to $1.54C_{L0}$. For $\rho = 1$ kg/m^3, this results in a wingtip static displacement as high as 30% of the semispan (the vertical displacement at $t = 0$ in Figure 9.5f), which is well within the geometrically nonlinear regime. As the sections toward the wing tip now have a significant effective dihedral from the elastic deformations, the effect of the gust velocity, which is always oriented toward the vertical direction of the Earth axes, is now much reduced.[1] As a result, the maximum incremental lift due to the gust is around $0.35C_{L0}$, compared with $0.49C_{L0}$ of Figure 9.5c. Finally, the static deformation of the wing also shifts its natural vibration characteristics. Here, this results in a higher amplitude vibrations for the shorter gusts when the wing is moved from $\alpha_0 = 0°$ to $10°$. Larger oscillations are indeed apparent after the gust has disappeared for cases $H = c$ and $2c$ in Figure 9.5e, even though the vertical gust effectively has a smaller strength due to the dihedral effects of the highly deformed wing.

9.2.4 Linearized Dynamics Around Nonlinear Trim

The coupled discrete-time system defined by Equation (9.5) can be linearized around any equilibrium point. This describes the perturbation dynamics of a very flexible aircraft in trimmed flight with, in general, large structural deflections. The process is entirely analogous to the one followed in Section 6.3.3, but here we use a different initial formulation. The analytical linearization of the structural equations in the displacement-based finite-element formulation has already been outlined in Section 8.4, while the linearization of the unsteady aerodynamics around a steady state yielded the DLTI UVLM in Section 7.3.3. The linearization of the coupled system dynamics also needs the linear form of the interface relations that were defined in Section 9.2.2. Here, particular attention needs to be paid to the transfer of aerodynamic forces onto the structure, since steady follower forces contribute to the effective stiffness of the linearized structure. This effect is akin to that of follower forces in buckling analysis, namely small changes in the geometry of the structure after static equilibrium produce a small but nonnegligible reorientation of the internal stresses that contribute to the tangent stiffness.

---

[1] Note that a gust loading in a very flexible wing results in a hybrid situation between the dead and follower forces that were illustrated in Figure 8.3. The background velocity field, or upwash, is given in the inertial axis, but the resulting pressure field is still normal to the deformed aerodynamic surfaces.

The process of assembly of the linearized system around a nonlinear equilibrium is almost identical to that followed in Section 6.6.1 for a general linear problem. The main difference is that the DLTI UVLM aerodynamics results in EoMs written in discrete time. Therefore, we write here the state-space form of the discrete-time linear time-invariant coupled system as

$$\Delta \mathbf{x}_{\mathrm{ae}}^{n+1} = \mathbf{A}_{\mathrm{ae}} \Delta \mathbf{x}_{\mathrm{ae}}^{n} + \mathbf{B}_{\mathrm{ae}} \Delta \mathbf{u}_{\mathrm{ae}}^{n},$$
$$\Delta \mathbf{y}_{\mathrm{ae}}^{n} = \mathbf{C}_{\mathrm{ae}} \Delta \mathbf{x}_{\mathrm{ae}}^{n} + \mathbf{D}_{\mathrm{ae}} \Delta \mathbf{u}_{\mathrm{ae}}^{n}, \tag{9.8}$$

where we have used the definition of the input vector of Equation (6.125), namely

$$\mathbf{u}_{\mathrm{ae}} = \left\{ \begin{matrix} \mathbf{f}_{qt}^{\top} & \delta_c^{\top} & \frac{1}{V} \mathbf{w}_{g0}^{\top} \end{matrix} \right\}^{\top}. \tag{9.9}$$

The state vector of the coupled system is defined as the concatenation of the fluid and structural states defined in Equations (9.4) and (9.6), respectively, as

$$\mathbf{x}_{\mathrm{ae}} = \left\{ \begin{matrix} \boldsymbol{\Gamma}_b^{\top} & \boldsymbol{\Gamma}_w^{\top} & \dot{\boldsymbol{\Gamma}}_b^{\top} \end{matrix} \middle| \begin{matrix} \boldsymbol{\xi}^{\top} & \dot{\boldsymbol{\xi}}^{\top} & \mathbf{v}_B^{\top} & \boldsymbol{\omega}_B^{\top} & \theta & \phi \end{matrix} \right\}^{\top}. \tag{9.10}$$

As we have discussed in Section 7.3.3, the DLTI UVLM assumes a prescribed wake model and therefore, the coordinates of the wake panels are not included in the states of the linearized system. Equation (9.8) defines a monolithic formulation for the linear dynamics of a flexible aircraft around an arbitrary static reference condition. It has been built directly on the spatial discretization of the structural and aerodynamic states (i.e., finite-element DoFs and vortex ring circulations). This typically yields models with tens of thousands of states for convergence. However, the size of the model can be dramatically reduced in a two-step process (Hesse and Palacios, 2014; Maraniello and Palacios, 2020): First, the structural equations are projected on the linear normal modes at the equilibrium point, following the solution process of Section 6.4; and second, the frequency-limited balancing of Section 7.4.2 is used to obtained a reduced-order approximation of the DLTI UVLM.

The eigenvalues of the discrete-time system matrix $\mathbf{A}_{\mathrm{ae}}$ determine the dynamic stability of the vehicle, including both aeroelastic and flight-dynamic characteristics (and couplings between them). Note that if λ_i are the discrete-time eigenvalues, then their continuous-time approximation is obtained as $\log(\lambda_i)/\Delta t$ (see Section 1.4.2). The process is similar to the one defined in Section 6.6.2, with the difference being that the state matrix depends here on the reference conditions, and therefore we need to reevaluate the system dynamics at, for example, each freestream velocity. This defines an iterative method that results in *matched conditions*, in which the computed flutter speed is computed on the dynamic system resulting from the equilibrium point at that speed. A flowchart for this solution process is included in Figure 9.6.

Finally, it is worth mentioning that while the linearization about a nonlinear operating point is very useful for the calculation of the flutter boundary, the same cannot be said for a generic dynamic simulation of the very flexible aircraft. Unless it is guaranteed that the dynamic motion is bounded under low-amplitude vibrations, something that depends on vehicle parameters (intrinsic flexibility and mass condition) and the level of excitation (control inputs and gust excitation), the linearization of the flight

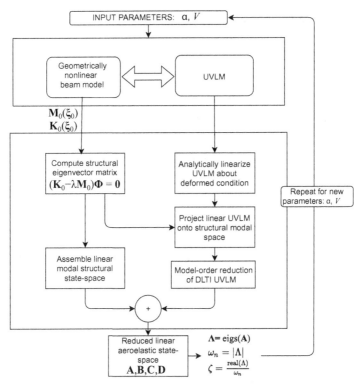

Figure 9.6 Solution process to obtain dynamic stability boundaries with matched conditions on a very flexible wing (Goizueta et al., 2022).

dynamics of the aircraft would lead to higher errors in the coupled aeroelastic–flight-dynamics solution than if the aeroelastic effects were neglected altogether (noting that both are incorrect when compared with the full nonlinear dynamics). The reason is that resulting errors tend to accumulate in time and will clearly show on the resulting vehicle trajectory, as shown by Shearer and Cesnik (2007). In those cases, a nonlinear rigid-linear elastic formulation, such as that introduced in Section 6.3.2, typically offers a suitable compromise. That formulation has been extended to very flexible aircraft dynamics around large-deflection trim equilibrium by Hesse and Palacios (2014), and further details can be found there. A second computationally efficient alternative is the nonlinear modal formulation with linearized aerodynamics that is described in the following section.

9.3   A Nonlinear Modal Formulation

A simplified description of the coupled flight dynamics–aeroelastic response of a very flexible aircraft can be derived if the aerodynamic forces are obtained on the undeformed geometry. This was the main assumption in the generation of the DLTI UVLM in Section 7.3.3, and it implies that the relative distance between the origin and target points in the Biot–Savart law, Equation (7.21), is not updated as the aircraft deforms. As the aerodynamic interference between panels goes as the square of the distance, such an approximation can give good results for high-aspect-ratio wings, even with rather large deformations, as shown in Example 9.2 below. Note that this is also the implicit assumption in thin strip aerodynamics, which we discuss in Section 9.4, where spanwise interactions are neglected and wingtip corrections are first computed on the undeformed geometry and then applied on the deformed aircraft.

In Section 7.5, we have seen that the DLTI UVLM, combined with the model-order reduction schemes of Section 7.4, results in a compact description of the unsteady aerodynamics for linear aeroelastic analysis. The structural dynamics has been given there in modal coordinates, Equation (6.51). We have later seen in Section 8.6.1 that the intrinsic formulation facilitates the description of geometrically nonlinear beam dynamics using the LNMs of the structure as a basis. In this section, we combine both formulations to construct a coupled flight dynamics–aeroelastic formulation for very flexible aircraft in modal coordinates that captures the effects of the large displacements on the structural dynamics but does not actually update the geometry for the computation of aerodynamic forces. With reference to Figure 5.3, such a coupled model would appropriately describe geometric stiffening of the structure gyroscopic couplings (including the rigid-body dynamics), the dependency of global inertia on instantaneous vehicle shapes, and follower force effects on the aerodynamic loading. While the simplification of the aerodynamics would bring some loss of accuracy, this strategy has excellent computational performance, and it will be used as the internal model for nonlinear control design in Chapter 10. The main features of the approach have been proposed by Artola et al. (2021a) and Cea and Palacios (2021) for incompressible and compressible flows, respectively, and a summary of the solution process for low-speed flight is included next.

The nonlinear modal EoMs of a geometrically nonlinear beam in intrinsic coordinates, Equation (8.57) are used to describe the airframe dynamics. They have been written in terms of the velocity and internal force (strain) of its linear normal modes, whose amplitudes have been defined as $\mathbf{q}_1(t)$ and $\mathbf{q}_2(t)$, respectively. The generalized aerodynamic forces, Equation (7.66), also depend on the displacement amplitude of the modes. Therefore, for the purpose of establishing the instantaneous shape of the linearized aerodynamic model, we define a displacement amplitude vector, $\mathbf{q}_0(t)$, from the integration of the modal velocities, that is, $\dot{\mathbf{q}}_0 = \mathbf{q}_1$. With those definitions, Equation (7.66) can be written as

$$\mathbf{f}_q(t) = \tfrac{1}{2}\rho V^2 \left[\mathcal{A}_{q0}\mathbf{q}_0 + \mathcal{A}_{\delta 0}\boldsymbol{\delta} + \tfrac{c}{2V}\left(\mathcal{A}_{q1}\mathbf{q}_1 + \mathcal{A}_{\delta 1}\dot{\boldsymbol{\delta}}\right) + \mathcal{C}_a\mathbf{x}_a + \tfrac{1}{V}\mathcal{C}_g\mathbf{x}_g \right]. \quad (9.11)$$

In general, the modal and gust aerodynamic states, $\mathbf{x}_a$ and $\mathbf{x}_g$, respectively, are obtained from the solution of Equations (7.41) and (7.64). As was seen in Section 7.4 however, a ROM with relatively few balanced states can be constructed for the DLTI UVLM with minimal penalty in accuracy. As the size of the resulting system is small, a robust discrete-to-continuous transformation can be finally introduced using the methods of Appendix A.3. The resulting linear ODEs will be written as

$$\frac{c}{2V}\dot{\mathbf{x}}_a = \widetilde{\mathbf{A}}_a\mathbf{x}_a + \widetilde{\mathbf{B}}_0\mathbf{q}_0 + \widetilde{\mathbf{B}}_\delta\boldsymbol{\delta} + \frac{c}{2V}\left(\widetilde{\mathbf{B}}_1\mathbf{q}_1 + \widetilde{\mathbf{B}}_{\dot{\delta}}\dot{\boldsymbol{\delta}}\right),$$

$$\frac{c}{2V}\dot{\mathbf{x}}_g = \widetilde{\mathbf{A}}_g\mathbf{x}_g + \widetilde{\mathbf{B}}_g\mathbf{w}_{g0}. \tag{9.12}$$

An even smaller ROM size can be obtained if the balanced realization is defined on the combined aerodynamic system simultaneously considering modal, control surfaces, and gust inputs, as done by Artola et al. (2021a), but this is not pursued here.

Substituting the aerodynamic forces given by Equation (9.11) into Equation (8.57) and introducing the modal gravitational stiffness matrix defined in Equation (6.51), we have

$$\dot{\mathbf{q}}_0 = \dot{\mathbf{q}}_1,$$

$$\dot{\mathbf{q}}_1 = \boldsymbol{\Omega}\mathbf{q}_2 - \sum_\ell q_{1\ell}\boldsymbol{\Gamma}_1^\ell\mathbf{q}_1 - \sum_\ell q_{2\ell}\boldsymbol{\Gamma}_2^\ell\mathbf{q}_2 + \mathbf{G}_q\mathbf{q}_0$$

$$+ \tfrac{1}{2}\rho V^2\left(\mathcal{A}_{q0}\mathbf{q}_0 + \mathcal{A}_{\delta0}\boldsymbol{\delta} + \mathcal{C}_a\mathbf{x}_a\right) + \tfrac{1}{4}\rho cV\left(\mathcal{A}_{q1}\mathbf{q}_1 + \mathcal{A}_{\delta1}\dot{\boldsymbol{\delta}}\right) + \tfrac{1}{2}\rho V\mathcal{C}_g\mathbf{x}_g,$$

$$\dot{\mathbf{q}}_2 = -\boldsymbol{\Omega}\mathbf{q}_1 + \sum_\ell q_{2\ell}\boldsymbol{\Gamma}_2^{\ell\top}\mathbf{q}_1. \tag{9.13}$$

For a given control surface deflection, which may be determined as the output of an actuator model (as in Equation (6.106)) and gust profile, Equations (9.12) and (9.13) determine the coupled flight-dynamics–aeroelastic response of a flexible vehicle. Note that the modal basis may include the rigid-body DoFs. In Equation (9.13), we have also introduced the gravitational forces using a linear approximation. This implies that they are obtained on the undeformed vehicle at the reference attitude, which may need updating under complex kinematics. A computationally efficient approach to compute the geometrically nonlinear effects in the instantaneous gravitational forces has been proposed by Artola et al. (2022), but it is not discussed here.

Example 9.2 High-Aspect-Ratio Very Flexible Wing. An illustrative example of the nonlinear modal formulation, which was first considered in Artola et al. (2021a), is included next. A cantilever wing of a very high aspect ratio is considered with the geometrical and structural properties of Table 9.1. There are no rigid-body modes, and the wing root is taken as a reference for the formulation.

The nonlinear modal equations of Equation (9.13) are considered with 12 natural modes for the structure. The aerodynamic model employed in the nonlinear modal formulation is based on the DTLI UVLM for the undeformed wing at angle of attack $\alpha_0 = 5°$. The discretization for the aerodynamic grid and structural elements is chosen to ensure convergence on tip deflection and total aerodynamic forces. It includes 8 vortex rings along the chord, 64 along the span (resulting in 32 structural beam elements), and a 10-chord wake length. This equates to a linear system with 13,321 states, including the gust system states. A Krylov subspace model-order reduction is performed to reduce it as little as 36 states. Figure 9.7 shows the static aeroelastic equilibrium shape of the wing for $\rho = 1.02$ kg/m^3 and $V = 35.35$ m/s. As can be seen, this wing is very flexible in bending, and the maximum tip displacement reaches nearly 50% of the span.

Table 9.1 High-aspect-ratio wing properties.

Span	L	45.72 m
Chord	$2b$	1.8288 m
Mass per unit length	m	20.71 kg/m
Mass moments of inertia per unit length	I_{yy}	0.86 kg·m
	I_{zz}	7.77 kg·m
Out-of-plane bending stiffness	EI_{yy}	9.77 MN·m^2
In-plane bending stiffness	EI_{zz}	977 MN·m^2
Torsional stiffness	GJ	98.7 MN·m^2
Axial stiffness	EA	1 MN
Shear stiffness	GA	10 MN
Center of mass offset (in body axes)	η_{cm}	$[0, -0.1829, 0]^\top$ m
Elastic axis from leading edge (in body axes)	η_{ea}	$[0, -0.6035, 0]^\top$ m

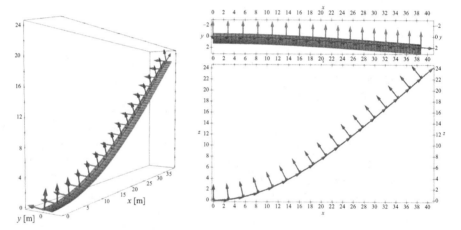

Figure 9.7 Nonlinear aeroelastic equilibrium of the very flexible wing at $\alpha = 5°$. At each beam node, the local frame of reference is shown, with y pointing upstream.

The interest here is in the assessment of the relative importance of different non-linear effects for such large-amplitude deformations. For that problem, we consider a step change of attack from an initially undeformed wing and neglecting gravity effects. The wing will display the transient dynamics associated with both the built-up of lift (i.e., Wagner's effect, Section 3.4.3) and the structure vibrations. However, this second effect is negligible for this configuration, as the natural frequencies of the wing are much larger than the convective timescale of the fluid. This can be seen in Figure 9.8 that shows the time history of the vertical displacement at the wing tip for three different airspeeds, namely 17.67, 25.00, and 35.35 m/s. Each increase in airspeed corresponds to the doubling of the dynamic pressure for easy assessment of the geometrically nonlinear effects. Four different modeling fidelities are considered: (1) a fully nonlinear aeroelastic simulation, using SHARPy; (2) a nonlinear structural model in modal coordinates, with the 36-state linear aerodynamic model described earlier (NL$_{struct}$ + L$_{aero}$); (3) a structural model solving the linearized EoMs in intrinsic

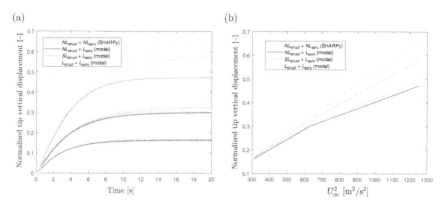

Figure 9.8 (a) Response of the wing released from an undeformed configuration at $\alpha_0 = 5°$ at 17.67, 25, and 35.35 m/s, using three different structural models coupled with reduced linear aerodynamics. (b) Steady-state values against the square of the airspeed (Artola et al., 2021a).

variables, but retaining the nonlinear kinematic relations in the evaluation of displacements and rotations ($SL_{\text{struct}} + L_{\text{aero}}$); and (4) a fully linearized aeroelastic system.

As can be seen, the tip deflection ranges from 17% of the span at the lowest airspeed to 48% at the highest one. At the lowest airspeed, $V = 17.67$ m/s, all four models give essentially the same results, which indicates that no significant nonlinear effects are present. As the airspeed is increased, clear differences are observed between the three approximations to the intrinsic beam equations. The linear model significantly overpredicts the structural deformations, while introducing the nonlinear kinematic relations (a postprocessing step here) reduces that discrepancy by nearly half. Finally, the model including all the nonlinear terms in the structure, but linearized aerodynamics, shows only small errors (of around 4% of the wingspan) compared with the fully nonlinear formulation in SHARPy (see Section 9.4). This indicates that the effects of the change of geometry in the aerodynamic model are mostly negligible even with very large wing bending and justify the use of the DLTI UVLM for very flexible aircraft applications. The computational cost of that formulation, particularly if model reduction is also included, is almost negligible compared with the direct solution of the full UVLM equations.

9.4 Some Representative Examples

At the time of writing this book, there are several software implementations in the open literature of the basic methods introduced in Section 9.2. A comprehensive review is beyond the ambition of this book, as new software packages regularly appear, but a very good introduction to some common implementations can be found in the work of Afonso et al. (2017). The numerical examples in this chapter use two of the most mature software implementations, namely SHARPy and UM/NAST, which are briefly described first.

SHARPy (del Carre et al., 2019a) is an open-source[2] Python implementation of a coupled flight dynamics–aeroelastic formulation built on the displacement-based geometrically nonlinear composite beam structure and UVLM aerodynamics. It includes a fully nonlinear time-domain formulation with a block Gauss–Seidel coupling algorithm, linearized solutions around arbitrary equilibrium points, as well as a nonlinear rigid-body–linear elastic formulation about at arbitrary reference. It has a very efficient numerical implementation that has permitted the analysis of optimal maneuvers of very flexible aircraft (del Carre and Palacios, 2020; Maraniello and Palacios, 2017). The aerodynamic model includes nonlifting bodies to account for interactions between wings and fuselage, and a variable discretization for the wake panel (Duessler et al., 2022). The dimensionality (and cost) of the DLTI UVLM models for large problems is reduced using the balancing strategies of Section 7.4 (Maraniello and Palacios, 2020), as well as iterative moment-matching methods (Goizueta et al., 2021). A generic flexible multibody dynamics implementation expands the solution of Section 8.4 to problems with internal joints (e.g., hinges) and mixed-boundary conditions, which has facilitated application in aeroelastic control of floating wind turbines (Ng et al., 2015, 2017). Finally, SHARPy is equipped with an interface for arbitrary 3-D velocity fields obtained from atmospheric simulation (Deskos et al., 2020) and native nonlinear model-predictive control algorithms (Artola et al., 2021a).

UM/NAST is the University of Michigan's Nonlinear Aeroelastic Simulation Toolbox that includes the implementation of the strain-based geometrically nonlinear composite beam formulation of Section 8.6 with four elastic DoFs per element (axial, torsion and in- and out-of-plane bending). Started in the late 1990s, it has been used to study multiple cases related to very flexible wings, for example, conventional subsonic configuration (Shearer and Cesnik, 2007), flying wings (Su and Cesnik, 2011a), blended wing body (BWB) (Jones and Cesnik, 2016; Su and Cesnik, 2010), joined wing (Cesnik and Brown, 2003), active wing warping for roll control (Cesnik and Brown, 2002), trajectory control (Shearer and Cesnik, 2008), high-aspect-ratio-wing transport aircraft (Kitson et al., 2016; Riso et al., 2020), loads control (Hansen et al., 2020; Pereira et al., 2019a), etc. Regarding the types of aerodynamics, UM/NAST was originally built with a potential-flow finite-state 2-D unsteady aerodynamic model (strip theory) with 3-D corrections (Cesnik and Brown, 2002). Along the years, many different aerodynamic formulations were included to represent the flight regime being considered. More recently, steady and unsteady vortex-lattice solvers (Teixeira and Cesnik, 2019), as well as ROMs for transonic regimes, were also added as aerodynamic options in the aeroelastic framework. Finally, propeller effects can be accounted for in the unsteady aerodynamics through the viscous vortex-particle method (Teixeira and Cesnik, 2020), as well as sectional viscous corrections. The framework written in C++ and Python includes solutions for trim, statics, modal, linearization, stability, fully nonlinear time simulation, and sensitivity evaluation for integration into multidisciplinary design optimization codes.

---

[2] The software, with extensive documentation, is available from www.imperial.ac.uk/aeroelastics/sharpy.

Table 9.2 Pazy wing: Stiffness properties of the constant section model.

Axial stiffness	EA	7.16 MN
Out-of-plane shear stiffness	GA_y	1 MN
In-plane shear stiffness	GA_z	3.31 MN
Torsional stiffness	GJ	7.2 N·m^2
Out-of-plane bending stiffness	EI_{yy}	4.67 N·m^2
In-plane bending stiffness	EI_{zz}	3.31 kN·m^2

Using both tools, the rest of this section will present three numerical examples of increasing complexity that illustrate typical situations found in the dynamics of very flexible aircraft. The first example, in Section 9.4.1, demonstrates geometrically nonlinear effects in the aeroelastic response of the very flexible wing. The second one, Section 9.4.2, illustrates the coupled flight-dynamics–aeroelastic response of a BWB aircraft subject to moderate wing deformations and complex (nonlinear) rigid-body dynamics. Finally, in Section 9.4.3, we explore the dynamic response to atmospheric gusts of a prototype air vehicle designed to display very large wing deformations.

9.4.1 Technion's Pazy Wing

The Pazy wing (Avin et al., 2022) is a very flexible wind-tunnel model designed to assess geometrically nonlinear effects on wing flutter. This is often referred to as *nonlinear flutter*, although it should be remarked that the nonlinearity is in the equilibrium conditions and whether flutter occurs is still established from linear stability analysis, as shown in Figure 9.6. Several wings have been built with slightly different properties, and the one considered here is the first prototype, referred to as the (pre)Pazy wing, which is shown in Figure 9.9. It has span $b = 0.55$ m and constant chord $c = 0.10$ m. The wing is built on NACA 0018 Nylon ribs mounted on an aluminum spar and covered by an Oralight skin, as shown in Figure 9.9a. A wingtip rod is also included to reduce the natural frequency of the first torsional mode. Figure 9.9b shows the wing displaying large deformations during a wind-tunnel test.

An approximated beam model of the wing has been assembled in SHARPy, and it is used here to assess its dynamic aeroelastic stability. The beam model assumes constant properties along the span and that there exists an *elastic axis* that diagonalizes the sectional stiffness. The beam properties are given in a frame of reference with x going along the elastic axis pointing towards the wing tip z pointing upward and y toward the leading edge. The corresponding sectional stiffness and inertial properties are listed in Tables 9.2 and 9.3, respectively. The leading edge of the prismatic wing is $0.4475c$ ahead of the elastic axis, while the CM of the constant section model is only slightly behind the reference line ($-3.5 \times 10^{-3}c$). The wingtip mass is modeled as a lumped mass located 5 mm aft the elastic axis. Its inertia properties are also listed in Table 9.3. A more detailed model has been constructed by Riso and Cesnik (2021) and has been used also by Goizueta et al. (2022), but this simplified approximation is preferred here for clarity.

Table 9.3 Pazy Wing: Mass properties in the constant section model.

Wing mass per unit length	m	$0.55\,\mathrm{kg/m}$
Wing inertia per unit length[a]	I_{xx}	$3.030 \times 10^{-4}\,\mathrm{kg \cdot m}$
	I_{yy}	$1.515 \times 10^{-4}\,\mathrm{kg \cdot m}$
	I_{zz}	$3.636 \times 10^{-3}\,\mathrm{kg \cdot m}$
Wingtip rod mass	m_{rod}	$19.95\,\mathrm{g}$
Wingtip rod inertia	I_{xx}	$1.2815 \times 10^{-4}\,\mathrm{kg \cdot m^2}$
	I_{yy}	$2.87 \times 10^{-7}\,\mathrm{kg \cdot m^2}$
	I_{zz}	$1.17 \times 10^{-4}\,\mathrm{kg \cdot m^2}$

[a]Note that these are identified coefficients that do not satisfy the sectional constraint that $I_{xx} = I_{yy} + I_{zz}$.

(a)   (b)

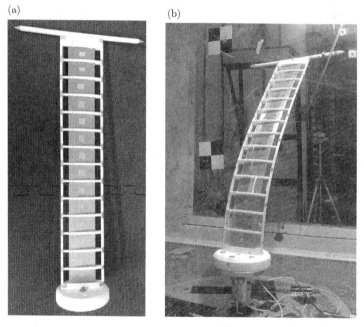

Figure 9.9 Pazy wing experimental model. (a) Substructure without skin and (b) wing in the wind tunnel (reprinted from Avin et al. (2022) with permission from the authors).

Flutter onset strongly depends on the wing deflection, which is a unique function of the root angle of attack, α, and the dynamic pressure. Therefore, dynamic stability analysis for the wing is performed by first setting α and then increasing the freestream velocity. This has been done numerically using SHARPy. Figure 9.10 shows the evolution of the first three aeroelastic modes in a V–g diagram for $\alpha = 1°$ and $4°$ at sea-level conditions, $\rho = 1.225$ kg/m^3. The unstable region is shown in gray in the figure.

Consider first the wing at a small angle of attack, $\alpha = 1°$. As we have seen in Equation (8.66), for long flexible beams, the natural frequencies of the bending modes are proportional to $\left(\frac{1}{b}\right)^2$, while those of the torsional modes to $\left(\frac{1}{b}\right)$. As a result, it is

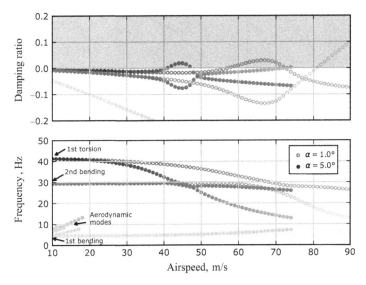

Figure 9.10 V–g diagram of the Pazy wing for two root angles of attack. Modes at very low speeds have been identified, but their features change with the airspeed.

common for very-high-aspect-ratio wings to have their first torsional frequency above that of the second out-of-plane bending mode, as is the case here. Flutter then occurs through the coalescence of the torsion and second out-of-plane modes. This is a conventional flutter mechanism driven by the loss of stiffness of the torsional (aeroelastic) mode with the dynamic pressure, which has been described in Section 3.2.2. The instability occurs when the torsional mode approaches the second bending mode, just above 60 m/s. It can also be seen in the evolution of the damping ratio in Figure 9.10 that the unstable mode for $\alpha = 1°$ recovers stlability at around 73 m/s. Such instability is known as a *hump mode* and, as the unstable eigenvalue remains close to the imaginary axes, it is typically a target for flutter suppression either via wing redesign (i.e., mass balancing) or via active methods, as will be seen in Chapter 10.

After the first flutter instability, there is a redistribution of the contribution of the torsional and second-bending mode shapes to the aeroelastic eigenvectors. The second aeroelastic mode now has a larger torsional component and, as the freestream velocity increases further, shows a large drop in frequency from around 70 m/s onward. As it approaches the first bending mode, this results in a second flutter mechanism, at 82 m/s for $\alpha = 1°$.

The wing is very flexible in out-of-plane bending, and increasing its root angle of attack results in very large displacements. The contour lines of wingtip displacements (normalized with the span) in static aeroelastic equilibrium for varying freestream velocities are shown in Figure 9.11. As an example, the wingtip displacement for $V = 50$ m/s is $0.07b$ for $\alpha = 1°$, but as much as $0.26b$ for $\alpha = 4°$. As the wing deforms, the natural frequency of the torsional (aeroelastic) mode is reduced, as can be seen in Figure 9.10. The physical mechanism that drives this is the coupling between torsional and in-plane bending DoFs that occurs as the wing bends up. This is a

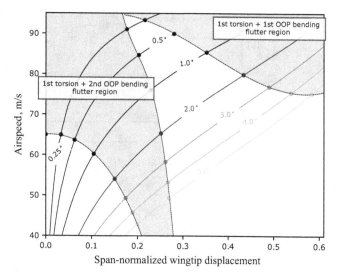

Figure 9.11 Flutter speed vs. wing displacement of the Pazy wing. Contour lines correspond to the equilibrium points with constant angle of attack (Goizueta et al., 2022).

finite-rotation effect, as a small twisting around a large out-of-plane curvature creates in-plane dynamics. Note that this always occurs in cantilever beams under large bending displacements, regardless of the type of loading (aerodynamic or otherwise). This additional softening of the third aeroelastic mode brings forward the first flutter mechanism, which for $\alpha = 4°$ now occurs at 45 m/s. As can be seen in Figure 9.10, it is still a hump mode that recovers stability at 52 m/s, while the second flutter mechanism now occurs at around 71 m/s.

A more detailed exploration of the relation between aeroelastic stability and wing deformations is included in Figure 9.11. The figure shows the stability boundaries as a function of the wingtip deflection and the freestream velocity. The static aeroelastic equilibrium conditions are then first obtained for varying root angle of attack and freestream velocity, and the contour lines for discrete values of α between 0.25° and 5° are included in the figure. For each equilibrium point, dynamic stability is assessed, and regions of flutter instability are then identified. For example, if we follow the line for $\alpha = 1°$ in the figure, we can identify the two stability regions of Figure 9.11, while the horizontal axis in the figure gives the wing deflection. For $\alpha = 1°$, the first instability occurs with a vertical displacement of the wing tip of around 10%, while the displacement in the second stability is above 35%. Results show a relatively narrow region of instability for the first flutter mode, in which the wing becomes stable again as its bending increases further. The strong dependency of the flutter characteristics on the wing displacement could be exploited by a closed-loop system that deforms the wing as a mechanism to alter its stability properties. This will be numerically demonstrated in the nonlinear control design studies of Section 10.6.2, in which the controller effectively moves the wing's state along the horizontal (constant velocity) lines, as shown in Figure 9.11.

Figure 9.12 HiLDA BWB wing experimental model in the NASA Langley's Transonic Dynamics Tunnel (photo by NASA).

Further investigations by Goizueta et al. (2022) have given further insights into the main nonlinear mechanisms that arise as the wing deforms. For wingtip displacements until around 25% of the span, the dominant effect determining the wing stability is the change of natural frequencies as the structure deforms, in particular, the drop in the frequency of the first torsional mode that we have just discussed earlier. As the displacement increases beyond that point, two additional nonlinear aerodynamic effects also become relevant: First, the changes in the wing geometry modify the relative distance between the vortex rings in the Biot–Savart law, Equation (7.21). This effect is, however, relatively modest, as the interference between panels decays with the distance and nearby panels do not significantly change their relative position. The second and more important effect is associated with the steady aerodynamics forces on the wing. As was seen in Example 7.3, steady aerodynamics are follower forces on the structure, and perturbations of the geometry may introduce additional couplings between the aeroelastic modes. For this wing, those effects are stabilizing, thus delaying the onset of instability as the deformation increases.

9.4.2   HiLDA Blended Wing Body

As an example of complex rigid-body–aeroelastic coupling, we consider next a full vehicle configuration with a *blended wing body* (*BWB*) that has been proposed for a "flying antenna" (Martinez et al., 2008). It is based on an experimental wind-tunnel model developed by Northrop Grumman, under the U.S. Air Force's High Lift-over-Drag Active (HiLDA) Wing program (Bartley-Cho and Henderson, 2008; Scott et al., 2008). A half-vehicle model, which is shown in Figure 9.12, was tested in NASA Langley's Transonic Dynamics Tunnel, and tests showed a very lightly damped in-plane bending model of the wing (Lockyer et al., 2005) that had not been predicted in the pretest linear aeroelastic analysis.

In order to understand that problem, a simulation model based on the wind-tunnel model has been built in UM/NAST. Its main properties are shown in Table 9.4. Besides

Table 9.4 Elastic and inertia properties of the BWB model.

	Body	Wing	Units
Beam axis from L.E. (root/tip)	64.38/45.60	45.60/45.60	% chord
Extension stiffness	1.69×10^8	1.55×10^8	N
Out-of-plane bending stiffness	7.50×10^5	1.14×10^4	$N \cdot m^2$
In-plane bending stiffness	3.50×10^7	1.30×10^5	$N \cdot m^2$
Torsion stiffness	2.25×10^6	1.10×10^4	$N \cdot m^2$
Mass per unit length	50	6.2	kg/m
Flat bending inertia per unit length	0.7	5.00×10^{-4}	$kg \cdot m$
Edge bending inertia per unit length	22	4.63×10^{-3}	$kg \cdot m$
Torsional inertia per unit length	4.5	5.08×10^{-3}	$kg \cdot m$

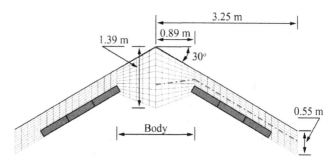

Figure 9.13 Geometry of the BWB model (Su and Cesnik, 2010).

the half-vehicle model, we created a full-vehicle representation to investigate the effects of the rigid-body DoFs. The results in this section have been presented in Su and Cesnik (2010).

Figure 9.13 shows the geometry of the full vehicle. A strain-based geometrically nonlinear beam model is used for the structure, and the location of the reference line is shown as dashed-dotted line in the figure. The local CM for the beam is at its axis. The main payload is modeled as an 80-kg concentrated mass in the symmetry plane of the aircraft and 0.89 m ahead of the beam root. Nine 2-kg concentrated masses are also evenly distributed along the beam reference line of each wing. The control surfaces are three elevons on each wing, which are also shown in Figure 9.13 and can be independently actuated. They cover 75% of the wingspan, starting from the root, and their chord is 25% of the wing chord. The unsteady aerodynamic model is based on *thin-strip aerodynamics* with sectional forces defined as in Section 3.4 and the steady aerodynamic coefficients obtained from the NACA 0012 airfoil with the Prandtl–Glauert compressibility correction. The same airfoil is chosen throughout the body and wing members.

The stability boundary is then calculated from a sequence of linearized solutions around varying equilibrium conditions, as was done in Section 9.4.1. For each flight condition, the trimmed state of the aircraft is first found using nonlinear static aeroelastic simulation. A linearized model for the vehicle dynamics is then built around

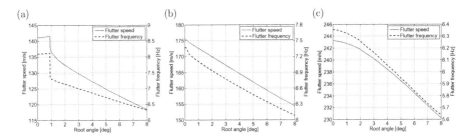

Figure 9.14 BWB flutter speed and frequency with respect to changes in the root angle of attack at three altitudes. (a) Sea level, (b) 6,096-m altitude, and (c) 15,000-m altitude (Su and Cesnik, 2010).

that equilibrium point, and eigenvalue analysis is finally performed to assess its stability characteristics. The process is repeated for increasing flight speed until either trim becomes impossible (e.g., from aeroelastic divergence) or a root has positive real part. This process has been performed under different modeling assumptions to consider free-flight flexible and rigid aircraft, and the wall-attached wind-tunnel conditions. They are discussed in the following section.

Flutter Boundary of a Half-Vehicle Model without Rigid-Body Motion

As in the wind-tunnel setup, the root angle of attack, α_0, of the half-vehicle model is varied from $0°$ to $8°$, elevons are retracted, and clamped boundary conditions are imposed (i.e., no rigid-body DoFs). Figure 9.14 shows the flutter boundaries obtained from simulations at three different air densities (corresponding to altitudes varying from sea level to 15,000 m). First, note that changes in density modify the dynamic pressure and, therefore, under constant α_0, lower altitudes correspond to a lower flutter speed. Moreover, since the effective mass, damping, and stiffness from the unsteady aerodynamics scale differently with changes in density and airspeed, Equation (6.123), a rather different evolution is observed for the flutter speed at each altitude. At sea level, the flutter speed initially increases with α_0, but a change of flutter mechanism is clearly observed for $\alpha_0 \approx 0.92°$, similar to the one discussed in Figure 9.11. The flutter modes are shown in Figure 9.15 for two different values of the root angle of attack. As can be seen in the figure, for small values of α_0, flutter results from the coupling of the first out-of-plane bending and torsion modes of the wing. However, at sufficiently large values of α_0, the first in-plane bending mode also contributes to the flutter instability. In Figure 9.14, we can also see that, at the higher altitudes (6,096 and 15,000 m), the flutter speed monotonically decreases with α_0. This is mainly because, at higher α_0, the geometrically nonlinear effects associated with an increased out-of-plane bending facilitate the coupling between the torsional and in-plane bending deformations (i.e., rotations in two axes produce a rotation on the third axis), which results in a larger in-plane bending of the wing.

In order to visualize the time response at particular conditions, consider the aircraft at 6,096-m altitude, $8°$ root angle of attack, and two airspeeds, namely, 147 m/s (below flutter) and 162 m/s (above flutter). Figure 9.16 shows the wingtip vertical

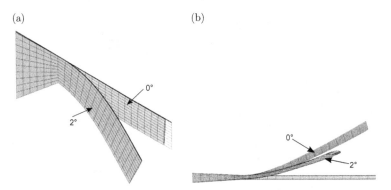

Figure 9.15 BWB flutter mode (at the maximum out-of-plane amplitude) of the half-wing model at sea level and for two root angles of attack. (a) Top view and (b) rear view (Su and Cesnik, 2010).

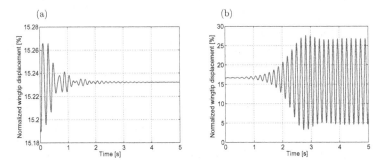

Figure 9.16 Wingtip displacements of BWB, normalized by the semispan, at 6,096-m altitude and 8° root angle. (a) Speed: 147 m/s and (b) Speed: 162 m/s (Su and Cesnik, 2010).

displacement of the half-model subjected to a small disturbance. In Figure 9.16(a), we can see that the disturbance dies out as expected from a stable system. However, in Figure 9.16(b), we can see that the disturbance grows when the flight speed is higher than the flutter speed as expected. Moreover, due to the nonlinearities present in the system, the motion reaches a saturation point and subsequently displays a limit cycle oscillation (LCO) with fixed frequency and large amplitudes of the wingtip displacement.

Flutter Boundary of the Full Vehicle in Free Flight

When we look at the complete vehicle in free flight, now with the inclusion of the rigid-body modes, the situation changes. Before calculating flutter, let us first look at the basic flight dynamic and elastic frequencies of this vehicle. By trimming the vehicle at every flight condition, we can obtain the linearized longitudinal flight modes at different altitudes and flight speeds. The phugoid and short-period modes are shown in Figure 9.17, where it can be seen that they are always stable in the range 0–15,000 m and 0.2–0.7 Ma. It is important to note that, contrary to the cantilever condition above, each altitude and flight velocity pair determines now the root angle of attack at which lift and weight balance each other. For the phugoid mode, an increase in Mach brings

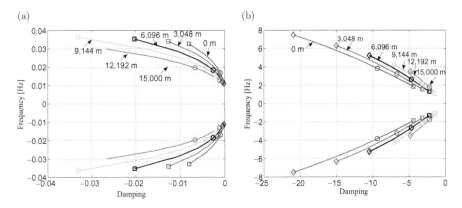

Figure 9.17 Effect of altitude in the phugoid and short-period modes of the BWB configuration. Airspeed is varied from 0.2 Ma (square) to 0.7 Ma (diamond). (a) Phugoid mode and (b) short-period mode.

a reduction on both frequency and damping (a destabilizing effect) of the eigenvalues. The contrary behavior is observed on the short-period mode in Figure 9.17b. Also, for a given flight speed, the effect of increasing the altitude is to increase the damping of the phugoid mode (a larger stability margin), while this is again reversed for the short-period mode.

Table 9.5 contains the first natural frequencies of the elastic vehicle in vacuum. Combining that with Figure 9.17, it can be seen that the frequency of the first out-of-plane bending mode is in the same range as the frequency of the short-period mode, which implies that a strong coupling between both modes may appear through the aerodynamic loading.[3] To investigate the coupled aeroelastic–flight dynamics problem, consider first the vehicle flying at 6,096 m altitude and trimmed for level flight at each flight speed where the stability is evaluated. In order to assess which rigid-body motion has the highest influence on the aeroelastic stability, multiple constraints on the rigid-body DoFs are imposed when computing the flutter boundary. The flutter onset conditions for each of those conditions are included in Table 9.6. Case 1 corresponds to clamped boundary conditions on the aircraft half-model (i.e., rigid-body motions are removed). Case 2 corresponds to a model that is free to displace in the vertical direction, while Case 3 is free in both plunging and pitching. Finally, Case 4 has no constraints on the rigid-body modes.

The root loci of Cases 1 and 4 are plotted in Figure 9.18, where an identification has been included for the various dynamic modes of the system (by analyzing the features of the eigenvalue Φ obtained in the vibration analysis step in Figure 9.6). Recall that they correspond to the dominant feature in vacuum, as the shape of the

---

[3] Recall that the vibration modes and the rigid-body modes are linearly orthogonal, and under small-amplitude oscillations, they behave as independent oscillators. Nonlinear couplings between modes are possible under large-amplitude vibrations. Aerodynamic forces do introduce linear (as well as nonlinear) couplings as, for example, a change in the rigid pitch generates lift that deforms the wing, thus increasing its bending.

Table 9.5 BWB fundamental modes and frequencies.

No.	Mode	Frequency (Hz)
1	First out-of-plane bending	3.3
2	First in-plane bending	10.9
3	Second out-of-plane bending	20.4
4	Third out-of-plane bending	55.6
5	Second in-plane bending	69.5
6	Fourth out-of-plane bending	82.0

Table 9.6 Flutter boundaries of the BWB for different rigid-body motion constraints.

Case	D.O.F.	Speed (m/s)	Frequency (Hz)
1	Fully constrained	172.52	7.30
2	Free plunging only	164.17	7.07
3	Free pitching-plunging	123.17	3.32
4	Free flight	123.2	3.32

modes changes with the airspeed. The elastic modes are identified in Figure 9.18a, and they are also present in Figure 9.18b with small shift toward lower frequencies, and so more elastic shapes are visible on the free flight-model (despite the slightly larger frequency range). Figure 9.18b also identifies the longitudinal flight-dynamic modes that are not present on the cantilever model. The instability mechanism for the aircraft in free flight is a *body-freedom flutter* (BFF), which has previously appeared in another flying wing in Example 6.6. The flutter mode is a complex eigenvector, and a snapshot of it, corresponding to the time instant with minimum pitch angle, is shown in Figure 9.19, where the vehicle in wireframe denotes the reference shape and position. From this figure, one may find that the rigid-body component of the flutter mode is a coupled pitching–plunging, while the elastic component consists of both out-of-plane bending and twist. Note finally that as this is a longitudinal motion, there are no differences in stability characteristics between Cases 3 and 4. Finally, when only plunging is allowed (Case 2), the main coupling mechanism between the elastic deformations and the rigid-body DoFs is removed and the flutter speed is very close to that of the fully constrained model (Case 1).

Response of Full Vehicle to a Discrete Gust

The dynamics of the free-flying aircraft in a gust encounter are finally considered. To investigate nonlinear effects in the vehicle stability, a flying condition just before the flutter speed is considered. In particular, simulations are carried out at 6,096 m and Ma = 0.39, corresponding to a speed of 122 m/s, which is approximately 1% below the flutter onset. A uniform "1-cosine" discrete gust, defined as in Section 2.4.2, is

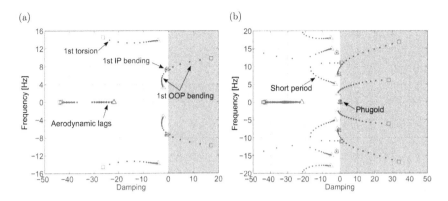

Figure 9.18 Stability diagram of the BWB configuration at 6,096-m altitude. Airspeed is varied from 100 m/s (triangle) to 200 m/s (square). (a) Fully constrained and (b) free flight (Su and Cesnik, 2010).

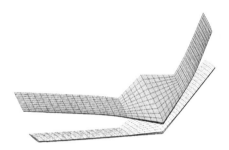

Figure 9.19 BWB flutter mode shape of the free-flight case at 6,096-m altitude.

used as a disturbance. The gust intensity is $w_0 = 15.24$ m/s (50 ft./sec.), and its length is $H = 6.86$ m, which corresponds to 12.5 times the wing chord. The simplified stall model of Su and Cesnik (2011a) has been considered in these results, in which the sectional lift is kept constant for local angles of attack above a critical stall angle.

Figure 9.20 shows the response of the vehicle to the gust and compares it to the response in calm air. As expected, the vehicle is stable in calm air since the flight speed is below the flutter limit. However, the response of the vehicle to the discrete gust is clearly unstable, which is something that would not be captured by a linear model. When encountering the gust, the aircraft sinks, which increases its flight speed. After around 15 s, the instantaneous flight speed is greater than the flutter onset and the aircraft displays growing oscillations on both the pitch rate and the wing's elastic displacements (shown in Figure 9.20c and Figure 9.20d, respectively). This motion eventually develops into an LCO with a low-frequency amplitude modulation, or beating. This behavior is typical of subcritical Hopf bifurcation (Hassard et al., 1981) and can be particularly detrimental when a certain flutter clearance must be guaranteed for certification.

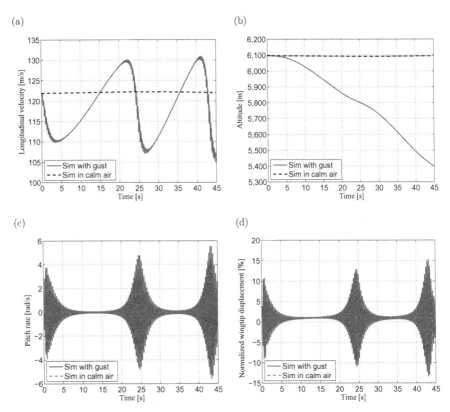

Figure 9.20 BWB dynamics in calm air and under gusts at the subcritical condition at 6,096-m altitude. (a) Body-attached longitudinal velocity, (b) altitude, (c) body-attached pitch rate, and (d) wingtip displacement (Su and Cesnik, 2010).

9.4.3   University of Michigan's X-HALE UAS

We consider next the University of Michigan (U-M) X-HALE unmanned aerial system (UAS) (Cesnik et al., 2012; Jones and Cesnik, 2015). This aircraft has been developed at the U-M's Active Aeroelasticity and Structures Research Laboratory as a nonlinear aeroelastic testbed for model validation and for the development and testing of nonlinear control laws for very flexible aircraft. Its design was conducted using UM/NAST. Snapshots of the vehicle on the ground and in flight are illustrated in Figure 9.21 and show the very large deflections that appear under aerodynamic loading (around 18% of the semispan in cruise conditions). The main wing of the aircraft has an aspect ratio of 30 and is composed of six panels of one-meter span each with five pods in the corresponding junctions. Each wing panel has constant structural and aerodynamic properties, which are shown in Tables 9.7–9.10. The outer panels are mounted on a 10° dihedral, as shown in Figure 9.21a. The center tail is actuated along the longitudinal axis, and it can be rotated to act either as a horizontal or vertical tail. The remaining four tails (two on each semispan) are movable and act as elevators. Three ventral fins, between the wing and the center and inboard tails, can also be attached to

(a)   (b)

Figure 9.21 (a) X-HALE UAS on the ground and (b) shortly after takeoff.

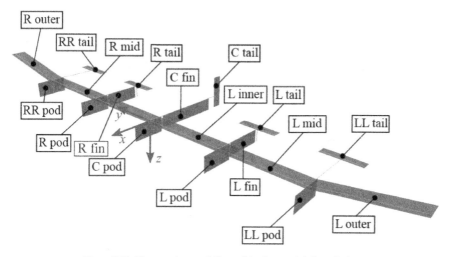

Figure 9.22 Nomenclature followed in the model description.

augment the lateral stability of the vehicle, although they are not shown in the prototype pictured in Figure 9.21. Finally, five motor-propeller combinations, one located at
each pod, provide thrust. The pods also contain the avionics equipment, battery packs,
instrumentation, as well as the landing gear.

The X-HALE poses multiple challenges for aeroelastic modeling and simulation.
First, it has a relatively complex geometry with multiple intersecting lifting surfaces.
Second, all its primary subcomponents are very lightweight and display both elastic
and inertia coupling (i.e., full $\mathcal{M}$ and $\mathcal{C}$ matrices in Equation (8.5)). Third, in order
to achieve the span loading of the wing (see Figure 5.2), it has a rather involved mass
distribution. Furthermore, the first bending mode of the aircraft has a natural frequency
close to 0.6 Hz, and as a result, the vehicle displays strong aeroelastic and flight-
dynamics coupling. The wing also has relatively low stiffness in torsion, which means
that the moments from the tails affect the lift distribution on the wing by modifying
its twist.

The full-vehicle simulations in this section have been carried out with both SHARPy
and UM/NAST applied to the configuration shown in Figure 9.22. The figure also shows
the naming conventions for the different lifting surfaces and the global reference

Table 9.7 Stiffness distribution of the X-HALE model [i: inner; m: mid; o: outer].

Section [-]	EA [N]	GJ [N · m^2]	EI_{yy} [N · m^2]	EI_{zz} [N · m^2]
R i/m/o	2.14×10^6	5.93×10^1	1.12×10^2	6.35×10^3
Fuselage/fin	5.39×10^7	5.39×10^7	5.39×10^7	5.39×10^7
Tail	3.21×10^6	2.14×10^1	9.10×10^1	4.27×10^3

Table 9.8 Distributed mass of the X-HALE model in local coordinates for each beam segment [x: torsional; y: in-plane; z: out-of-plane].

Section [-]	Mass [kg/m]	I_{xx} [kg · m]	I_{yy} [kg · m]	I_{zz} [kg · m]	I_{yz} [kg · m]	x_{cg} [mm]	y_{cg} [mm]	z_{cg} [mm]
R i/m	3.94×10^{-1}	8.1×10^{-4}	1.22×10^{-5}	7.97×10^{-4}	6.5×10^{-6}	0	29.4	0
R outer	5.0×10^{-1}	8.1×10^{-4}	1.22×10^{-5}	7.97×10^{-4}	6.5×10^{-6}	0	21.4	0
Fuselage	0.0429	2.91×10^{-9}	1.46×10^{-9}	1.46×10^{-9}	0	0	0	0
Tail	0.2614	1.6×10^{-4}	2.910×10^{-6}	1.57×10^{-4}	0	0	14.4	0
C fin	0.5092	3.19×10^{-3}	9.34×10^{-5}	3.28×10^{-3}	0	0	0	0
L/R fin	0.3208	8.17×10^{-4}	5.88×10^{-5}	8.76×10^{-4}	0	0	0	0

frame. The stiffness and mass distributions of the aircraft are given in Tables 9.7 and 9.8, respectively. Only right-wing information is given, as the stiffness and distributed mass data are symmetrical for this model (although not exactly so in the actual UAS). All structural properties are given in the local beam axes (frame of reference S) and are identified using the notation of Equation (8.7). In particular, EI_{yy} represents here the out-of-plane bending stiffness of the wing, and EI_{zz} represents the bending stiffness in the plane of each lifting surface. A constant coupling term between both, $k_{yz} = 4.63 \times 10^1$ N · m^2, also appears in the main wing.

The inertia of the pods is defined in the form of three lump masses with nonzero moments of inertia rigidly attached to a node along the beam reference line of the wing. The inertia components of each pod, which are identified using the notation of Figure 9.22, are given in Table 9.9. The table includes the location of the center of gravity for each pod (given with respect to the local cross section in the wing reference line), its mass, and its components of the inertia tensor, Equation (4.21), written in the local structural reference frame (frame S).

Finally, the geometric parameters needed by the aerodynamic model are given in Table 9.10, including the distance between the leading edge of the lifting surfaces and the elastic axis where the structural properties given in Tables 9.7–9.8 are defined. A typical aerodynamic discretization is shown in Figure 9.23. For the results in this chapter, reasonable convergence was obtained with a model with over 600 bound vortex rings and wakes on all lifting surfaces that extend to 4 m (14-chord length) downstream. All results are obtained at a reference flight speed of $V = 14$ m/s, with air density $\rho = 1.225$ kg/m^3 and gravitational acceleration $g = 9.807$ m/s^2.

Table 9.9 Lumped mass of the X-HALE pods given in the wing structural reference frame [x: torsional; y: in-plane (positive upstream); z: out-of-plane (positive up)].

Mass [kg]	x_{cg} [m]	y_{cg} [m]	z_{cg} [m]	I_{xx} [kg·m²]	I_{xy} [kg·m²]	I_{xz} [kg·m²]	I_{yy} [kg·m²]	I_{yz} [kg·m²]	I_{zz} [kg·m²]
C pod									
0.3746	0	0.1	0	1.15×10^{-3}	0	0	8.90×10^{-4}	0	8.90×10^{-4}
1.0462	3.97×10^{-3}	0.0612	-0.0168	1.48×10^{-2}	2.32×10^{-4}	2.27×10^{-5}	2.82×10^{-3}	4.50×10^{-4}	2.50×10^{-4}
0.023	0	0.260	-0.023	0	0	0	0	0	0
L/R pod									
0.548	-0.01	0.090	0	1.54×10^{-3}	0	0	8.90×10^{-4}	0	8.90×10^{-4}
0.929	2.14×10^{-3}	0.04	1.39×10^{-2}	1.13×10^{-2}	-1.21×10^{-3}	1.06×10^{-5}	3.21×10^{-3}	4.60×10^{-5}	8.48×10^{-3}
0.023	0	0.259	-0.023	0	0	0	0	0	0
LL/RR pod									
0.571	-0.01	0.091	0	1.54×10^{-3}	0	0	8.90×10^{-4}	0	8.90×10^{-4}
0.929	2.14×10^{-3}	0.04	-1.39×10^{-2}	1.13×10^{-2}	-1.21×10^{-3}	1.06×10^{-5}	3.21×10^{-3}	4.60×10^{-5}	8.48×10^{-3}
0.023	0	0.259	-0.023	0	0	0	0	0	0

Table 9.10 X-HALE aerodynamic model description.

Part	Span	Chord	Elastic axis	Dihedral	Airfoil
[–]	[m]	[m]	[% chord]	[deg]	[–]
Inner/mid wing panel	1	0.2	28.8	0	EMX-07
Outer wing panel	1	0.2	28.8	10	EMX-07
C tail	0.385	0.11	32.35	0	Flat plate
LL/L/R/RR tail	0.48	0.11	32.35	0	Flat plate
C fin	0.15	0.78	122.56	0	Flat plate
L/C/R pod	0.184	0.38	60.93	0	Flat plate

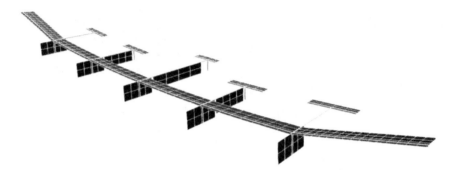

Figure 9.23 Aerodynamic model of the X-HALE.

As can be seen in Figure 9.21, the X-HALE is a very flexible aircraft with large shape changes between its *trim shape*, that is, its aeroelastic equilibrium condition in steady forward flight, and its *jig shape*, its geometry on the ground. As a result, stability and dynamic response in gust encounters or maneuvers need to be assessed around the trim point of interest. In the rest of this section, we will first numerically investigate the static aeroelastic characteristics of the aircraft, followed by a study on the dynamic response to a vertical gust.

Static Aeroelastic Results

The static aeroelastic equilibrium conditions have been found for the aircraft clamped at the central pod/main spar intersection and are shown in Figure 9.24 for three differ-ent body angles of attack, namely $0°$, $3°$ and $5°$. The body angle of attack is defined here as the angle between the freestream and the undeformed central pod. The thrust has been set to zero, and the tails are aligned with the pods ($\delta = 0°$). Results are written in terms of the coordinates in the body-attached frame of reference of Figure 9.22. For convenience, the vertical displacement is shown in reverse scale so that the wing bends upward. The deformed aircraft shapes are obtained with both SHARPy and UM/NAST and show almost negligible differences between both solvers for deflections that go as high as 20% of the wing semispan.

Trimming a very flexible aircraft is significantly more complex than doing so in a rigid one. In addition, the unconventional configuration of the X-HALE contributes to this difficulty. Given the relatively low torsional stiffness of the wing spar, a positive

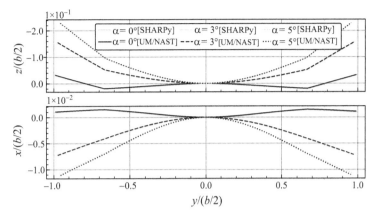

Figure 9.24 Spar deformation for different root angles of attack.

increase in control surface deflection at any of the tails has two effects: First, it produces a nose-down pitching moment and a momentary positive contribution from the tail to the overall lift; and second, the main wing eventually suffers a drop in lift, especially in the outer parts of the wing, due to the negative twisting moment produced by the tail.

The X-HALE has some characteristics that, when considered in the trim routine, simplify the numerical process. First, the four outer tails are actuated using a single input. Second, the five motors are controlled using a *symmetric* and *antisymmetric* thrust in order to provide lateral control. Finally, for our current modeling accuracy, the aircraft can be considered to be symmetric, and only the longitudinal equations are used by the trim algorithm. Rigid aircraft trim algorithms assume constant aerodynamic derivatives near the trimmed state (De Marco et al., 2007). This is not necessarily valid here since the aerodynamic surfaces change shape between updates of the trim variables, such as tail-induced wing twist or lift-distribution variations due to structural deformations. The approach implemented in both SHARPy and UM/NAST is based on a modified Newton–Raphson with approximated Jacobians obtained by first-order finite differences. In SHARPy, a diagonal Jacobian is updated at every iteration, while in UM/NAST, the Jacobian is calculated just in the first step and then updated at next iterations using Broyden's method (Broyden, 1965). Both trim algorithms compute the resultant loads at the origin of the body-attached frame B, which is modeled as clamped, and iterate on the trim parameters until they are smaller than a given tolerance.

The computed trim conditions, in terms of the root angle of attack of the main wing (measured as before as the body angle at the center of the fuselage), α, deflection on the control surfaces, δ, and thrust on the five engines, T, are included in Table 9.11. Results are presented for both SHARPy and UM/NAST for the central tail in both vertical and horizontal configurations. A slightly smaller angle of attack is needed with the vertical tail configuration, mainly to compensate for the lack of lift from the central tail. The agreement for α and δ between both codes is good, while the small difference in thrust reflects varying convergence rates between codes in induced drag

Table 9.11 Trim results for cruise flight.

Code	Tail orientation	α [°]	δ [°]	T [N]
SHARPy	Vertical	2.64	1.19	0.223
	Horizontal	2.21	0.52	0.213
UM/NAST	Vertical	2.59	1.15	0.179
	Horizontal	2.38	0.66	0.179

Table 9.12 Natural vibration frequencies (in Hz) of the aeroelastically trimmed X-HALE.

Mode	Mode type	SHARPy	UM/NAST
1	1st Out-of-plane	0.57	0.59
2	1st Torsion	2.50	2.57
3	2nd Out-of-plane	3.62	3.70
4	1st In-plane	4.58	4.45
5	2nd Torsion	6.59	6.57

prediction. Finally, after the trim of the flexible aircraft has been obtained, the linear normal modes of the structural subsystem are computed to assess the natural vibration characteristics in flight. This is done from the generalized eigenvalue problem defined with the structural mass and the stiffness matrices around the equilibrium shape, as in Equation (8.27). The first five natural frequencies, obtained again using both SHARPy and UM/NAST, are presented in Table 9.12.

Vertical Gusts

Next, we consider the response of the aircraft to spanwise-homogeneous "1-cosine" gusts, defined in Equation (2.54). As the aircraft have multiple lifting surfaces, the gust encounters them at different times as it travels through the airfoil. This is illustrated in Figure 9.25, which presents the time history (in the dimensionless time $s = 2Vt/c$) of the normalized velocities at the reference axis of three different lifting surfaces (defined in Figure 9.22) and for two gust lengths. Note that the shorter gust ($2H = 15c$) has nearly left the main wing when its intensity is maximum in the central tail. As will be shown next, this has a substantial impact on the dynamics of the vehicle, given the strong flight dynamics–aeroelastic couplings.

For a discrete vertical gust with $2H = 15c$, the instantaneous vertical displacement of the wing spar center is shown in Figure 9.26, while the changes to the longitudinal and vertical velocities at that point are shown in Figure 9.27. Results are presented for the central tail in vertical and horizontal configurations, and again for both SHARPy and UM/NAST. Figure 9.26 shows results for a very long simulation time (10 s in physical time). From Figure 9.25, it can be seen that the gust vanishes at $s = 50$, but this initiates a very slowly damped phugoid mode with a period of around 7.5 s, which dominates the long-term response of the aircraft.

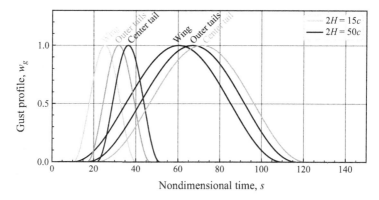

Figure 9.25 Gust velocity time histories for different aerodynamic surfaces (as seen from the elastic axis in steady forward flight).

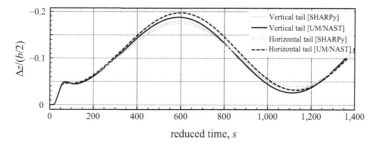

Figure 9.26 X-HALE under a vertical gust, $2H = 15c$. Vertical displacement at the wing midpoint.

The current formulation uses a material point on the structure as a reference for rigid-body dynamics of the vehicle (the frame of reference B, defined in Section 8.3). This implies that the kinematics of that reference include both local and global effects. An example of this can be seen in the longitudinal velocity in Figure 9.27, which includes high-frequency vibrations associated with the first in-plane mode of the wing.

Further insights into the vehicle dynamics can be obtained from the time history of the pitch angle of the body-attached frame B. Changes to this angle from the reference condition are shown in Figure 9.28. From Figure 9.25 we can see that the outer tails enter the gust around $s \approx 20$ and reach the peak velocity at $s = 35$. This can be observed in Figure 9.28, where the pitch reaches the maximum due to the wing-only excess of lift, and then the increasingly stronger tail influence causes it to pitch down heavily. Due to the span loading of this wing, the influence of the main wing on the pitch-up part of the response is weaker than could be expected from conventional aircraft. Since the main contribution to both aircraft lift and mass comes from the wing, the aerodynamic center and the CM are located close to each other. Thus, a lift increment in the wing generates a vertical velocity increase, but not a strong pitch-up. Note also that having the center of lift and gravity so closely located also results in low damping of the phugoid mode.

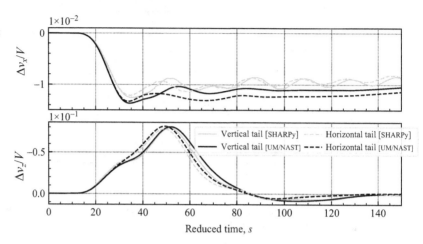

Figure 9.27 X-HALE under a vertical gust, $2H = 15c$. Inertial velocity at the wing midpoint.

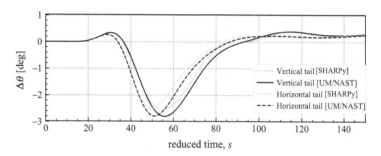

Figure 9.28 Pitch response of the X-HALE to a vertical gust $2H = 15c$.

The tail orientation also has an effect on the pitch-down part of the response ($30 \lesssim s \lesssim 50$). The peak value of the pitch is similar for both tail orientations. In contrast, the peak is reached earlier in time for the horizontal tail. This is a predictable result, as the increased horizontal wetted area in the aft part of the aircraft will make the aircraft pitching characteristics more responsive to perturbations in the angle of attack.

Finally, the time history of the wing-root deformations during the gust encounter is shown in Figure 9.29. The out-of-plane bending and torsional moment strains have been included, which correspond to the first and second components of the curvature vector in the local material frame. Results are qualitatively similar to those obtained for a single wing in Figure 9.5, with the difference that the rigid-body inertia includes a negative bending curvature after around $s = 35$ that gives the maximum loading on the wing.

9.5 Design Strategies for Very Flexible Aircraft

So far in this chapter, we have outlined the main features for modeling and simulation of the dynamics of very flexible aircraft and have investigated some typical scenarios

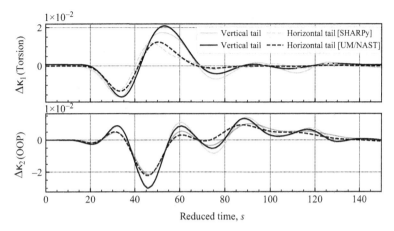

Figure 9.29 Right wing-root element strain components due to gust response for $H = 15$ chords.

through numerical examples. To conclude it, it is interesting to assess the impact that the introduction of more flexible airframes is having on the aircraft design process.

The design methods adopted by the aircraft industry stem from the cumulative knowledge gained through many previous successful aircraft design exercises. They are also continuously evolving as new features are sought and innovative technologies are available, and yet the vast majority of aircraft in operation are built with a relatively stiff airframe that can be analyzed using linear models. However, as we have already discussed in Section 9.1, the quest for additional fuel efficiency demands air vehicles with lighter structures and higher aspect-ratio wings. This often results in very flexible-aircraft concepts, whose primary structures may display large deformations under loading, and therefore need to be studied including geometrically nonlinear effects. Their structural vibrations will also have lower natural frequencies, thus resulting in strong couplings between the rigid-body and flexible characteristics of the aircraft, as we have already discussed in Chapter 6.

A very flexible airframe is essentially the result of a design choice. The designers seeking the additional performance need to jump outside the comfort zone of previous vehicles and regression charts and need to substantiate their studies with suitable analysis methods from the *conceptual design* phase. As we have seen in the numerical studies earlier, the resulting aircraft are likely to display strong (and often nonlinear) interactions between structural, aerodynamics, and flight dynamics, which need to be harnessed in the design optimization to achieve a high-performing configuration. A major implication of this is that some of the existing design procedures, built on civil (FAA, 2011) or military (MIL-HDBK-516B, 2005) standards, will need to be updated to a more generic, and essentially nonlinear, analysis paradigm. As a minimum, the new design processes should be built on the following premises (Cesnik et al., 2014): (1) the deformed aircraft geometry, which will depend on the operating (trim) condition, should now be the baseline in weight, structural, and stability analyses; (2) transient dynamic simulations (e.g., gust response) should include large nonlinear displacements of the aircraft; (3) aeroelastic models should incorporate the rigid-body motion of the vehicle; and (4) flight-dynamics models should incorporate nonlinear aeroelastic effects. We further expand on them in the following sections.

In the rest of the chapter, we present first a review of some aspects of the standard airframe design process, which is geared toward relatively stiff airframes. We then provide some recommendations for the development of tools and frameworks suitable for very flexible aircraft design. The focus will be on geometrically nonlinear structural analysis and coupled flight-dynamics–aeroelastic effects and their impact on current design practices. Finally, Section 9.5.2 shows a numerical example for a very flexible unmanned aerial vehicle (UAV).

9.5.1   Critical Review of Some Standard Analysis Procedures

As we have mentioned in the introduction, several aspects of the design process need to be reassessed for very flexible aircraft. We discuss here some important ones in aeroelastic stability analysis, response to dynamic loads, and flight stability and control. For each discipline, we will first briefly summarize the current practice, before outlining the key challenges associated with the design of very flexible aircraft.

Aeroelastic Stability Evaluation

Aeroelastic stability analysis is concerned with the identification (and prevention) of any occurrence of flutter or divergence within the extended flight envelope. In particular, flutter clearance is performed through a combination of experimental testing campaigns and linear analysis methods, as we discuss in some detail in Chapter 11. Cheap computational power has enabled the routine use of higher fidelity unsteady aerodynamic models, using methods such as those described in Section 6.5.4. However, the vast number of simulations that are needed for certification still makes the potential flow aerodynamics of Section 6.5.1 very difficult to replace. Instead, corrections to the linear aerodynamic analysis predictions are generally utilized, using either some targeted higher fidelity aerodynamic simulations or wind-tunnel measurements, or both. Correction methods based on either steady (Palacios et al., 2001) or unsteady (Nguyen et al., 2019) high-fidelity data are widely used, although their performance needs to be carefully assessed for each particular configuration. Some large development programs also include aeroelastic wind-tunnel tests as a derisking strategy. A dynamically scaled model of the full-scale vehicle is built and nonlinear unsteady aerodynamics are investigated in the wind tunnel. The results are used to calibrate the simulation model on which the type certification is based.

A typical flutter analysis process for aircraft certification is then mainly based on linear aeroelastic analysis methods, possibly corrected with data from wind-tunnel tests and higher fidelity aerodynamic models, and supplemented by some higher fidelity aeroelastic simulations at the most critical flight conditions. Here, the assumption has been that the timescales in the flight dynamics are much larger than the aeroelastic timescales (i.e., with modal frequencies at least one order of magnitude apart) and therefore that both analyses can be effectively decoupled. However, as we have shown in Figure 5.3, very flexible aircraft are likely to display some very low natural vibration frequencies, which prevent this decoupling. The resulting stability analysis then needs the framework of Section 6.6, with the linear normal modes including both rigid and elastic components and the aerodynamics including steady and unsteady contributions. In particular, BFF is more likely to occur from the coupling between the first bending frequency and the short-period rigid-body mode. As was discussed in Section 6.6, BFF typically appears on vehicles with low pitch inertia (e.g., flying wings and BWB body configurations). For long-endurance aircraft, it is also strongly affected by the fuel fraction, which modifies the inertia, and therefore the vibration characteristics, of the wings between different flight segments. Removing BFF from the flight envelope via redesign then becomes very challenging without also penalizing performance, and active flutter suppression strategies have been suggested instead (Love et al., 2005).

The strong dependency on wing deformations brings an additional complication in flutter prediction for very flexible aircraft. The geometry, inertia, and stiffness become dependent on the flight condition, and therefore the various parameters that determine flutter onset (the CM of the aircraft, structural frequencies of the wing, flight speed, etc.) are no longer independent from each other. However, for supercritical bifurcation, the flutter onset conditions can still be determined by linear stability analysis about the (nonlinear) reference equilibrium. As we have seen in Section 9.4.1, the aeroelastic stability characteristics of a wing can be very sensitive to the reference geometry. This may still be the case, under relatively small deformations, as shown for joined wings by Cesnik and Brown (2003) and Strong et al. (2005), and on T-tails by Murua et al. (2014). In Example 8.3, Section 8.6, we have also seen that a geometrically nonlinear structure displays couplings between bending in both axes and the torsional DoF (a finite-rotation effect). In particular, in-plane displacements of a wing with out-of-plane steady forces produce an incremental torsional moment. This will affect the linearized solutions and, therefore, would need to be included in flutter analysis around a nonzero equilibrium (in a T-tail, for a nonzero incidence of the horizontal tail).

The geometrically nonlinear model for the structure needs to be coupled with a suitable unsteady aerodynamic solver, that is, a potential flow solver of the type discussed in Chapter 7, or, for transonic flight, a numerical solver of the Euler or Navier–Stokes equations (e.g., Bendiksen, 2006; Strganac et al., 2005). This may capture new physical phenomena that do not appear in linear studies. A good example is the wash-out effect on a sweptback wing, that is, the reduction in the effective angle of incidence along the span due to wing bending (Torenbeek, 1982). It is easy to show that it has a strong beneficial effect on delaying wing divergence (Bisplinghoff et al., 1955). However, Bendiksen (2006) has shown that it also has a stabilizing effect on the dynamic aeroelastic behavior of wings of a high aspect ratio in the transonic regime. In particular, the wash-out modifies the location of the shock waves toward the wing tip in such a way that the resulting instantaneous pressure distribution reduces the amplitudes of the oscillations. In general, LCOs on high-aspect-ratio wings depend both on structural and aerodynamic nonlinearities and may display, as we have just seen, new and interesting physics. Finally, it is possible, as shown by Patil et al. (2001) and Kim and Strganac (2003), as well as in our example in Section 9.4.2, that the nonlinear stability corresponds to a subcritical LCO, which would correspond to an unstable flight points at speeds below the flutter boundary.

Structural Design Load Evaluation

We have discussed dynamic loads in Section 6.4.4 using linear analysis methods. The actual approach used is, as always, tailored to each particular class of aircraft. For vehicles with a relatively rigid airframe, such as a fighter aircraft, design loads are often obtained from experimental testing and subsequent analytical corrections. In particular, a wind-tunnel model is built and instrumented to measure component loads (i.e., hinge moments on a control surface). Thus, a database on internal loads is built as a function of the flight speed and angle of attack. A linear aeroelastic model of the airframe is then built and used to compute flex-to-rigid ratios, as we have discussed in Section 5.3, which are then applied on the component loads. This results in a database of *elastified* loads, which is incorporated into a 6-DoF flight simulator that is used to estimate the structural loads on certain specified maneuvers. Any potential structural integrity risks resulting from the worst-case loads are then addressed by stress analysis.

For more flexible aerostructures, such as those currently found in large transport aircraft, the analysis is mostly based on linear aeroelastic models, with experimental corrections on the potential flow aerodynamics similar to those used for flutter analysis. This traditional approach is increasingly challenged by advances in higher fidelity aerodynamics (Kroll et al., 2016), which can provide sufficiently accurate loading directly from simulation, albeit at a higher computational cost. Loads at certain monitoring stations are then obtained on a variety of configurations and flight conditions, which may correspond to either trim flight or specific instants during a maneuver. Additional loading conditions are defined from the response to discrete gusts and continuous turbulence.

As has been seen, conventional practice relies on linear aeroelastic models calibrated with wind-tunnel data. Linear theory can then be directly used to obtain the power-spectral density of the expect loads, using strategies similar to those of Section 3.7. This radically changes, however, for very flexible aircraft. While flutter analysis was still possible on a lineared aircraft model (around a nonlinear equilibrium), the transient dynamics of a very flexible aircraft needs to be computed using nonlinear time-domain simulation. This was the approach followed in the examples of Section 9.4, which have showed strong couplings between the flight dynamics and the structural response. While those examples have only displayed moderate twist deformations, for design under critical loads, models may need to be augmented to account for local stall conditions (e.g., Su and Cesnik, 2011). Stall acts as a saturation on the loading and, therefore, reduces the maximum loading on the structure – while of course having an adverse effect, and possibly unacceptably so, on flight performance.

Finally, it is worth noting that geometrically nonlinear effects on the wing dynamics may substantially modify the response of the aircraft to ground loads. This is particularly relevant for landing loads, as a more elastic wing will absorb a larger part of the energy during the impact of the landing gear with the runway.

Flight Simulation Development

Flight simulation models often incorporate linear aeroelastic effects. The most common situation is that the frequency difference between rigid and flexible dynamics enables the use of *elastified* stability and control databases built from wind-tunnel models on a rigid aircraft model, and flex-to-rigid ratios determined from linear aeroelastic models. Those are then incorporated into a flight simulation model, using the strategies of Chapter 5, to assess the effect of airframe flexibility on stability, control, and aircraft handling.

The stability derivatives will change with the geometrical changes of a very flexible aircraft, in a similar manner to the flutter models. Therefore, they need to be estimated for the trimmed aircraft at each flight point and also during maneuvers. This may have a significant impact on the load and balance diagrams (Torenbeek, 1982) that are used to establish safety margins for payload and fuel distribution during operation. In general, shifting the CM of the aircraft results in a rebalancing of the loading between the front and rear landing gear, as well as forces appearing on the tail during maneuvers. For a very flexible aircraft, the instantaneous position of the CM, and also the aircraft moments of inertia, depend not only on the mass and distribution of the remaining fuel but also on the instantaneous geometry of the (deforming) airframe, which changes between flight points.

The effect of airframe flexibility on the aircraft balance can be addressed in at least two different ways. The simplest solution is to introduce additional safety margins in the allowed weight distribution that account for the larger shifts in the aircraft CM due to airframe deformations. While this would not essentially modify the current weight analysis procedures, it may be too restrictive. A second alternative is then to define the critical weight distribution for each operation condition, including the expected deformed airframe shape. To achieve this, for each payload, fuel distribution, and flight point, the shape of the deformed airframe would need to be estimated, and an analysis of the aircraft stability should be performed to establish the critical conditions (see, for example, its impact on a Helios-like aircraft as discussed in Su and Cesnik, 2011). The result is then either clearance or more restricted limits to the acceptable fuel and payload distributions. It is, therefore, clear that for very flexible aircraft, aeroelastic effects need to consider in the weight optimization process from early on in the design process (Grouas, 2001). If aeroelastic phenomena are only included once the layout is fixed, this may result in dramatic penalties in the performance of the final design.

Finally, the changes of location of the CM of the aircraft with the wing deformation have an important effect on the reference frames used for flight simulation. To address this, throughout this chapter, we have used a body-fixed reference frame to describe the dynamics of very flexible aircraft. The fixed point on the vehicle may coincide with either the initial location of the CM or that of the inertial measurement unit used by the flight control system.

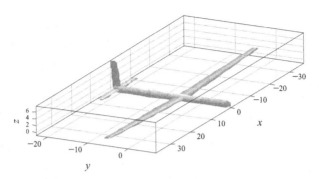

Figure 9.30 Basic geometry (jig shape) of the long-endurance UAV. Dimensions are in meters (Cesnik et al., 2014).

Flight Control Development

For aircraft with sufficient frequency separation between the rigid and elastic modes, such as fighter aircraft, two (mostly) independent control loops are defined for flight dynamics and aeroelastic purposes. If that separation is smaller, as in large transport aircraft, then aeroelastic models for control are needed that retain both the rigid and elastic modes. Those are derived as state-space models from a linear finite-element model of the aircraft, and linear unsteady aerodynamics approximated by rational functions, as we discussed in Chapter 6. Corrections to the model coefficients associated with the rigid-body dynamics are often included to achieve a better correlation wiht the 6-DoF flight simulation database.

Flight control laws are often designed with filters on the natural frequecies of the structure (Halsey et al., 2005). This attenuates any vibrations that may be induced by the actuation, without violating the phase margin requirements in the vehicle flight dynamics. This again fundamentally changes for very flexible aircraft, for which the most common approach is a local linearization around each trim condition (Shearer and Cesnik, 2008; Tuzcu et al., 2007), at most including some model updating (Wang et al., 2018). Some direct nonlinear approaches have been the topic of recent investigations (Pereira et al., 2019a,b), and some promising solutions are discussed in Section 10.5.

9.5.2 A Numerical Example: Long-Endurance UAV

We will illustrate the issues above on a very flexible aircraft configuration representative of a medium-altitude long-endurance UAV. The aircraft has the geometry as shown in Figure 9.30, and it is analyzed using UM/NAST, which has been introduced in Section 9.4. The fuselage is assumed to be rigid, while both the high-aspect-ratio wings and the tails are flexible structures. Tables 9.13–9.15 show the main dimensions of the long-endurance UAV, the properties of the beam sections, and the airfoil details, with linear interpolation used between the values in the tables. The aircraft has a basic empty mass of 4,000 kg, including a nonstructural mass of 220 kg that is uniformly distributed along the wing. We also consider a variable payload M_P of up to 1,000 kg,

Table 9.13 Geometric properties for the long-endurance UAV.

Wing		HTP		VTP	
Span	75 m	Span	18 m	Span	8 m
Root chord	3 m	Root chord	2 m	Root chord	2 m
Tip chord	1.5 m	Tip chord	1.2 m	Tip chord	1.5 m
Area	168.75 m^2	Wing-to-HTP dist.	15.5 m	Wing-to-nose	12 m

Table 9.14 Cross-sectional properties of the long-endurance UAV.

	Wing	HTP	VTP
Material	AL 6061-T6	AL 6061-T6	AL 6061-T6
Reference axis (chord fraction)	0.45	0.3	0.3
Root skin/spar thickness	1.8 mm	1.2 mm	1.2 mm
Tip skin/spar thickness	1.0 mm	1.0 mm	1.0 mm
Spar location (chord fraction)	0.45	0.45	0.45

Table 9.15 Aerodynamic properties of the long-endurance UAV.

	Wing	HTP	VTP
Airfoil	NACA 4415	NACA 0012	NACA 0012
Incidence angle	2.0°	−2.0°	0°
Control surface location (start–end)	(20.83–29.17) m	(1.8–9.0) m	(1.6–6.4) m
Control surface chord (chord fraction)	0.2	0.2	0.2

which is a point mass located 5 m ahead of the origin of the body-fixed frame along the fuselage, and a fuel mass M_F of up to 2,500 kg, which is linearly distributed from the root to the point at 40% of the wing semispan. The ratio of the fuel mass between both points is 10:7. All results in this section have been presented in Cesnik et al. (2014).

Each wing is modeled using nine strain-based finite elements, while five elements are used for each tail. Wings and tails are linked via a rigid fuselage and a thin-strip aerodynamic model is used, with 6-inflow states per airfoil colocated at each structural node. Stall is modeled by a cutoff angle at 12°. Propellers are not explicitly modeled, and, instead, a thrust force is included at the CM that is assumed to be a constant force that exactly cancels the horizontal resultant force for each trim value. For each combination of flight speed V, payload, and fuel mass, the flexible aircraft is trimmed for level flight by finding suitable values of the control surface deflections.

Figure 9.31 shows the normalized wingtip deflection in level flight as a function of the payload, for different flight speeds and two different fuel masses. Under full load (maximum values of M_F and payload), the wing shows a 25% deflection, for which strong geometrically nonlinear effects should be expected.

We can also compute the (longitudinal) stability derivatives for each trimmed condition. For a rigid aircraft, they have been defined in Equation (4.41), while for a

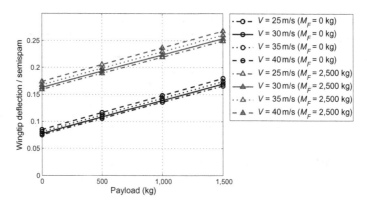

Figure 9.31 Wingtip deflection in trim of the long-endurance UAV for two different fuel fractions (Cesnik et al., 2014).

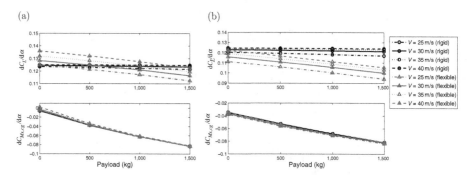

Figure 9.32 Trim stability derivatives of the long-endurance UAV with respect to the angle of attack and under different mass combinations. (a) Fuel mass, $M_F = 0$ kg and (b) fuel mass, $M_F = 2,500$ kg (Cesnik et al., 2014).

flexible aircraft, they are the result of the residualization process in Equation (5.11). Figures 9.32 and 9.33 show the derivatives of lift and pitch moment with body angle of attack and a symmetric aileron deflection, respectively. They compare the results for the flexible aircraft with the material properties of Table 9.14 with those of a rigid aircraft with the same mass and dimensions. It can be seen that increasing payload or fuel mass, thus increasing wing deformations, results in a reduction of $\frac{dC_L}{d\alpha}$ that would not appear on a rigid aircraft. Note also that the neutral stability $\left(\frac{dC_{Mx,cg}}{d\alpha} = 0\right)$ is constant for the rigid aircraft for all flight speeds, while it has a relatively wide range for the flexible aircraft. For example, the minimum payload to ensure statically stable aircraft at 25 m/s would be $M_F = -70.5$ kg, while at 40 m/s, it would be $M_F = 23.1$ kg.

The dynamic response to commanded actuation on the control surfaces is studied next. Three models are considered, corresponding to a geometrically nonlinear model, a linearized model about the trim shape, and a rigid aircraft model also with the shape similar to that of the flexible aircraft in trim. The aircraft is set at $V = 35$ m/s, with $M_P = 1,000$ kg and $M_F = 2,500$ kg. A doublet, defined as in Figure 9.34, is commanded on the control surfaces and is normalized by $\delta_{\max}$ and T, which are the parameters. A Newmark integration algorithm is used with time step $\Delta t = T/100$.

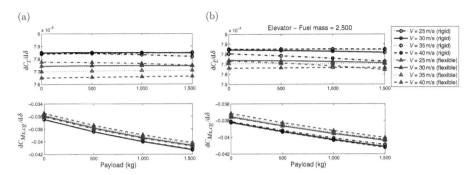

Figure 9.33 Trim stability derivatives of the long-endurance UAV with respect to the elevator deflection and under different mass combinations. (a) Fuel mass, $M_F = 0$ kg and (b) fuel mass, $M_F = 2,500$ kg (Cesnik et al., 2014).

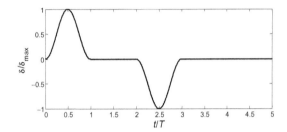

Figure 9.34 Control surface deflections in a "1-cosine" doublet input.

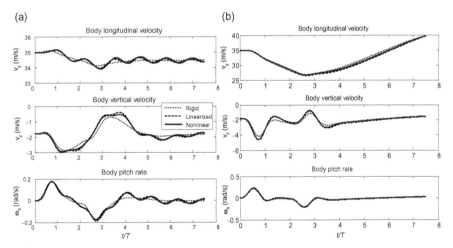

Figure 9.35 Body-frame longitudinal velocities for elevator deflection ($\delta_{max} = -20°$). (a) $T = 1$ s and (b) $T = 2$ s (Cesnik et al., 2014).

Consider first a symmetric actuation of the elevator with $\delta_{max} = -20°$ and two different actuation times, $T = 1$ s and 2 s. Figure 9.35 shows the resulting nonzero velocities, in the body axis, of the origin of the body-attached reference frame. There is a clear difference between the rigid and the flexible models, which increases as T decreases and gets into the range of the natural vibration characteristics of the

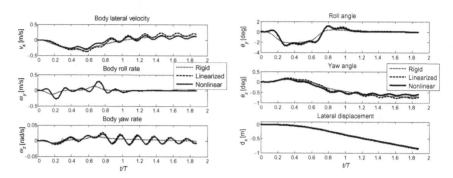

Figure 9.36 Lateral response to aileron deflection ($T = 0.5$ s, $\delta_{max} = -30°$) (Cesnik et al., 2014).

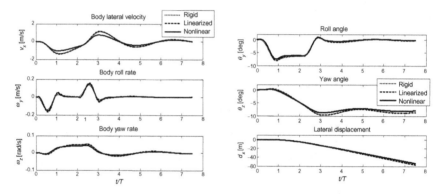

Figure 9.37 Lateral response to aileron deflection ($T = 2$ s, $\delta_{max} = 30°$) (Cesnik et al., 2014).

airframe. Note, however, that there is only a very small difference between the linear and the nonlinear models, since control surface actuation does not significantly deform the structure for this particular case.

Similar trends are obtained under an antisymmetric deflection of the ailerons, as shown in Figures 9.36 and 9.37. First, structural vibrations again become more apparent at higher actuation frequencies ($T = 0.5$ s). However, as the amplitude of the excitation is relatively small, there is little difference between the geometrically nonlinear and the linear structural models.

This changes when considering the response to a "1-cosine" discrete gust. The aircraft is in the same conditions as before, that is, trimmed for level flight at $M_P = 1,000$ kg, $M_F = 2,500$ kg, and $V = 35$ m/s. A vertical gust with the profile similar to that of Equation (2.54) is then defined along the flight path, with intensity $w_0 = 15$ m/s and varying gust length H. This gust intensity would be comparable to that in the FAR25 regulations. As in the previous cases, the aircraft is trimmed under large wing deformations, and then three different models are assumed for the subsequent dynamics (rigid, linear, and geometrically nonlinear). In all cases, Newmark integration is used with time step $\Delta t = \frac{1}{50}\frac{H}{V}$. The gust reaches the wing at $t = 0$, and the subsequent inertial

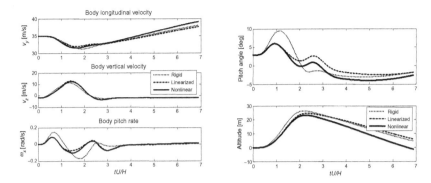

Figure 9.38 Longitudinal response to symmetric discrete gust ($V = 35$ m/s, $H = 50$ m) (Cesnik et al., 2014).

velocities, altitude, and pitch at the origin of the body-attached reference are shown in Figure 9.38 for the case $H = 50$ m. The excitation is large, and it clearly has a substantial effect on the aircraft dynamics, with some noticeable differences between the three models. In all cases, a large fraction of the gust energy is absorbed into kinetic energy, in a process known as *inertia relief*. For a vehicle with high wing flexibility, some energy is also absorbed by the wing deformations, thus resulting in smaller maximum values in pitch oscillations. Finally, Cesnik et al. (2014) have shown that this vehicle has an unstable phugoid mode, which explains why the aircraft does not go back to a level flight after the gust encounter.

The internal moments at the wing root are shown in Figure 9.39. The plots include the instantaneous values of both out-of-plane bending (M_y) and torsional (M_x) moments for three different gust lengths, and both the linear and the geometrically nonlinear aircraft models. The values of the moments for the aircraft in level flight ($t = 0$) are marked with a circle. While the moments obtained by both models are comparable, the maximum values are always obtained with the geometrically nonlinear model, which shows that the linear assumption would underpredict the worst-case scenarios in this configuration, and therefore result in a nonconservative design.

Gust loads are likely to define a critical structural design condition for wings of a very high aspect ratio. Certification requirements imply the analysis of loads under rather large gust intensities, which, as it was seen here, may result in complex flight dynamics and large-amplitude wing displacements. Importantly, for a coupled aeroelastic–flight dynamics problem, one cannot assure that the linearized structural model is going to provide a conservative estimate of the internal loads on the primary structures.

9.6   Summary

This chapter has introduced models and analysis tools for the study of dynamics of very flexible aircraft, which have been defined as air vehicles for which linear structural models are not sufficient to describe the deformations of the airframe. The focus

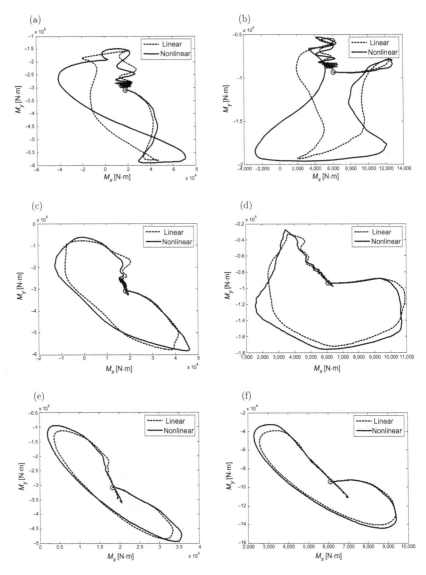

Figure 9.39 Bending-twist moment ($M_y - M_x$) diagrams in the response to symmetric discrete gust for $V = 35$ m/s. (a) Wing root ($H = 25$ m). (b) Wing midpoint ($H = 25$ m). (c) Wing root ($H = 50$ m). (d) Wing midpoint ($H = 50$ m). (e) Wing root ($H = 100$ m). (f) Wing midpoint ($H = 100$ m) (Cesnik et al., 2014).

has been put on numerical strategies that facilitate the prediction of the (nonlinear) coupled flight dynamics–aeroelastic features of those vehicles. This is necessary to: (1) identify trim conditions with aeroelastic effects; (2) describe their dynamics under commanded maneuvers, including linear (asymptotic) stability around nonlinear trim; and (3) predict their dynamic response to atmospheric turbulence. We have restricted our modeling exploration to geometrically nonlinear composite beams for the structural response, and potential flow unsteady aerodynamics, for which the underlying mathematical models have been introduced in Chapters 7 and 8. This modeling approach results in a suitable compromise between appropriate characterization of key physics on high-aspect-ratio wing vehicles under attached-flow conditions and a manageable computational cost of the fully nonlinear time-domain simulations. They are the current state of the art in the simulation of flexible vehicle dynamics and also give good insights into the key features of the problem when there is no flow separation.

Details on the time-domain unsteady aerodynamics models have been presented in Chapter 7, while different beam formulations were outlined in Chapter 8. Integration between both has resulted here in a medium-fidelity framework to explore the dynamics of very flexible aircraft. As the UVLM defines a discrete-time formulation for the unsteady aerodynamics, the aeroelastic coupling is introduced after discretization in time of the beam equations. Multiple strategies are available, and the focus has been on implicit Newmark integration methods. With respect to the spatial coupling, the instantaneous geometry of the lifting surfaces is obtained by assuming rigid cross sections attached to each point along the deforming reference line (Cosserat's *elastica* model), while the instantaneous forces and moments on the beam reference line are obtained from the sectional integration of the aerodynamic forces. This basic approach can then be used to define the aeroelastic equilibrium of the aircraft in steady flight, dynamic stability characteristics around that equilibrium, and any subsequent nonlinear dynamics.

An alternative formulation has also been introduced that preserves the geometric nonlinearities in the structure but assumes linear unsteady aerodynamics. This additional assumption can be used to reduce, quite dramatically, the computational cost of the problem by using (1) intrinsic beam equations in modal coordinates for the structure and (2) a frequency-limited balanced ROM of the DLTI UVLM for the unsteady aerodynamics. Results of a cantilever wing showing large bending deflections have shown that the loss in accuracy can be rather minimal by this approach. For this reason, it will be therefore used for the internal models for nonlinear control in Chapter 10.

The computational implementation for generic problems can be involved, and two different software packages (SHARPy and UM/NAST) have been introduced for this. They have been used to exemplify some typical scenarios where the dynamics of very flexible aircraft and wings differ from those predicted by the linear methods of Chapter 6. Three problems of increased complexity have been chosen for which the computer models replicate the properties and characteristics of real experimental prototypes.

Some guidelines have finally been produced to update existing airframe design processes so that more flexible vehicle concepts can be more easily accommodated. The main drive needs a shift toward multidisciplinary thinking using nonlinear analysis methods. That brings complexity but offers the promise of a much greater efficiency in future air vehicles.

10 Feedback Control

10.1 Introduction

This book has only considered so far the open-loop dynamics of flexible aircraft. However, as we discussed in Chapter 1, the flight control system (FCS) of virtually all modern aircraft integrates *feedback control* in its design architecture. Feedback control fundamentally achieves one or more of the following three objectives: (1) it stabilizes unstable (or marginally stable) dynamics, (2) it compensates for external (exogeneous) disturbances, and (3) it eases the design effort by correcting for unmodelled/unknown dynamics and/or model uncertainty.[1] The FCS of modern aircraft, therefore, consists of a set of physical sensors and actuators distributed throughout the vehicle, as well as the signal conditioning and control logic, which are implemented in on-board computing systems. Note that for some unmanned vehicles, ground-based computers are also an integral part of the FCS.

Sensors, such as Pitot tubes or those in the inertial measurement unit (IMU), collect information about the incoming wind and the instantaneous state of the aircraft, respectively. They unavoidably include a certain amount of measurement noise, and the resulting signals often need to be preconditioned before they can be employed to *estimate* the current state of the aircraft and, possibly, the wind conditions. As we have seen in this book, the definition of physical DoFs that determine the aircraft state can vary quite considerably between vehicles, from those that can be analyzed by rigid aircraft models to the very flexible ones we have considered in Chapter 9. Therefore, the number of sensors and the associated signal processing requirements are different for each aircraft. As a typical example, an IMU on a rigid aircraft (from a flight mechanics point of view) can be used to directly estimate the rigid-body velocities at the CM; however, the same IMU signals on an aircraft with flexible wings would be measuring both rigid and flexible contributions. Remember that the CM moves with respect to the aircraft as it deforms, but the IMU is in a fixed location on the aircraft. Additional sensors that estimate the elastic state of the aircraft are then also needed to determine instantaneous position and velocity of the CM. In other words, it is virtually impossible

---

[1] This is most apparent in the rapid development of small quadcopters, whose detailed physics, such as the unsteady aerodynamics in hovering flight, are rather challenging to model. This has been compensated for with large actuation power and feedback control.

to estimate the kinematics of the CM of a flexible aircraft without also estimating its instantaneous deformed shape.[2]

The flight control system then modifies the state of the system, as required by some predetermined control objectives, by means of commands on the relevant actuators. Typical actuators are the control surfaces and engine throttle setting that we have first defined in Section 4.4.1 and have utilized throughout this book. Other systems are available. In Section 10.2, we outline some of the sensors and actuators that have particularly facilitated the development of more flexible aircraft.

The information needed for the design of the control logic, also known as *controller synthesis*, includes: (1) a sufficiently detailed knowledge of the relevant aircraft dynamics, which can be given by the dynamic models introduced in this book, and possibly also augmented with any additional dynamics of the available sensors and actuators (which may be strongly nonlinear); (2) some knowledge (often qualitative, but increasingly grounded on physics-based modeling) of any external disturbances acting on the aircraft; and (3) the definition of a suitable performance objective for the controller. All of the above has developed into different control methods and system architectures.

Consider first the control objectives. In Section 1.2, we have introduced the handling qualities of an aircraft as the assessment of the required commands, and the associated effort, that the pilot needs to perform to maneuver the vehicle. The FCS can be designed to improve the handling qualities of an aircraft by for instance, providing *stability augmentation*. This is mainly achieved by placing the closed-loop system eigenvalues at more "pilot-friendly" locations in the complex plane (see the discussion by Stevens et al. (2016, pp. 274–287)) than where they would be for operating the aircraft without feedback control. The FCS can also be designed to keep the aircraft in a certain trajectory by compensating against disturbances (e.g., atmospheric turbulence). In this case, the control system acts as a *regulator*. It can also be tailored to reduce the loading on critical structural components under dynamic events (i.e., a gust encounter), in which case it is known as a *load alleviation system*.

The details of the control design process depend on the specific aircraft and performance objectives, but it can be outlined in general as follows. Once the vehicle and the control objectives are known, the analysis of the open-loop response of the aircraft to sizing maneuvers under typical disturbances is used to establish requirements for control authority (which determines the size of the actuator) and frequency bandwidth (which determines sampling and actuator rates). Next step is to size and locate the actuators and sensors on the aircraft, to ensure adequate *controllability* and *observability* of the vehicle. Controllability and observability have been defined in Sections 5.4.2 and 7.4.1, respectively, and they are not discussed further here, but it is assumed in what follows that there are sufficient sensors and actuators on the aircraft to guarantee them on the relevant states needed to describe its flight dynamics. A dynamic model of the vehicle with the selected sensors and actuators is then typically built. When the

---

[2] A different question, however, is whether knowledge of the velocity at the CM is strictly necessary to steer an air vehicle. Birds and insects have long showed to us that this is not at all the case.

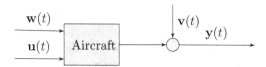

Figure 10.1 Block diagram for the open-loop aircraft dynamics with input and measurement noise.

vehicle dynamics can be described by a linear model, we use the standard notation of Section 1.4.2, namely

$$\dot{\mathbf{x}} = \mathbf{A}\mathbf{x} + \mathbf{B}_u\mathbf{u} + \mathbf{B}_w\mathbf{w},$$
$$\mathbf{y} = \mathbf{C}\mathbf{x} + \mathbf{D}\mathbf{u} + \mathbf{v},$$

(10.1)

where **u** refers to the control inputs, **w** to any disturbance inputs into the system, including in particular those resulting from the non stationary atmosphere, and **v** is any measurement noise in the output signal, as shown in Figure 10.1. While the main disturbance may come from atmospheric turbulence, it also includes errors in actuator inputs, when, for example, the actual aileron deflection differs from the commanded one due to mechanical losses. In rare situations, there is no *a priori* information about the values of the disturbance, **w**(*t*), but we typically know in advance some general characteristics of it, such as its frequency bandwidth, and those are used to improve the effectiveness of the controller.

Dynamic models of a flexible aircraft need to include elastic (structural) DoFs, and this typically results in descriptions of very large dimensionality for control design purposes, as we have seen, for example, in Equation (9.5). Thus, ROMs of a suitable size are often first derived. As it is always the case with computer models, this brings a trade-off between modeling accuracy and numerical performance, which is different for control design (where models need to be small) and for flight simulation and analysis (where they need to be accurate).

In applications to flexible structures, a key problem related to this model mismatch is the potential for *spillover*. Structures are distributed parameter systems which, theoretically, have infinite DoFs. In practice, this means that they may display vibrations with low damping and relatively high frequency. If the controller is designed for an exceedingly narrow frequency bandwidth, the actuators may unwittingly amplify high-frequency dynamics and lead to feedback instability (Preumont, 2018, Ch. 11).

The final stage is to define a suitable feedback control logic to achieve the performance objectives. This results in a set of *control laws*, or CLAWS. As it is possible to include sensors for the incoming wind velocity, the FCS architecture can also include feedforward blocks in addition to the feedback blocks for disturbance rejection scenarios (Hesse and Palacios, 2016). There are multiple strategies that have been successfully applied for both flight and aeroelastic control (see Schmidt (2012), Lavretsky and Wise (2013), or Livne (2018) for some good examples), and the focus in this book is on controller design using *optimal control theory*. This is a rather generic class of methods seeking the "best" set of control actions **u**(*t*) that can be selected to achieve a certain preset objective, which is mathematically defined by means of a metric of performance. Optimal control is, thus, concerned with establishing those best-performing control signals that minimize (or maximize) that performance criterion, while meeting any physical or operational constraints that may exist in the system. Adopting optimal control, however, also implies a departure from traditional flight control strategies based on individually shaping the closed-loop transfer functions of interest. That frequency-domain approach is indeed very well suited for the flight control of rigid aircraft, whose dynamics of interest can be generally analyzed as single-input single-output (SISO) systems. Optimal control methods require less manual tuning and have found wide application in more general multi-input multi-output (MIMO) problems (Skogestad and Postlethwaite, 2005), such as those appearing in the description of the dynamics of aircraft displaying rigid-elastic couplings.

Optimal control can be applied to linear or nonlinear systems with time-invariant or time-dependent dynamics, and with various levels of complexity in the definitions of the problem constraints. Here, we start on the simplest optimal control problem, known as the Linear Quadratic Regulator (LQR). LQR control assumes full-state feedback, which is rarely available, and therefore we introduce the associated linear state estimation problem, namely the Kalman filter. The combination of Kalman filter estimation and LQR control results in Linear Quadratic Gaussian (LQG) control, which is then outlined. This strategy is finally generalized to nonlinear systems by moving from offline to online (real-time) solutions of the optimal control problem. For the control problem, this results in a model predictive control (MPC) strategy, while the generalization of the Kalman filter results in moving horizon estimation (MHE). Several classical books, such as those of Kirk (1970) and Burl (1999), include excellent introductions to optimal control methods. The MPC/MHE strategy can deal with saturation constraints and nonlinear physics at the expense of much more demanding computational resources. For aircraft applications, most modern textbooks on flight dynamics include descriptions of the key elements of a modern flight control system (Cook, 2013; Schmidt, 2012). This chapter offers only a brief introduction of the underlying theories and their application to typical problems in aeroelasticity and flexible aircraft dynamics.

Although we have left control aspects to a late chapter, we need to emphasize that control design is now an integral part of the vehicle design from the conceptual phase. The ability of modern aircraft to meet their performance requirements, to operate safely, or to provide superior ride performance is often heavily dependent on rather sophisticated control systems. For flexible aircraft dynamics, there is a clear trend to replace hardware (e.g., stiffness of the wing) by software solutions (e.g., a gust load alleviation system), as exemplified in the NASA X-56A program (Burnett et al., 2016). Recent studies have also been shown that flutter clearance across the flight envelope may be achieved with active suppression systems (Livne, 2018; Ricci et al., 2021), although the industrial adoption of such solutions will require changes on the certification process.

This chapter continues with a discussion on some of the devices used in flexible aircraft control, with a particular focus on those used for gust load alleviation. We proceed then to review linear optimal control methods (LQR and LQG) before introducing nonlinear MPC strategies. Simple aeroservoelastic examples, building on the models of Chapter 3, are used to exemplify the linear control methods. More complex wing and aircraft models, described using the nonlinear simulation framework of Chapter 9, are finally introduced to demonstrate nonlinear control strategies for both stability augmentation and disturbance rejection. A final note of caution: At the time of writing, model-based nonlinear control is still in an early stage of maturation for application as the main on-board controller in air vehicles. We present the latest research in the area as a demonstration of what is possible at the time of writing, noting that several hurdles, such as functional acceptance testing across the entire flight envelope, still need to be overcome for the safe adoption of this technology.

10.2   Sensors and Actuators

In Chapter 4, we have introduced the conventional actuators that provide controllability and maneuverability to most air vehicles, namely throttle settings on the engine, and trailing-edge control surfaces on the wing and tails. We have seen that the resultant quasi-steady aerodynamic forces and moments can be included in the first approximation in flight-dynamic models through steady gains known as *control derivatives*. In Section 6.5, we have also seen that the control derivatives need to be further expanded to include unsteady physics for aeroelastic applications, thus resulting in frequency-domain transfer functions. In that case, the dynamics of servomotors often need to be included as well in the analysis, as was done in Equation (6.106). In addition to this, modern aircraft have a relatively large number of sensors to determine the vehicle state, either for flight control purposes or for health checks on critical subsystems. In particular, the development of more flexible aircraft is being facilitated by some specific hardware solutions, and the most important of them are outlined in this section.

10.2.1   Sensors

Sensors determine the measurements in Equation (10.1). They can give information about the instantaneous aerodynamic, the structural and/or the inertial state of the aircraft. In particular, for flexible aircraft applications, we are often concerned with distributed information (e.g., to estimate the spanwise lift distribution, see Figure 5.1). This is a richer dataset than the one needed for rigid aircraft control, which only needs aggregate metrics (e.g., knowledge of vertical acceleration is sufficient to estimate changes in lift). Several important sensing technologies are briefly discussed in the rest of this section.

Wind Velocity Estimation
Measuring the instantaneous wind velocity provides the FCS with information about a key external disturbance during flight besides the airspeed itself. In particular, due to the relatively high-frequency content in the wind velocity impinging on the aircraft,

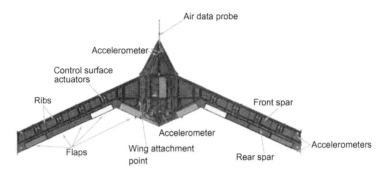

Figure 10.2 Basic layout of the NASA X-56A prototype showing the forward location of the Pitot tube (adapted from Pankonien et al. (2020)).

gust alleviation methods benefit from the direct use of those measurements. This is usually done in an indirect manner, by subtracting measurements on flow probes from measurements of the vehicle velocity.

The first requirement is then to determine the *inertial* rigid-body velocities of the aircraft. This is achieved by the IMU, which uses accelerometers and gyroscopes to estimate the instantaneous velocity vector and aircraft orientation by means of time integration. To avoid cumulative errors in the integral, which would create an artificial drift, they are aided – and sometimes replaced – by the Global Positioning System devices that determine the position by triangulation from several reference satellites. If B is a reference frame attached to the IMU (and aligned with the vehicle stability axis in some nominal conditions), the instantaneous inertial translational and angular velocities will be $\mathbf{v}_B$ and $\boldsymbol{\omega}_B$, respectively.

As discussed in Section 2.3, the typical length scale of turbulence in the atmosphere at altitude is around 2500 ft., which means that most aircraft will see a uniform gust along its span. As a result, one single probe can be used to measure airspeed. To measure the direction and orientation of the *true airspeed* vector, $\mathbf{v}_{\text{TAS}}$, a five-hole Pitot tube is typically used. Its magnitude is the relative airspeed between the aircraft and the wind, V_{rel}, and its direction angles are the instantaneous angles of attack, α, and sideslip, β, as was defined in Equation (4.34). To compensate for delays in signal processing, and also to obtain "cleaner" measurements, the probe is typically located at the front of the aircraft. A clear example of this is provided by the NASA X-56A prototype shown in Figure 10.2, which has already appeared in Example 6.6 on page 270. The air data probe has been identified at the wing apex.

If the position vector of the airspeed probe with respect to the reference frame attached to the IMU is $\mathbf{r}_{\text{probe}}$, then the wing velocity vector is obtained simply as

$$\mathbf{v}_g = \left(\mathbf{v}_B + \boldsymbol{\omega}_B \times \mathbf{r}_{\text{probe}} - \mathbf{v}_{\text{TAS}} \right). \tag{10.2}$$

Measurements of wind speed at the wing, however, give very limited time to the control system to act against any disturbances. An alternative approach, which is under study at the time of writing, uses remote sensing of the wing speed using a light detection and ranging (LIDAR) system. LIDAR devices emit a laser beam of a certain

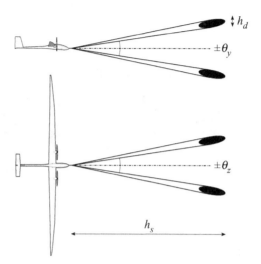

Figure 10.3 Typical LIDAR setup to measure three components of velocity.

wavelength λ, which is reflected by either the flow molecules (the resulting device is known as a Rayleigh-type LIDAR) or aerosol particles in the atmosphere (a Mie-type LIDAR). Measurement of the wavelength shift of the backscattered light gives the relative velocity of the aircraft with respect to the air. Aerosol particles provide a much stronger backscatter signal, but they are concentrated on the lower atmosphere. As a result, Mie-type LIDARs are used for measurements near the ground (e.g., to map wind conditions in wind farm siting), while Rayleigh-type LIDARs are considered for airborne applications.

LIDAR systems can only determine the component of the flow velocity along the line of sight. Since the most critical gust conditions are normally along the vertical and lateral directions, the 3-D velocity vector needs to be determined. This is done by generating LIDAR beams along various directions, as shown in Figure 10.3. At least three beams are necessary to determine the full velocity vector, although additional beams can be added to improve the measurement accuracy (four beams are used in the example in the figure).

As LIDAR systems provide preview information of the disturbance, they naturally fit within feedforward control architectures. The forward-looking range is limited by both the technology to generate very short wavelength laser beams necessary to illuminate the flow molecules and also by the timescale in the fluctuations of the wind velocity (so that the measured speed of the flow particles does not change by the time they reach the vehicle). For a commercial transport aircraft flying at 250 m/s, an appropriate forward-looking distance is about $h_s = 75$ m ahead of the wings. The resulting beam has a diameter of about $h_d = 15$ cm (Rabadan et al., 2010).

Stagnation/Separation Point Sensors

Airspeed sensors do not give distributed information of the flow field along the wings. They also have a rather low bandwidth and only give average values of the velocity. Both aspects can be improved by directly measuring some flow features along the

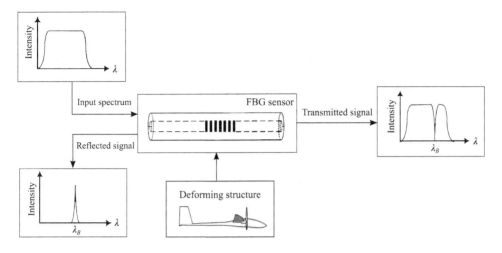

Figure 10.4 Fiber Bragg grating sensor.

wing. Two measurements have been found to be particularly suitable: the location of the local leading-stagnation point (LESP) and the flow separation point (FSP). For subsonic 2-D flows on airfoils, as the angle of attack increases, the LESP moves rearward, while the FSP moves upstream once the flow separates. The position of both points uniquely determines the average circulation on the airfoil. LESP sensors are typically hot-film flow sensors installed as a compact array perpendicular to the flow direction near the leading edge (Lee and Basu, 1998). Hot-film sensors calculate shear stress by measuring the heat dissipated in the sensor. They have a large frequency bandwidth and can therefore be used to estimate the instantaneous local forces. Importantly, they can also be used to provide an early warning of changes in the airflow velocity, thus indicating the presence of gusts (Ryan et al., 2014; Suryakumar et al., 2016).

Strain Sensors

Conventional strain-gauge sensors have long been used to measure structural deformation in ground and test flights, as we will see in Chapter 11. Strain gauges present, however, problems of reliability and electromagnetic interference, and alternative methods have been developed for in-flight wing shape estimation. In particular, fiber-optic strain sensors (FOSS) are increasingly used (Derkevorkian et al., 2013). Those sensors measure the change on the characteristics of the transmitted (or reflected) light beam in the optical fiber as it is subject to external actions. They are, therefore, insensitive to electromagnetic fields, but they are also substantially more expensive than foil strain gauges.

The most common FOSS technology is the fiber Bragg grating (FBG) sensor (Kang et al., 2007). As schematically shown in Figure 10.4, it works by reflecting only a narrow band of wavelength of the light beam (while transmitting all others). This is achieved by introducing a periodic variation of its refractive index (known as "grating") on a relatively small part of the fiber. That segment of fiber then defines the strain sensor. As the fiber is deformed, the wavelength of the reflected light will change. The wavelength shift provides a direct measure of strain along the direction of the fiber as

$$\varepsilon = \frac{1}{1 - p_e} \frac{\Delta \lambda_B}{\lambda_B}, \tag{10.3}$$

where p_e is the strain-optic coefficient, λ_B is the nominal Bragg wavelength, and $\Delta \lambda_B$ is the change of the wavelength. This can be measured using an interferometer with a reference FBG sensor. This compensated sensor can also be designed to remove thermal effects from the measurements. Using multiple strain sensors on a structure (i.e., the wing), it is possible to reconstruct the displacement field.

Accelerometers

Finally, accelerometers may be additionally used to measure structural vibrations induced by any external excitation on the vehicle during flight. For that, it is critical to have first a good understanding of the resonant characteristics of the vehicle, that is, of its natural frequencies and the corresponding mode shapes. Recall, however, that the mode shapes measured on the ground change, sometimes quite significantly, during flight due to the apparent mass and stiffness effects generated by the changes on aerodynamic forces under wing deformations, as was seen in Section 6.6. Each mode shape is defined by a unique displacement field for a given flight condition. This includes *nodal lines*, which are the zero-amplitude points of that field. Therefore, to measure the response of a structure in a given resonant mode, accelerometers should be positioned away from the nodal lines of that mode. Accelerometers are extensively used in vibration testing of aircraft structures (both on the ground and in flight), and they will reappear in Chapter 11.

10.2.2   Actuators

Control surfaces installed on wings and tails are the most obvious actuation solution on the instantaneous distribution of aerodynamic forces. They offer the control authority needed to shift and reshape the aerodynamic loading on the lifting surfaces. However, the standard layout with longitudinal flight control using elevator commands on the horizontal tailplane is typically too slow to compensate for the angle of attack changes due to atmospheric gusts (Hesse and Palacios, 2016). Therefore, for gust alleviation purposes, it is often more efficient to use wing-mounted control surfaces that have a direct effect on the local aerodynamic forces. Depending on the vehicle (or sometimes on the flight speed for a given vehicle), there are two possible solutions.

The first solution is to use control surfaces located near the root of the wing, such as those used during take-off and landing. Those control surfaces have a direct effect on the wing aerodynamic forces and are located in the region when those forces are larger. They are, therefore, referred to as direct lift control (DLC) flaps (Hahn and Schwarz, 2008). A typical layout with ailerons for roll control and DLC flaps for gust load control was shown in Chapter 1. They are the most efficient to achieve a smooth ride, but as the additional lift due to the gust-induced velocity increases toward the wingtips, they do not necessarily reduce the bending moment at the wing root.

A second solution consists in utilizing control surfaces near the tip of the wing, such as those used for roll control. These control surfaces are less effective to reduce lift, but they can be designed to reduce the root bending moments. These control surfaces are currently investigated as key enablers for very high aspect wings (Riso et al., 2020; Sanghi et al., 2022), including nonconventional solutions, such as hinged wingtips that are left to float freely whenever atmospheric turbulence is sufficiently high (Castrichini et al., 2020; Sanghi et al., 2023).

In practice, the optimal actuator distribution for a given configuration may include a combination of inner and outer wing control surfaces, as shown by Stanford (2020). Furthermore, while the actuation of flaps for flight control is typically done at low frequencies, control surface deflection rates in the response to atmospheric turbulence need to be relatively fast. If the typical timescale of that actuation is comparable with the convective timescale in the fluid (see Section 3.3.3), then quasi-steady assumptions on the flap aerodynamics are no longer valid. As a result, unsteady aerodynamic models for control surface deflections are often necessary for control synthesis for gust load alleviation.

Finally, smaller flow control devices with very fast actuation times are also being considered. Both microtabs and microjects have gained some traction, particularly for applications to load control in wind turbines. Microtabs are small trailing-edge tabs with large deflection angles (Frederick et al., 2010), while microjets are small pneumatic jets blown normal to the blade surface from spanwise slots near the trailing edge (Blaylock et al., 2014). In both cases, the effect is to modify the Kutta–Joukowski condition at the trailing-edge and, therefore, such devices can be designed to have a relatively large effect on the total lift (Blaylock et al., 2014). They can be much cheaper to install than large flaps and can be even retrofitted on existing wings or turbine blades. However, their operational robustness is not yet guaranteed, and at the time of writing, they are still in the development phase.

10.3   Linear Quadratic Regulator

After our short review of the typical hardware elements in flexible aircraft control, we turn our attention to controller design using optimal control methods. As has been discussed in Section 10.1, controller synthesis is done on certain test conditions and on a suitable model of the physical system, which is often a simplified version of the one used for simulation. Needless to say, the controller is later expected to operate on the real physical system and under a wide range of conditions. We restrict ourselves first to linear methods, in which the system model used for control design is given in linear time-invariant (LTI) state-space form, Equation (10.1). Particular examples of this class of problems in this book are Equation (4.93), for a rigid aircraft, and Equation (6.125), for a flexible one.

We consider first the LQR, which, despite its limited application in practice, introduces most building blocks in optimal control. In essence, all optimal control

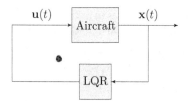

Figure 10.5 LQR control design problem. The outputn signal is the full state, and the problem is solved for the initial conditions $\mathbf{x}(0) = \mathbf{x}_0$.

problems achieve regulation by measuring error from a reference and penalizing it. This penalization is defined by an optimization problem performed on a quadratic, that is, nonnegative, index obtained from this error, which the controller is designed to minimize.

The control synthesis process for the LQR assumes a perfect system with no disturbances on which the full-state vector can be measured. This is known as *full-state feedback*. The resulting system model used for the control synthesis is, therefore,

$$\dot{\mathbf{x}} = \mathbf{Ax} + \mathbf{Bu},$$
$$\mathbf{y} = \mathbf{x}. \tag{10.4}$$

It is important to remark that this simplification is only used for the definition of the CLAWS and the controller will still have to operate on the original system described by Equation (10.1). The input signal is then defined by means of an optimization problem on this LTI state equations, with nonzero initial conditions, full-state feedback, and a given performance index (also known as the cost function). This results in a feedback loop such as the one schematically shown in Figure 10.5.

In particular, the LQR control minimizes a quadratic *cost function* of the inputs and system states over a given time horizon, τ_c. Without loss of generality, we also assume in our description here that the reference state is zero, which results in a cost function written as

$$J(\mathbf{x}, \mathbf{u}) = \frac{1}{2}\mathbf{x}_c^\top \mathbf{H}\mathbf{x}_c + \frac{1}{2}\int_0^{\tau_c} \left[\mathbf{x}^\top \mathbf{Q}\mathbf{x} + \mathbf{u}^\top \mathbf{R}\mathbf{u}\right] dt. \tag{10.5}$$

This expression has three weighting matrices that are chosen to be symmetric. Furthermore, $\mathbf{R}$ is chosen to be positive definite, and $\mathbf{Q}$ and $\mathbf{H}$ are positive semidefinite. This results in J being a nonnegative function regardless of the choice of the state and input signals, for which a minimum can then be sought. For simplicity of notation, we have defined $\mathbf{x}_c = \mathbf{x}(\tau_c)$ as the value of the state vector at the end of the control horizon.

The first term in Equation (10.5) is the *terminal cost*, which penalizes any deviations from zero on the state at the end of the time horizon. The first term inside the integral penalizes nonzero values of the state vector during the control horizon. Therefore, the relative magnitude between $\mathbf{H}$ and $\mathbf{Q}$ determines whether we are more concerned about keeping the state as close as possible to the reference trajectory at all times, or about achieving a certain state at τ_c. Finally, the second term within the integral puts a cost to the control inputs, and $\mathbf{R}$ is known as the control weighting matrix.[3] The values of the three weighting matrices are selected depending on the desired performance of the closed-loop system. A second optimization problem may be posed for the selection of the weights, in which the performance of the closed-loop system, particularly in terms of actuator demand, is analyzed for some limit-case scenarios. In practice, it is often simply done by trial and error until an acceptable design is obtained.

Equations (10.4) and (10.5), with the initial conditions $\mathbf{x}(0) = \mathbf{x}_0$, define a constrained minimization problem, that is, the minimization of the cost function subject to constraints on state and control (in particular, they need to satisfy the initial conditions at $t = 0$ and the EoMs for $0 \leq t \leq \tau_c$). Mathematically, we write this problem as

$$\min_{\mathbf{x},\mathbf{u}} \quad J(\mathbf{x},\mathbf{u})$$

$$\text{subject to} \quad \mathbf{Ax}+\mathbf{Bu}-\dot{\mathbf{x}} = \mathbf{0}, \quad \text{for} \quad 0 \leq t \leq \tau_c, \tag{10.6}$$

$$\mathbf{x}(0) = \mathbf{x}_0.$$

The solution to Equation (10.6) can be sought directly using numerical methods of dynamic programming (Kirk, 1970), and indeed this is the approach followed in Section 10.5, where we address a more general nonlinear optimization problem. However, the problem defined by Equation (10.6), with linear dynamics and no additional constraints on inputs or states, can be analytically solved using the methods of variational calculus. This is done in the following section.

10.3.1   Solution Using Lagrange Multipliers

A closed-form solution of constrained optimization problems such as the one in Equation (10.6) can be found by transforming them into an equivalent unconstrained optimization problem on a larger solution space. For that purpose, we define the augmented cost function

$$J_{\mathrm{aug}}(\mathbf{x},\mathbf{u},\boldsymbol{\lambda}) = J(\mathbf{x},\mathbf{u}) + \int_0^{\tau_c} \boldsymbol{\lambda}^\top [\mathbf{Ax}+\mathbf{Bu}-\dot{\mathbf{x}}] \, dt, \tag{10.7}$$

where $\boldsymbol{\lambda}(t)$ is the vector of Lagrange multipliers, also known as *co-states*, which has the same size as the state vector, $\mathbf{x}(t)$, and which also needs to be solved for. Note that, since the term between square brackets within the integral in Equation (10.7) is zero for all time, the integrand is always zero for any finite value of the Lagrange multipliers. As a result, the minimum of the original cost function is still the minimum of the augmented function for any finite value of the Lagrange multipliers. We will demonstrate now that when the Lagrange multipliers are also considered as variables to be minimized against, the resulting problem also enforces the constraints given by the state equation, Equation (10.4).

---

[3] The term $\frac{1}{2}\int_0^{\tau_c} \mathbf{u}^\top \mathbf{Ru}\, dt$ can then be interpreted as a measurement of the total energy consumed by the controller.

The minimum of J_{aug} is found by enforcing that its variations are equal to zero. Using that all weight matrices are symmetric, this gives

$$
\begin{aligned}
\delta J_{\text{aug}} = {}& \mathbf{x}_c^\top \mathbf{H} \delta \mathbf{x}_c + \int_0^{T_c} \left[\mathbf{x}^\top \mathbf{Q} \delta \mathbf{x} + \mathbf{u}^\top \mathbf{R} \delta \mathbf{u} \right] \mathrm{d}t \\
& + \int_0^{T_c} \boldsymbol{\lambda}^\top \left[\mathbf{A} \delta \mathbf{x} + \mathbf{B} \delta \mathbf{u} - \delta \dot{\mathbf{x}} \right] \mathrm{d}t + \int_0^{T_c} \delta \boldsymbol{\lambda}^\top \left[\mathbf{A}\mathbf{x} + \mathbf{B}\mathbf{u} - \dot{\mathbf{x}} \right] \mathrm{d}t = 0.
\end{aligned}
\tag{10.8}
$$

The time derivatives $\delta \dot{\mathbf{x}}$ are next removed by performing an integration by parts. Noting that $\delta \mathbf{x}(0) = \mathbf{0}$, since the initial constraint needs to be satisfied by the virtual displacements, and denoting $\boldsymbol{\lambda}_c = \boldsymbol{\lambda}(\tau_c)$, this results in

$$
\delta J_{\text{aug}} = \delta \mathbf{x}_c^\top \left[\mathbf{H}\mathbf{x}_c - \boldsymbol{\lambda}_c \right]
\tag{10.9a}
$$

$$
+ \int_0^{T_c} \delta \mathbf{x}^\top \left[\mathbf{Q}\mathbf{x} + \dot{\boldsymbol{\lambda}} + \mathbf{A}^\top \boldsymbol{\lambda} \right] \mathrm{d}t
\tag{10.9b}
$$

$$
+ \int_0^{T_c} \delta \mathbf{u}^\top \left[\mathbf{R}\mathbf{u} + \mathbf{B}^\top \boldsymbol{\lambda} \right] \mathrm{d}t
\tag{10.9c}
$$

$$
+ \int_0^{T_c} \delta \boldsymbol{\lambda}^\top \left[\mathbf{A}\mathbf{x} + \mathbf{B}\mathbf{u} - \dot{\mathbf{x}} \right] \mathrm{d}t = 0.
\tag{10.9d}
$$

Equation (10.9) has to be satisfied for any variations of the independent variables, that is, $\delta \mathbf{x}$, $\delta \mathbf{u}$, $\delta \boldsymbol{\lambda}$, and the terminal conditions, $\delta \mathbf{x}_c$. That results in four equations that will provide a solution to the optimization problem defined in Equation (10.6). Note in particular that the last integral, Equation (10.9d), corresponds to a weak enforcement of the original state equations. From the terms in brackets in Equation (10.9c), we obtain a relation for the input vector

$$
\mathbf{u} = -\mathbf{R}^{-1} \mathbf{B}^\top \boldsymbol{\lambda},
\tag{10.10}
$$

and the Lagrange multipliers are solved simultaneously with the state variable, by virtue of Equations (10.9b) and (10.9d). After substituting the input vector using Equation (10.10), we have

$$
\left\{ \begin{array}{c} \dot{\mathbf{x}} \\ \dot{\boldsymbol{\lambda}} \end{array} \right\} = \left[\begin{array}{cc} \mathbf{A} & -\mathbf{B}\mathbf{R}^{-1}\mathbf{B}^\top \\ -\mathbf{Q} & -\mathbf{A}^\top \end{array} \right] \left\{ \begin{array}{c} \mathbf{x} \\ \boldsymbol{\lambda} \end{array} \right\}.
\tag{10.11}
$$

These equations are known as the *Euler–Lagrange equations*, and the system matrix in Equation (10.11) is referred to as the *Hamiltonian matrix* associated with Equation (10.6). The equation is solved subject to initial conditions on the state variables, but the right-hand-side term in Equation (10.9a) imposes terminal conditions on the Lagrange multiplier, that is,

$$
\begin{aligned}
\mathbf{x}(0) &= \mathbf{x}_0, \\
\boldsymbol{\lambda}(\tau_c) &= \mathbf{H}\mathbf{x}_c.
\end{aligned}
\tag{10.12}
$$

As a result, solving for the state and co-state vectors implies marching forward and backward in time, respectively. This is a characteristic feature of optimal control problems that make their direct solution rather cumbersome. While specifically tailored numerical algorithms have been proposed to solve this general class of problems, under the assumptions of LQR control, there is a closed-form solution that is described next.

10.3.2 The Riccati Equation

In order to solve Equation (10.11), we postulate the existence of a symmetric matrix function $\mathbf{P}(t)$ that at all times satisfies $\boldsymbol{\lambda}(t) = \mathbf{P}(t)\mathbf{x}(t)$. Since $\boldsymbol{\lambda}(t)$ is algebraically related to the input vector $\mathbf{u}(t)$ from Equation (10.10), this assumption implies that the optimal control input can be obtained from the state vector as

$$\mathbf{u} = -\mathbf{R}^{-1}\mathbf{B}^{\top}\mathbf{P}\mathbf{x}, \tag{10.13}$$

where $\mathbf{R}$ and $\mathbf{B}$ are constant matrices, while $\mathbf{P}(t)$ is a function of time. An equation for $\mathbf{P}(t)$ is obtained by differentiating $\boldsymbol{\lambda}(t) = \mathbf{P}(t)\mathbf{x}(t)$ with time as

$$\dot{\boldsymbol{\lambda}} = \dot{\mathbf{P}}\mathbf{x} + \mathbf{P}\dot{\mathbf{x}}. \tag{10.14}$$

Substituting here $\dot{\mathbf{x}}$ and $\dot{\boldsymbol{\lambda}}$ from Equation (10.11) results in

$$\dot{\mathbf{P}}\mathbf{x} + \mathbf{P}\left(\mathbf{A}\mathbf{x} - \mathbf{B}\mathbf{R}^{-1}\mathbf{B}^{\top}\boldsymbol{\lambda}\right) - \left(-\mathbf{Q}\mathbf{x} - \mathbf{A}^{\top}\boldsymbol{\lambda}\right) = 0, \tag{10.15}$$

and if we finally replace $\boldsymbol{\lambda} = \mathbf{P}\mathbf{x}$, yields

$$\left(\dot{\mathbf{P}} + \mathbf{P}\mathbf{A} + \mathbf{A}^{\top}\mathbf{P} + \mathbf{Q} - \mathbf{P}\mathbf{B}\mathbf{R}^{-1}\mathbf{B}^{\top}\mathbf{P}\right)\mathbf{x} = 0. \tag{10.16}$$

This equation holds for any initial condition $\mathbf{x}(0) = \mathbf{x}_0$. Since by varying the initial condition we can obtain any value of $\mathbf{x}(t)$, the terms within brackets have to be identically zero, that is,

$$\dot{\mathbf{P}} + \mathbf{P}\mathbf{A} + \mathbf{A}^{\top}\mathbf{P} - \mathbf{P}\mathbf{B}\mathbf{R}^{-1}\mathbf{B}^{\top}\mathbf{P} = -\mathbf{Q}. \tag{10.17}$$

This is a matrix equation with as many independent equations as coefficients in the symmetric matrix $\mathbf{P}$, and it is known as the *Riccati equation*. Since $\boldsymbol{\lambda}$ and $\mathbf{x}$ have to satisfy the terminal condition at $t = \tau_c$ given in Equation (10.12), the Riccati equation needs to be integrated backward in time from the condition $\mathbf{P}(\tau_c) = \mathbf{H}$. The equation is nonlinear in $\mathbf{P}$, but it can be proved that it has a unique solution (Burl, 1999, p. 187). Finally, using Equation (10.13), we obtain the time-varying feedback gain matrix (the *optimal gains*) as

$$\mathbf{K}(t) = \mathbf{R}^{-1}\mathbf{B}^{\top}\mathbf{P}(t), \tag{10.18}$$

which closes the feedback loop in Figure 10.5. In Section 10.5, we will generalize the previous solution process to the online solution of the (in general, nonlinear) optimal problem over a fixed time horizon.

10.3.3   Steady-State LQR Control

In many applications, the global timescales of the processes to be controlled (e.g., the time it takes to perform a maneuver) are much longer than those of the controller actions, and in that case, we can assume a very long control horizon, τ_c. This is the situation most commonly assumed in LQR synthesis and significantly simplifies the control design problem. To study this, consider the particular case of the *infinite horizon problem* ($\tau_c \to \infty$). In this case, the cost function becomes

$$J(\mathbf{x}, \mathbf{u}) = \frac{1}{2} \int_0^\infty \left[\mathbf{x}^\top \mathbf{Q} \mathbf{x} + \mathbf{u}^\top \mathbf{R} \mathbf{u} \right] dt. \tag{10.19}$$

Since our concern is now the long-term steady-state response of the system, the problem of identifying the optimal gains becomes time independent, and it can be assumed that $\dot{\mathbf{P}} = \mathbf{0}$. Furthermore, the terminal cost becomes irrelevant, and it has been removed from the cost function. As a result, Equation (10.17) becomes

$$\mathbf{P} \mathbf{A} + \mathbf{A}^\top \mathbf{P} - \mathbf{P} \mathbf{B} \mathbf{R}^{-1} \mathbf{B}^\top \mathbf{P} = -\mathbf{Q}. \tag{10.20}$$

A nonlinear matrix equation of this form is referred to as *algebraic Riccati equation*. Equation (10.13) now defines a constant feedback gain matrix,

$$\mathbf{K} = \mathbf{R}^{-1} \mathbf{B}^\top \mathbf{P}, \tag{10.21}$$

which, when substituted into Equation (10.1) with $\mathbf{u} = -\mathbf{K}\mathbf{x}$, results in the closed-loop dynamics

$$\dot{\mathbf{x}} = (\mathbf{A} - \mathbf{B}\mathbf{K}) \mathbf{x} + \mathbf{B}_w \mathbf{w}. \tag{10.22}$$

A well-tuned steady-state LQR controller typically increases the stability of the system and can be included as an inner loop to the pilot commands. In that case, it is known as a *stability augmentation system* (SAS). Other feedback systems are often also introduced to improve the responsiveness of the aircraft to the pilot inputs, and they provide *command augmentation*. The CLAWS include a number of gains that are identified during the initial flight test campaigns to achieve desired handling qualities. Optimization-based controllers are only part of the general architecture of a modern FCS, whose design also needs to address robustness and off-design closed-loop performance (Lavretsky and Wise, 2013).

Example 10.1 An Analytical Example.   Consider an LTI system

$$\dot{\mathbf{x}}(t) = \mathbf{A}\mathbf{x} + \mathbf{B}\mathbf{u}(t), \tag{10.23}$$

with states x_1 and x_2, and state and input matrices given as

$$\mathbf{A} = \begin{bmatrix} -1 & 0 \\ 1 & 0 \end{bmatrix} \text{ and } \mathbf{B} = \begin{bmatrix} 1 \\ 0 \end{bmatrix}. \tag{10.24}$$

An equation of this form has appeared in Section 5.7 as a simplified model for the lateral dynamics. There x_1 is the roll rate, x_2 is the roll angle, and u is an antisymmetric aileron deflection. The cost function can be chosen as

$$J = \int_0^\infty \left(x_2^2 + Ru^2 \right) dt, \tag{10.25}$$

with $R > 0$, which penalizes the second state in the system (the roll angle) and the effort on the aileron, with a relative weight between the two given by R (e.g., if $R = 1$, they have equal weight). Given a value of R, the optimal gains for the closed-loop problem are obtained from the algebraic Riccati equation, Equation (10.20), which results here in the matrix equation

$$\begin{bmatrix} P_{11} & P_{12} \\ P_{12} & P_{22} \end{bmatrix} \begin{bmatrix} -1 & 0 \\ 1 & 0 \end{bmatrix} + \begin{bmatrix} -1 & 1 \\ 0 & 0 \end{bmatrix} \begin{bmatrix} P_{11} & P_{12} \\ P_{12} & P_{22} \end{bmatrix}$$
$$-\frac{1}{R} \begin{bmatrix} P_{11} & P_{12} \\ P_{12} & P_{22} \end{bmatrix} \begin{bmatrix} 1 \\ 0 \end{bmatrix} \begin{bmatrix} 1 & 0 \end{bmatrix} \begin{bmatrix} P_{11} & P_{12} \\ P_{12} & P_{22} \end{bmatrix} = - \begin{bmatrix} 0 & 0 \\ 0 & 1 \end{bmatrix}, \tag{10.26}$$

where we have used that the unknown matrix P is symmetric. It simplifies to

$$\begin{bmatrix} P_{11} & P_{12} \\ P_{12} & P_{22} \end{bmatrix} \begin{bmatrix} -R & 0 \\ R & 0 \end{bmatrix} + \begin{bmatrix} -R & R \\ 0 & 0 \end{bmatrix} \begin{bmatrix} P_{11} & P_{12} \\ P_{12} & P_{22} \end{bmatrix} - \begin{bmatrix} P_{11}^2 & P_{11}P_{12} \\ P_{11}P_{12} & P_{12}^2 \end{bmatrix} = \begin{bmatrix} 0 & 0 \\ 0 & -R \end{bmatrix}. \tag{10.27}$$

The unknowns are, therefore, P_{11}, P_{12}, and P_{22}, and their corresponding three (quadratic) equations are:

$$-2RP_{11} + 2RP_{12} - P_{11}^2 = 0,$$
$$-RP_{12} + RP_{22} - P_{11}P_{22} = 0, \tag{10.28}$$
$$P_{12}^2 = R.$$

From the last equation, we have $P_{12} = \sqrt{R}$. Then the first equation results in a second-order polynomial in P_{11} with roots, $P_{11} = -R \pm \sqrt{R^2 + 2R\sqrt{R}}$. Since $\mathbf{P}$ must be positive definite, only the positive root is considered, which gives $P_{11} = -R + R\sqrt{1 + 2/\sqrt{R}}$ (positive for all $R > 0$). Finally, substituting this result into the second equation gives $P_{22} = \sqrt{R}/\sqrt{1 + 2\sqrt{R}}$.

The optimal feedback gains are then given by Equation (10.21), which results in

$$\mathbf{K} = \frac{1}{R} \begin{bmatrix} 1 & 0 \end{bmatrix} \begin{bmatrix} P_{11} & P_{12} \\ P_{12} & P_{22} \end{bmatrix} = \begin{bmatrix} \sqrt{1 + 2/\sqrt{R}} - 1 & 2/\sqrt{R} \end{bmatrix} \tag{10.29}$$

The closed-loop system dynamics is finally obtained from replacing $u = -\mathbf{K}\mathbf{x}$ in the state equation, that is,

$$\dot{\mathbf{x}} = (\mathbf{A} - \mathbf{BK})\mathbf{x}. \tag{10.30}$$

By varying the control penalty R, we can then explore its effect on the eigenvalues of the closed-loop system. First of all, we can compute the open-loop eigenvalues

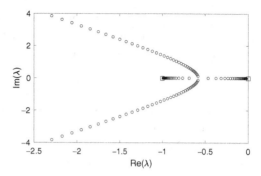

Figure 10.6 Closed-loop eigenvalues for $10^{-2} < R < 10^4$. The squares denote the open-loop eigenvalues.

of the system (those of **A**), which are $\lambda_1 = 0$ and $\lambda_2 = -1$. Note that, from Equation (10.29), when $R \to \infty$, then $\mathbf{K} = \mathbf{0}$, and in that limit case, the closed-loop system behaves as the open-loop system. This is because a very large penalty on the control u prevents it from acting outright. On the other extreme, the gains on the feedback matrix go to infinity as $R \to 0$, which means that if the penalty on the control is too small, we have very large (nonphysical) actuation. Figure 10.6 shows the eigenvalues of the closed-loop system for values of R ranging from 10^{-2} (where the system has two highly damped complex root pairs) to 10^4 (where the roots are very close to those in open loop). A logarithmic distribution of R is used within that range.

Example 10.2 Load Alleviation of a 2-DoF Airfoil in Continuous Turbulence.   Consider again the 2-DoF airfoil under continuous turbulence introduced in Example 3.3 and 3.4. The same properties are considered now, namely $\mu = 5$, $x_{ac} = 0.25c$, $x_{ea} = 0.35c$, $x_{cg} = 0.45c$, $r_\alpha = 0.25c$, and $\omega_h/\omega_\alpha = 0.5$, with the geometric description of Figure 3.1. As the interest here is now on closed-loop performance, a trailing-edge flap is included. The flap chord is $0.15c$, or equivalently $x_{fh} = 0.85c$. The dynamics of the airfoil is modeled again using Equation (3.92), with the aerodynamic forces and moments given by Equations (3.75) and (3.76) to include the flap actuation. The command is on the flap accelerations, with rates and deflections obtained through a double integrator. No actuator model is included and a fixed reference airspeed $V = 0.5c\omega_\alpha$ is considered in all simulations. The lift-deficiency function, Equation (3.47), is approximated with four aerodynamic lags, while the two-lag approximation of Table 3.2 is considered for Sears's function. This results in a system with 20 states,[4] which are ordered as

$$\mathbf{x} = \left\{ \eta \quad \alpha \quad \delta \quad \eta' \quad \alpha' \quad \delta' \quad \mathbf{x}_a \quad \mathbf{x}_g \right\}^\top \qquad (10.31)$$

---

[4] Recall that the number of aerodynamic states coming from the lift-deficiency function is the number of aerodynamic lags times the number of DoFs. This results in $\mathbf{x}_a$ being of dimension 12 here.

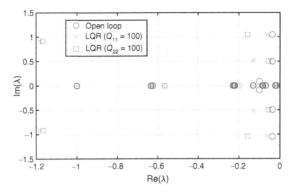

Figure 10.7 Open- and closed-loop dimensionless eigenvalues of the 2-DoF airfoil with LQR flap control.

using the notation of Chapter 3. First, we consider full-state feedback using LQR control. The state cost $\mathbf{Q}$ is chosen to be a diagonal matrix with all coefficients being equal to 1 except for the coefficient associated with either η or α, which is chosen to be 100. Figure 10.7 shows all the open- and closed-loop eigenvalues for this problem with both controllers. As the system is written in terms of the (nondimensional) reduced frequency, the eigenvalues are scaled with the convective timescale of the flow, $c/(2V)$. The aerodynamic lags in the open-loop system result in eigenvalues close to the negative real axis. The LQR controller increases damping on the mode shapes associated with the pitch and plunge DoFs, with higher damping on the DoF with higher cost in the objective function.

This additional system damping is most clearly seen in Figure 10.8 that shows the Bode plots (in linear scale) of the FRFs between the normalized leading-edge gust profile w_g/V and the plunge, pitch, and flap deflections (the last two given in radians). The two resonant modes are clearly visible in the open-loop plunge and pitch responses. In the open loop, however, the flap is locked, and δ is identically zero. Closing the loop defines a feedback command on the flap. Depending on whether the control effort is put on the plunge mode, $Q_{11} = 100$, or the pitch mode, $Q_{22} = 100$, the flap has its peak magnitude at the corresponding resonance frequency, which results in higher reductions of the peak value in the admittance functions for the pitch and plunge DoFs.

The airfoil is next subject to a turbulent inflow that is described by the 1-D von Kármán model, Equation (2.44). Only the vertical component of the velocity is considered, which results in a turbulence filter (written in terms of the reduced frequency) of the form

$$G_{\text{wn}}(ik) = \sigma_w \sqrt{\frac{2\ell}{\pi c}} \frac{1 + ik\sqrt{\frac{8}{3}\frac{2a\ell}{c}}}{1 + ik\frac{2a\ell}{c}} \frac{1}{\left(1 + ik\frac{2a\ell}{c}\right)^{5/6}}, \qquad (10.32)$$

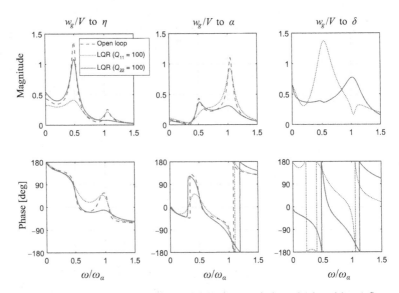

Figure 10.8 FRFs between gust velocity and airfoil plunge, pitch, and (closed-loop) flap deflection. [10Control/turb2dof_lqr.m]

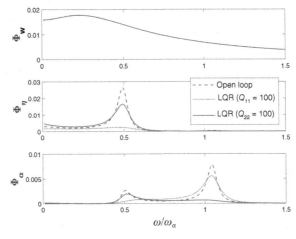

Figure 10.9 PSD in the open- and closed-loop response to von Kármán isotropic turbulence with $\ell = c$ and $\sigma_w = 0.1V$. [10Control/turb2dof_lqr.m]

where the last term is approximated by Equation (2.49). The turbulent length scale is chosen to be $\ell = c$, which would be a typical value if the airfoil was mounted in a wind tunnel, and the turbulence intensity $\sigma_w = 0.1V$. The open-loop system corresponds to the problem outlined in Figure 3.27, while in the closed-loop system, the flap is actuated, with acceleration given by the LQR feedback gain. The open- and closed-loop power spectral density of the simulated airfoil response is shown in Figure 10.9. It can be seen how the peak values of the FRFs obtained before have now been modulated by the spectral content of the turbulent inflow.

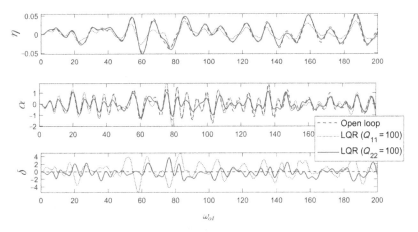

Figure 10.10 Sample time histories of plunge, pitch, and flap command under 1-D von Kármán turbulence with $\ell = c$ and $\sigma_w = 0.1V$. [10Control/turb2dof_lqr.m]

Similar spectral content is obtained for the flap deflections, but it has not been included in the figure. Instead, we present in Figure 10.10 the time history of plunge, pitch, and flap deflections under a realization of turbulence inflow for $\sigma_w = 0.1V$ and $\ell = c$. This is necessary to quantify the expected maximum deflection and the maximum rate of the flap for the chosen weights in the control system. For the 10% gust intensity and the LQR controllers described earlier, the amplitude of the flap deflection in Figure 10.10 never exceeds $5°$, which can be assumed to be an acceptable value. Refined values of the weights in Equation (10.19) are typically obtained through an optimization on representative situations of a worst-case scenario and under saturation constraints on the actuator commands. Finally, the typical metric used to report load levels is the RMS, defined as in Equation (2.10), of the relevant DoFs. This can be done from either the integration of the area under the curves in Figure 10.9 or from sufficiently long (i.e., statistically relevant) time series, using Equation (2.10). For this problem, a 42% reduction in the RMS of the plunge DoF is achieved with a controller with $Q_{11} = 100$. The second controller, with $Q_{22} = 100$, reduces the pitch RMS by 31%.

10.4   Linear Quadratic Gaussian Control

A major limitation of LQR control is that the full state of the system must be known (i.e., measured) in real time, as it is the input to the controller. This information is, however, very rarely available and particularly so for larger systems, such as those considered in this book. This lack of actual data can be overcome by introducing state estimation strategies, in which only partial information of the state is measured over time, potentially also including measurement noise. The instantaneous state is then approximately reconstructed from those noisy measurements. Here, we consider the Kalman filter, which estimates the state of a linear system given by Equation (10.1) assuming a Gaussian distribution of the disturbance noise. This is particularly relevant in applications to load control of flexible aircraft since, as we saw in Section 2.3, atmospheric turbulence is reasonably well approximated as a Gaussian process. A very good introduction to Kalman filters is given by Welch and Bishop (2001). As with LQR control, there exists a general theory for finite-horizon observer models, but our discussion will be restricted here to infinite-horizon, or *steady-state*, Kalman filters. In what follows, it is also assumed, although this is not strictly necessary, that there is no direct feedthrough between inputs and outputs (i.e., the system to be controlled is strictly proper). Multiple extensions of the Kalman filter exist for nonlinear systems or for colored-noise disturbance, among others (Lewis et al., 2008). Moreover, our focus is on the continuous-time version of the filter, often referred to as the *Kalman–Bucy* filter. The combination of a Kalman filter and LQR control results in the LQG control, and an outline of the relevant theory, with an example of a typical aeroelastic application, is included next.

10.4.1   The Kalman Filter

The steady-state Kalman filter estimates the instantaneous state of a dynamic system, given the values of its inputs and measured outputs. It assumes that the system dynamics are well approximated by an LTI state-space model with known coefficients as

$$
\begin{aligned}
\dot{\mathbf{x}} &= \mathbf{A}\mathbf{x} + \mathbf{B}_u\mathbf{u} + \mathbf{B}_w\mathbf{w}, \\
\mathbf{y} &= \mathbf{C}\mathbf{x} + \mathbf{v},
\end{aligned}
\tag{10.33}
$$

that is, as in Equation (10.1) but without feedthrough. Furthermore, for the design of the Kalman filter, both $\mathbf{v}(t)$ and $\mathbf{w}(t)$, the measurement and input (or *process*) noise, respectively, are assumed to be (uncorrelated) white noise signals with zero mean and known covariance matrices $\boldsymbol{\Sigma}_v$ and $\boldsymbol{\Sigma}_w$, respectively. Recall that the covariance was defined in Equation (2.11). An additional condition is that there is measurement noise in all channels, which implies that $\boldsymbol{\Sigma}_v$ is positive definitive. The Kalman filter seeks an estimation of the state, $\hat{\mathbf{x}}(t)$, that minimizes the mean square value, defined in Equation (2.10), of the estimation error $\mathbf{e} = \mathbf{x} - \hat{\mathbf{x}}$, namely $\sigma_e^2 = \mathrm{E}[\mathbf{e}^\top(t)\mathbf{e}(t)]$. This is achieved (Burl, 1999, Ch. 7) by means of a linear observer equation (the *Kalman state equation*) of the form

$$
\begin{aligned}
\dot{\hat{\mathbf{x}}} &= \mathbf{A}\hat{\mathbf{x}} + \mathbf{B}_u\mathbf{u} + \mathbf{L}\left(\mathbf{y} - \mathbf{C}\hat{\mathbf{x}}\right), \\
\hat{\mathbf{x}}(0) &= \mathbf{0},
\end{aligned}
\tag{10.34}
$$

where the constant gain matrix $\mathbf{L}$ is known as the *Kalman gain*. Although it will not be proved here, it can be shown (Anderson and Moore, 1979, Ch. 3) that the optimal gain that minimizes the estimation error is

$$
\mathbf{L} = \boldsymbol{\Sigma}_e\mathbf{C}^\top\boldsymbol{\Sigma}_v^{-1},
\tag{10.35}
$$

where $\boldsymbol{\Sigma}_e$ is the covariance matrix of $\mathbf{e}(t)$, which will be determined next. Note first that Equation (10.34) retrieves the solution of Equation (10.33) with no disturbances if the difference between the estimated, $\mathbf{C}\hat{\mathbf{x}}$, and the actual, $\mathbf{y}$, output signals is sufficiently small. Combining Equations (10.33) and (10.34) results in the following evolution equation for the error signal:

$$\dot{\mathbf{e}} = (\mathbf{A} - \mathbf{LC})\,\mathbf{e} + \mathbf{B}_w \mathbf{w} - \mathbf{Lv},$$
$$\mathbf{e}(0) = \mathbf{x}(0). \tag{10.36}$$

Therefore, for the estimation error to decrease with time, we need a stable matrix $\mathbf{A} - \mathbf{LC}$, which imposes a constraint to the measurement noise. The (steady-state) covariance matrix of the state error, $\boldsymbol{\Sigma}_e$, can now be obtained from the associated Lyapunov equation to Equation (10.36). From Equation (2.27), it can be written as

$$(\mathbf{A} - \mathbf{LC})\,\boldsymbol{\Sigma}_e + \boldsymbol{\Sigma}_e\,(\mathbf{A} - \mathbf{LC})^\top + \mathbf{B}_w \boldsymbol{\Sigma}_w \mathbf{B}_w^\top - \mathbf{L}\boldsymbol{\Sigma}_v \mathbf{L}^\top = \mathbf{0}, \tag{10.37}$$

where we have used that the input and measurement noise are uncorrelated signals. Finally, substituting the optimal Kalman gain and simplifying the algebra, we obtain the nonlinear equation that determines the covariance of the estimation error as

$$\mathbf{A}\boldsymbol{\Sigma}_e + \boldsymbol{\Sigma}_e \mathbf{A}^\top - \boldsymbol{\Sigma}_e \mathbf{C}^\top \boldsymbol{\Sigma}_v^{-1} \mathbf{C}\boldsymbol{\Sigma}_e = -\mathbf{B}_w \boldsymbol{\Sigma}_w \mathbf{B}_w^\top, \tag{10.38}$$

which has the same structure as Equation (10.20) and, therefore, can be identified as an *algebraic Riccati equation*. Note that, if there are no measurements, that is, $\mathbf{C} = \mathbf{0}$, Equation (10.38) gives the propagation of the covariance of the error to input noise, Equation (2.27). It is common to refer to Equations (10.20) and (10.38) as the controller and observer algebraic Riccati equations, respectively. Indeed, the optimal estimation problem with the known covariance of the measurement noise is the dual (or adjoint) problem to the evaluation of the optimal quadratic control with full-state feedback (Kwakernaak and Sivan, 1972). This duality will be used again when considering the nonlinear control and estimation problems in Section 10.5. Finally, after the covariance matrix of the estimation error is computed from Equation (10.38), the Kalman gain in Equation (10.35) can be obtained. As it can be seen, $\mathbf{L}$ is uniquely determined for a given system, Equation (10.33), and given values of the covariance of the process and measurement noise, $\boldsymbol{\Sigma}_w$ and $\boldsymbol{\Sigma}_v$, respectively.

Example 10.3 A Kalman Filter for a 1-DoF Oscillator.   Consider the 1-DoF damped oscillator under the stochastic excitation of Equation (2.37), with noisy velocity measurements. Its dynamics are described by

$$\dot{\mathbf{x}} = \begin{bmatrix} 0 & 1 \\ -\omega_0^2 & -2\omega_0\xi \end{bmatrix} + \begin{bmatrix} 0 \\ \omega_0^2 \end{bmatrix} w,$$
$$y = \begin{bmatrix} 0 & 1 \end{bmatrix} \mathbf{x} + v, \tag{10.39}$$

where the state vector $\mathbf{x}$ contains the displacement and velocity of the oscillator, the excitation $w(t)$ is a white noise with RMS σ_w, and the RMS of the measurement noise is σ_v. If we write the error covariance, which is a symmetric matrix, as

$$\Sigma_e = \begin{bmatrix} p_{11} & p_{12} \\ p_{12} & p_{22} \end{bmatrix}, \tag{10.40}$$

the unknown coefficients are given by Equation (10.38), which results in the following three independent equations:

$$2p_{12} - \frac{p_{12}^2}{\sigma_v^2} = 0,$$

$$p_{22} - \omega_0^2 p_{11} - 2\xi\omega_0 p_{12} - \frac{p_{12}p_{22}}{\sigma_v^2} = 0, \tag{10.41}$$

$$-2\omega_0^2 p_{12} - 4\xi\omega_0 p_{22} - \frac{p_{22}^2}{\sigma_v} + \omega_0^4\sigma_w^2 = 0.$$

From the first equation, we have $p_{12} = 0$. Next, the last equation gives a quadratic polynomial in p_{22}, from which only the positive root needs to be considered. Therefore, $p_{22} = 2\xi\omega_0\sigma_v^2\left(\sqrt{1+\lambda^2}-1\right)$, where $\lambda = \frac{\omega_0\sigma_w}{2\xi\sigma_v}$ is a normalized signal-to-noise ratio (Bryson and Ho, 1979). Finally, the second equation gives $p_{11} = \frac{p_{22}}{\omega_0^2}$. Substituting into Equation (10.35) gives the Kalman gain for this problem as

$$\mathbf{L} = \begin{bmatrix} 0 \\ 2\xi\omega_0\left(\sqrt{1+\lambda^2}-1\right) \end{bmatrix}. \tag{10.42}$$

The corresponding steady-state Kalman filter, Equation (10.34), is then

$$\dot{\tilde{\mathbf{x}}} = \begin{bmatrix} 0 & 1 \\ -\omega_0^2 & -2\xi\omega_0\sqrt{1+\lambda^2} \end{bmatrix}\tilde{\mathbf{x}} + \begin{bmatrix} 0 \\ 2\xi\omega_0\left(\sqrt{1+\lambda^2}-1\right) \end{bmatrix}y. \tag{10.43}$$

Comparing the state matrices of Equations (10.39) and (10.43), we can see that both systems are oscillators with natural frequency ω_0, but the damping in the observer grows with the normalized signal-to-noise ratio, λ. It is then illustrative to write the FRF between the measured and the estimated velocity, which becomes

$$H(i\omega) = \frac{2i\xi\left(\sqrt{1+\lambda^2}-1\right)\frac{\omega}{\omega_0}}{1+2i\xi\sqrt{1+\lambda^2}\frac{\omega}{\omega_0} - \frac{\omega^2}{\omega_0^2}}. \tag{10.44}$$

Ideally, we would like the estimator to result in $H(i\omega) = 1$, that is, equal values for the estimated and measured velocity. Figure 10.11 shows the amplitude and phase of $H(i\omega)$ for two different levels of damping on the original system, and increasing values of the signal-to-noise ratio, λ. As the noise increases (corresponding to a smaller λ), a mismatch between both signals appears around the natural frequency ω_0. For very low noise ($\lambda \to 0$), on the other hand, the observer becomes an *all-pass filter* with unit amplitude, since good quality measurements are available.

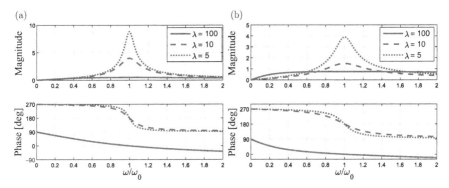

Figure 10.11 Transfer function from measured to estimated velocity. (a) $\xi = 0.01$ and (b) $\xi = 0.02$.

10.4.2   Closed-Loop System

We have now all the ingredients to formulate the LQG control problem. For a system such as the one defined in Equation (10.33), the LQG problem is posed as a steady-state LQR control problem where the system state is estimated by a Kalman filter from noisy measurements, as shown in Figure 10.12. As a result, the closed-loop input signal is given by

$$\mathbf{u} = -\mathbf{K}\hat{\mathbf{x}}, \tag{10.45}$$

with the feedback matrix given by Equation (10.21). Combining the state and error equations, Equation (10.36) finally results in the closed-loop system

$$\left\{ \begin{matrix} \dot{\mathbf{x}} \\ \dot{\mathbf{e}} \end{matrix} \right\} = \begin{bmatrix} \mathbf{A} - \mathbf{B}\mathbf{K} & \mathbf{B}\mathbf{K} \\ \mathbf{0} & \mathbf{A} - \mathbf{L}\mathbf{C} \end{bmatrix} \left\{ \begin{matrix} \mathbf{x} \\ \mathbf{e} \end{matrix} \right\} + \begin{bmatrix} \mathbf{B}_w & \mathbf{0} \\ \mathbf{B}_w & -\mathbf{L} \end{bmatrix} \left\{ \begin{matrix} \mathbf{w} \\ \mathbf{v} \end{matrix} \right\}, \tag{10.46}$$

where the estimation error has already been defined above as $\mathbf{e} = \mathbf{x} - \hat{\mathbf{x}}$. Note that, while the infinite-horizon LQR results in a static feedback gain that did not add new states in Equation (10.22), LQG control needs to solve both the system and estimator dynamics, which doubles the size of the problem. An important characteristic of this system of equations is that the system matrix has a block triangular form, which implies that the eigenvalues of the regulator and the observer problem remain unaltered when assembling the feedback system. This is known as the *separation principle* (Anderson and Moore, 1990), and it justifies *a posteriori* our independent evaluation of the CLAW gain matrix $\mathbf{K}$ and the Kalman gain $\mathbf{L}$.

Waite et al. (2021) have recently demonstrated LQG for flexible aircraft control. In particular, an active flutter suppression solution was devised on a half-model of NASA's Common Research Model. Actuation is based on three trailing-edge control surfaces, one in board and two outboard, while 22 accelerometers over the wing provide the output measurements used by the Kalman filter to estimate the internal state. The control synthesis is carried out on an identified aeroelastic model built using

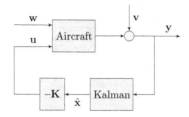

Figure 10.12 Block diagram for the LQG controller.

ERA algorithms from unsteady RANS (Section 6.5.4) and the linear normal modes of the structure. Other relevant control architecture for gust load alleviation built upon the LQG are those of Gangsaas et al. (1981), Matsuzaki et al. (1989), Dillsaver et al. (2011), and Christhilf et al. (2012). While LQG design produces a controller for which stability can be guaranteed, its robustness, that is, its performance under modeling errors, may still be an issue and needs to be checked *a posteriori* (Doyle, 1978). Actual implementations have then often included additional features, such as loop transfer recovery (Christhilf et al., 2012), to improve robustness. An alternative, and increasingly common approach, is to build the design using $\mathcal{H}_\infty$ control, in which the optimization problem is posed on an infinite norm of the closed-system model under perturbations. This results in a constrained optimization problem, which is in general more complex to solve than the LQG (and for which there may not be a solution), but for which some measures of robustness are available. Good introductions to robust control can be found in the books of Zhou et al. (1995), for the general theory, and Lavretsky and Wise (2013), for applications to aerospace engineering. Recent applications of $\mathcal{H}_\infty$ control to flexible aircraft are those of Cook et al. (2013), Hesse and Palacios (2016), and Fournier et al. (2022).

Example 10.4 Load Alleviation of a 2-DoF Airfoil in Continuous Turbulence.   In Example 10.2, we have considered the feedback control of a 2-DoF airfoil under continuous turbulence using LQR control. LQR, however, uses full-state feedback that would not be available in a practical implementation of the problem (in particular, it is not possible to measure the aerodynamic states in Equation (10.31)). This is addressed here by means of a Kalman filter for state estimation. For that purpose, we first augment the system given by Equation (3.92) with the turbulence filter of Equation (10.32) and use it as the internal model for the Kalman state equation. This gives information about the frequency content of interest to the state estimator. The Kalman filter defined here is driven by the turbulence disturbance, which defines the process noise, and uses as input only the pitch rate. A turbulent gust intensity equal to 10% of the airspeed is assumed to obtain the estimator, resulting in a process noise of 0.01. As Σ_w needs to be nonsingular in Equation (10.38), a covariance noise of 10^{-8} is also assumed on the flap input signal. A small output noise $\Sigma_v = 10^{-6}$ is finally assumed on the pitch-rate measurements.

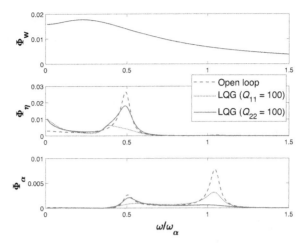

Figure 10.13 PSD in the open- and closed-loop (LQG) response to von Kármán isotropic turbulence with $\ell = c$ and $\sigma_w = 0.1V$. [10Control/turb2dof_lqg.m]

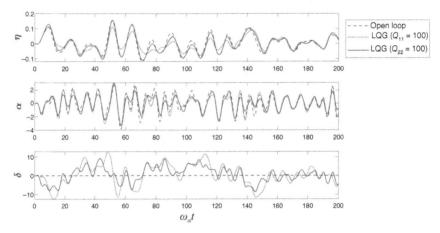

Figure 10.14 Sample time histories of plunge, pitch, and flap command for LQG control under 1-D von Kármán turbulence with $\ell = c$ and $\sigma_w = 0.1V$. [10Control/turb2dof_lqg.m]

The same weights as in Example 10.2 are used to determine the LQR gains. Closing the loop with the resulting LQG controller gives the PSD of Figure 10.13. This is a very similar PSD to that obtained by the LQR (Figure 10.9), meaning that a Kalman filter with pitch-rate measurements can produce an acceptable estimate (for control purposes) of the instantaneous state of the system. The main difference between the LQR and LQG PSDs is that the latter displays a small drop in performance on the estimated DoF (plunge, η) at low frequencies. A second difference can be observed in the time histories on a sample realization of the gust, which are shown in Figure 10.14 and can be compared with the LQR results in Figure 10.10. The results show that the input command is now in the range of $\pm 10°$, that is, a larger effort is required from the controller to achieve similar performance. It would be necessary to increase the penalty on the controller in the LQR used here to obtain a similar level of actuation to the problem with full-state feedback at the expense of poorer performance.

10.5   Nonlinear Optimal Control and Estimation

The methods of linear optimal control defined so far in this chapter can be extended to systems with more complex, in general, nonlinear dynamics if the evaluation of the optimal input signals is done in real time by an on-board computing system. The resulting strategy is known as MPC and defines an *online* solution process in contrast to the *offline* definition of the optimal gains of, for example, Section 10.3. To achieve this, MPC relies on an *internal model* of the system dynamics, known to the controller and used to produce real-time predictions of the response to both exogenous disturbance and commanded inputs, using an estimation of the states from measurements as initial conditions. The literature on MPC is vast, and good introductions to both the theoretical background and the numerical implementations are given by Magni et al. (2009) and Rawlings et al. (2017).

A key challenge in the development of MPC-based solutions is to generate an internal model that has sufficiently low dimensionality for real-time optimization but that still adequately captures the key physical process of interest. This is particularly challenging in flexible aircraft applications, which as we have already seen in this book, display complex dynamics involving unsteady aerodynamics, structural, and flight dynamics, as well as saturation limits that may act as nonlinear constraints on the system dynamics. The coupled flight dynamics–aeroelastic description of Section 9.3 gives an excellent starting point to address this problem, and a summary of the current state of the art is included in this section. To simplify the description, we restrict ourselves to linearized gravitational effects, as was done in Equation (9.13). A more general solution, including the finite rotations of the structure, has been recently presented by Artola et al. (2022) for nonlinear structural control and Artola et al. (2021a) and Pereira et al. (2019a, 2022a, 2022b) for the nonlinear aeroelastic control of very flexible wings.

The objective is to determine the optimal control inputs at a given (constant) sampling rate, Δt. Thus, given past measurements up to the current time t_k, the controller needs to compute the inputs on the system at the next sampling time, $t_{k+1} = t_k + \Delta t$. This is achieved here by repetitively solving an optimization problem defined by a quadratic cost function such as that of Equation (10.5) and the governing equations, Equation (9.13), with potentially additional constraints on input and/or states. Even though the objective is to compute only the control inputs one time step at a time, the governing equations need to be solved for a longer time horizon, τ_c, to inform the controller of probable future events and avoid constraint violations. As this process is done online on the real system, the MPC problem needs to be solved in a much faster time than Δt, which puts constraints on the maximum value of τ_c.

At the kth sampling point, the MPC problem is written as

$$\min_{\mathbf{x}(t),\mathbf{u}(t)} \quad J(\mathbf{x}-\mathbf{x}_r,\mathbf{u}-\mathbf{u}_r)$$

subject to

$$\mathbf{h}(\dot{\mathbf{x}},\mathbf{x},\mathbf{u},\mathbf{w}) = 0 \quad \text{for} \quad t_k \le t \le t_k + \tau_c, \tag{10.47}$$

$$\mathbf{x}(t_k) = \hat{\mathbf{x}}_k,$$

$$\mathbf{x}_{\min} \le \mathbf{x}(t) \le \mathbf{x}_{\max},$$

$$\mathbf{u}_{\min} \le \mathbf{u}(t) \le \mathbf{u}_{\max},$$

where $\mathbf{x}(t) = \left\{ \mathbf{q}_0^\top \quad \mathbf{q}_1^\top \quad \mathbf{q}_2^\top \quad \mathbf{x}_a^\top \quad \mathbf{x}_g^\top \right\}^\top$ is the aeroelastic state vector, $\mathbf{u}$ defines the system control inputs (in particular, δ in Equation (9.13)), and $\mathbf{w}$ includes the external, typically atmospheric, disturbances on the system ($\mathbf{w}_{g0}$ in Equation (9.12)). The cost function is written with respect to a predetermined nonzero reference state $\mathbf{x}_r$ and control $\mathbf{u}_r$, which may correspond to a trim state or to a set point in a maneuver (which would be determined by an additional navigation loop in autonomous vehicles). Finally, the constraints on inputs and states are restricted here to be *box constraints*, that is, saturation constraints on both minimum and maximum values that a variable can take, although that is not strictly necessary. The solution of this system is then repeated after a time Δt, which must be higher than the computational time required to solve the optimization problem.

Equation (10.47) defines a nonlinear extension to the LQR problem of Section 10.3. The nonlinearities arise from the constraints on control input (e.g., saturation levels on the aileron deflection) and possibly on some of the states, as well as from the nonlinear functional $\mathbf{h}$, which here represents Equations (9.12) and (9.13). Finally, $\hat{\mathbf{x}}_k$ is an estimate of the system state at $t = t_k$, which defines the initial condition for the internal time marching of the system dynamics within the MPC problem. The estimated state is obtained by solving a second optimization problem, known as *moving horizon estimation* (MHE), which is an extension to nonlinear systems of the Kalman filter of Section 10.4.1.

MHE defines a nonlinear optimization problem to estimate the state of the system at t_k, given past measurements over a recent estimation horizon τ_e, typically much larger than the sampling rate Δt, and an internal model for the system dynamics. MHE can also simultaneously estimate any (unknown) disturbance signals of interest that may have acted on the system over the horizon window. It poses the following problem at the current time t_k:

$$\min_{\mathbf{w}(t),\hat{\mathbf{x}}_e} \quad J_e(\mathbf{y}-\hat{\mathbf{y}}(\mathbf{w},\hat{\mathbf{x}}_e))$$

subject to

$$\mathbf{h}(\dot{\hat{\mathbf{x}}},\hat{\mathbf{x}},\mathbf{u},\mathbf{w}) = 0 \quad \text{for} \quad t_k - \tau_e \le t \le t_k, \tag{10.48}$$

$$\hat{\mathbf{x}}(t_k - \tau_e) = \hat{\mathbf{x}}_e,$$

where the cost function is defined with a positive semidefinite weighting matrix $\mathbf{Q}_e$ as

$$J_e = \int_{t_k-\tau_e}^{t_k} \frac{1}{2}(\mathbf{y}-\hat{\mathbf{y}})^\top \mathbf{Q}_e (\mathbf{y}-\hat{\mathbf{y}}) \, dt. \tag{10.49}$$

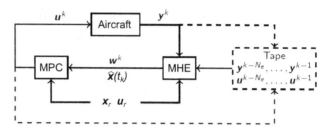

Figure 10.15 MPC-MHE framework for control of very flexible aircraft (Artola et al., 2021b).

This cost function penalizes the difference between the observed measurements $\mathbf{y}(t)$ over the estimation horizon τ_e and the estimated values $\hat{\mathbf{y}}(t)$, which are obtained from the solution of the governing equations with the known past control inputs, $\mathbf{u}(t)$, but unknown disturbance signals, $\mathbf{w}(t)$. In Equation (10.48), the unknown parameters are the external disturbance, $\mathbf{w}(t)$, for which additional box constraints my be needed, and the initial state of the system, $\hat{\mathbf{x}}_e$, at the beginning of the estimation horizon is $t_k - \tau_e$. The estimated state between $t_k - \tau_e$ and t_k is then obtained from the integration of the governing equation $\mathbf{h}$, given by Equations (9.12) and (9.13). Once Equation (10.48) is solved, the final state in the estimation window $\hat{\mathbf{x}}(t_k)$ defines the initial conditions for the MPC problem, Equation (10.47). All other intermediate solutions can be discarded, although, as it will be seen later in this section, they can be used to accelerate the converge in future time steps. This problem is repeated at time $t_k + \Delta t$.

The resulting control architecture is schematically shown in Figure 10.15. Current inputs and measurements on the aircraft are fed into the MHE, while the most recent $N_e = \tau_e / \Delta t$ samples are stored to accelerate the convergence of the MPC-MHE solutions. Note that, if the governing equations were linear, Equation (10.48) would simplify to the optimization problem for the Kalman filter over a finite horizon. As it should be expected by now, a major challenge for the practical implementation of both MPC and MHE is to solve the optimization problems at a faster rate than that of the control actuation Δt. Consequently, this approach has gained a foothold on the dynamical process with either slow timescales (Verwaal et al., 2015), simple dynamics (Diehl et al., 2004), or ideally both. Advances in computer power have made it increasingly attractive for aircraft load alleviation (Kopf et al., 2018), and a successful wind-tunnel implementation on a relatively stiff wing, with saturation constraints, has been demonstrated by Barzgaran et al. (2021). An application with a very flexible wing, also in the wind tunnel, was recent achieved by Pereira et al. (2022c). Application of MPC-MHE in nonlinear control of very flexible aircraft has been recently shown via simulation by Pereira et al. (2021) and Artola et al. (2021b). These recent developments have benefited from bespoke numerical acceleration strategies to satisfy the extreme computational demands on real-time optimization. Two of them are outlined next.

The time-marching integration of the internal model is the most expensive part in the solution of an optimal control problem. This is a nonlinear problem that needs an iterative solution with multiple evaluations until convergence. In practice, however, there are small differences between those solutions and this can be exploited by the control algorithm to obtain large reductions in the computational cost. A classical solution strategy consists in splitting the prediction or estimation horizons in multiple subintervals, which can then be solved in parallel and initialized with the previous results. The remainder of this section discusses the resulting *multiple shooting method* (Bock and Krämer-Eis, 1984) and a numerical acceleration process based an efficient data transfer between subiterations and between the estimation and prediction problems. As in the linear case, the nonlinear problems for estimation and control are dual descriptions that can be solved in almost identical procedures. Consequently, only the nonlinear MPC optimization problem is considered here.

In a multiple shooting solution, the prediction horizon τ_c of the MPC problem is split into M_c time subintervals, and the initial conditions to each of the subintervals are then added as optimization parameters. In general, those are longer than the sampling rate, that is, $M_c < \tau_c / \Delta t$. *Matching conditions* are then included in the optimization to enforce that the states at the end of one subinterval are the same as the start of the following one. We identify the subintervals with superscript $m = 1, \ldots, M_c$. The control input and disturbance vectors are then defined by a piecewise constant parameterization, where $\mathbf{u}^m$ and $\mathbf{w}^m$ are their values at the mth subinterval. Note that this parametrization can be chosen independently of the multiple shooting discretization, such that within a shooting subinterval the control and disturbance signals can be given with a faster sampling. The initial and final conditions at the subintervals are denoted as $\mathbf{x}_0^m$ and $\mathbf{x}_f^m$, respectively, and we define the functional $\mathbf{f}(\mathbf{x}_0^m, \mathbf{u}^m)$ from the time integration of Equations (9.12) and (9.13) within the mth interval. With those definitions, Equation (10.47) can be rewritten as

$$\min_{\mathbf{x}_0, \mathbf{u}} \sum_{m=1}^{M_c} J_m(\mathbf{x}^m - \mathbf{x}_r^m, \mathbf{u}^m - \mathbf{u}_r^m) \tag{10.50a}$$

$$\text{subject to} \quad \mathbf{x}_0^1 = \hat{\mathbf{x}}_i, \tag{10.50b}$$

$$\mathbf{x}_f^m = \mathbf{f}(\mathbf{x}_0^m, \mathbf{u}^m, \mathbf{w}^m), \quad m = 1, \ldots, M_c - 1, \tag{10.50c}$$

$$\mathbf{x}_0^{m+1} = \mathbf{x}_f^m, \tag{10.50d}$$

$$\mathbf{x}_{\min} \le \mathbf{x}_f^m \le \mathbf{x}_{\max}, \quad m = 1, \ldots, M_c, \tag{10.50e}$$

$$\mathbf{u}_{\min} \le \mathbf{u}^m \le \mathbf{u}_{\max}, \tag{10.50f}$$

where J_m is the cost function of the mth subinterval. Equation (10.50d) defines the matching conditions between the subintervals. The inequality constraints on the state vector, Equation (10.50e), are only enforced at the final state $\mathbf{x}_f^m$ of each subintervals. To achieve a computationally efficient implementation, Equation (10.50c) is linearized at each optimization iteration, which leads to nonphysical intermediate solutions, possibly including discontinuties, before convergence is achieved. In the numerical examples in Section 10.6, the solution to Equation (10.50) is iteratively sought using a sequential quadratic programming (SQP) algorithm with analytical adjoints, and the Hessian of the system is approximated using the Broyden–Fletcher–Goldfarb–Shanno (BFGS) update formula. See Chapter 18 of Nocedal and Wright (2006) for details of the methods and Artola et al. (2021a) for their application.

10.6   Numerical Examples

10.6.1   Gust Load Alleviation in a Cantilever Wing with Saturation Constraints

As a first numerical example, consider the closed-loop gust load alleviation of the flexible cantilever wing originally defined by Goland (1945). This is the "unstretched" version of the test case studied in Example 6.3 on page 257. Flutter at sea level occurs at 170 m/s, and a numerical study of its open- and closed-loop response at 140 m/s was carried by Simpson et al. (2014), and it is summarized here. The wing aerodynamics is modeled using a UVLM discretization with $N = 16$ chordwise and $M = 40$ span-wise vortex rings and a 20-chord wake. This ensures the convergence of the results. As this is a relatively stiff wing, a linearized aeroelastic model of the wing has been considered, using the approach of Section 9.2.4. Finally, an 8-state ROM is obtained through balanced truncation of this linearized system, which is finally transformed to continuous time using a bilinear transformation, Equation (A.12). Two optimal control architectures for state regulation are considered: MPC with full-state feedback, but including saturation constraints, and an LQR control scheme. The closed-loop response of the plant under MPC uses 100 time steps to define the prediction horizon. The input constraints correspond to flap deflections in the $\pm 10°$ range. The wing is excited by a vertical "1-cosine" gust, and the critical gust length (see Section 3.6.2) is first searched from the open-loop response.

The nominal control, for both LQR and MPC, is based on unit weights for all the plant states and inputs, that is, the matrices $\mathbf{H}$, $\mathbf{Q}$, and $\mathbf{R}$ in Equation (10.5) are identity matrices. The open- and closed-loop responses, using both controllers, to the critical gust length are shown in Figure 10.16a. As can be seen, this choice of weights reduces the maximum torsional strain with both MPC and LQR by approximately 25%, while increasing the maximum bending strain by around 13%. This is a typical trade-off in gust load alleviation using trailing-edge flaps. The control action for both closed-loop results is shown in Figure 10.16b, with the LQR reaching saturation during the gust encounter. The MPC, on the other hand, anticipates the constraint violations and achieves a similar performance to the LQR without saturating the inputs. The maximum torsional deformation with the MPC controller is slightly larger than that with the LQR, but also smoother, which may reduce fatigue risks.

Changing the weights in the control system allows us to shape the response in the closed-loop system. As an example, Figure 10.17 shows again the torsional and bending strains at the wing root under a "1-cosine" gust. In the first case (*output weighting* in the figure), unit weights are assigned to both bending and torsion strains, while in the second case, only the bending strain is penalized by the controller. The first case achieves a mild reduction in both torsional and bending deformations, while the second results in a much higher reduction of bending strains at the expense of higher torsional strains. The choice of which load component needs to be targeted depends in practice on which resulting stress state is effectively sizing that wing structure.

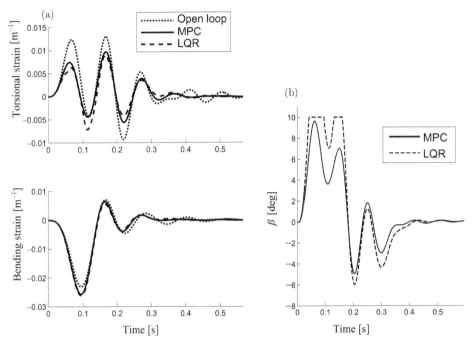

Figure 10.16 Response of the Goland wing to a "1-cosine" gust. Closed-loop results with unit weight on states and inputs. (a) Wing root strains and (b) commanded control input (Simpson et al., 2014).

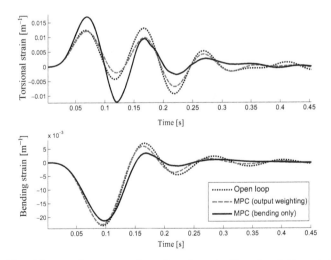

Figure 10.17 Closed-loop performance under a discrete gust with unit weights on both bending and torsion strains (output weighting) and with only on the bending strains (Simpson et al., 2014).

10.6.2   Nonlinear Control of the Pazy Wing

The Pazy wing has been introduced in Section 9.4.1, where it has been used to demonstrate the effect on large (geometrically nonlinear) deformations on wing flutter. This section explores active flutter suppression mechanisms on that configuration using the nonlinear control strategies of Section 10.5. The following results have first appeared in Artola et al. (2021a), where further details of this problem can be found.

The problem to be solved is the stabilization of the wing at a root angle of incidence $\alpha_0 = 4°$, $V = 50$ m/s, and $\rho = 1.225$ kg/m^3, which correspond to a point in the first flutter region in Figure 9.11. The nonlinear modal formulation of Equations (9.12) and (9.13) is used as an internal model for both estimation and control. The structural model includes nine LNMs of the underformed structure (the first three out-of-plane bending, the first two in-plane bending, the first three torsional, and the first axial modes). The aerodynamic model is a DLTI UVLM of the undeformed wing at $\alpha_0 = 4°$ angle of attack, with 128 elements in the spanwise discretization, 8 elements chordwise, and an 8-chord long wake. The resulting DLTI system has 11,273 states, which is clearly unmanageable for nonlinear MPC-MHE, and thus model-reduction methods are employed. In particular, a Krylov-subspace projection (Goizueta et al., 2021) has been used here. The simulation model on which the controller is exercised is built from a finite-element discretization of Equations (8.42) and (8.50) coupled to the full-order DLTI UVLM on the undeformed wing at $\alpha_0 = 4°$.

In each simulation, the nonlinear optimal control of Equations (10.47) and (10.48) is solved to convergence, with a relative tolerance of 10^{-4} for both the MPC and MHE problems. A key parameter in the definition of the MPC-MHE is the control horizon. If this horizon is chosen too short, the controller's predictions of the state of the system may be insufficient to anticipate key future events, while if it is too long, the prediction may be too inaccurate due to model mismatch and/or the computational cost of the real-time optimizations may be unaffordable. This choice is explored next for the flutter suppression problem using two distinct actuation mechanisms, which we have called *direct actuation* and *deformation control*. In all cases, the MHE measures the local wing velocity at four equally spaced spanwise locations, and has a weighting matrix $\mathbf{Q}_e = 0.1\mathcal{I}$. Its cost function, Equation (10.48), is augmented to reduce the error between the initial conditions in the estimation problem, $\hat{\mathbf{x}}_e$, and the best available prediction, which have been shown to improve convergence (Artola et al., 2021a). The cost function of the MPC penalizes wing velocities, with unit weight. The MPC also includes saturation constraints on the flap deflection, for which two values are considered, namely $\|\delta\| < \delta_{\max}$ with both a stricter constraint, $\delta_{\max} = 5°$, and a more permissive one, $\delta_{\max} = 15°$.

Flutter Suppression via Direct Actuation

First, consider the case of a relatively short control horizon. Equal estimation and control horizons, τ_e and τ_c, respectively, are selected, with $\tau_e = \tau_c = 0.06$ s, which correspond to approximately 1.8 times the flutter oscillation period at the nominal conditions, τ_f. A small sampling time, $\tau_s = 0.0025$ s ($\approx 0.075\tau_f$), is also considered for

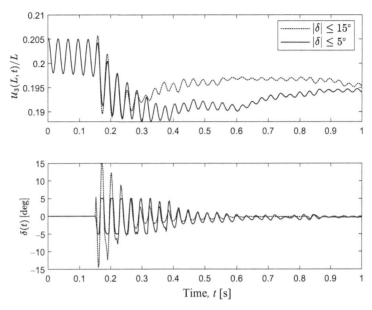

Figure 10.18 Time history of the normalized wingtip displacement and aileron deflection of the Pazy wing with direct actuation mechanism (Artola et al., 2021a).

the internal model. Figure 10.18 shows the closed-loop results for this setup, including both the normalized wingtip displacement and the flap-deflection time histories, for both saturation constraints on the controller. As can be seen, the controller is effective in suppressing the oscillations, and it does so by generating an oscillatory signal at the flutter frequency that counteracts the wing motions, that is, it has the effect of a viscous damper on the system dynamics. This would be a standard solution on a linear regulator, and therefore we refer to it as the *direct actuation* mechanism. It can be further observed in the figure that changing the saturation constraint on the flap actuation only has a marginal effect on the results, even though the saturation is active in both cases. The only differences appear, as expected, shortly after the activation of the control loop, where the amplitude of the oscillations is higher. Note also how the initial effect of the controller, while it learns about the system dynamics, is to increase the amplitude of the wing vibrations. Finally, a small offset in the mean value of the wing deflections can be observed before and after the closed-loop system is active. This is the result of the linearization around a nonzero reference: The steady component of the lift generates in-plane forces with nonzero mean under the vertical motions (see Section 3.3.5), which changes the equilibrium shape.

Flutter Suppression via Deformation Control

Much longer estimation and control horizons are considered now. The control horizon is $\tau_c = 0.12$ s ($\approx 3.6\tau_f$), and three values for the estimation horizon are investigated, namely $\tau_e = 0.12, 0.24$, and 0.48 s. For the internal model, a suitable larger sampling time is chosen, $\tau_s = 0.01$ s, to reduce the computational burden. The implication of this is that there are higher modeling errors for the MPC-MHE to look further into

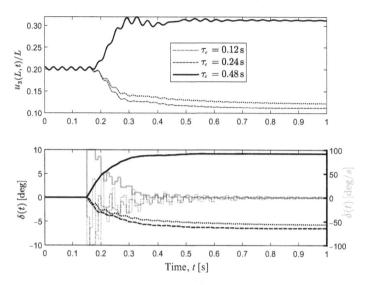

Figure 10.19 Normalized wingtip displacement, aileron deflection, and deflection rate (Artola et al., 2021a).

the future, but as will be seen, the resulting controller still displays good performance. As the interest is on the longer term wing response, an additional terminal penalty, $\mathbf{H} = 5\mathbf{Q}$, is included to the controller objective function. Saturation constraints are now $\pm15°$ for aileron deflection and $\mp10°$/s for the deflection rate.

This combination of longer prediction and lighter actuator penalty results in a very different flutter suppression mechanism than the first controller. The closed-loop system steers now the wing toward a different deformation state with a nonzero flap deflection in the steady state, as shown in Figure 10.19. Those steady states effectively correspond to the wing equilibrium moving along the horizontal line in Figure 9.11. This is because the extended estimation horizon is sufficiently long to capture the transient to the stable equilibrium point and identifies it as a solution with a minimum cost. Interestingly, for smaller values of τ_e, the wing leaves the instability region in Figure 9.11 toward the left (smaller deflection), but with the longest horizon, it moves toward the right in the stability plot (larger deflections).

Figure 10.19 also includes the time histories of the flap deflections and rates. Comparison of those signals with those in Figure 10.18 makes evident the advantages of this actuator strategy. For example, the case $\tau_e = 0.12$ s shows a very smooth evolution of the flap deflection that "gently" takes the vibrating wing (it is initially in an LCO) into a stable condition where the vibrations damp naturally. The flap reaches a steady state at an angle of $9.2°$, which results in wingtip deflections of 31% of the span. This is just outside the unstable region in Figure 9.11.[5] Finally, Figure 10.20 shows a 3-D

---

[5] It should be noted, however, that the equilibrium shape on the wing with a deflected flap is different from that resulting from a change of angle of attack. As a result, the stability boundaries would be slightly modified between both problems.

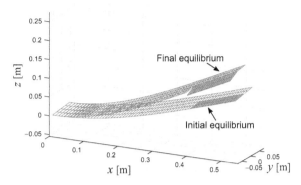

Figure 10.20 3-D view of the initial and final shapes of the wing for $\tau_e = 0.48$ s (Artola et al., 2021a).

view of the bound aerodynamic lattice for the initial and final equilibrium conditions corresponding to the longest estimation horizon.

To conclude this section, we should remark that this control strategy has automatically exploited the nonlinearities in the aeroelastic system to achieve a nonconventional and more efficient solution to a classical problem. This was made possible by the efficient embedding of key system nonlinearities in the internal model of the controller using Equation (9.13). To put this in context, the most common method for nonlinear MPC in large systems is a strategy known as *successive linearization*, in which every control action is defined from an optimal problem on a linearized system around the current (estimated) state. Such an approach would, however, miss here the presence of the stable region with an increase in deformation and therefore would always yield a direct actuation strategy. A recent development on addressing nonlinear constraint aggregation in MPC can be found in Pereira et al. (2022a), allowing for an effective way of dealing with large nonlinear problems while keeping reasonable computational cost.

10.6.3 Nonlinear Control of a Very Flexible Aircraft

To demonstrate the capabilities of the MPC-MHE framework of Section 10.5 in a very flexible aircraft, we finally consider the HALE configuration studied in Deskos et al. (2020). The aircraft is shown schematically in Figure 10.21 and has a conventional configuration with high-aspect-ratio wings and a T-tail. The weight of the aircraft is 78.25 kg, including a 50-kg payload that is modeled as a concentrated mass at the wing–fuselage intersection. The cruise speed for this aircraft is $V = 10$ m/s. The control inputs are the thrust on a single propeller, which is located at the wing root, and the deflections of a rudder and all-moving elevator on the tail. Nonlinear control studies for this aircraft were carried out by Artola et al. (2021b), and the main results are included here.

The control problem is the stabilization of the aircraft after a payload drop. For that purpose, the aircraft is first trimmed for level flight. At $t_0 = 0.125$ s, half of the

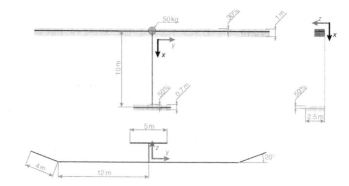

Figure 10.21 3-D view of the HALE aircraft model (Artola et al., 2021b).

Figure 10.22 Equilibrium wing shape before and after the payload drop (Artola et al., 2021b).

payload (31.95% of the aircraft weight) is released, which is a sudden, intense perturbation after which the aircraft undergoes a large reduction in the angle of attack,
while it quickly gains altitude in a response. Another consequence of the mass release
is that the distributed inertia of the resulting aircraft is severely modified, as observed
in Figure 10.22, including a CM that is dangerously displaced toward the tail, thus
posing a challenging control problem. The affected dynamics of the aircraft, as well
as the new stability properties, must be managed properly by the control framework
for successful stabilization.

Before presenting the MPC closed-loop results, it is interesting to assess the performance of the internal controller model and compare it to a full nonlinear simulation
using SHARPy, which has been introduced in Section 9.4. For this purpose, the aeroelastic system in modal coordinates, Equations (9.12) and (9.13), is constructed using
the first 17 normal modes of the aircraft. The model includes the 6 rigid-body modes
and 11 elastic modes, of which 6 are longitudinal modes, while the rest are lateral
ones. They have been computed for the total payload of 50 kg, while the mass that
will be ejected has been internally modeled as a forcing term, as described by Artola
et al. (2021b). A DLTI UVLM is used for the aerodynamics model, with linearization
around the undeformed configuration with an angle of attack of $4°$. The open-loop
results from both full-order (SHARPy) and low-order (modal) models are shown in
Figure 10.23. It includes the instantaneous airspeed V and the angle of attack α of the
aircraft, which are both defined from the velocity at the origin of the body-attached
frame B, and the pitch angle θ, defined between the x axis of the B frame and the horizontal direction. Releasing the payload produces an excess of lift that accelerates the
aircraft in the upward direction, thus producing a sudden drop in the angle of attack.
At the same time, the drastic change in the aircraft CM leads to an increase in

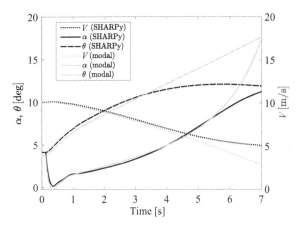

Figure 10.23 HALE aircraft: Open-loop results to payload drop (Artola et al., 2021b).

the aircraft pitch, which ends up stabilizing after the first 5 s of simulation. Finally, as the aircraft gains altitude, there is a continuous reduction in its airspeed.

The results in Figure 10.23 show that the nonlinear aeroelastic ROM is in very good agreement with the full-order nonlinear simulation for up to around 5 s. This is sufficient for the use of the nonlinear ROM as an internal model for control. Recall that the internal model in the MCP-MHE only needs to provide simulations over a finite-time horizon, which cannot be chosen to be very large for computational reasons. After the first 5 s, the results start to diverge, which is likely caused by the decrease in velocity magnitude, V, which is not captured in the nonlinear ROM, and of the linearization approximations introduced by the aerodynamic model. Note also that both simulations start from slightly different values of α and θ due to the difference in trim solutions between both solvers.

Importantly, the internal simulations required for the control framework are performed at a fraction of the computational cost of higher fidelity models. As a comparison, 1 s of physical simulation, using the model discretization given in Table 10.1, takes SHARPy around 2,200 s of CPU time (taking an average of two iterations per time step) to run on an Intel Xeon Silver 4114 CPU at 2.20 GHz, while the same time lapse is simulated in roughly 2.5 s (1,000 times faster) with the nonlinear ROM implemented in Matlab R2018b and running on an Intel Xeon CPU E5-2630 v3 2.40 GHz machine. This implies that, if minimal parallelization was included, each MPC-MHE iteration will run well below real time even using this nondedicated software.

For the control problem, the HALE aircraft is initially trimmed for a level flight at $V = 10$ m/s, and 50% of the payload located at the intersection between the wings and the fuselage is released shortly after the start of the simulation. The objective of the MPC-MHE framework is then to stabilize the aircraft after the payload drop, about which it does not have any direct information. This problem is representative of more general scenarios, such as a sudden relocation of the aircraft CM due to payload loosening. The estimator is responsible for providing an estimate of the state at the

Table 10.1 SHARPy model of HALE aircraft numerical discretization.

3-noded beam elements	Wing	16
	Fuselage	8
	Vertical tailplane	8
	Horizontal tailplane	8
UVLM	Chordwise panels	16
	Wake length	10 chords
	Total bound panels	1,792
	Total wake panels	17,920
Time step		6.25 ms

current sampling point together with an estimate of the external disturbances to the system. Here, the lost mass is the only one considered.

The internal model is constructed using the same number of modes as described earlier and simulated with a sampling time of $\tau_s = 0.2$ s. The estimator horizon is $\tau_e = 1.2$ s, which corresponds to 59% of the period of the first bending mode. Wing-mounted sensors are assumed to give the local translational velocites at five spanwise locations along the wing. All sensor measurements are given the same weight in the estimator, and $Q_e = \mathcal{I}$, $P_e = \mathcal{I}$, and $R_e = 1$. For the control problem, a longer time horizon with $\tau_c = 2.4$ s is chosen, which represents a window slightly longer than the period of the first bending mode. Control is slightly penalized with $R_c = 0.05\mathcal{I}$ while $Q_c = \text{blkdiag}(\mathcal{I}_6, 0)$, meaning that we mainly choose to penalize rigid-body perturbations with the aim to maintain a leveled flight. That avoids the drastic drop in speed that is observed in the open-loop response in Figure 10.23, and the sudden increase in the angle of attack that may lead to stall. Thrust is constrained to be $0 \le T \le 15$ N, while elevator deflection and deflection rates are constrained to be $\pm 15°$ and $\pm 100°/$s, respectively.

The closed-loop values of the pitch angle, angle of attack, and velocity magnitude can be seen in Figure 10.24a, where they have also been compared with the open-loop response that was presented above. First, a very successful control of the flight speed is achieved, only slightly deviating from the initial 10 m/s just after the payload drop. The angle of attack experiences a similar sudden drop at the beginning, due to the impossibility to counteract the initial upward velocity induced by the mass release, followed by some low-amplitude mild oscillations. However, its subsequent increase in value has been effectively suppressed. Regarding the pitch of the aircraft, it experiences some oscillations with higher amplitude, but it is also seen to be controlled with high efficacy. This is, perhaps, the most visual effect of the controller, since the nose-up configuration adopted by the vehicle in the open-loop case is suppressed. This is more clearly seen in Figure 10.25, which shows snapshots of the vehicle in the open and closed loops for the first 30 s after the payload drop. Overall, the nonlinear controller effectively stabilizes the aircraft within the first 5 s after the payload drop.

The control inputs are shown in Figure 10.24b. The figure also includes the estimation of the payload, which is treated as a disturbance by our MHE. The input signals are given as increments with respect to the trim values and have been normalized by

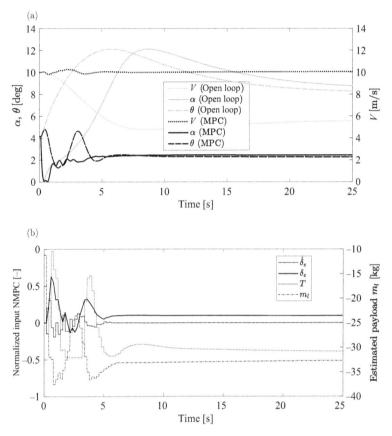

Figure 10.24 Open- and closed-loop response after the payload drop. (a) Angle of attack, pitch angle, and airspeed, and (b) normalized control inputs and value of mass estimated by the MHE (Artola et al., 2021b).

their maximum values within the saturation limits. Both aileron and thrust signals are oscillatory, responding to the corrections needed to maintain the flight velocity, and stabilize around different steady-state values to account for the new trim conditions of the lighter aircraft configuration. Note that the pitch and angle of attack do not end up coinciding exactly, and therefore the aircraft is left with a residual vertical velocity. This is explained by the plant-model mismatch and the different steady solutions that are obtained with both models. This mismatch is also believed to explain the offset and the fluctuations of the estimated payload. Nevertheless, this does not hinder the closed-loop performance, as suggested by the results.

Finally, it is worth mentioning that these results are only explained by the control framework being able to exploit the nonlinear knowledge of the system. As shown in Figure 10.22, the release of 50% of the aircraft's payload severely affects the wing loading, which adopts a much less deformed configuration. Hence, the controller has to be able to drive the system to a different nonlinear steady solution, while the estimator needs to adapt to the changing internal dynamics resulting from the large variations of the vehicle mass. The trajectory of the aircraft is shown in Figure 10.25.

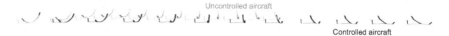

Uncontrolled aircraft

Controlled aircraft

Figure 10.25 Snapshots of the controlled–uncontrolled aircraft for 30 s after the payload drop, shown with 2.5 s intervals. The viewpoint is that of an inertial observer at a small distance from the center of the page (Artola et al., 2021b).

10.7   Summary

This chapter has outlined some common strategies for feedback control of the dynamics of flexible aircraft. Previous chapters have described air vehicles as dynamical systems defined by their inputs and outputs, and subject to external disturbances, and this has provided a suitable framework for the development of a flight control system. The discussion in this chapter started with an overview of the process for control synthesis, right from an understanding (via mathematical models) of the system dynamics to knowledge of its environment to specifications of performance objectives required from the controller. The control system has a logical component but also a hardware component, and we have consequently reviewed some of the most common sensors and actuators that are currently found in, or are being considered for, modern aircraft displaying coupled flight dynamic–aeroelastic couplings.

The focus then shifted to optimal control theory as it provides a suitable framework (albeit not the only one) for the initial development of feedback systems for trajectory control, stabilization, and active load alleviation of a flexible aircraft. A brief summary of the linear methods, that is, LQR and LQG control, has introduced the main concepts of optimal control, and their practical application to aeroelasticity has been demonstrated using the 2-DoF aeroelastic system of Chapter 3. The biggest control challenge occurs on very flexible aircraft with potentially large wing deformations: First, those vehicles are the most susceptible to atmospheric disturbances, and their successful performance is highly dependent on high-authority control; second, the dynamics may be highly nonlinear that pushes the limits of applicability of linear control methods. That double challenge is addressed here by the *online* version of the linear optimal control methods, namely model-based predictive control strategies, in which the optimization problem is solved in real time by the fligth controller. We have finalized this chapter with an overview of how a working MPC-MHE framework can be implemented for nonlinear aeroelastic control of very flexible wings and aircraft.

10.8   Problems

Exercise 10.1. Given the cost function $J(x_1, x_2) = x_1^2 - 2x_1 + 1 + x_2^2$ and the constraint $x_2 = x_1^2$, find the optimum solution using Lagrange multipliers.

Exercise 10.2. Consider an LTI system with both measurement and process noise, as in Equation (10.33). Obtain the optimal Kalman gains for the associate observer equation if there is a nonnegligible correlation between the measurement and process noise, Σ_{wv}, which are still assumed to be white noise signals.

Exercise 10.3. Write the LQG closed-loop system in terms of a new state vector defined as $\left\{ \mathbf{x}(t)^\top \quad \hat{\mathbf{x}}(t)^\top \right\}^\top$.

Exercise 10.4. For an F-16 aircraft flying at 550 km/h, the state and input matrices are given on pages 714–722 of Stevens et al. (2016) as

$$
\mathbf{A}_d = \begin{bmatrix} -0.0322 & 0.064 & 0.0364 & -0.9917 \\ 0 & 0 & 1 & 0.0037 \\ -30.6492 & 0 & -3.6784 & 0.6646 \\ 8.5396 & 0 & -0.0254 & -0.4764 \end{bmatrix} \quad \text{and}
$$

$$
\mathbf{B}_d = \begin{bmatrix} -0.0003 & -0.0008 \\ 0 & 0 \\ 0.7333 & -0.1315 \\ 0.0319 & 0.0620 \end{bmatrix}.
$$

(i) Compute the open-loop eigenvalues of the system and identify the Dutch roll, roll subsidence, and spiral modes.
(ii) Build steady-state LQR controllers for this system with $\mathbf{R} = \mathcal{I}$ and a diagonal $\mathbf{Q}$, and identify numerically combinations of state weights that improve the Dutch roll damping.

Exercise 10.5. Consider the system given by the state equation

$$
\dot{\mathbf{x}} = \begin{bmatrix} 0 & 1 \\ -10 & -1 \end{bmatrix} \mathbf{x} + \begin{Bmatrix} 0 \\ 1 \end{Bmatrix} u.
$$

We seek a controller able to track a constant signal $x_1(t) = r$ using the integral error $\dot{e} = r - x_1$ and the cost function $J = \int_0^t e^2 \, dt + Ru^2$.

(i) Derive the augmented state-space equations in terms of $\left\{ \mathbf{x}^\top \quad e \right\}^\top$.
(ii) Compute the optimal LQR feedback gains for $r = 1$ and $R = 10^{-4}$.
(iii) Simulate the open-loop response starting from zero initial conditions.

Exercise 10.6. Consider again the LQR and LQG controllers for the 2-DoF airfoil derived in Examples 10.2 and 10.4, respectively. Compute the closed-loop response of the airfoil to a "1-cosine" vertical gust, defined as in Equation (2.54), with gust gradient $H = 2c$ and intensity $w_g = 0.1V$.

11 Experimental Testing

11.1 Introduction

Aircraft are very complex machines, and extensive experimental testing is a necessary step to achieve airworthiness certification. Such test campaigns have two main objectives: First, they ensure safe operation, which is achieved by a systematic approach in which incremental checks are made on increasingly complex features (from the ultimate stress test of a single component to high-speed maneuvers of the aircraft in flight). Second, they validate the extensive numerical analysis that supports the design. This is important because there are hundreds of thousands of combinations of configurations, loadings, and flight conditions that the aircraft may encounter during its operational life, and it is materially impossible to test the prototype vehicles under all those scenarios. Those are assessed instead by analysis, using computer models that are refined as experimental data become available along the aircraft design process.

Experimental testing is performed on aerodynamic features, using mostly wind-tunnel tests; on the strength and fatigue behavior of both the materials and the assembled structures; on the control systems, using first a ground-based *iron bird* and then through flight testing; on each major subsystem, for example, landing gear, de-icing, hydraulics, and so on. The list is long, and the focus in this chapter is on the main experimental tests that are necessary to assess the vibration characteristics of a flexible aircraft, both on the ground and in flight. Section 11.2 describes a typical ground vibration test (GVT), in which an aircraft vibration characteristics are measured once the aircraft is fully assembled for the first time. Section 11.3 then outlines the flight vibration test, which is one of the most dangerous phases in the flight testing of any vehicle, as it involves taking the aircraft to points at the edge of flight envelope for the first (and often hopefully last) time. Both ground and flight vibration tests of an aircraft are particular forms of *experimental modal analysis*, in which the *test specimen* (the aircraft) is excited at the desired frequency range, its oscillatory response is measured using appropriate transducers, and the measured datasets are analyzed to identify the vibration and damping characteristics of the vehicle. Critically, both tests need to be conducted on the full aircraft, which brings substantial schedule and cost constraints as both tests typically become project bottlenecks at a critical time in the certification process. Substantial resources are often put on both to ensure a timely and satisfactory outcome that does not delay the entry of service of a new aircraft. We note finally that most of the description in this chapter is based on conventional vehicles that can be described using the linear methods of Chapter 6. The equivalent process for a very flexible aircraft is still an active topic of research (see, for example, Sharqi and Cesnik, 2020, 2022).

11.2   Ground Vibration Tests

11.2.1   Introduction

In Section 6.4.1, we have introduced the normal modes of a structure as its resonant states, and we have discussed how they are an intrinsic property of a built-up structure. Here we discuss how the LNMs are measured on large aircraft, for which we rely on the descriptions of Oliver et al. (2009) and Rodriguez Ahlquist et al. (2010). The GVT is a nondestructive structural test that is performed on the ground on a complete aircraft with the aim of experimentally obtaining its normal modes. The GVT is carried out on a vehicle that is "ready for flight," in the sense that both its airframe and mass configuration should be those of the operational aircraft (including, e.g., ballast weights to account for any relevant payload), so that the GVT measures the actual LNMs in those conditions. During a GVT, the aircraft is first dynamically excited, typically with known excitation sources, in one or several points, and the subsequent response of the airframe is measured in hundreds of positions covering the entire structure. A standard configuration is shown in Figure 11.1. Typically, the GVT uses as test specimen the first aircraft prototype, and it is performed just a few weeks before its first flight.

GVTs, therefore, enable the experimental modal analysis of the aircraft in the frequency range where either potentially dynamic amplifications in the aircraft response or destructive aeroelastic instabilities, such as flutter, may occur. The first objective of

Figure 11.1 Typical aircraft GVT configuration (reprinted from Oliver et al. (2009) with permission from the authors).

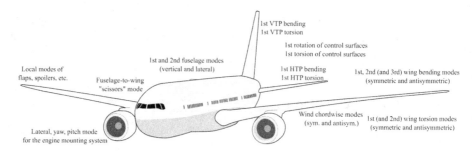

Figure 11.2 Typical normal modes that are measured in a GVT.

the GVT is to obtain the natural frequencies, mode shapes, and modal mass of each LNM of the aircraft within that frequency range. Those can be used to update the finite-element model of the aircraft to match the experimental predictions (Pak and Truong, 2015). Importantly, GVTs also measure modal damping, which is very difficult to estimate without experiments, and which can then be included for the first time in subsequent studies of the vehicle dynamics. Finally, the GVT can also be used to assess some nonlinear features of the normal modes by, for example, measuring changes on modal frequencies with the amplitude of the excitation (Fellowes et al., 2011; Rodriguez Ahlquist et al., 2010).

The range of frequencies varies depending on the type of aircraft or the components considered. For large aircraft, normal modes typically need to be investigated up to at least 30 Hz. For medium and light aircraft, fighters, trainers, etc., this number goes up to around 50 Hz. Finally, for certain critical components, such as the canard or the flaperon of a fighter aircraft, or the tailplane of a business jet, the frequencies of interest span up to 70–80 Hz. The typical normal modes that are sought in the GVT of a commercial aircraft are schematically shown in Figure 11.2. For other aircraft, they may also include lateral, pitch, and yaw modes of any external stores, or of their air-to-air refueling probe.

As mentioned above, the experimental natural frequencies and mode shapes are then used to update the finite-element model used for aircraft dynamics and, therefore, the distributed inertia and the stiffness in Equation (6.35). This updated model is later used to determine flutter and dynamic load clearances for the aircraft first flight, flight envelope expansion, and the subsequent flight test campaign. Also, the experimental validation of the models is needed for certification purposes, and after the GVT, the final certification tasks, corresponding to the response and stability of the aircraft, may be initiated.

11.2.2 Components of the GVT

The structure to be investigated is typically referred to as the *specimen*. In a GVT, we need to determine the characteristics of the specimen, as well as suitable actuators and sensors. Those are described in the following section.

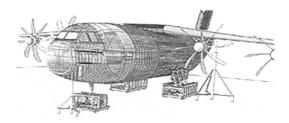

Figure 11.3 Typical supporting system using bungees (reprinted from Oliver et al. (2009) with permission from the authors).

Specimen

The specimen is the aircraft in a particular configuration. As mentioned in the earlier section, the aircraft needs to be in a configuration "ready for flight," and, therefore, the tests can only take place after the first prototype is fully assembled. The first configuration to test is the aircraft with a *clean wing*, that is, empty of fuel. All other configurations with significant differences in dynamic behavior should also be included in the performance of a GVT (full fuel, open rear ramp on a cargo airplane, mounted underwing pods on a fighter plane, flaps deployed, etc.). The way the aircraft is supported determines its *boundary conditions*, which need to be as close as possible to a *free–free* condition to simulate the real aircraft in flight. To realize it, ideally the natural frequency of the support should be separated from the lowest elastic mode of interest by at least one order of magnitude. There are different support solutions depending on the aircraft weight, allotted time and budget, which include:

1. *Aircraft on bungees*. For large airplanes, the aircraft is supported on bungees, which are elastic supports with very low natural frequencies. The bungees are attached on the landing gear wheel axes, as shown in Figure 11.3. Note the jacking towers in the figure, which are located below the three lifting points (two under each wing and the third in the nose fuselage) and protect the aircraft in the case of bungees failure. For smaller aircraft, the bungees are directly attached on the aircraft lifting points.
2. *Pneumatic platforms*. They are less expensive but may add some uncertainty to the damping measurement of rigid-body modes.
3. *Half-deflated tyres*. This is the cheapest and quickest solution and may be acceptable for testing of aircraft modifications of well-known designs.
4. *Aircraft on jacks*. This is usually the last option to be considered because of the uncertainty introduced by the jack structure on the measured normal modes.

Finally, a *scaffolding* is necessary to allow access to the aircraft (e.g., for installation of instrumentation), to support the instrumentation cables, and to locate the dynamic shakers. It must be sufficiently stiff to avoid undesired vibrations when exciting the aircraft. A typical scaffolding is shown in Figure 11.4.

Actuation

The aircraft structure needs to be dynamically excited to measure its subsequent response. There are multiple actuation methods for aircraft dynamic testing, but the

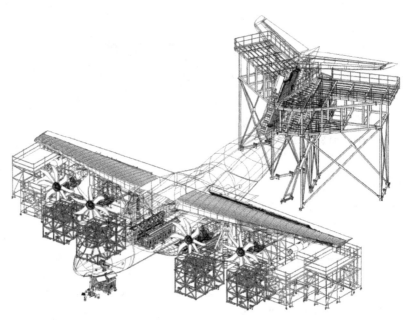

Figure 11.4 Typical scaffolding for a GVT (reprinted from Oliver et al. (2009) with permission from the authors).

most common actuation for a GVT is based on *electrodynamic shakers*. These devices are suitable to introduce a desired controlled excitation such as frequency sweeps and dewells, as well as, random excitations, and pulses (Friswell and Mottershead, 1995, Ch. 3). The excitation level needs to be large enough to obtain a good signal-to-noise quality, but also not so large that it may go above the load levels that impact fatigue life. This is done to avoid the GVT consuming any significant amount of fatigue life of the prototype. We also want to avoid exciting nonlinearities in the structure at this point.

Other excitation means may complement the shakers in some particular applications. For example, to identify the low-frequency rigid-body modes in the supported specimen,[1] excitation by hand can be reasonably adequate. In other situations, such as a test at the small component level, an instrumented impact hammer can be convenient. Finally, for high loads and large strokes, hydraulic exciters may be necessary.

The distribution of shakers around the aircraft should guarantee that all the relevant normal modes are properly excited. As an example, close to each wing tip, the shakers have to be located to excite both vertically and horizontally, which means two shakers for each wing. In this way, both bending and chordwise (in-plane) modes can

---

[1] Note that the frequencies of the rigid-body modes will never be exactly zero in the test as the aircraft needs support.

be excited, as well as torsion (as long as the shakers are not aligned with the elastic axis of the wing). Equally, two additional shakers are needed for each horizontal tailplane and for the vertical tailplane. Other shaker arrangements are also necessary to measure local engine and engine-mounting system modes, as well as to measure the normal modes of each control surface, among others.

Finally, sine sweeps are typically performed with the shakers symmetrically arranged over the aircraft. Sweeps are performed at either $0°$ or $180°$ phase difference between the shakers on the left and on the right of the aircraft to seek excitation of both the symmetric and antisymmetric modes.

Sensing

Airframe vibrations are typically measured using *accelerometers*. A sufficient number of accelerometers needs to be selected to reproduce the mode shapes of all the normal modes to be measured. This usually results in between 400 and 600 accelerometers, depending on aircraft size.

Accelerometers are distributed through the entire aircraft structure, thus covering wings, HTP, VTP, fuselage, control surfaces, engine-mounting system, engines, and landing gears. Importantly, they must be located in hard points of the airframe, so as to measure global aircraft modes instead of local vibrations.

Finally, cables are deployed throughout the structure for the connection of all the accelerometers to the data-acquisition (DAQ) system. Special attention must be paid to minimize added weight onto the structure. Once the installation is complete, a number of systematic checks are necessary to ensure that all accelerometers are correctly installed and connected, and that they are supplied with adequate levels of excitation. Checks are also made on *reciprocity*, that is, the response at point B from an excitation at point A needs to be very similar (in theory, identical) to the response at point A from an excitation at point B, as well as on the *coherence* between the input and the output. Signal coherence was defined in Equation (2.14), and it is used here to assess the quality of measurements, since a good test setup results in high coherence across the relevant frequency spectrum. Finally, potential digital signal-processing issues, such as aliasing, leading, and/or a poor noise-to-signal ratio, are investigated and addressed as necessary before the GVT can take place.

11.2.3 Completing the GVT

The duration of a typical aircraft GVT campaign is in the range of two to four weeks. As it is performed very close to its first flight, the testing campaign is very intense, and any delays come at a large financial cost to the aircraft program. Two teams work in parallel: one for data acquisition and another for in-site experimental modal analysis. The second team assesses the test performance as it progresses to produce immediate feedback on its results.

Despite its very short duration, the GVT needs to be planned years in advance. For a new aircraft, in general, the Request for Test (RFT) is first released about two

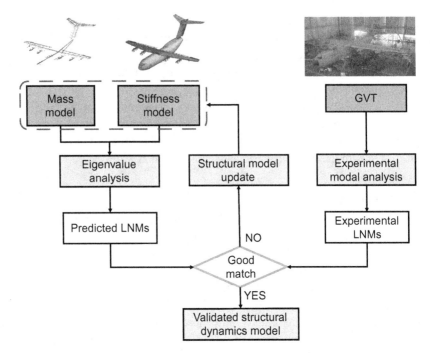

Figure 11.5 GVT model validation flowchart.

years before the tests take place, and it is updated several times before the actual GVT. After the GVT, the normal modes obtained from the experiments are compared with pretest normal modes analyses to decide whether the pretest finite-element model is suitable for the certification tasks or should be updated. Figure 11.5 shows a typical validation flowchart for the finite-element model using the experimental modes from the GVT. The acceptable mismatch between experiments and predictions depends on the particular regulations that apply to a given aircraft. For example, the US military standard (MIL-STD-1540C, 1994, Sec. 6.2.10) requires that, within the bandwidth of interest, the predicted and experimental frequencies are within 3% and that the cross-correlation of the eigenvectors is at least 95%.

11.3   Flight Vibration Tests

11.3.1   Introduction

As has been seen in this book, aeroelastic instabilities can produce severe damage, often resulting in vehicle loss, and therefore the stability boundaries of new aircraft are thoroughly assessed in their flight test campaign. Indeed, airworthiness regulations (EASA, 2017) establish that full-scale flight flutter tests must be conducted for new type designs, as well as for modifications to a type design that may have a nonnegligible effect on the aircraft aeroelastic characteristics. This results in the the *flight flutter test* or *flight vibration test* (FVT).[2] The objectives of the FVT are, first, to ensure that the aircraft is free from flutter in its entire flight envelope; and second, to obtain the experimental evolution of frequency and damping of the (aeroelastic) normal modes with the flight speed. As we have seen in Section 6.6.1, the feedback interactions between the structure and the unsteady aerodynamics substantially modify the vibration characteristics of an aircraft in flight. With reference to Figure 6.20, the GVT has measured the normal modes of the airframe in the absence of the aerodynamic forces.[3] Since the structural properties are the same on the ground and in flight, the FVT can be used to perform a validation of the unsteady aerodynamic model for the vehicle (i.e., the lower block in Figure 6.20).

The flight flutter test uses as test specimen the first aircraft prototype that takes place at the very beginning of the certification flight test campaign. In a modern FVT, the aircraft is dynamically excited in a controlled way using its control surfaces, and the response of the aircraft is measured in various positions covering the entire vehicle. The FVT is recognized as one of the most dangerous tests in the entire flight test campaign of an aircraft and, as a consequence, the greatest possible precautionary measures are taken, such as flying with a minimum crew (fully equipped with parachutes) and monitoring the flight using *telemetry*. The FVT progresses sequentially from flight points at low dynamic pressure and Mach number to points at increasingly higher dynamic pressure and Mach number, up to the *design dive speeds*. A typical set of flight test points for an FVT is shown in Figure 11.6. The isolines indicate the Mach number and calibrated airspeed (in knots) for each combination of altitude and (true) airspeed. The calibrated airspeed is the airspeed directly computed from the dynamic pressure measured at the Pitot tube, and corrected to include instrument errors. It differs from the EAS, which was defined in Section 2.2, in that it does not account for compressibility effects. The gray dots in the figure indicate the test points, with the number showing the order of the tests. At each Mach number and altitude, the following processes take place:

1. The pilot trims the aircraft at the current flight point.
2. Rapid, symmetric and antisymmetric, excitations are introduced using successively ailerons, elevators, and rudder.
3. The response to each excitation run is measured with the onboard sensors. The data is sent to a ground station via telemetry where an online experimental modal analysis (Taylor et al., 2017) is performed to ensure that all modes are properly damped at that particular flight point.
4. Only when the stability of all modes is verified, the aircraft will be authorized to proceed to the next flight test point at a higher speed/dynamic pressure.

---

[2] The acronym FFT is seldom used to avoid confusion with the Fast Fourier Transforms that are used in the postprocessing on the flight data.

[3] Strictly speaking, the GVT only removes the circulatory terms of the aerodynamic forces, and it would need to be performed in vacuum to avoid apparent mass effects. For most aircraft, apparent mass is negligible in the frequency range of interest and this distinction is rarely made. For very low wing-loading aircraft, like Solar Impulse, that could not be neglected (Böswald et al., 2010).

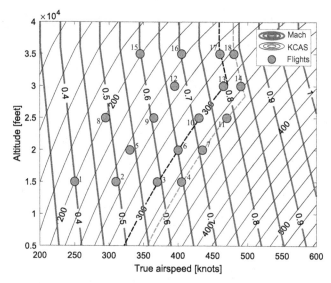

Figure 11.6 Typical flight test points to perform excitation runs in an FVT. The dotted lines correspond to the design cruise speed, VC (dark gray) and the design dive speed, VD (light gray).

FVTs enable experimental modal analysis (EMA) of the aircraft in the frequency range where potential aeroelastic instabilities, such as flutter, can take place. It needs to meet two objectives at each test point before the test pilot can proceed to the following one: (1) ensure that all normal modes are properly damped and (2) obtain, as accurately as possible, frequency and damping for each relevant normal mode of the aircraft at that flight condition. The FVT finalizes when the design dive speed (V_D) is reached. The progressive flight envelope expansion is typically achieved by a combination of flutter and the handling qualities tests.

As the frequency bandwidth of actuation of the control surfaces is smaller than that of the shakers in the ground, the range of frequencies that can be excited in flight are necessarily lower than those tested in the GVT. Nevertheless, all potentially critical flutter mechanisms found in the pretest numerical simulations should be tested in flight to confirm the predicted frequency and damping values. Results from the FVT are critical to validate the dynamic aeroelastic model of the aircraft. This model is later used in two certification tasks: (1) to perform stability analysis and to ensure that the aircraft is free from aeroelastic instabilities, especially from any component failure that cannot be tested in flight (e.g., after sustaining damage); and (2) to perform in-flight dynamic response analysis of the conditions defined in Chapter 2, that is, discrete tuned gusts, continuous turbulence, buffet, wake encounter, etc.

11.3.2   Components of the FVT

Similar to the GVT, the FVT is composed of a specimen, actuators, and sensors that are described next.

Specimen

The specimen in the FVT is the aircraft in the mass configuration that is deemed to be more critical in the pretest flutter numerical simulations. The flight conditions should be as close as possible to 1-g flight condition, and therefore most of the FVT is carried out on a straight trajectory with leveled wings. There is an exception here, however, for very high speeds and dynamic pressures that can only be achieved by diving.

Actuation

The preferred option for actuation is to use the control surfaces already existing on the aircraft. For aircraft with actuators and FCS, the regular way is to introduce dedicated signals for symmetric and antisymmetric pulses and frequency sweeps applied sequentially to ailerons, elevators, and rudder. For small and medium aircraft (up to $\approx$ 20 tons of maximum take-off weight), which usually have manual controls, the excitation may be introduced through *stick raps* (Silva and Dunn, 2005) sequentially in the three axes.

Similar to the GVT, the excitation levels should be large enough to obtain a good signal-to-noise ratio but ensuring always that the level of response is below the fatigue life threshold (i.e., the FVT should not consume the fatigue life of the prototype), remains in the liinear regime, and does not adversely affects the controllability of the aircraft.

Before the development of fast electromechanically actuated control surfaces, other means of excitations were used for the FVT, although they are now mostly discontinued. They include (Kehoe, 1995):

1. Thrusters, also aptly known as *bonkers*, which are pyrotechnic devices located at the tip of each lifting surface. This option is no longer in use, first, for obvious safety reasons, and second, because only one shot per lifting surface could be used at a time what makes the duration of the FVT unnecessarily long. Sometimes the weight of the bonker itself (typically 8–10 kg) could act as a mass balance and hide a real aeroelastic problem in the aircraft once the bonker weight is removed.
2. *External oscillating vanes*, which are small aerodynamic vanes located at the tip of the lifting surfaces. This approach requires the installation of a hydraulic system able to move harmonically the vanes.
3. *External vanes with rotating trailing edge slots*, which are fixed aerodynamic vanes located at the tip of the lifting surfaces with a rotating cylinder in the trailing edge. The cylinder has a rotating slot that alternatively allows the flow to pass through it or restricts it. When the flow is allowed to pass the vane, the lift is produced, and when the flow is restricted, the lift vanishes. The frequency of the rotation of the cylinder will provide the frequency of the excitation force. This approach eliminates the need of a complex hydraulic system like the one in the oscillating vanes.

Sensing

Again similar to the GVT, the aircraft response in the FVT is typically measured using *accelerometers*. However, as the aircraft structural modes have already been measured in the previous GVT, the number of accelerometers in the FVT can be reduced. A

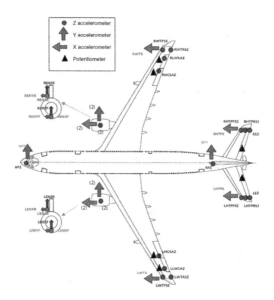

Figure 11.7 Typical arrangement of accelerometers for an FVT (courtesy of Airbus).

typical FVT uses 40–80 accelerometers, depending on aircraft size and the number of underwing external stores. The accelerometers are usually located at the tip of all lifting surfaces (wings, HTPs, and VTP), at the trailing edge of all control surfaces (ailerons, elevators, and rudder), and at the engine and engine-mounting system. A typical layout is shown in Figure 11.7. For fighter aircraft, they are also positioned on the underwing external stores. In all cases, and same as the GVT, accelerometers must be located at hard points of the structure. Indeed, accelerometers in the FVT are typically at locations that were already used in the GVT (although no external, surface-mounted accelerometers can be used for FVT).

Cables need to be deployed internally throughout the structure, which can be considered as a major difference with the GVT. The cables connect the accelerometers to an on-board DAQ system normally located in the fuselage. The DAQ system immediately transmits the accelerometers and excitation signals to the on-ground telemetry room for on-line experimental modal analysis. In addition, sensor data are also recorded and stored in the aircraft to protect against potential loss of the telemetry signal, which may happen, in particular, if the aircraft is flying at a low altitude.

Finally, the excitation signal on the control surface rotation is measured using potentiometers, rotary encoders, or other types of angular measurement devices. Checks are then carried out on the telemetry system to verify that the rigid-body modes are captured and that the quality of the sensing signal is acceptable. The sampling ratio is at least in the range of 128–512 samples per second.

11.3.3   Completing the FVT

An FVT campaign typically consists of two to four flights. It involves the aircraft flying at very high speeds, which may make the testing rather expensive in terms of fuel. Similar to the GVT, it needs to be planned early with an RFT that is first released two years in advance of initiating the tests. A typical selection of FVT points is shown in Figure 11.6. There the dark gray dotted line corresponds to the *design cruise speed* (VC) envelope, and it can be generally reached at straight and level 1-g flight. Frequency sweeps and pulses can be measured used up to these flight points. The light gray dotted line in the figure corresponds to the *design dive speed* envelope, and those points can only be reached during a dive of the aircraft. As a result, only for a few seconds, the aircraft is in the neighborhood of the target flight test point, and only pulses or natural turbulence can be used as excitations at these points.

The FVT produces two types of results. The first batch of results, obtained through experimental modal analysis, is the frequency and damping of the excited aeroelastic normal modes. This information is obtained online, and the results are assessed before authorizing to proceed to the next flight test point. In particular, it must be ensured that all modes are properly damped and that there is not a significant drop of damping for any mode. Those results are also compared with the frequency and damping, which are estimated through numerical simulation using the pretest model. This will enable to identify any major divergence between the predicted and the actual performance of the aircraft and to decide whether the aeroelastic model is suitable for the certification tasks or whether it should be updated. A typical validation flowchart is shown in Figure 11.8.

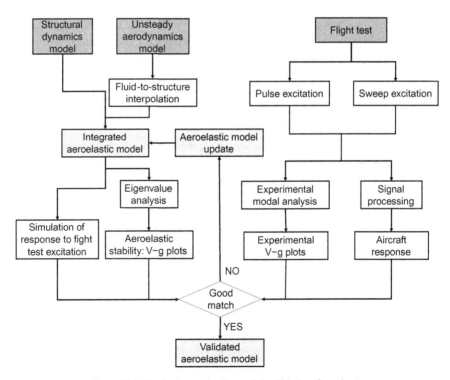

Figure 11.8 Typical aeroelastic model validation flowchart.

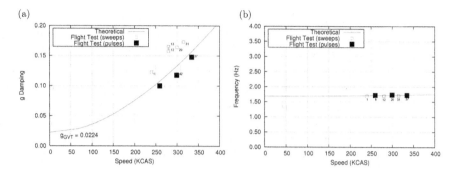

Figure 11.9 Typical results of an FVT. Numbers identify the flight tests. (a) Damping of the first bending mode and (b) frequency of the first bending mode (courtesy of Airbus).

The second batch of results is the time histories of accelerometers or *dynamic loads* at certain monitoring stations (bending moment, shear force, and/or torque, as discussed in Chapter 1). This is done offline after finishing the FVT, and it is another way of validating the aeroelastic model. For that purpose, the measured time histories are compared with the time histories of the same magnitudes of the aeroelastic model obtained by the numerical simulation of the response to the same excitation (sweeps or pulses). As an example, Figure 11.9 shows the comparison, for different flight speeds, between the predicted and the measured damping and frequency of the first symmetric wing bending mode of an aircraft. Note that the numerical results start at $V = 0$, which corresponds to the GVT measured damping. Note also the slight difference in terms of measured damping between two types of excitations: frequency sweeps and pulses. Finally, the numbers in the plots correspond to different excitation runs.

11.4   Summary

In this final chapter, we have introduced a quick summary of the processes and objectives of the ground vibration and flight flutter tests of large aircraft. Both GVT and FVT give critical experimental information about the dynamics of a new air vehicle, which informs the final stages in the design process while being mandatory for airworthiness certification. The GVT determines the normal modes of the airframe on the ground, and the FVT determines how those modes change with the airspeed. The FVT also ensures that the aircraft is free from flutter in its entire flight envelope.

For certification purposes, the GVT and FVT are typically performed on the first full prototype of a new aircraft type, which brings significant constraints on the available time for the tests. Both are nondestructive tests that need strict risk-management strategies to limit the possibility of accidental damage (particularly in the in-flight tests) but also to avoid a negative impact on the fatigue life of the vehicle. During the tests, information is constantly monitored and postprocessed to ensure that data conform to the numerical predictions. When critical decisions are needed, they must be taken within hours for the GVT and within seconds for the FVT.

The experimental data collected from both the GVT and the FVT are finally used to validate, and update when necessary, the finite-element model used for the modal analysis of the airframe and the aeroelastic model used for flutter analysis and computation of in-flight dynamic loads.

As we could see in this chapter, the process of validation of numerical aeroelastic models and the path to aircraft certification involves complex ground and flight tests. As the numerical methods and tools become more complex for very flexible aircraft as discussed in Chapter 9, additional challenges will appear in the corresponding GVT and FVT as well. The issue of the calibration of the FE model from GVT data is reasonably well established for more traditional aircraft and numerical models of the type described in this chapter and Chapter 5, respectively. But when we move to more flexible aircraft, even the ones that can be modeled with the methods described in Chapter 6, the experimental processes become more complex and difficult to execute. When we move to fully geometrically nonlinear aircraft as discussed in Chapter 9, those experimental and numerical updating processes require a new approach altogether. These are on-going research topics reflected, for example, in the X-56A and the X-HALE examples introduced in Chapters 6 and 9, respectively. Further developments in this area should be expected in the years to come, and they will be an essential component to support the design and certification of highly-efficient aircraft for sustainable aviation.

Appendix A The Fourier Transform

A.1   Fourier Transform Pairs

A function $g(t)$ is said to be *absolutely integrable* if, for every T, there is a finite value k for which $\int_{-T/2}^{T/2} |g(t)| \, dt < k$. Let $g(t)$ be an absolutely integrable function of time with finite discontinuities. Its Fourier transform is defined as

$$G(i\omega) = \frac{1}{2\pi} \int_{-\infty}^{\infty} g(t) e^{-i\omega t} \, dt. \tag{A.1}$$

The corresponding inverse Fourier transform is

$$g(t) = \int_{-\infty}^{\infty} G(i\omega) e^{i\omega t} \, d\omega. \tag{A.2}$$

In this book, we use capitalization, $G(i\omega)$, and also overbars, $\overline{g}(i\omega)$, depending on the context, to indicate the Fourier transform of a time-domain function $g(t)$. Functions $g(t)$ and $G(i\omega)$ (or $\overline{g}(i\omega)$) are known as a Fourier transform pair. The definitions above assume that the time is given in seconds and ω is given in rad/s. Alternative definitions of the Fourier transform with different normalizations (e.g., without the $(2\pi)^{-1}$ coefficient in Equation (A.1)) can often be found in the literature, including expressions where the frequency is given in Hertz, but this one is used throughout this book.

A.2   Correlation and Convolution Functions

Of particular importance in this book is the Fourier transform of the correlation and convolution operations between two real functions, $f(t)$ and $g(t)$. In continuous time, the correlation between two scalar functions is defined as[1]

$$\phi(\tau) = (f * g)(\tau) = \int_{-\infty}^{\infty} f(t+\tau) g(t) \, dt. \tag{A.3}$$

The correlation function has been extensively used in Section 2.3.1 to characterize the statistical properties of random processes. Its Fourier transform is

---

[1] The correlation is sometimes defined as $\phi^-(\tau) = \int_{-\infty}^{\infty} f(t) g(t+\tau) \, dt$, so that $\phi(\tau) = \phi^-(-\tau)$ as in Equation (2.6), but the definition used here is consistent with that of Chapter 2.

$$
\begin{aligned}
\Phi(i\omega) &= \frac{1}{2\pi} \int_{-\infty}^{\infty} \left(\int_{-\infty}^{\infty} f(t+\tau)g(t)\,dt \right) e^{-i\omega\tau}\,d\tau \\
&= \int_{-\infty}^{\infty} g(t) \left(\frac{1}{2\pi} \int_{-\infty}^{\infty} f(t+\tau)e^{-i\omega\tau}\,d\tau \right) dt \\
&= \int_{-\infty}^{\infty} g(t) \left(\frac{1}{2\pi} \int_{-\infty}^{\infty} f(\tau)e^{-i\omega(\tau-t)}\,d\tau \right) dt \\
&= \int_{-\infty}^{\infty} F(i\omega)g(t)e^{i\omega t}\,dt = 2\pi F(i\omega)G(-i\omega).
\end{aligned}
\tag{A.4}
$$

Noting finally the Hermitian symmetry of real functions, it is

$$
\Phi(i\omega) = 2\pi F(i\omega)G^*(i\omega).
\tag{A.5}
$$

A particular situation arises when $f(t) = g(t)$; in that case, Equation (A.3) becomes the autocorrelation function, for which the Fourier transform is $2\pi |F|^2$. In other words, the Fourier transform of the autocorrelation function of a given signal goes as the squared modulus of the Fourier transform of that signal. Recall that the modulus of a complex number is $|a+ib| = \sqrt{a^2+b^2}$.

It is also interesting to consider the case with a zero time lag, similarly to Equation (2.11). From Equation (A.3), we have $\phi(0) = \int_{-\infty}^{\infty} f(t)g(t)\,dt$, while from the inverse Fourier transform, Equation (A.2), at $t = 0$, we then get $\phi(0) = \int_{-\infty}^{\infty} \Phi(i\omega)\,d\omega$. Equating both expressions and introducing Equation (A.5) result in the following relation between the time and frequency integrals of two functions:

$$
\int_{-\infty}^{\infty} f(t)g(t)\,dt = 2\pi \int_{-\infty}^{\infty} F(i\omega)G^*(i\omega)\,d\omega.
\tag{A.6}
$$

This is the well-known *Plancherel's theorem*, which has been used in Section 7.4.2.

The second operation of interest is the convolution integral, often referred to as *Duhamel's integral* in structural vibrations (Shabana, 1991b, Ch. 5). The convolution integral has appeared in linear dynamic analyses to construct forced responses from an impulse response (see Equation (1.14)). It is defined as

$$
h(t) = (f \star g)(t) = \int_{-\infty}^{\infty} f(t-\tau)g(\tau)\,d\tau.
\tag{A.7}
$$

The Fourier transform of the convolution function is obtained similarly as above and results in

$$
H(i\omega) = 2\pi F(i\omega)G(i\omega).
\tag{A.8}
$$

Therefore, convolution in the frequency domain corresponds to the series connection between the admittance of the system and the frequency content of the input. Note finally that all the definitions in this section are easily expanded to vectors or matrix arrays of functions in either the time or frequency domains, and this is how they have often appeared in this book.

A.3   The Z-Transform

Consider a discrete-time signal $\{\mathbf{x}_n\}$ of finite length, that is, with $n = 0, 1, 2, \ldots, N$, sampled at a constant time step Δt. We assume here that all $\mathbf{x}_n$ are real numbers, although this is not strictly necessary. We define its *discrete Fourier transform* (DFT) as the new series

$$\mathbf{X}_m = \sum_{n=0}^{N} e^{-2\pi i \frac{m}{N} n} \mathbf{x}_n, \tag{A.9}$$

for $m = 0, 1, 2, \ldots, N$. It is easy to see that $e^{-2\pi i \frac{m}{N}}$ defines points in the complex plane located along a single loop on the unit circle. In the same similar manner as the Laplace transform expands the Fourier transform from the imaginary axis to the complex domain on continuous fractions, we introduce the *Z-transform* as the discrete-time equivalent for the DFT (Franklin et al., 1998).

The formal definition of the Z-transform is on a discrete-time signal $\{\mathbf{x}_n\}$ of infinite length, that is, $N \to \infty$, for which its (one-sided) Z-transform is then given by

$$\mathbf{X}(z) = \sum_{n=0}^{\infty} z^{-n} \mathbf{x}_n, \tag{A.10}$$

where z is a complex variable. A two-sided (or bilateral) transform can also be defined and is preferred here for consistency with Equation (A.1) as

$$\mathbf{X}(z) = \sum_{n=-\infty}^{\infty} z^{-n} \mathbf{x}_n. \tag{A.11}$$

In general, Equation (A.11) may not be a convergent series. The *region of convergence* of a Z-transform are the points in the complex plane for which $\mathbf{X}(z)$ takes finite values. Some important properties of the Z-transform are as follows:

1. *Linearity.* From the definition in Equation (A.11), it is easy to see that the Z-transform of $\alpha \{\mathbf{x}_n\} + \beta \{\mathbf{y}_n\}$ is $\alpha \mathbf{X}(z) + \beta \mathbf{Y}(z)$.
2. *Time delay.* If $\mathbf{X}(z)$ is the Z-transform of $\{\mathbf{x}_n\}$, then the Z-transform of $\{\mathbf{x}_{n-k}\}$ for $k > 0$, with $\mathbf{x}_{n-k} = 0$ if $n < k$, is $z^{-k} \mathbf{X}(z)$.
3. *Forward differences.* The Z-transform of $\left\{ \mathbf{x}^{n+1} - \mathbf{x}^n \right\}$ is $(z-1)\mathbf{X}(z) - z\mathbf{x}(0)$.

In general, a function in the Z-domain can be mapped into a function in the Laplace domain using the *Tustin transformation*

$$z = e^{is} \approx \frac{1 + s\Delta t/2}{1 - s\Delta t/2}, \tag{A.12}$$

where s is the Laplace variable. The inverse operation is the bilinear transformation, namely

$$s = \frac{2}{T} \frac{z-1}{z+1}. \tag{A.13}$$

These transformations map the left half of the complex plane in continuous time into a unit circle in discrete time. Importantly, they also preserve the observability and controllability Gramians between the discrete and continuous representations of a given system (Antoulas, 2005, Ch. 4).

To complete this brief summary, consider again Plancherel's theorem, Equation (A.6). For discrete-time signals, it becomes

$$\sum_{n=-\infty}^{\infty} \mathbf{x}_n^\top \mathbf{y}_n = \frac{1}{2\pi} \int_{-\pi}^{\pi} \mathbf{X}(e^{-i\theta})^\top \mathbf{Y}(e^{i\theta}) \, d\theta, \tag{A.14}$$

where the integral in the frequency domain is solved for convenience in the phase angle $\theta = \omega \Delta t$ and integrated up to the Nyquist angular frequency, $\omega_{\text{nyq}} = \pi/\Delta t$. The $1/2\pi$ on the right-hand side comes from the choice of normalization in the definition of the Z-transform, which is missing the 2π factor of Equation (A.1). Note that the discrete version of Equation (A.6) would be $2\pi \int_{-\pi}^{\pi} \frac{\mathbf{X}^*}{2\pi} \frac{\mathbf{Y}}{2\pi} \, d\theta$, which results in Equation (A.14).

Appendix B Von Kármán Model for Isotropic Turbulence

We have used von Kármán's classical description of isotropic turbulence to model atmospheric turbulence above the Earth's boundary layer in Chapter 2. This appendix derives the expressions used there for the point (1-D) assumption, which has been adopted by the airworthiness certification rules, as well as for 2-D conditions, which may be given consideration for aircraft with very large wingspan.

We first define the turbulent velocity as the difference between the instantaneous and the mean velocity at a point in a turbulent fluid. Let $u_i'(\mathbf{x},t)$ be the component of the turbulent fluid velocity vector at position $\mathbf{x}$ along the ith direction at time t. The total turbulent kinetic energy (TKE) per unit mass of the fluid is defined as half the covariance of turbulent velocity, that is, $\frac{1}{2}\sigma_{u'}^2 = \frac{1}{2}\mathrm{E}[u_1'^2 + u_2'^2 + u_3'^2]$, using the definition of Equation (2.10). The wavenumber-based spectral density of the TKE, $E(\kappa)$, known as the *energy spectrum function*, will be such that $\frac{1}{2}\sigma_{u'}^2 = \int_0^\infty E(\kappa)\,\mathrm{d}\kappa$. It represents the contribution to the TKE from all fluid spatial features with wavelength κ.

In Section 2.3.3, we have defined the 1-D velocity spectrum tensor $\hat{\mathbf{\Phi}}_w(\kappa)$. More generally, one can define the 3-D spectrum tensor $\hat{\mathbf{\Phi}}_w(\boldsymbol{\kappa})$ in which the wavenumber vector $\boldsymbol{\kappa}$ also includes directional information. Physically, it corresponds to the correlation being defined between the instantaneous velocities at two different points in space, which brings the effect of their relative position into the definition. Under the assumption of isotropic turbulence, there exists a closed-form relation between the energy spectrum function and the 3-D velocity spectrum tensor (Batchelor, 1953, Eq. (3.4.12)), namely[1]

$$\hat{\Phi}_{ij}(\boldsymbol{\kappa}) = \frac{E(\kappa)}{2\pi\kappa^2}\left(\delta_{ij} - \frac{\kappa_i\kappa_j}{\kappa^2}\right), \tag{B.1}$$

with $\kappa = \|\boldsymbol{\kappa}\|$. Note that, as one would expect from an isotropic fluid, this expression is invariant with a change of coordinate system.

So far, we have only introduced definitions for the statistical characterization of the turbulent field. Next, we introduce a predictive model based on semi-empirical analysis. Von Kármán proposed the following expression for the energy spectrum function of isotropic turbulence in the inertial subrange (which is the one of interest to us):

---

[1] Note that we have introduced a factor of 2 in this expression, for consistency with the definitions in Section 2.3.

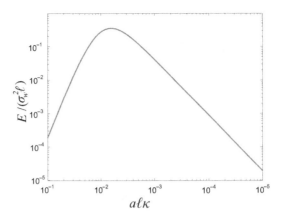

Figure B.1 The von Kármán model for the energy of isotropic turbulence.

$$E(\kappa) = \frac{55}{9\pi}\sigma_w^2 \ell \frac{(a\ell\kappa)^4}{\left[1 + (a\ell\kappa)^2\right]^{17/6}}, \tag{B.2}$$

where we have used the parameters defined in Section 2.3.3. This function is shown in Figure B.1. Substituting Equation (B.2) into Equation (B.1) results in a 3-D velocity spectrum tensor given as

$$\hat{\Phi}_{ij}(\bar{\kappa}) = \frac{55\sigma_w^2 a^2 \ell^3}{18\pi^2} \frac{\bar{\kappa}^2 \delta_{ij} - \bar{\kappa}_i \bar{\kappa}_j}{\left(1 + \bar{\kappa}^2\right)^{17/6}}, \tag{B.3}$$

with the dimensionless wavenumber $\bar{\kappa}_i = a\ell\kappa_i$. The 1-D spectrum is obtained by integration along both lateral directions:

$$\hat{\Phi}_{ij}(\kappa_1) = \int_0^\infty \int_0^\infty \hat{\Phi}_{ij}(\kappa_1, \kappa_2, \kappa_3)\, d\kappa_2\, d\kappa_3. \tag{B.4}$$

For the longitudinal component, this yields after transformation to cylindrical coordinates (Pope, 2000, Eq. (6.214)):

$$\hat{\Phi}_{11}(\kappa_1) = \int_{\kappa_1}^\infty \frac{E(\bar{\kappa})}{\bar{\kappa}}\left(1 - \frac{\kappa_1^2}{\bar{\kappa}^2}\right) d\bar{\kappa}, \tag{B.5}$$

while for the lateral component, it is

$$\hat{\Phi}_{22}(\kappa_1) = \frac{1}{2}\int_{\kappa_1}^\infty \frac{E(\bar{\kappa})}{\bar{\kappa}}\left(1 + \frac{\kappa_1^2}{\bar{\kappa}^2}\right) d\bar{\kappa}. \tag{B.6}$$

Note that they are related as in Equation (2.42). Integrating both equations gives the 1-D spectra of Equations (2.43) and (2.44).

Integrating along only one dimension gives the two-wavenumber spectrum, that is,

$$\hat{\Phi}_{ij}(\kappa_1, \kappa_2) = \int_0^\infty \hat{\Phi}_{ij}(\kappa_1, \kappa_2, \kappa_3)\, d\kappa_3, \tag{B.7}$$

which results in

$$\hat{\Phi}_{11}(\kappa_1,\kappa_2) = \frac{55\sigma_w^2 a\ell^2}{72\pi^{3/2}} \frac{\Gamma(\frac{4}{3})\left(1+\bar{\kappa}_1^2+\bar{\kappa}_2^2\right)+2\Gamma(\frac{7}{3})\bar{\kappa}_2^2}{\Gamma(\frac{17}{6})\left(1+\bar{\kappa}_1^2+\bar{\kappa}_2^2\right)^{7/3}},$$

$$\hat{\Phi}_{22}(\kappa_1,\kappa_2) = \frac{55\sigma_w^2 a\ell^2}{72\pi^{3/2}} \frac{\Gamma(\frac{4}{3})\left(1+\bar{\kappa}_1^2+\bar{\kappa}_2^2\right)+2\Gamma(\frac{7}{3})\bar{\kappa}_1^2}{\Gamma(\frac{17}{6})\left(1+\bar{\kappa}_1^2+\bar{\kappa}_2^2\right)^{7/3}}, \tag{B.8}$$

$$\hat{\Phi}_{33}(\kappa_1,\kappa_2) = \frac{55\sigma_w^2 a\ell^2}{36\pi^{3/2}} \frac{\Gamma(\frac{7}{3})\left(\bar{\kappa}_1^2+\bar{\kappa}_2^2\right)}{\Gamma(\frac{17}{6})\left(1+\bar{\kappa}_1^2+\bar{\kappa}_2^2\right)^{7/3}}.$$

Noting that $a = \dfrac{\Gamma(\frac{1}{3})}{\sqrt{\pi}\Gamma(\frac{5}{6})} \approx 1.339$, and using the properties of the Γ function, we finally have

$$\hat{\Phi}_{11}(\kappa_1,\kappa_2) = \frac{\sigma_w^2 a^2\ell^2}{6\pi} \frac{1+\bar{\kappa}_1^2+\frac{11}{3}\bar{\kappa}_2^2}{\left(1+\bar{\kappa}_1^2+\bar{\kappa}_2^2\right)^{7/3}},$$

$$\hat{\Phi}_{22}(\kappa_1,\kappa_2) = \frac{\sigma_w^2 a^2\ell^2}{6\pi} \frac{1+\frac{11}{3}\bar{\kappa}_1^2+\bar{\kappa}_2^2}{\left(1+\bar{\kappa}_1^2+\bar{\kappa}_2^2\right)^{7/3}}, \tag{B.9}$$

$$\hat{\Phi}_{33}(\kappa_1,\kappa_2) = \frac{4\sigma_w^2 a^2\ell^2}{9\pi} \frac{\bar{\kappa}_1^2+\bar{\kappa}_2^2}{\left(1+\bar{\kappa}_1^2+\bar{\kappa}_2^2\right)^{7/3}}.$$

This result has been reported by Etkin (1972, Eq. (13.2,17)), among others. Finally, we can obtain the cross-correlation spectra between two points at a distance Δy by introducing an inverse Fourier transform in the lateral direction, that is,

$$\Phi_{ij}(\omega,\Delta y) = \int_0^\infty \hat{\Phi}_{ij}(\omega/V,\kappa_2)\cos(\kappa_2\Delta y)\,\mathrm{d}\kappa_2. \tag{B.10}$$

For the vertical component of the turbulent velocity, this results in the cross-spectrum of Equation (2.50).

Appendix C   Finite Rotations

C.1   Rodrigues Rotation Formula

Given an Euclidean vector in 3-D space, we transform its components between two different frames of reference B and A as

$$\mathbf{v}_A = \mathbf{R}_{AB}\mathbf{v}_B, \tag{C.1}$$

where $\mathbf{R}_{AB}$ is the coordinate transformation matrix from B to A, which has been introduced in Section 4.2. This change of basis can also be seen as a *finite rotation* between both frames in 3-D space, and $\mathbf{R}_{AB}$ is then describing the relative orientation of frame B with respect to A. Some properties of the coordinate transformation matrix have already been discussed in Section 4.2.2, where we have seen in particular that it can be parameterized using the Euler angles describing the relative orientation between both frames.

Here, the link between coordinate transformation matrices and finite rotations is established using *Euler's rotation theorem*. It is known that the kinematics of a rigid body can be decomposed into a translation and a rotation, and that the rotation component always defines points that remain fixed in space. Euler's rotation theorem states that the rotation field can be uniquely described by a rotation angle ψ, with $0 \leq \psi < 2\pi$, around a fixed axis, defined by a unit vector $\mathbf{n}$, with origin in a fixed point. Following on that, the relative rotations between frames A and B can be described by a rotation ψ about the axis defined by the unit vector $\mathbf{n}$ (which has the same components in both A and B and, therefore, it is written here with no subindex, i.e., $\mathbf{n} = \mathbf{n}_A = \mathbf{n}_B$). These definitions are sketched in Figure C.1.

Given a rotation defined by the pair $(\psi, \mathbf{n})$, we can decompose a generic vector, given in B frame as $\mathbf{v}_B$, into its components in the parallel and normal directions to $\mathbf{n}$. The parallel component is the *vector projection*, which we can write as $\left(\mathbf{v}_B^\top \mathbf{n}\right)\mathbf{n}$, while the perpendicular component, which is sometimes called for consistency the *vector rejection* (Figure C.2), can be written as $\mathbf{P}_n\mathbf{v}_B$, where $\mathbf{P}_n$ is the *orthogonal projection operator*, given by

$$\mathbf{P}_n = \mathcal{I} - \mathbf{n}\mathbf{n}^\top. \tag{C.2}$$

It is straightforward to see that the projection and rejection vectors add up to the original vector, that is,

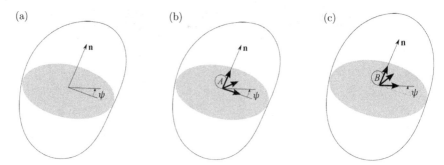

Figure C.1 Definitions in Euler rotation's theorem. Frame B results from frame A after the rigid-body rotation (for simplicity, we have chosen an axis of A to be along $\mathbf{n}$). (a) Rigid body before a rotation ψ along $\mathbf{n}$. (b) Frame A, defined on a rigid body before rotation. (c) Frame B, defined on the rigid body after rotation.

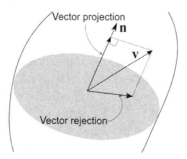

Figure C.2 Geometric definition of vector projection and vector rejection.

$$\left(\mathbf{v}_B^\top\mathbf{n}\right)\mathbf{n} + \mathbf{P}_n\mathbf{v}_B = \left(\mathbf{v}_B^\top\mathbf{n}\right)\mathbf{n} + \mathbf{v}_B - \mathbf{n}\mathbf{n}^\top\mathbf{v}_B = \mathbf{v}_B + \left(\mathbf{v}_B^\top\mathbf{n} - \mathbf{v}_B^\top\mathbf{n}\right)\mathbf{n} = \mathbf{v}_B. \quad \text{(C.3)}$$

In a finite rotation about $\mathbf{n}$, the component parallel to the rotation axis remains invariant, while the perpendicular component rotates an angle ψ in the normal plane (which is defined by $\mathbf{P}_n\mathbf{v}_B$ and its cross product with $\mathbf{n}$, i.e., $\tilde{\mathbf{n}}\mathbf{P}_n\mathbf{v}_B$). The components of the vector in the reference frame before the rotation, A, can then be obtained simply as

$$\mathbf{v}_A = \left(\mathbf{v}_B^\top\mathbf{n}\right)\mathbf{n} + (\mathbf{P}_n\mathbf{v}_B)\cos\psi + (\tilde{\mathbf{n}}\mathbf{P}_n\mathbf{v}_B)\sin\psi. \quad \text{(C.4)}$$

Comparing this with Equation (C.1) results in the following expression for the coordinate transformation matrix:

$$\begin{aligned}
\mathbf{R}_{AB} &= \mathbf{n}\mathbf{n}^\top + \cos\psi\,\mathbf{P}_n + \sin\psi\,\tilde{\mathbf{n}}\mathbf{P}_n \\
&= \mathcal{I}\cos\psi + \mathbf{n}\mathbf{n}^\top(1 - \cos\psi) + \tilde{\mathbf{n}}\sin\psi,
\end{aligned} \quad \text{(C.5)}$$

where we have used in the last term that $\tilde{\mathbf{n}}\mathbf{n} = \mathbf{0}$. This last expression, therefore, serves to write the coordinate transformation matrix in terms of the two parameters in Euler's rotation theorem. Equation (C.5) is known as *Rodrigues' rotation formula*.

From the first line of Equation (C.5), it can be seen that the coordinate transformation matrix satisfies

$$\frac{\partial \mathbf{R}_{AB}}{\partial \psi} = \tilde{\mathbf{n}} \mathbf{R}_{AB}, \tag{C.6}$$

and, integrating from $\psi = 0$, for which the transformation is the identity, results in the rotation matrix given by the exponential map

$$\mathbf{R}_{AB} = \exp(\tilde{\mathbf{n}}\psi). \tag{C.7}$$

This is an alternative expression to Equation (C.5) to compute the rotation matrix corresponding to the finite rotation defined by $(\psi, \mathbf{n})$.

Example C.1   Consider a vector along the x axis of the B frame that results from a rotation ψ around the z axis of the A frame. Then its components in the B and A frames are given, respectively, as $\mathbf{v}_B = (1,0,0)^\top$ and $\mathbf{v}_A = (\cos\psi, \sin\psi, 0)^\top$, and the unit normal $\mathbf{n} = (0,0,1)$. In that case, the vector projection of $\mathbf{v}_B$ is $\left(\mathbf{v}_B^\top \mathbf{n}\right)\mathbf{n} = (0,0,0)^\top$ and the orthogonal projection is $\mathbf{P}_n \mathbf{v}_B = (1,0,0)$, with the orthogonal projection operator

$$\mathbf{P}_n = \begin{bmatrix} 1 & 0 & 0 \\ 0 & 1 & 0 \\ 0 & 0 & 0 \end{bmatrix}.$$

We then finally have $\tilde{\mathbf{n}}\mathbf{P}_n\mathbf{v}_B = (0,1,0)^\top$, which defines all the terms needed to obtain again $\mathbf{v}_A$ from $\mathbf{v}_B$, $\mathbf{n}$ and ψ using Equation (C.4).

C.2   Cartesian Rotation Vector

An important parameterization for finite rotations, which has been used in the finite-element solution for composite beams in Section 8.4, is the Cartesian rotation vector, sometimes also referred to as the *rotational pseudovector* (Argyris, 1982). It is defined as a vector with the direction given by $\mathbf{n}$ and the magnitude given by ψ, that is,

$$\boldsymbol{\psi} = \psi \mathbf{n}, \tag{C.8}$$

where $\mathbf{n}$ and ψ have been defined in Euler's rotation theorem in the previous section. This is the rotation associated with the coordinate transformation from B to A, while a rotation $-\psi$ around the same axis, thus giving a Cartesian rotation vector $-\boldsymbol{\psi}$, defines the transformation from A to B. Since the normal vector is unitary, then $\psi = \|\boldsymbol{\psi}\|$. The exponential map derived in Equation (C.7) gives now a direct relation between a Cartesian rotation vector and its corresponding coordinate transformation matrix. We write it as

$$\mathbf{R}_{AB} = \mathbf{R}(\boldsymbol{\psi}) = \exp(\tilde{\boldsymbol{\psi}}) = \mathcal{I} + \sum_{k=1}^{\infty} \frac{\tilde{\boldsymbol{\psi}}^k}{k!}. \tag{C.9}$$

For practical purposes, the last expression is written as a finite series using Cayley–Hamilton's theorem. It results in

$$\mathbf{R}_{AB} = \mathcal{I} + \frac{\sin\psi}{\psi}\tilde{\boldsymbol{\psi}} + \frac{1-\cos\psi}{\psi^2}\tilde{\boldsymbol{\psi}}\tilde{\boldsymbol{\psi}}. \tag{C.10}$$

Note that Equation (C.10) can also be obtained from the direct manipulation of Equation (C.5) with the definition given in Equation (C.8). This expression is preferable to Equation (C.9) for numerical efficiency, while Equation (C.9) is very useful for analysis. However, for small angles, the singularity in the denominators needs to be removed, and this is achieved by taking a truncated series expansion of the exponential map as

$$\mathbf{R}_{AB} = \mathcal{I} + \tilde{\boldsymbol{\psi}} + \frac{1}{2}\tilde{\boldsymbol{\psi}}\tilde{\boldsymbol{\psi}} + \mathcal{O}\left(\psi^3\right). \tag{C.11}$$

C.3   Infinitesimal Rotations

We have seen in Section 4.2.3 that angular velocities describe the rate of change of rotation in the relative motion between two frames A and B. The angular velocity, expressed in its components of the B frame, has been defined as

$$\tilde{\boldsymbol{\omega}}_B = \mathbf{R}_{BA}\frac{\mathrm{d}}{\mathrm{d}t}\mathbf{R}_{AB}. \tag{C.12}$$

An analogous relation has been obtained when the change of rotation occurs along a space curve, which defines the curvature as

$$\tilde{\mathbf{k}}_B = \mathbf{R}_{BA}\frac{\mathrm{d}}{\mathrm{d}s}\mathbf{R}_{AB}. \tag{C.13}$$

This can be generalized to any variation of the rotation by means of a virtual rotation, defined as

$$\widetilde{\delta\boldsymbol{\varphi}}_B = \mathbf{R}_{BA}\delta\mathbf{R}_{AB}. \tag{C.14}$$

Virtual rotations defined in this way are known as *material* (or right) virtual rotations. Importantly, they only measure infinitesimal changes and do not integrate into a finite rotation. Transformation to the origin frame defines the *spatial* (or left) virtual rotation as

$$\delta\boldsymbol{\varphi}_A = \mathbf{R}_{AB}\delta\boldsymbol{\varphi}_B, \tag{C.15}$$

which results in

$$\widetilde{\delta\boldsymbol{\varphi}}_A = \delta\mathbf{R}_{AB}\mathbf{R}_{BA}. \tag{C.16}$$

Consider now a parameterization of the rotation matrix. One such parameterization is the Cartesian rotation vector $\mathbf{R}_{AB} = \mathbf{R}(\psi)$, where $\mathbf{R}(\psi)$ is the exponential map, Equation (C.10). The material virtual rotation of Equation (C.14) can then be written as $\widetilde{\delta\varphi}_B = \mathbf{R}^\top(\boldsymbol{\psi})\delta\mathbf{R}(\boldsymbol{\psi})$. Introducing the Levi–Civita symbol, ε_{ijk}, for the skew-symmetric operator, this expression can also be written in an index form as

$$\varepsilon_{ijk}\delta\varphi_{B,k} = R_{ki}(\boldsymbol{\psi})\delta R_{kj}(\boldsymbol{\psi}) = R_{ki}(\boldsymbol{\psi})\frac{\partial R_{kj}(\boldsymbol{\psi})}{\partial \psi_l}\delta\psi_l. \tag{C.17}$$

Premultiplying this expression by ε_{ijn} (also summing through both i and j) and using the property that $\varepsilon_{ijk}\varepsilon_{ijn} = 2\delta_{kn}$, we finally obtain a linear relation between the virtual rotation and the variations of the rotation parameter as

$$\delta\varphi_B = \mathbf{T}(\boldsymbol{\psi})\delta\boldsymbol{\psi}, \tag{C.18}$$

where the new matrix is known as the *tangential rotational operator*, and it is defined as

$$T_{nl}(\boldsymbol{\psi}) = \frac{1}{2}\varepsilon_{ijn}R_{ki}(\boldsymbol{\psi})\frac{\partial R_{kj}(\boldsymbol{\psi})}{\partial \psi_l}. \tag{C.19}$$

So far, we have not used any properties of the exponential map, and this relation is valid for any parameterization of rotations (by simply replacing it instead of $\boldsymbol{\psi}$ in Equation (C.19)). Once the parameterization is chosen, a closed-form analytical expression for the tangential operator can be obtained. The expression in terms of Euler angles was given in Equation (4.15). For the Cartesian rotation vector, it yields

$$\mathbf{T}(\boldsymbol{\psi}) = \sum_{k=0}^{\infty}\frac{(-1)^k}{(k+1)!}\tilde{\psi}^k. \tag{C.20}$$

As with the coordinate transformation matrix, the infinite series expansion can be simplified using Cayley–Hamilton's theorem. It results in

$$\mathbf{T}(\boldsymbol{\psi}) = \boldsymbol{\mathcal{I}} + \frac{\cos\psi - 1}{\psi^2}\tilde{\psi} + \frac{\psi - \sin\psi}{\psi^3}\tilde{\psi}\tilde{\psi}. \tag{C.21}$$

This expression has also a singularity in the denominators near $\psi = 0$, and it is substituted near the origin by a small angle approximation of the infinite series defined by Equation (C.20).

C.4   Skew-Symmetric Matrices

In the manipulation of finite rotations, we have often found the skew-symmetric operator, which, from Equation (C.11), is associated with infinitesimal rotations around the origin. Given a vector in a 3-D Euclidean space defined by its components on a certain orthonormal basis, $\boldsymbol{\alpha} = (\alpha_1, \alpha_2, \alpha_3)^\top$, we have:

$$\tilde{\boldsymbol{\alpha}} = \begin{bmatrix} 0 & -\alpha_3 & \alpha_2 \\ \alpha_3 & 0 & -\alpha_1 \\ -\alpha_2 & \alpha_1 & 0 \end{bmatrix}. \tag{C.22}$$

Skew-symmetric matrices have the following properties:

$$
\begin{aligned}
&\tilde{\alpha}^\top = -\tilde{\alpha} && \tilde{\alpha}\beta = -\tilde{\beta}\alpha \\
&\tilde{\alpha}\alpha = \mathbf{0} && \mathbf{R}\tilde{\alpha}\mathbf{R}^\top = \widetilde{\mathbf{R}\alpha} \\
&\alpha^\top\beta = -\tfrac{1}{2}\mathrm{trace}(\tilde{\alpha}\tilde{\beta}) && \tilde{\alpha}\tilde{\beta} = \beta\alpha^\top - \left(\alpha^\top\beta\right)\mathcal{I} \\
&(\tilde{\alpha}\beta)^\top = -\alpha^\top\tilde{\beta} && \widetilde{\tilde{\alpha}\beta} = \tilde{\alpha}\tilde{\beta} - \tilde{\beta}\tilde{\alpha} = \beta\alpha^\top - \alpha\beta^\top ,
\end{aligned}
\tag{C.23}
$$

where $\mathbf{R}$ is a generic coordinate transformation matrix.

References

Abramowitz, M. and I. A. Stegun. *Handbook of Mathematical Functions*. Dover, 9th edition, New York, 1972.

Abzug, M. J. and E. E. Larrabee. *Airplane Stability and Control, A History of the Technologies That Made Aviation Possible*. Cambridge University Press, Cambridge, UK, 2nd edition, 2002.

Afonso, F., J. Vale, E. Oliveira, F. Lau, and A. Suleman. A review on non-linear aeroelasticity of high aspect-ratio wings. *Progress in Aerospace Sciences*, 89(3):40–57, February 2017. doi:10.1016/j.paerosci.2016.12.004.

Ahmad, N. and F. Proctor. Mesoscale simulation data for initializing fast-time wake transport and decay models. In *51st AIAA Aerospace Sciences Meeting*, Grapevine, Texas, USA, January 2013. doi:10.2514/6.2013-510.

Albano, E. and W. P. Rodden. A doublet-lattice method for calculating lift distributions on oscillating surfaces in subsonic flow. *AIAA Journal*, 7(2):279–285, 1969. doi:10.2514/3.5086.

Anderson, B. D. O. and J. B. Moore. *Optimal Filtering*. Prentice-Hall, Englewood Cliffs, New Jersey, USA, 1979.

Anderson, B. D. O. and J. B. Moore. *Optimal Control*. Dover Publications, Mineola, New York, USA, 1990.

Anon. Aeroelastic effects from a flight mechanics standpoint. In *34th Meeting of the AGARD Flight Mechanics Panel*, Marseille, France, 21–25 April 1969. NATO Advisory Group for Aerospace Research and Development.

Antoulas, A. C. *Approximation of Large-Scale Dynamical Systems*. Society for Industrial and Applied Mathematics, 2005. doi:10.1137/1.9780898718713.

Argentina, M., J. Skotheim, and L. Mahadevan. Settling and swimming of flexible fluid-lubricated foils. *Physical Review Letters*, 99(22):224503–4, 2007. doi:10.1103/PhysRevLett.99.224503.

Argyris, J. An excursion into large rotations. *Computers Methods in Applied Mechanics and Engineering*, 32:85–155, 1982. doi:10.1016/0045-7825(82)90069-X.

Artola, M., N. Goizueta, A. Wynn, and R. Palacios. Aeroelastic control and estimation with a minimal nonlinear modal description. *AIAA Journal*, 59(7):2697–2713, 2021a. doi:10.2514/1.J060018.

Artola, M., N. Goizueta, A. Wynn, and R. Palacios. Proof of concept for a hardware-in-the-loop nonlinear control framework for very flexible aircraft. In *AIAA Science and Technology Forum and Exposition*, Nashville, Tennessee, USA, 4–8 January 2021b. doi:10.2514/6.2021-1392.

Artola, M., A. Wynn, and R. Palacios. A generalised Kelvin-Voigt damping model for geometrically-nonlinear beams. *AIAA Journal*, 59(1):356–365, 2021c. doi:10.2514/1.J059767.

Artola, M., A. Wynn, and R. Palacios. Modal-based nonlinear model predictive control for 3D very flexible structures. *IEEE Transactions in Automatic Control*, 2022. doi:10.1109/TAC.2021.3071326.

Ashley, H. H. *Engineering Analysis of Flight Vehicles*. Addison-Wesley Aerospace Series. Addison-Wesley Publishing Co., Reading, Massachusetts, USA, 1974.

Avin, O., D. E. Raveh, A. Drachinsky, Y. Ben-Shmuel, and M. Tur. Experimental aeroelastic benchmark of a very flexible wing. *AIAA Journal*, 60(3):1745–1768, 2022. doi:10.2514/1.J060621.

Bairstow, L. and A. Fage. Oscillations of the tail plane and body of an aeroplane in flight. Reports and Memoranda 276, Part II, British Advisory Committee for Aeronautics, 1916.

Bairstow, L., B. M. Jones, and A. W. H. Thomson. Investigation into the stability of an aeroplane, with an examination into the conditions necessary in order that the symmetric and antisymmetric oscillations can be considered independently. Reports and Memoranda 7, British Advisory Committee for Aeronautics, 1913.

Baker, M. L., D. L. Mingori, and P. J. Goggin. Approximate subspace iteration for constructing internally balanced reduced order models of unsteady aerodynamic systems. In *37th AIAA/ASME/ASCE/AHS/ASC Structures, Structural Dynamics, and Materials Conference*, Salt Lake City, Utah, USA, 15–17 April 1996. doi:10.2514/6.1996-1441.

Balachandran, B. and E. B. Magrab. *Vibrations*. Cambridge University Press, Cambridge, UK, 3rd edition, 2018.

Baldelli, D. H., P. C. Chen, and J. Panza. Unified aeroelastic and flight dynamic formulation via rational function approximations. *Journal of Aircraft*, 43(3):763–772, 2006. doi:10.2514/1.16620.

Ballhaus, W. F. and P. M. Goorjian. Computation of unsteady transonic flows by the indicial method. *AIAA Journal*, 16(2):117–124, 1978. doi:10.2514/3.60868.

Bartley-Cho, J. D. and A. Henderson. Design and analysis of HiLDA/AEI aeroelastic wind tunnel model. In *AIAA Applied Aerodynamics Conference*, Honolulu, Hawaii, USA, 18–21 August 2008. doi:10.2514/6.2008-7191.

Barzgaran, B., J. D. Quenzer, M. Mesbahi, K. A. Morgansen, and E. Livne. Real-time model predictive control for gust load alleviation on an aeroelastic wind tunnel test article. In *AIAA Science and Technology Forum and Exposition*, Nashville, Tennessee, USA, 4–8 January 2021. doi:10.2514/6.2021-0500.

Batchelor, G. K. *The Theory of Homogeneous Turbulence*. Cambridge University Press, Cambridge, UK, 1953.

Bauchau, O. A., A. Epple, and S. Heo. Interpolation of finite rotations in flexible multi-body dynamics simulations. *Proceedings of the Institution of Mechanical Engineers, Part K: Journal of Multi-body Dynamics*, 222(4):353–366, 2008. doi:10.1243/14644193JMBD155.

Bauchau, O. A. *Flexible Multibody Dynamics*, volume 176 of *Solid Mechanics and Its Applications*. Springer Netherlands, Dordrecht, The Netherlands, 2011. doi:10.1007/978-94-007-0335-3.

Bazilevs, Y., K. Takizawa, and T. E. Tezduyar. *Computational Fluid-Structure Interaction: Methods and Applications*. John Wiley & Sons, Chichester, UK, 2013.

Beal, T. R. Digital simulation of atmospheric turbulence for Dryden and von Kármán models. *Journal of Guidance, Control, and Dynamics*, 16(1):132–138, 1993. doi:10.2514/3.11437.

Beck, M. S. Correlation in instruments: cross correlation flowmeters. *Journal of Physics E: Scientific Instruments*, 14(1):7–19, 1981. doi:10.1088/0022-3735/14/1/001.

Bell, T. M., P. M. Klein, J. K. Lundquist, and S. Waugh. Remote sensing and radiosonde datasets collected in the San Luis Valley during the LAPSE-RATE campaign. *Earth System Science Data*, 13(3):1041–1051, March 2021. doi:10.5194/essd-13-1041-2021.

Bendiksen, O. O. Transonic limit cycle flutter of high-aspect-ratio swept wings. In *47th AIAA/ASME/ASCE/AHS/ASC Structures, Structural Dynamics, and Materials Conference*, Newport, Rhode Island, USA, 1–4 May 2006. doi:10.2514/1.29547.

Benner, P., P. Kürschner, and J. Saak. Frequency-limited balanced truncation with low-rank approximations. *SIAM Journal on Numerical Analysis*, 38(1):A471–A499, 2016. doi:10.1137/15M1030911.

Beran, P. S., N. S. Khot, F. E. Eastep, R. D. Snyder, and J. V. Zweber. Numerical analysis of store-induced limit-cycle oscillation. *Journal of Aircraft*, 41(6):1315–1326, 2004.

Beranek, J., L. Nicolai, M. Buonanno, E. Burnett, C. Atkinson, B. Holm-Hansen, and P. M. Flick. Conceptual design of a multi-utility aeroelastic demonstrator. In *13th AIAA/ISSMO Multidisciplinary Analysis Optimization Conference*, Fort Worth, Texas, USA, 13–15 September 2010. doi:10.2514/6.2010-9350.

Bhardwaj, M. K., R. K. Kapania, E. Y. Reichenbach, and G. P. Guruswamy. Computational fluid dynamics/computational structural dynamics interaction methodology for aircraft wings. *AIAA Journal*, 36(12):2179–2186, 1998.

Bieniek, D. and R. Luckner. Simulation of aircraft encounters with perturbed vortices considering unsteady aerodynamic effects. *Journal of Aircraft*, 51(3):705–718, 2014. doi:10.2514/1.C032383.

Birnbaum, W. Die tragende Wirbelfläche als Hilfsmittel zur Behandlung des ebenen Problems der Tragflügeltheorie. *ZAMM*, 3(4):290–297, 1923.

Birnbaum, W. Das ebene Problem des schlagenden Flügels (The Plane Problem of the Flapping Wing–also NACA TM 1364, 1954). *ZAMM*, 4:277–298, 1924.

Bisplinghoff, R. L., H. Ashley, and R. L. Halfman. *Aeroelasticity*. Addison-Wesley, Reading, Massachusetts, USA, 1955.

Blair, M. A compilation of the mathematics leading to the doublet-lattice method. Report 9227812, Air Force Research Laboratories, Dayton, Ohio, USA, March 1992.

Blaylock, M., R. Chow, A. Cooperman, and C. P. van Dam. Comparison of pneumatic jets and tabs for active aerodynamic load control. *Wind Energy*, 17(9):1365–1384, 2014. doi:10.1002/we.1638.

Block, J. J. and T. W. Strganac. Applied active control for a nonlinear aeroelastic structure. *Journal of Guidance, Control and Dynamics*, 21(6), 1998.

Bock, H. G. and P. Krämer-Eis. A multiple shooting method for numerical computation of open and closed loop controls in nonlinear systems. *IFAC Proceedings*, 17(2):411–415, July 1984. doi:10.1016/S1474-6670(17)61005-X.

Böswald, M., Y. Govers, A. Vollan, and M. Basien. Solar Impulse – How to validate the numerical model of a superlight aircraft with A340 dimensions! In *Proceedings of ISMA 2010–International Conference on Noise and Vibration Engineering*, Katholieke Universiteit Leuven, Leuven, Belgium, 2010, pp. 2451–2466.

Boyd, S. and L. Vandenberghe. *Convex Optimization*. Cambridge University Press, Cambridge, UK, 2004.

Brandt, S. A., J. J. Bertin, R. J. Stiles, and R. Whitford. *Introduction to Aeronautics*. AIAA Educational Series, Reston, Virginia, USA, 2nd edition, 2006. doi:10.2514/4.862007.

Breitsamter, C. Wake vortex characteristics of transport aircraft. *Progress in Aerospace Sciences*, 47(2):89–134, 2011. doi:10.1016/j.paerosci.2010.09.002.

Britt, R. T., S. B. Jacobson, and T. D. Arthurs. Aeroservoelastic analysis of the B-2 bomber. *Journal of Aircraft*, 37(5):745–752, 2000. doi:10.2514/2.2674.

Brown, S. A. Displacement extrapolations for CFD + CSM aeroelastic analysis. In *38th AIAA/ASME/ASCE/AHS/ASC Structures, Structural Dynamics, and Materials Conference*, Kissimmee, Florida, USA, 07–10 April 1997.

Broyden, C. G. A class of methods for solving nonlinear simultaneous equations. *Mathematics of Computation*, 19(92):577–593, 1965.

Brunton, S. L. and J. N. Kutz. *Data-Driven Science and Engineering*. Cambridge University Press, Cambridge, UK, 2019. doi:10.1017/9781108380690.

Brunton, S. L. and C. W. Rowley. Empirical state-space representations for Theodorsen's lift model. *Journal of Fluids and Structures*, 38:174–186, 2013.

Bryan, G. H. *Stability in Aviation*. MacMillian, London, UK, 1911.

Bryan, G. H. and W. E. Williams. The longitudinal stability of aerial gliders. In *Proceedings of the Royal Society of London, Series A*, volume 73, London, UK, 1904.

Bryant, L. W. and A. G. Pugsley. The lateral stability of highly loaded aeroplanes. Reports and Memoranda 1840, British Advisory Committee for Aeronautics, 1936.

Bryson, A. E. and Y. C. Ho. *Applied Optimal Control: Optimization, Estimation, and Control*. Hemisphere Publishing Corporation, London, UK, 1979.

Buck, B. K. and B. A. Newman. Aircraft acceleration prediction due to atmospheric disturbances with flight data validation. *Journal of Aircraft*, 43(1):72–81, 2006. doi:10.2514/1.12074.

Burl, J. B. *Linear Optimal Control – $\mathcal{H}_2$ and $\mathcal{H}_\infty$ Methods*. Addison-Wesley, Menlo Park, California, USA, 1999.

Burnett, E. L., J. A. Beranek, B. T. Holm-Hansen, C. J. Atkinson, and P. M. Flick. Design and flight test of active flutter suppression on the X-56A multi-utility technology test-bed aircraft. *The Aeronautical Journal*, 120(1228):893–909, 2016. doi:10.1017/aer.2016.41.

Burnham, D. C. and J. N. Hallock. Chicago monostatic acoustic vortex sensing system. Technical report, National Information Service, Springfield, Virginia, USA, 1982.

Burris, P. M. and M. A. Bender. Aircraft load alleviation and mode stabilization (LAMS) - B-52 systems analysis, synthesis, and design. Technical report AFFDL-TR-68-161, Air Force Flight Dynamics Laboratory, Wright-Patterson Air Force Base, Ohio, USA, November 1969.

Campbell, C. W. Monte Carlo turbulence simulation using rational approximations to von Kármán spectra. *AIAA Journal*, 24(1):62–66, 1986. doi:10.2514/3.9223.

Castrichini, A., T. Wilson, F. Saltari, F. Mastroddi, N. Viceconti, and J. E. Cooper. Aeroelastics flight dynamics coupling effects of the semi-aeroelastic hinge device. *Journal of Aircraft*, 57(2):1–9, 2020. doi:10.2514/1.C035602.

Cea, A. and R. Palacios. A non-intrusive geometrically nonlinear augmentation to generic linear aeroelastic models. *Journal of Fluids and Structures*, 101:103222, 2021. doi:10.1016/j.jfluidstructs.2021.103222.

Cea, A. and R. Palacios. Assessment of geometrically nonlinear effects on the aeroelastic response of a transport aircraft configuration. *Journal of Aircraft*, 60(1):205–220, January 2023. doi:10.2514/1.C036740.

Cesnik, C. E. S., Aeroelasticity of very flexible aircraft: Prof. Dewey Hodges' three-decade contributions to the field. In Proc. *AIAA Science and Technology Forum and Exposition (SciTech2023)*, National Harbor, Maryland, USA, 23–27 January 2023.

Cesnik, C. E. S. and E. L. Brown. Modeling of high aspect ratio active flexible wings for roll control. In *43rd AIAA/ASME/ASCE/AHS/ASC Structures, Structural Dynamics, and Materials Conference*, Denver, Colorado, USA, 22–25 April 2002. doi:10.2514/6.2002-1719.

Cesnik, C. E. S. and E. L. Brown. Active warping control of a joined-wing airplane configuration. In *44th AIAA/ASME/ASCE/AHS/ASC Structures, Structural Dynamics, and Materials Conference*, Norfolk, Virginia, USA, 7–10 April 2003. doi:10.2514/6.2003-1715.

Cesnik, C. E. S. and W. Su. Nonlinear aeroelastic modeling and analysis of fully flexible aircraft. In *46th AIAA/ASME/ASCE/AHS/ASC Structures, Structural Dynamics, and Materials Conference*, Austin, Texas, 18–21 April 2005. doi:10.2514/6.2005-2169.

Cesnik, C. E. S. and D. H. Hodges. VABS: A new concept for composite rotor blade cross-sectional modeling. *Journal of the American Helicopter Society*, 42(1):27–38, 1997.

Cesnik, C. E. S., D. H. Hodges, and V. G. Sutyrin. Cross-sectional analysis of composite beams including large initial twist and curvature effects. *AIAA Journal*, 34(9):1913–1920, 1996.

Cesnik, C. E. S., P. J. Senatore, W. Su, E. M. Atkins, and C. M. Shearer. X-HALE: A very flexible unmanned aerial vehicle for nonlinear aeroelastic tests. *AIAA Journal*, 50(12):2820–2833, 2012. doi:10.2514/1.J051392.

Cesnik, C. E. S., R. Palacios, and E. Y. Reichenbach. Re-examined structural design procedures for very flexible aircraft. *Journal of Aircraft*, 51(5):1580–1591, 2014. doi:10.2514/1.C032464.

Chevalier, H. L., G. M. Dornfeld, and R. C. Schwanz. An analytical method for predicting the stability and control characteristics of large elastic airplanes at subsonic and supersonic speeds, Part II – Application. In *34th Meeting of the AGARD Flight Mechanics Panel*, Marseille, France, 21–25 April 1969. NATO Advisory Group for Aerospace Research and Development.

Christhilf, D. M., B. Moulin, E. Ritz, P. C. Chen, K. M. Roughen, and B. Perry III. Characteristics of control laws tested on the semi-span supersonic transport (S4T) wind-tunnel model. In *53rd AIAA/ASME/ASCE/AHS/ASC Structures, Structural Dynamics, and Materials Conference*, Waikiki, Hawaii, 23–26 April 2012. doi:10.2514/6.2012-1555.

Cicala, P. Aerodynamic forces on an oscillating profile in uniform stream. *Memorie della Reale Accademia delle Scienze*, II-68:73–98, 1935.

Claverias, S., J. Cerezo, M. A. Torralba, M. Reyes, H. Climent, and M. Karpel. Wake vortex encounter loads numerical simulation. In *International Forum on Aeroelasticity and Structural Dynamics*, Bristol, UK, June 2013.

Collar, A. R. The expanding domain of aeroelasticity, *The Journal of the Royal Aeronautical Society*, 50(428):613–636, August 1946. doi:10.1017/S0368393100120358.

Collar, A. R. The Second Lanchester Memorial Lecture: Aeroelasticity, retrospect and prospect. *The Journal of the Royal Aeronautical Society*, 63(577):1–15, 1959. doi:10.1017/S0368393100070450.

Collar, A. R. The first fifty years of aeroelasticity. *Aerospace*, 5(2):12–20, 1978.

Connolly, J. W., G. Kopasakis, P. Chwalowski, J.R. Carlson, M. D. Sanetrik, W. A. Silva, and J. McNamara. Aero-propulso-elastic analysis of a supersonic transport. *Journal of Aircraft*, 57(4):569–585, July 2020. ISSN 1533-3868. doi:10.2514/1.C035531.

Cook, M. V. *Flight Dynamics Principles: A Linear Systems Approach to Aircraft Stability and Control*. Butterworth-Heinemann, Waltham, Massachusetts, USA, 3rd edition, 2013.

Cook, R. G., R. Palacios, and P. J. Goulart. Robust gust alleviation and stabilization of very flexible aircraft. *AIAA Journal*, 51(2):330–340, 2013. doi:10.2514/1.j051697.

Cook, W. H. *The Road to the 707*. TYC Publishing Co., Bellevue, Washington, USA, 1991.

Cornman, L. B., C. S. Morse, and G. Cunning. Real-time estimation of atmospheric turbulence severity from in-situ aircraft measurements. *Journal of Aircraft*, 31(1):171–177, 1995. doi:10.2514/3.46697.

Cottet, G.-H. and P. D. Koumoutsakos. *Vortex Methods: Theory and Practice*. Cambridge University Press, Cambridge, UK, 2000.

Cox, H. R. and A. G. Pugsley. Theory of loss of lateral control due to wing twisting. Reports and Memoranda 1506, British Advisory Committee for Aeronautics, 1932.

Crisfield, M. A. and G. Jelenic. Objectivity of strain measures in the geometrically exact three-dimensional beam theory and its finite-element implementation. *Proceedings of the Royal Society of London. Series A: Mathematical, Physical and Engineering Sciences*, 455(1983): 1125–1147, 1999. doi:10.1098/rspa.1999.0352.

Da Ronch, A., A. J. McCracken, K. J. Badcock, M. Widhalm, and M. S. Campobasso. Linear frequency domain and harmonic balance predictions of dynamic derivatives. *Journal of Aircraft*, 50(3):694–707, 2013. doi:10.2514/1.C031674.

Dawson, S. T. M. and S. L. Brunton. Improved approximations to Wagner function using sparse identification of nonlinear dynamics. *AIAA Journal*, 60(3):1–17, 2021. doi:10.2514/1 .j060863.

De Marco, A., E. Duke, and J. Berndt. A general solution to the aircraft trim problem. In *AIAA Modeling and Simulation Technologies Conference and Exhibit*, Hilton Head, South Carolina, USA, 20–23 August 2007. doi:10.2514/6.2007-6703.

del Carre, A. and R. Palacios. Low-altitude dynamics of very flexible aircraft. In *AIAA Science and Technology Forum and Exposition*, San Diego, California, USA, 7–11 January 2019. AIAA Paper No. 2019–2038.

del Carre, A. and R. Palacios. Simulation and optimization of takeoff maneuvers of very flexible aircraft. *Journal of Aircraft*, 57(6):1097–1110, November 2020. doi:10.2514/1.C035901.

del Carre, A., A. Muñoz-Simón, N. Goizueta, and R. Palacios. SHARPy: A dynamic aeroelastic simulation toolbox for very flexible aircraft and wind turbines. *Journal of Open Source Software*, 4(44):1885, December 2019a. doi:10.21105/joss.01885.

del Carre, A., P. Teixeira, R. Palacios, and C. E. S. Cesnik. Nonlinear response of a very flexible aircraft under lateral gust. In *International Forum on Aeroelasticity and Structural Dynamics*, Savannah, Georgia, USA, 10–13 June 2019b. IFASD Paper 2019-090.

Derkevorkian, A., S. F. Masri, J. Alvarenga, H. Boussalis, J. Bakalyar, and W. L. Richards. Strain-based deformation shape-estimation algorithm for control and monitoring applications. *AIAA Journal*, 51(9):2231–2240, 2013. doi:10.2514/1.J052215.

Deskos, G., A. del Carre, and R. Palacios. Assessment of low-altitude atmospheric turbulence models for aircraft aeroelasticity. *Journal of Fluids and Structures*, 95:102981, May 2020. doi:10.1016/j.jfluidstructs.2020.102981.

Diehl, M., L. Magni, and G. De Nicolao. Efficient NMPC of unstable periodic systems using approximate infinite horizon closed loop costing. *Annual Reviews in Control*, 28(1):37–45, January 2004. doi:10.1016/j.arcontrol.2004.01.011.

Dillsaver, M. J., C. E. S. Cesnik, and I. V. Kolmanovsky. Gust load alleviation control for very flexible aircraft. In *AIAA Atmospheric Flight Mechanics Conference*, Portland, Oregon, USA, August 2011. doi:10.2514/6.2011-6368.

Dillsaver, M. J., C. E. S. Cesnik, and I. V. Kolmanovsky. Gust response sensitivity characteristics of very flexible aircraft. In *AIAA Atmospheric Flight Mechanics Conference*, Minneapolis, Minnesota, 08–11 August 2012. doi:10.2514/6.2012-4576.

Disney, T. E. The C-5A active load alleviation system. In *AIAA Aircraft Systems and Technology Meeting*, Los Angeles, California, USA, 4–7 August 1975. American Institute of Aeronautics and Astronautics. AIAA Paper 75–991.

Dizy, J., R. Palacios, and S. T. Pinho. Homogenisation of slender periodic composite structures. *International Journal of Solids and Structures*, 50:1473–1481, May 2013. doi:10.1016/j.ijsolstr.2013.01.017.

Dowell, E. H., A. R. Dusto, and K. C. Hall. Eigenmode analysis in unsteady aerodynamics: Reduced order models. *Applied Mechanics Reviews*, 50(6):371–387, 1997. doi:10.1115/1.3101718.

Dowell, E. H. and D. Tang. *Dynamics of Very High Dimensional Systems*. World Scientific Publishing Company, Singapore, 2003.

Dowell, E. H., R. Clark, D. Cox, H. C. Curtis Jr., J. W. Edwards, K. H. Hall, D. A. Peters, R. Scanlan, E. Simiu, F. Sisto, and T. W. Strganac. *A Modern Course in Aeroelasticity*. Kluwer Academic Publishers, Dordrecht, The Netherlands, 4th edition, 2004.

Doyle, J. Guaranteed margins for LQG regulators. *IEEE Transactions on Automatic Control*, 23 (4):756–757, 1978. doi:10.1109/TAC.1978.1101812.

Drela, M. Integrated simulation model for preliminary aerodynamic, structural, and control-law design of aircraft. In *40th AIAA/ASME/ASCE/AHS/ASC Structures, Structural Dynamics, and Materials Conference*, St. Louis, Missouri, USA, 12–15 April 1999. doi:10.2514/6.1999-1394.

Drela, M. *Flight Vehicle Aerodynamics*. The MIT Press, Cambridge, Massachusetts, USA, 2014.

Drewiacki, D., F. J. Silvestre, and A. B. Guimarães Neto. Influence of airframe flexibility on pilot-induced oscillations. *Journal of Guidance, Control, and Dynamics*, 42(7):1537–1550, 2019. doi:10.2514/1.G004024.

Duessler, S., N. Goizueta, A. Muñoz-Simón, and R. Palacios. Modelling and numerical enhancements on a UVLM for nonlinear aeroelastic simulation. In *AIAA Science and Technology Forum and Exposition*, San Diego, California, USA, 3–7 January 2022. doi:10.2514/6.2021-0363.

Dullerud, G. E. and F. Paganini. *A Course in Robust Control Theory: A Convex Approach*, volume 36. Springer Science & Business Media, New York, USA, 2013.

Duncan, W. J. A suggested investigation on wing flutter. A.R.C. Report 4281 (O.159), November 1939.

Duncan, W. J. and A. R. Collar. Calculations of the resistance derivatives of flutter theory. Reports and Memoranda 1500, British Advisory Committee for Aeronautics, 1932.

Duncan, W. J. and G. A. McMillan. Reversal of aileron control due to wing twist. Reports and Memoranda 1499, British Advisory Committee for Aeronautics, 1932.

Dusto, A. R. An analytical method for predicting the stability and control characteristics of large elastic airplanes at subsonic and supersonic speeds, part I – Analysis. In *34th Meeting of the AGARD Flight Mechanics Panel*, Marseille, France, 21–25 April 1969. NATO Advisory Group for Aerospace Research and Development.

EASA. *Certification Specifications for Large Aeroplanes (CS-25)*, 2017. European Aviation Safety Agency.

Eichenbaum, F. D. A general theory of aircraft response to three-dimensional turbulence. *Journal of Aircraft*, 8(5):353–360, 1971. doi:10.2514/3.59108.

Enns, D. F. Model reduction with balanced realization: An error bound and a frequency weighted generalization. In *Proceedings of the 23rd Conference on Decision and Control*, Las Vegas, Nevada, USA, December 1984. doi:10.1109/CDC.1984.272286.

Erickson, A. L. and R. L. Mannes. Wind-tunnel investigation on transonic aileron flutter. Technical report, National Advisory Commitee for Aeronautics, Moffett Field, California, USA, 1949. NACA-RM-A9B28.

Etkin, B. *Dynamics of Atmospheric Flight*. Addison-Wesly Aerospace Series. John Wiley & Sons, New York, New York, USA, 1972.

Etkin, B. Turbulent wind and its effect on flight. *Journal of Aircraft*, 18(5):327–345, 1981. doi:10.2514/3.57498.

Eversman, W. and A. Tewari. Consistent rational-function approximation for unsteady aerodynamics. *Journal of Aircraft*, 28(9):545–552, September 1991. doi:10.2514/3.46062.

FAA. *Airworthiness Standards: Transport Category Airplanes. U.S. Code of Federal Regulations, 14 CFR Part 25, Appendix G*. Federal Aviation Administration, Government Printing Office, Washington, DC, USA, 2011.

Farhat, C. CFD-based nonlinear computational aeroelasticity. In E. Stein, R. de Borst, and T. J. R. Hughes, editors, *Encyclopedia of Computational Mechanics*, chapter 13. John Wiley & Sons, Chichester, England, UK, November 2004. doi:10.1002/0470091355.ecm063.

Farhat, C., K. G. van der Zee, and P. Geuzaine. Provably second-order time-accurate loosely-coupled solution algorithms for transient nonlinear computational aeroelasticity. *Computer Methods in Applied Mechanics and Engineering*, 195(17–18):1973–2001, 2006. doi:10.1016/j.cma.2004.11.031.

Fellowes, A., T. Wilson, G. Kemble, C. Havill, and J. Wright. Wing box non-linear structural damping. In *International Forum on Aeroelasticity and Structural Dynamics*, Paris, France, 27–29 June 2011. IFASD Paper 2011–11.

Fournier, H., P. Massioni, M. Tu Pham, L. Bako, R. Vernay, and M. Colombo. Robust gust load alleviation of flexible aircraft equipped with LIDAR. *Journal of Guidance, Control, and Dynamics*, 45(1):58–72, 2022. doi:10.2514/1.G006084.

Franklin, G. F., J. D. Powell, and M. L. Workman. *Digital Control of Dynamic Systems*. Addison-Wesley, Menlo Park, California, USA, 3rd edition, 1998.

Frazer, R. A. and W. J. Duncan. The flutter of aeroplane wings. Reports and Memoranda 1155, British Advisory Committee for Aeronautics, August 1928.

Frederick, M., E. C. Kerrigan, and J. M. R. Graham. Gust alleviation using rapidly deployed trailing-edge flaps. *Journal of Wind Engineering and Industrial Aerodynamics*, 98(12):712–723, 2010. doi:10.1016/j.jweia.2010.06.005.

Friedmann, P. P. The renaissance of aeroelasticity and its future. *Journal of Aircraft*, 36(1):105–121, 1999.

Friswell, M. I. and J. E. Mottershead. *Finite Element Model Updating in Structural Dynamics*. Kluwer Academic Publishers, Dordrecht, The Netherlands, 1995.

Fung, Y. C. *An Introduction to the Theory of Aeroelasticity*. Courier Dover Publications, Mineola, New York, USA, 2008.

Gage, S. Creating a unified graphical wind turbulence model from multiple specifications. In *AIAA Modeling and Simulation Technologies Conference and Exhibit*, Austin, Texas, USA, August 2003. doi:10.2514/6.2003-5529.

Gangsaas, D., U. Ly, and D. Norman. Practical gust load alleviation and flutter suppression control laws based on a LQG methodology. In *19th Aerospace Sciences Meeting*, St Louis, Missouri, USA, January 1981. doi:10.2514/6.1981-21.

Garrick, I. E. On some reciprocal relations in the theory of nonstationary flows. Technical Note TN 629, N.A.C.A., 1938.

Garrick, I. E. *Nonsteady Wing Characteristics, Division F., Vol. VII High Speed Aerodynamics and Jet Propulsion; Aerodynamic Components of Aircraft at High Speeds, Eds. Donovan, AF and Lawrence, HR*. Princeton University Press, Princeton, New Jersey, USA, 1957.

Garrick, I. E. and W. H. Reed. Historical development of aircraft flutter. *Journal of Aircraft*, 18(11):897–912, 1981.

Gaunaa, M. Unsteady 2D potential-flow forces on a thin variable geometry airfoil undergoing arbitrary motion. Technical report Risø-R-1478(EN), Risø National Laboratory, Denmark, 2006.

Gawronski, W. K. *Advanced Structural Dynamics and Active Control of Structures*. Springer-Verlag, New York, New York, USA, 2004.

Gawronski, W. K. and J. N. Juang. Model reduction in limited time and frequency intervals. *International Journal of Systems Science*, 21(2):349–376, 1990. doi:10.1080/00207729008910366.

Géradin, M. and A. Cardona. *Flexible Multibody Dynamics: A Finite Element Approach*. Chichester, UK John Wiley & Sons, 2001.

Géradin, M. and D. Rixen. *Mechanical Vibrations – Theory and Applications to Structural Dynamics*. John Wiley & Sons, Chichester, UK 2nd edition, 1997.

Glauert, H. Theoretical relationships for the lift and drag of an aerofoil structure. *The Journal of the Royal Aeronautical Society*, 27:512–518, 1923.

Glauert, H. The force and moment of an oscillating aerofoil. Reports and Memoranda 1242, British Advisory Committee for Aeronautics, 1929.

Goizueta, N., A. Wynn, and R. Palacios. Parametric Krylov-based order reduction of aircraft aeroelastic models. In *AIAA Science and Technology Forum and Exposition*, Nashville, Tennessee, USA, 4–8 January 2021.

Goizueta, N., A. Wynn, R. Palacios, A. Drachinsky, and D. E. Raveh. Flutter predictions for very flexible wing wind tunnel test. *Journal of Aircraft*, 59(4):1082–1097, July 2022, doi:10.2514/1.C036710.

Goland, M. The flutter of a uniform cantilever wing. *Journal of Applied Mechanics*, 12(4):197–208, December 1945.

Goldstein, H., C. P. Poole, and J. Safko. *Classical Mechanics*. Pearson Education, Third Edition, 2014, London, UK, 2011.

Gonzalez-Salcedo, A., M. Aparicio-Sanchez, X. Munduate, R. Palacios, J. M. R. Graham, O. Pires, and B. Mendez. A computationally-efficient panel code for unsteady airfoil modelling including dynamic stall. In *35th Wind Energy Symposium*, Grapevine, Texas, USA, 9–13 January 2017. doi:10.2514/6.2017-2000.

Grauer, J. A. and M. Boucher. System identification of flexible aircraft: lessons learned from the X-56A phase 1 flight tests. In *AIAA Science and Technology Forum and Exposition*, Orlando, Florida, USA, 6–10 January 2020. doi:10.2514/6.2020-1017.

Greenwood, D. T. *Principles of Dynamics*. Prentice-Hall, Upper Saddle River, New Jersey, USA, 1988.

Grouas, J. A very large aircraft, a challenging project for aeroelastics and loads. In *International Forum on Aeroelasticity and Structural Dynamics*, Madrid, Spain, 5–7 June 2001.

Gugercin, S. and A. C. Antoulas. A survey of model reduction by balanced truncation and some new results. *International Journal of Control*, 77(8):748–766, 2004. doi:10.1080/00207170410001713448.

Guimarães Neto, A. B., R. G. A. Silva, P. Paglione, and F. J. Silvestre. Formulation of the flight dynamics of flexible aircraft using general body axes. *AIAA Journal*, 54(11):3516–3534, 2016. doi:10.2514/1.j054752.

Haghighat, S., H. H. T. Liu and J. R. R. A. Martins. Model-predictive gust load alleviation controller for a highly flexible aircraft. *Journal of Guidance, Control, and Dynamics*, 35(6):1751–1766, November 2012a. doi:10.2514/1.57013.

Haghighat, S., J. R. R. A. Martins, and H. H. T. Liu. Aeroservoelastic design optimization of a flexible wing. *Journal of Aircraft*, 49(2):432–443, 2012b. doi:10.2514/1.C031344.

Hahn, K.-U. and R. König. ATTAS flight test and simulation results of the advanced gust management system LARS. In *AIAA Guidance, Navigation, and Control Conference*, Hilton Head, South Carolina, USA, 10–12 August 1992, AIAA Paper 92-4343-CP.

Hahn, K.-W. and C. Schwarz. Alleviation of atmospheric flow disturbance effects on aircraft response. In *26th International Congress of the Aeronautical Sciences (ICAS)*, Anchorage, Alaska, USA, 14–19 September 2008.

Hall, K. C. Eigenanalysis of unsteady flows about airfoils, cascades and wings. *AIAA Journal*, 32(12):2426–2432, December 1994. doi:10.2514/3.12309.

Halsey, S. A., R. M. Goodall, B. D. Caldwell, and J. T. Pearson. Filtering structural modes in aircraft: Notch filters vs. Kalman filters. In *16th IFAC World Congress*, 38(1):205–210, 2005. doi:10.3182/20050703-6-CZ-1902.00255.

Hansen, J. H., M. Duan, I. Kolmanovsky, and C. E. S. Cesnik. Control allocation for maneuver and gust load alleviation of flexible aircraft. In *AIAA Science and Technology Forum and Exposition*, Orlando, Florida, USA, 6–10 January 2020. doi:10.2514/6.2020-1186.

Hariharan, N. and J. G. Leishman. Unsteady aerodynamics of a flapped airfoil in subsonic flow by indicial concepts. *Journal of Aircraft*, 33(5):855–868, 1996.

Hassard, B. D., N.D. Kazarinoff, and Y. H. Wan. *Theory and Applications of Hopf Bifurcation*. Cambridge Tracts in Mathematics. Cambridge University Press, Cambridge, UK, 1981.

Hassig, H. J. An approximate true damping solution of the flutter equation by determinant iteration. *Journal of Aircraft*, 8(11):885–889, 1971. doi:10.2514/3.44311.

Hesse, H. *Consistent Aeroelastic Linearisation and Reduced-order Modelling in the Dynamics of Manoeuvring Flexible Aircraft*. PhD thesis, Imperial College London, UK, 2013.

Hesse, H. and R. Palacios. Consistent structural linearisation in flexible-body dynamics with large rigid-body motion. *Computers & Structures*, 110–111:1–14, November 2012. doi:10.1016/j.compstruc.2012.05.011.

Hesse, H. and R. Palacios. Reduced-order aeroelastic models for dynamics of maneuvering flexible aircraft. *AIAA Journal*, 52(8):1717–1732, 2014. doi:10.2514/1.j052684.

Hesse, H. and R. Palacios. Dynamic load alleviation in wake vortex encounters. *Journal of Guidance, Control, and Dynamics*, 39(4):801–813, 2016. doi:10.2514/1.G000715.

Hoadley, S. T. and M. Karpel. Application of aeroservoelastic modeling using minimum-state unsteady aerodynamic approximations. *Journal of Guidance, Control, and Dynamics*, 14(6): 1267–1276, November 1991. doi:10.2514/3.20783.

Hoblit, F. M. *Gust Loads on Aircraft: Concepts and Applications*. AIAA Education Series, Reston, Virginia, USA, 1988.

Hodges, D. H. A mixed variational formulation based on exact intrinsic equations for dynamics of moving beams. *International Journal of Solids and Structures*, 26(11):1253–1273, 1990. doi:10.1016/0020-7683(90)90060-9.

Hodges, D. H. *Nonlinear Composite Beam Theory*. American Institute of Aeronautics and Astronautics, Reston, Virginia, USA, January 2006. doi:10.2514/4.866821.

Hodges, D. H. and G. A. Pierce. *Introduction to Structural Dynamics and Aeroelasticity*. Cambridge University Press, New York, New York, USA, 2nd edition, 2014.

Hodges, D. H. Geometrically exact, intrinsic theory for dynamics of curved and twisted anisotropic beams. *AIAA Journal*, 41(6):1131–7, 2003. doi:10.2514/2.2054.

Hönlinger, H., H. Zimmermann, O. Sensburg, and J.. Becker. Structural aspects of active control technology. In *AGARD-CP-560*, Turin, Italy, 9–13 May 1994. NATO Advisory Group for Aerospace Research and Development.

Horn, R. A. and C. R. Johnson. *Matrix Analysis*. Cambridge University Press, Cambridge, UK, 2012.

Houbolt, J. C. Atmospheric turbulence. *AIAA Journal*, 11(4):421–437, 1973.

Hsiao, K. M., J. Y. Lin, and W. Y. Lin. A consistent co-rotational finite element formulation for geometrically nonlinear dynamic analysis of 3-D beams. *Computer Methods in Applied Mechanics and Engineering*, 169(1–2):1–18, 1999. doi:10.1016/S0045-7825(98)00152-2.

Hunsaker, J. C. and E. B. Wilson. Report on behavior of aeroplanes in gusts. Technical report, National Advisory Committee for Aeronautics, 1917.

ICAO. *Manual of the ICAO Standard Atmosphere*, 3rd edition, 1993. International Civil Aviation Organization, Doc 7488-CD.

Jacobson, S., R. T. Britt, D. Freim, and P. Kelly. Residual pitch oscillation (RPO) flight test and analysis on the B-2 bomber. In *39th AIAA/ASME/ASCE/AHS/ASC Structures, Structural Dynamics, and Materials Conference*, Long Beach, California, USA, 20–23 April 1998. doi:10.2514/6.1998-1805.

Johnson, W. *Theory of Helicopter*. Princeton University Press, Mineola, New York, USA, 1980.

Jones, J. G. Measured statistics of multicomponent gust patterns in atmospheric turbulence. *Journal of Aircraft*, 44(5):1559–1567, 2007.

Jones, J. R. and C. E. S. Cesnik. Nonlinear aeroelastic analysis of the X-56A multi-utility aeroelastic demonstrator. In *15th AIAA Dynamics Specialists Conference*, San Diego, California, USA, 4–8 January 2016. doi:10.2514/6.2016-1799.

Jones, J. R. *Development of a Very Flexible Testbed Aircraft for the Validation of Nonlinear Aeroelastic Codes*. PhD thesis, University of Michigan, Ann Arbor, Michigan, USA 2017.

Jones, J. R. and C. E. S. Cesnik. Preliminary flight test correlations of the X-HALE aeroelastic experiment. *Aeronautical Journal*, 119(1217):855–870, 2015. doi:10.1017/S0001924000010952.

Jones, R. T. Operational treatment of the nonuniform lift theory to airplane dynamics. Technical Note 667, N.A.C.A., 1938.

Joseph, C. and R. Mohan. Closed-form expressions of lift and moment coefficients for generalized camber using thin-airfoil theory. *AIAA Journal*, 59(10):1–7, 2021. doi:10.2514/1.j060859.

Joukowski, N. On the flight of birds (in Russian). *Obshchestvo liubitelei estestvoznaniia, antropologit i etnografii*, 73:29–43, 1891.

Juang, J. N. and R. S. Pappa. An eigensystem realization algorithm for modal parameter identification and model reduction. *Journal of Guidance, Control, and Dynamics*, 8(5):620–627, 1985. doi:10.2514/3.20031.

Kaimal, J. C., J. C. Wyngaard, Y. Izumi, and O. R. Coté. Spectral characteristics of surface-layer turbulence. *Quarterly Journal of the Royal Meteorological Society*, 98(417):563–589, 1972. doi:10.1002/qj.49709841707.

Kang, L. H., D. K. Kim, and J. H. Han. Estimation of dynamic structural displacements using fiber Bragg grating strain sensors. *Journal of Sound and Vibration*, 305(3):534–542, 2007.

Karpel, M. Design for active flutter suppression and gust alleviation using state-space aeroelastic modeling. *Journal of Aircraft*, 19(3):221–227, 1982. doi:10.2514/3.57379.

Karpel, M. Time-domain aeroservoelasticity modeling using weighted unsteady aerodynamic forces. *Journal of Guidance, Navigation and Control*, 13(1):30–37, January 1990.

Karpel, M. Procedures and models for aeroservoelastic analysis and design. *Zeitschrift fur Angewandte Mathematik und Mechanik*, 81(9):579–92, 2001. doi:10.1002/1521-4001(200109)81:9<579::AID-ZAMM579>3.0.CO;2-Z.

Karpel, M. and L. Brainin. Stress considerations in reduced-size aeroelastic optimization. *AIAA Journal*, 33(4):716–722, 1995. doi:10.2514/3.12447.

Karpel, M. and Z. Sheena. Structural optimization for aeroelastic control effectiveness. *Journal of Aircraft*, 26(5):493–495, May 1989. doi:10.2514/3.45791.

Karpel, M., S. Yaniv, and D. S. Livshits. Integrated solution for computational static aeroelastic problems. In *AIAA, NASA, and ISSMO Symposium on Multidisciplinary Analysis and Optimization*, Bellevue, Washington, USA, 1996. doi:10.2514/6.1996-4012.

Karpel, M., B. Moulin, E. Presente, L. Anguita, C. Maderuelo, and H. Climent. Dynamic gust loads analysis for transport aircraft with nonlinear control effects. In *49th AIAA/ASME/ASCE/AHS/ASC Structures, Structural Dynamics, and Materials Conference*, Schaumburg, Illinois, USA, 7–10 April 2008. doi:10.2514/6.2008-1994.

Katz, J. and A. Plotkin. *Low-Speed Aerodynamics*. Cambridge Aerospace Series. Cambridge University Press, New York, New York, USA, 2nd edition, 2001.

Kayran, A. Küssner's function in the sharp-edged gust problem–A correction. *Journal of Aircraft*, 43(5):1596–1598, 2006. doi:10.2514/1.2029.

Kehoe, M. W. A historical overview of flight flutter testing. Technical Memorandum TM-4720, NASA, 1995.

Kelley, N. D. and B. J. Jonkman. Overview of the TurbSim stochastic inflow turbulence simulator. Technical report NREL/TP-500-41137, National Renewable Energy Laboratory 2007.

Kennedy, G. J., G. K. W. Kenway, and J. R. R. A. Martins. High aspect ratio wing design: Optimal aerostructural tradeoffs for the next generation of materials. In *AIAA Science and Technology Forum and Exposition*, National Harbor, Maryland, USA, 13–17 January 2014. doi:10.2514/6.2014-0596.

Kier, T. M. An integrated loads analysis model including unsteady aerodynamic effects for position and attitude dependent gust fields. In *International Forum on Aeroelasticity and Structural Dynamics*, Paris, France, 27–29 June 2011. IFASD Paper 2011-052.

Kier, T. M. An integrated loads analysis model for wake vortex encounters. In *International Forum on Aeroelasticity and Structural Dynamics*, Bristol, UK, 24–26 June 2013.

Kim, K. and T. W. Strganac. Nonlinear responses of a cantilever wing with a external store. In *44th AIAA/ASME/ASCE/AHS Structures, Structural Dynamics, and Materials Conference*, Norfolk, Virginia, USA, 7–10 April 2003. doi:10.2514/6.2003-1708.

Kirk, D. E. *Optimal Control Theory*. Prentice-Hall, Englewood Cliffs, New Jersey, USA, 1970.

Kitson, R. C., C. A. Lupp, and C. E. S. Cesnik. Modeling and simulation of flexible jet transport aircraft with high-aspect-ratio wings. In *15th Dynamics Specialists Conference*, San Diego, California, USA, 4–8 January 2016. doi:10.2514/6.2016-2046.

Klöckner, A., M. Leitner, D. Schlabe, and G. Looye. Integrated modelling of an unmanned high-altitude solar-powered aircraft for control law design analysis. In *2nd CEAS Specialist Conference on Guidance, Navigation & Control*, The Netherlands, 12–14 April 2013.

Ko, J., T. W. Strganac, and A. J. Kurdila. Stability and control of a structurally nonlinear aeroelastic system. *Journal of Guidance, Control, and Dynamics*, 21(5):718–725, 1998.

Kopf, M., E. Bullinger, H. G. Giesseler, S. Adden, and R. Findeisen. Model predictive control for aircraft load alleviation: Opportunities and challenges. In *2018 Annual American Control Conference*, pages 2417–2424, Milwaukee, Minnesota, USA, 27–29 June 2018. doi:10.23919/ACC.2018.8430956.

Kraft Jr, C. C. Initial results of a flight investigation on a gust-alleviation system. N.A.C.A. Technical Note 3612, Washington, DC, USA, April 1956.

Kroll, N., M. Abu-Zurayk, D. Dimitrov, T. Franz, T. Führer, T. Gerhold, S. Görtz, R. Heinrich, C. Ilic, J. Jepsen, J. Jägersküpper, M. Kruse, A. Krumbein, S. Langer, D. Liu, R. Liepelt, L. Reimer, M. Ritter, A. Schwöppe, J. Scherer, F. Spiering, R. Thormann, V. Togiti,

D. Vollmer, and J. H. Wendisch. DLR project Digital-X: Towards virtual aircraft design and flight testing based on high-fidelity methods. *CEAS Aeronautical Journal*, 7(1):3–27, 2016. doi:10.1007/s13272-015-0179-7.

Krzysiak, A. and J. Narkiewicz. Aerodynamic loads on airfoil with trailing-edge flap pitching with different frequencies. *Journal of Aircraft*, 43(2):407–418, 2006.

Kuchemann, D. A simple method for calculating the span and chordwise loading on straight and swept wings of any given aspect ratio at subsonic speeds. Technical report, Aeronautical Research Council, 1952. R.A.E. Report No. 2476.

Kumar, D. and C. E. S. Cesnik. Performance enhancement in dynamic stall condition using active camber deformation. *Journal of American Helicopter Society*, 60(2):1–12, 2015. doi:10.4050/JAHS.60.022001.

Küssner, H. G. Schwingungen von Flugzeugflugeln (Flutter of aircraft wings). *Luftfahrt-forschung*, 4:41–62, June 1929.

Küssner, H. G. Zusammenfassender Bericht Über den instationären Auftrieb von Flügeln (Comprehensive report on the non-stationary lift of wings). *Luftfahrtforschung*, 13(12):410–424, 1936.

Kwakernaak, H. and R. Sivan. *Linear Optimal Control Systems*. John Wiley & Sons, New York, New York, USA, 1972.

Lahooti, M., R. Palacios, and S. J. Sherwin. Thick strip method for efficient large-eddy simulations of flexible wings in stall. In *AIAA Science and Technology Forum and Exposition*, Nashville, Tennessee, USA, 4–8 January 2021. doi:10.2514/6.2021-0363.

Lanchester, F. W. *Aerodonetics*. Archibald Constable & Co. Ltd., London, 1908.

Lanchester, F. W. Torsional vibrations of the tail of aeroplane. Reports and Memoranda 276, British Advisory Committee for Aeronautics, 1916.

Lavretsky, E. and K. A. Wise. *Robust and Adaptive Control*. Advanced Textbooks in Control and Signal Processing. Springer London, London, UK 2013. doi:10.1007/978-1-4471-4396-3.

Le, K. C. *Vibrations of Shells and Rods*. Springer-Verlag, Berlin, Germany 1999.

Lee, T. and S. Basu. Measurement of unsteady boundary layer developed on an oscillating airfoil using multiple hot-film sensors. *Experiments in Fluids*, 25(2):108–117, 1998. doi:10.1007/s003480050214.

Leishman, J. G. *Principles of Helicopter Aerodynamics*. Cambridge Aerospace Series. Cambridge University Press, New York, New York USA, 2nd edition, 2006.

Levien, R. The elastica: A mathematical history. Technical report UCB/EECS-2008-103, EECS Department, University of California, Berkeley, California, USA August 2008.

Lewis, F. L., L. Xie, and D. Popa. *Optimal and Robust Estimation: With an Introduction to Stochastic Control Theory*. Automation and Control Engineering. Taylor & Francis, Boca Raton, Florida, USA 2nd edition, 2008.

Li, D., S. Guo, and J. Xiang. Aeroelastic dynamic response and control of an airfoil section with control surface nonlinearities. *Journal of Sound and Vibration*, 329(22):4756–4771, 2010. doi:10.1016/j.jsv.2010.06.006.

Li, H. and K. Ekici. A novel approach for flutter prediction of pitch–plunge airfoils using an efficient one-shot method. *Journal of Fluids and Structures*, 82:651–671, 2018. doi:10.1016/j.jfluidstructs.2018.08.012.

Li, H. and K. Ekici. Aeroelastic modeling of the AGARD 445.6 wing using the harmonic-balance-based one-shot method. *AIAA Journal*, 57(11):4885–4902, 2019. doi:10.2514/1.J058363.

Lind, R. and M. Brenner. *Robust Aeroservoelastic Stability Analysis: Flight Test Applications*. Advances in Industrial Control. Springer-Verlag, London, UK, 2012.

Livne, E. Aircraft active flutter suppression: State of the art and technology maturation needs. *Journal of Aircraft*, 55(1):410–452, 2018. doi:10.2514/1.C034442.

Livne, E. and T. Weisshaar. Aeroelasticity of nonconventional airplane configurations–Past and future. *Journal of Aircraft*, 40(6):1047–1065, 2003. doi:10.2514/2.7217.

Lockyer, A. J., A. Drake, J. Bartley-Cho, E. Vartio, D. Solomon, and T. Shimko. High lift over drag active (HiLDA) wing. Technical report, U.S. Air Force Research Laboratories, Wright Patterson Air Force Base, Ohio, USA 2005. AFRL-VA-WP-TR-2005-3066.

Lomax, T. L. *Structural Loads Analysis for Commercial Transport Aircraft: Theory and Practice*. AIAA Education Series. American Institute of Aeronautics and Astronautics, Reston, Virginia, USA, 1996.

Love, A. E. H. *A Treatise on the Mathematical Theory of Elasticity*. Cambridge University Press, Cambridge, UK, 1927.

Love, M. H., P. S. Zink, P. A. Wieselmann, and H. Youngren. Body freedom flutter of high aspect ratio flying wings. In *46th AIAA/ASME/ASCE/AHS/ASC Structures, Structural Dynamics, and Materials Conference*, Austin, Texas, USA, 18–21 April 2005. doi:10.2514/6.2005-1947.

Lu, K. J. and S. Kota. Design of compliant mechanisms for morphing structural shapes. *Journal of Intelligent Material Systems and Structures*, 14(6):379–391, 2003. doi:10.1177/1045389X03035563.

Lucia, D. J. The SensorCraft configurations: A non-linear aeroservoelastic challenge for aviation. In *46th AIAA/ASME/ASCE/AHS/ASC Structures, Structural Dynamics, and Materials Conference*, pages 1768–1774, Austin, Texas, USA, 18–21 April 2005. AIAA Paper 2005-1943.

Lupp, C. A. and C. E. S. Cesnik. A gradient-based flutter constraint including geometrically nonlinear deformations. In *AIAA Science and Technology Forum and Exposition*, San Diego, California, USA, 7–11 April 2019. doi:10.2514/6.2019-1212.

MacNeal, R. H. and C. W. McCormick. The NASTRAN computer program for structural analysis. *Computers & Structures*, 1:389–412, 1972.

Magni, L., D. M. Raimondo, and F. Allgöwer, editors. *Nonlinear Model Predictive Control*, volume 384 of *Lecture Notes in Control and Information Sciences*. Springer, Berlin, Germany, 2009. doi:10.1007/978-3-642-01094-1.

Mann, J. The spatial structure of neutral atmospheric surface-layer turbulence. *Journal of Fluid Mechanics*, 273:141–168, 1994. doi:10.1017/S0022112094001886.

Mann, J. Wind field simulation. *Probabilistic Engineering Mechanics*, 13(4):269–282, 1998. doi:10.1016/S0266-8920(97)00036-2.

Manwell, J. F., J. G. McGowan, and L. Anthony. *Wind Energy Explained: Theory, Design and Application*. John Wiley & Sons, Chichester, UK, 2nd edition, 2009.

Maraniello S. and R. Palacios. Optimal rolling maneuvers with very flexible wings. *AIAA Journal*, 55(9):2964–2979, 2017. doi:10.2514/1.J055721.

Maraniello, S. and R. Palacios. State-space realizations and internal balancing in potential-flow aerodynamics with arbitrary kinematics. *AIAA Journal*, 57(6):2308–2321, 2019. doi:10.2514/1.J058153.

Maraniello, S. and R. Palacios. Parametric reduced-order modeling of the unsteady vortex-lattice method. *AIAA Journal*, 58(5):2206–2220, 2020. doi:10.2514/1.j058894.

Marsden, J. E. and T. J. R. Hughes. *Mathematical Foundations of Elasticity*. Prentice-Hall, Englewood Cliffs, New Jersey, USA, 1983

Martinez, J. R., P. M. Flick, J. Perdzock, G. Dale, and M. B. Davis. An overview of SensorCraft capabilities and key enabling technologies. In *AIAA Applied Aerodynamics Conference*, Honolulu, Hawaii, USA, August 2008. doi:10.2514/6.2008-7185.

Matsuzaki, Y., T. Ueda, Y. Miyazawa, and H. Matsushita. Gust load alleviation of a transport-type wing: Test and analysis. *Journal of Aircraft*, 26(4):322–327, 1989. doi:10.2514/3.45763.

Mayo, A. J. and A. C. Antoulas. A framework for the solution of the generalized realization problem. *Linear Algebra and Its Applications*, 425(2–3):634–662, 2007. doi:10.1016/j.laa.2007.03.008.

McCroskey, W. J. Unsteady airfoils. *Annual Review of Fluid Mechanics*, 14(1):285–311, 1982.

Meirovitch, L. Hybrid state equations of motion for flexible bodies in terms of quasi-coordinates. *Journal of Guidance, Control, and Dynamics*, 14(5):1008–1013, 1991. doi:10.2514/3.20743.

Meirovitch, L. and I. Tuzcu. Unified theory for the dynamics and control of maneuvering flexible aircraft. *AIAA Journal*, 42(4):714–727, April 2004. doi:10.2514/1.1489.

MIL-HDBK-1797. *Flying Qualities of Piloted Aircraft*. U.S. Department of Defense, December 1997.

MIL-HDBK-516B. *Airworthiness Certification Criteria*. U.S. Department of Defense, 2005.

MIL-STD-1540C. *Test Requirements for Launch, Upper-Stage, and Space Vehicles*. U.S. Department of Defense, September 1994.

Millikan, W. F. Progress in dynamic stability and control research. *Journal of the Aeronautical Sciences*, 14(9):493–519, 1947. doi:10.2514/8.1434.

Milne, R. D. Dynamics of the deformable aeroplane, Parts I and II. Report 3345, Aeronautical Research Council, London, UK 1962.

Milne, R. D. Some remarks on the dynamics of deformable bodies. *AIAA Journal*, 6(3):556–558, 1968.

Moore, B. Principal component analysis in linear systems: Controllability, observability, and model reduction. *IEEE Transactions on Automatic Control*, 26(1):17–32, 1981. doi:10.1109/TAC.1981.1102568.

Morino, L. and G. Bernardini. Singularities in BIEs for the Laplace equation; Joukowski trailing-edge conjecture revisited. *Engineering Analysis with Boundary Elements*, 25(9): 805–818, October 2001. doi:10.1016/S0955-7997(01)00063-7.

Mouyon, P. and N. Imbert. Identification of a 2D turbulent wind spectrum. *Aerospace Science and Technology*, 6(8):599–605, 2002. doi:10.1016/S1270-9638(02)01198-7.

Muñoz Simón, A., R. Palacios, and A. Wynn. Some modelling improvements for prediction of wind turbine rotor loads in turbulent wind. *Wind Energy*, 22(2):333–353, 2022. doi:10.1002/we.2675.

Munk, M. M. General theory of thin wing sections. N.A.C.A. Report No. 142, 1922.

Muñoz-Esparza, D., B. Kosović, J. van Beeck, and J. Mirocha. A stochastic perturbation method to generate inflow turbulence in large-eddy simulation models: Application to neutrally stratified atmospheric boundary layers. *Physics of Fluids*, 27(3):035102, 2015. doi:10.1063/1.4913572.

Murrow, H. N., K. M. Pratt, and J. C. Houbolt. N.A.C.A./NASA research related to evolution of U.S. gust design criteria. In *30th AIAA/ASME/ASCE/AHS/ASC Structures, Structural Dynamics, and Materials Conference*, Mobile, Alabama, USA, 03–05 April 1989. doi:10.2514/6.1989-1373.

Murua, J., R. Palacios, and J. Peiro. Camber effects in the dynamic aeroelasticity of compliant airfoils. *Journal of Fluids and Structures*, 26:527–543, 2010. doi:10.1016/j.jfluidstructs.2010.01.009.

Murua, J., R. Palacios, and J. M. R. Graham. Applications of the unsteady vortex-lattice method in aircraft aeroelasticity and flight dynamics. *Progress in Aerospace Sciences*, 55:46–72, November 2012. doi:10.1016/j.paerosci.2012.06.001.

Murua, J., P. Martínez, H. Climent, L. H. van Zyl, and R. Palacios. T-tail flutter: Potential-flow modelling, experimental validation and flight tests. *Progress in Aerospace Sciences*, 71: 54–84, November 2014. doi:10.1016/j.paerosci.2014.07.002.

Nelson, R. C. *Flight Stability and Automatic Control*. McGraw-Hill, Boston, Massachusetts, USA, 2nd edition, 1998.

Ng, B. F., R. Palacios, E. C. Kerrigan, J. M. R. Graham, and H. Hesse. Aerodynamic load control in horizontal axis wind turbines with combined aeroelastic tailoring and trailing-edge flaps. *Wind Energy*, 19(2):243–263, 2015. doi:10.1002/we.1830.

Ng, B. F., R. Palacios, and J. M. R. Graham. Model-based aeroelastic analysis and blade load alleviation of offshore wind turbines. *International Journal of Control*, 90(1):15–36, 2017. doi:10.1080/00207179.2015.1068456.

Nguyen, N., K. Trinh, D. Nguyen, and I. Tuzcu. Nonlinear aeroelasticity of flexible wing structure coupled with aircraft flight dynamics. In *53rd AIAA/ASME/ASCE/AHS/ASC Structures, Structural Dynamics, and Materials Conference*, Honolulu, Hawaii, USA, 23–26 April 2012. doi:10.2514/6.2012-1792.

Nguyen, N., J. Fugate, J. Xiong, and U. Kaul. Flutter analysis of the transonic truss-braced wing aircraft using transonic correction. In *AIAA Science and Technology Forum and Exposition*, San Diego, California, USA, 7–11 January 2019. doi:10.2514/6.2019-0217.

Nickel, K. and M. Wohlfahrt. *Tailless Aircraft in Theory and Practice*. Butterworth-Heinemann, Oxford, England, 1994.

Niu, M. C. Y. *Airframe Structural Design: Practical Design Information*. Hong Kong Conmilit Press, Hong Kong, 2nd edition, 1999.

Nocedal, J. and S. J. Wright. *Numerical Optimization*. Springer, New York, New York, USA, 2nd edition, 2006.

Noll, T. E., J. M Brown, M. E Perez-Davis, S. D Ishmael, G. C Tiffany, and M. Gaier. Investigation of the Helios prototype aircraft mishap. Mishap Report: Volume 1, NASA, January 2004.

Norris, G. and M. Wagner. *Boeing 787 Dreamliner*. Zenith Press, Minneapolis, Minnesota, USA, 2009.

Oliver, M., J. Rodriguez Ahlquist, J. M. Carreno, H. Climent, R. De Diego, and J. De Alba. A400M GVT: The challenge of nonlinear modes in very large GVT. In *International Forum on Aeroelasticity and Structural Dynamics*, Seattle, Washington, USA, June 22–25 2009.

Ouellette, J. A. and F. D. Valdez. Generation and calibration of linear models of aircraft with highly coupled aeroelastic and flight dynamics. In *AIAA Science and Technology Forum and Exposition*, Orlando, Florida, USA, 6–10 January 2020. doi:10.2514/6.2020-1016.

Pak, C. and S. Truong. Creating a test-validated finite-element model of the X-56A aircraft structure. *Journal of Aircraft*, 52(5):1644–1667, 2015. doi:10.2514/1.C033043.

Palacios, R. Nonlinear normal modes in an intrinsic theory of anisotropic beams. *Journal of Sound and Vibration*, 330(8):1772–1792, 2011.

Palacios, R. and A. Cea. Nonlinear modal condensation of large finite element models: Application of Hodges's intrinsic theory. *AIAA Journal*, 57(10):4255–4268, 2019. doi:10.2514/1.J057556.

Palacios, R. and C. E. S. Cesnik. Cross-sectional analysis of non-homogeneous anisotropic active slender structures. *AIAA Journal*, 43(12):2624–2638, 2005.

Palacios, R., H. Climent, A. Karlsson, and B. Winzell. Assessment of strategies for correcting linear unsteady aerodynamics using CFD or experimental results. In *International Forum on Aeroelasiticy and Structural Dynamics*, Madrid, Spain, 5–7 June 2001.

Palacios, R., J. Murua, and R. Cook. Structural and aerodynamic models in the nonlinear flight dynamics of very flexible aircraft. *AIAA Journal*, 48(11):2648–2659, November 2010. doi:10.2514/1.J050513.

Panchal, J. and H. Benaroya. Review of control surface freeplay. *Progress in Aerospace Sciences*, 127:100729, November 2021. doi:10.1016/j.paerosci.2021.100729.

Pankonien, A. M., P. M. Suh, J. R. Schaefer, and R. M. Mitchell. Deadbands tell no tails: X-56A dynamic actuation requirements. In *ASME 2020 Conference on Smart Materials, Adaptive Structures and Intelligent Systems*, Virtual conference, 15 September 2020. doi:10.1115/smasis2020-2427.

Parenteau, M. and E. Laurendeau. A general modal frequency-domain vortex lattice method for aeroelastic analyses. *Journal of Fluids and Structures*, 99:103146, 2020. doi:10.1016/j.jfluidstructs.2020.103146.

Pasinetti, G. and P. Mantegazza. Single finite states modeling of aerodynamic forces related to structural motions and gusts. *AIAA Journal*, 37(5):604–612, 1999. doi:10.2514/2.760.

Patil, M. J. From fluttering wings to flapping flight: The energy connection. *Journal of Aircraft*, 40(2):270–276, 2003. doi:10.2514/2.3119.

Patil, M. J. and D. H. Hodges. Flight dynamics of highly flexible flying wings. *Journal of Aircraft*, 43(6):1790–1798, 2006. doi:10.2514/1.17640.

Patil, M. J., D. H. Hodges, and C. E. S. Cesnik. Nonlinear aeroelastic analysis of aircraft with high-aspect-ratio wings. In *39th AIAA/ASME/ASCE/AHS/ASC Structures, Structural Dynamics, and Materials Conference*, Long Beach, California, USA, 20–23 April 1998. doi:10.2514/6.1998-1955.

Patil, M. J., D. H. Hodges, and C. E. S. Cesnik. Nonlinear aeroelasticity and flight dynamics of high-altitude long-endurance aircraft. *Journal of Aircraft*, 38(1):88–94, 2001. doi:10.2514/2.2738.

Payne, C. B. A flight investigation of some effects of automatic control on gust loads. N.A.C.A. Research Memorandum L53E14a, Washington, DC, USA, July 1953.

Penzl, T. Numerical solution of generalized Lyapunov equations. *Advances in Computational Mathematics*, 8(1):33–48, 1998. doi:10.1023/A:1018979826766.

Penzl, T. Algorithms for model reduction of large dynamical systems. *Linear Algebra and Its Applications*, 415(2–3):322–343, 2006. doi:10.1016/j.laa.2006.01.007.

Pereira, M. F. V., I. Kolmanovsky, C. E. S. Cesnik, and F. Vetrano. Model predictive control for maneuver load alleviation in flexible airliners. In *International Forum on Aeroelasticity and Structural Dynamics*, Savannah, Georgia, USA, 9–13 June 2019a. IFASD Paper No 2019-018.

Pereira, M. F. V., I. Kolmanovsky, C. E. S. Cesnik, and F. Vetrano. Model predictive control architectures for maneuver load alleviation in very flexible aircraft. In *AIAA Science and Technology Forum and Exposition*, San Diego, California, USA, 7–11 January 2019b. doi:10.2514/6.2019-159.

Pereira, M. F. V., I. Kolmanovsky, C. E. S. Cesnik, and F. Vetrano. Time-distributed scenario-based model predictive control approach for flexible aircraft. In *AIAA Science and Technology Forum and Exposition*, Nashville, Tennessee, USA, 4–8 January 2021. doi:10.2514/6.2021-0502.

Pereira, M. F. V., I. Kolmanovsky, and C. E. S. Cesnik. Nonlinear model predictive control with aggregated constraints, *Automatica*, 146:1106–1149, December 2022a. doi:10.1016/j.automatica.2022.110649.

Pereira, M. F. V., M. Duan, C. E. S. Cesnik, I. Kolmanovsky, and F. Vetrano. Model predictive control for very flexible aircraft based on linear parameter varying reduced-order models, In *International Forum on Structural Dynamics and Aeroelasticity*, Madrid, Spain, 13–17 June 2022b.

Pereira, M. F. V., G. B. Chaves, R. M. Bertolin, I. Kolmanovsky, and C. E. S. Cesnik. Experimental validation of model predictive controllers for load alleviation in very flexible aircraft, In *International Forum on Structural Dynamics and Aeroelasticity*, Madrid, Spain, 13–17 June 2022c.

Perkins, C. D. Development of airplane stability and control technology. *Journal of Aircraft*, 7(4):290–301, 1970. doi:10.2514/3.44167.

Pesmajoglou, S. D. and J. M. R. Graham. Prediction of aerodynamic forces on horizontal axis wind turbines in free yaw and turbulence. *Journal of Wind Engineering and Industrial Aerodynamics*, 86(1):1–14, 2000. doi:10.1016/S0167-6105(99)00125-7.

Peters, D. A. and M. J. Johnson. Finite-state airloads for deformable airfoils on fixed and rotating wings. In *ASME Symposium on Aeroelasticity and Fluid-Structure Interaction*, volume 44, page 1–28, Fairfield, New Jersey, USA, November 1994.

Phillips, W. B. Loads implications of gust-alleviation systems. N.A.C.A. Technical Note 4056, Washington, DC, USA, June 1957.

Pope, S. B. *Turbulent Flows*. Cambridge University Press, New York, New York, USA, 2000.

Possio, C. L'azione aerodinamica sul profilo oscillante in un fluido compressibile a velocita iposonora. *L'Aerotecnica*, 18(4):441–458, 1938.

Pratt, K. G. A revised formula for the calculation of gust loads. Technical Note 2964, National Advisory Committee for Aeronautics, Langley Aeronautical Laboratory, 1953.

Press, H., M. T. Meadows, and I. Hadlock. Estimates of probability distribution of root-mean-square gust velocity of atmospheric turbulence from operational gust-load data by random-process theory. Technical Note 3362, N.A.C.A. 1955.

Preumont, A. *Vibration Control of Active Structures: An Introduction*. Springer International Publishing, Berlin, Germany, 4th edition, 2018.

Pugsley, A. G. The influence of wing elasticity upon longitudinal stability. Reports and Memoranda 1548, British Advisory Committee for Aeronautics, January 1933.

Quero, D., P. Vuillemin, and C. Poussot-Vassal. A generalized state-space aeroservoelastic model based on tangential interpolation. *Aerospace*, 6(1):9, 2019. doi:10.3390/aerospace6010009.

Quero, D., P. Vuillemin, and C. Poussot-Vassal. A generalized eigenvalue solution to the flutter stability problem with true damping: The p-L method. *Journal of Fluids and Structures*, 103:103266, 2021. doi:10.1016/j.jfluidstructs.2021.103266.

Rabadan, G. J., N. P. Schmitt, T. Pistner, and W. Rehm. Airborne LIDAR for automatic feedforward control of turbulent in-flight phenomena. *Journal of Aircraft*, 47(2):392–403, March–April 2010.

Ramesh, K., A. Gopalarathnam, K. Granlund, M. V. Ol, and J. R. Edwards. Discrete-vortex method with novel shedding criterion for unsteady aerofoil flows with intermittent leading-edge vortex shedding. *Journal of Fluid Mechanics*, 751:500–538, 2014. doi:10.1017/jfm.2014.297.

Raveh, D. E., Y. Levy, and M. Karpel. Efficient aeroelastic analysis using computational unsteady aerodynamics. *Journal of Aircraft*, 38(3):547–556, 2001. doi:10.2514/2.2795.

Rawlings, J. B., D. Q. Mayne, and M. Diehl. *Model Predictive Control: Theory, Computation, and Design*. Nob Hill Publishing, Madison, Wisconsin, USA, 2017.

Rea, J. B. Aeroelasticity and stability and control. Wright Air Development Center, WADC TR 55–173, Ohio, USA, 1957.

Reddy, J. N. *Energy Principles and Variational Methods in Applied Mechanics*. John Wiley & Sons, Hoboken, New Jersey, USA, 2nd edition, 2002.

Reeh, A. D. and C. Tropea. Behaviour of a natural laminar flow aerofoil in flight through atmospheric turbulence. *Journal of Fluid Mechanics*, 767:394–429, 2015. doi:10.1017/jfm.2015.49.

Regan, C. D. and C. V. Jutte. Survey of applications of active control technology for gust alleviation and new challenges for lighter-weight aircraft. Technical report, NASA Dryden Flight Research Center, Edwards, California, USA, April 2012.

Reimer, L., M. Ritter, R. Heinrich, and W. R. Krüger. CFD-based gust load analysis for a free-flying flexible passenger aircraft in comparison to a DLM-based approach. In *22nd AIAA Computational Fluid Dynamics Conference*, Dallas, Texas, USA, June 2015. doi:10.2514/6.2015-2455.

Reissner, E. On one-dimensional large-displacement finite-strain beam theory. *Studies in Applied Mathematics*, 52(2):87–95, June 1973.

Reissner, H. Neurere probleme aus der flugzeugstatik. *Zeitschrift für Flugtechnik und Motorluftschiffahrt*, 17:137–146, April 1926.

Ricci, S., L. Marchetti, L. Riccobene, A. De Gaspari, F. Toffol, F. Fonte, P. Mantegazza, J. Berg, K. A. Morgansen, and E. Livne. An active flutter suppression (AFS) project: Overview, results and lessons learned. In *AIAA Science and Technology Forum and Exposition*, Virtual Event, 11–21 January 2021. doi:10.2514/6.2021-0908.

Ricciardi, A. P., M. J. Patil, R. A. Canfield, and N. Lindsley. Evaluation of quasi-static gust loads certification methods for high-altitude long-endurance aircraft. *Journal of Aircraft*, 50 (2):457–468, 2013. doi:10.2514/1.C031872.

Richardson, J. R., E. M. Atkins, P. T. Kabamba, and A. R. Girard. Envelopes for flight through stochastic gusts. *Journal of Guidance Control and Dynamics*, 36(5):1464–1476, 2013. doi:10.2514/1.57849.

Ripepi, M. and P. Mantegazza. Improved matrix fraction approximation of aerodynamic transfer matrices. *AIAA Journal*, 51(5):1156–1173, 2013. doi:10.2514/1.J052009.

Riso, C. and C. E. S. Cesnik. Correlations between UM/NAST nonlinear aeroelastic simulations and the pre-Pazy wing experiment. In *AIAA Science and Technology Forum and Exposition*, Nashville, Tennessee, USA, 4–8 January 2021. doi:10.2514/6.2021-1712.

Riso, C. and C. E. S. Cesnik. Investigation of geometrically nonlinear effects in the aeroelastic behavior of a very flexible wing, In *AIAA Science and Technology Forum and Exposition (SciTech2023)*, National Harbor, Maryland, USA, 23–27 January 2023.

Riso, C., D. Sanghi, C. E. S. Cesnik, F. Vetrano, and P. Teufel. Parametric roll maneuverability analysis of a high-aspect-ratio-wing civil transport aircraft, In *AIAA Science and Technology Forum and Exposition (SciTech2020), 61st AIAA/ASCE/AHS/ASC Structures, Structural Dynamics, and Materials Conference*, Orlando, Florida, USA, 6–10 January 2020. AIAA–2020–1191. doi:10.2514/6.2020-1191.

Rodden, W. P. The development of the doublet-lattice method. In *International Forum on Aeroelasticity and Structural Dynamics*, Rome, Italy, 17–20 June 1997.

Rodriguez Ahlquist, J., J. M. Carreno, H. Climent, R. De Diego, and J. De Alba. Assessment of nonlinear structural response in A400M GVT. In *28th International Modal Analysis Conference (IMAC 28)*, Jacksonville, Florida, USA, 2010.

Roesler, B. T. and B. P. Epps. Discretization requirements for vortex lattice methods to match unsteady aerodynamics theory. *AIAA Journal*, 56(6):2478–2483, 2018. doi:10.2514/1.j056400.

Roger, K. L. Airplane math modelling and active aeroelastic control design. In *AGARD Structures and Materials Panel*, Loughton, Essex, UK, 1977. AGARD-CP-228.

Roger, K. L. and G. E. Hodges. Active flutter suppression–A flight test demonstration. *Journal of Aircraft*, 12(6):410–450, June 1975.

Routh, E. J. *A Treatise on the Stability of a Given State of Motion, Particularly Steady Motion*. MacMillian, London, UK, 1877.

Rule, J. A., D. E. Cox, and R. L. Clark. Aerodynamic model reduction through balanced realization. *AIAA Journal*, 42(5):1045–1048, May 2004. doi:10.2514/1.9596.

Ryan, J. J., J. T. Bosworth, J. J. Burken, and P. M. Suh. Current and future research in active control of lightweight, flexible structures using the X-56 aircraft. In *52nd Aerospace Sciences Meeting*, National Harbor, Maryland, USA, 13–17 January 2014. doi:10.2514/6.2014-0597.

Sahoo, D. and C. E. S. Cesnik. Roll maneuver control of UCAV wing using anisotropic piezoelectric actuators. In *43rd AIAA/ASME/ASCE/AHS/ASC Structures, Structural Dynamics, and Materials Conference*, Denver, Colorado, USA, 22–25 April 2002. doi:10.2514/6.2002-1720.

Saltari, F., C. Riso, G. Matteis, and F. Mastroddi. Finite-element-based modeling for flight dynamics and aeroelasticity of flexible aircraft. *Journal of Aircraft*, 54(6):2350–2366, 2017. doi:10.2514/1.c034159.

Sanghi, D., C. Riso, and C. E. S. Cesnik. Conventional and unconventional control effectors for load alleviation in high-aspect-ratio-wing aircraft, In *AIAA Science and Technology Forum and Exposition (SciTech2022)*, San Diego, California, USA, 3–7 January 2022.

Sanghi, D., C. E. S. Cesnik, and C. Riso. Roll maneuvers of very flexible aircraft with flared folding wingtips, In *AIAA Science and Technology Forum and Exposition (SciTech2023)*, National Harbor, Maryland, USA, 23–27 January 2023.

Sarpkaya, T., R. E. Robins, and D. P. Delisi. Wake-vortex eddy-dissipation model predictions compared with observations. *Journal of Aircraft*, 38(4):687–692, July 2001. doi:10.2514/2.2820.

Schmidt, D. K. *Modern Flight Dynamics*. Mc-Graw Hill, New York, New York, USA, 2012.

Schmidt, D. K. Discussion: "The Lure of the Mean Axes" (Meirovitch, L., and Tuzcu, I., *ASME J. Appl. Mech.*, 74(3), pp. 497–504). *Journal of Applied Mechanics*, 82(12):125501, December 2015. doi:10.1115/1.4031567.

Scott, R. C., M. A. Castelluccio, D. A. Coulson, and J. Heeg. Aeroservoelastic wind-tunnel tests of a free-flying, joined-wing SensorCraft model for gust load alleviation. In *52nd AIAA/ASME/ASCE/AHS/ASC Structures, Structural Dynamics, and Materials Conference*, pages 1–36, Denver, Colorado, USA, 4–7 April 2011. doi:10.2514/6.2011-1960.

Scott, R. C., T. K. Vetter, K. B. Penning, D. A. Coulson, and J. Heeg. Aeroservoelastic testing of a sidewall mounted free flying wind-tunnel model. In *AIAA Applied Aerodynamics Conference*, Honolulu, Hawaii, USA, 18–21 August 2008. doi:10.2514/6.2008-7186.

Sears, W. R. Some aspects of non-stationary airfoil theory and its practical application. *Journal of the Aeronautical Sciences*, 8(3):104–108, 1941.

Sears, W. R. *Stories from a 20th Century Life*. Parabolic Press Ltd, Stanford, California, USA, 1994.

Sears, W. R. and T. von Kármán. Airfoil theory for non-uniform motion. *Journal of the Aeronautical Sciences*, 5:379–390, 1938.

Shabana, A. A. *Theory of Vibration. Volume II: Discrete and Continuous Systems.* Mechanical Engineering Series. Springer US, New York, New York, USA, 1991a. doi:10.1007/978-1-4684-0380-0.

Shabana, A. A. *Theory of Vibration. Volume I: Introduction.* Mechanical Engineering Series. Springer US, New York, New York, USA, 1991b. doi:10.1007/978-3-319-94271-1.

Shabana, A. A. Flexible multibody dynamics: Review of past and recent developments. *Multibody System Dynamics*, 1(2):189–222, 1997. doi:10.1023/A:1009773505418.

Sharman, R. D., S. B. Trier, L. P. Lane, and J. D. Doyle. Sources and dynamics of turbulence in the upper troposphere and lower stratosphere: A review. *Geophysical Research Letters*, 39, 2012. doi:10.1029/2012GL051996.

Sharqi, B. and C. E. S. Cesnik. Ground vibration testing on very flexible aircraft, In *AIAA Science and Technology Forum and Exposition (SciTech2020), 61st AIAA/ASCE/AHS/ASC Structures, Structural Dynamics, and Materials Conference*, Orlando, Florida, USA, 6–10 January 2020.

Sharqi, B. and C. E. S. Cesnik. Finite element model updating for very flexible aircraft, *Journal of Aircraft*, 14 pages, October 2022. doi:10.2514/1.C036894

Shearer, C. M. *Coupled Nonlinear Flight Dynamics, Aeroelasticity, and Control of Very Flexible Aircraft.* PhD thesis, Aerospace Engineering, The University of Michigan, Ann Arbor, Michigan, USA, 2006.

Shearer, C. M. and C. E. S. Cesnik. Nonlinear flight dynamics of very flexible aircraft. *Journal of Aircraft*, 44(5):1528–1545, 2007. doi:10.2514/1.27606.

Shearer, C. M. and C. E. S. Cesnik. Trajectory control for very flexible aircraft. *Journal of Guidance, Control, and Dynamics*, 31(2):340–357, March–April 2008. doi:10.2514/1.29335.

Silva, W. A. Identification of linear and nonlinear aerodynamic impulse responses using digital filter techniques. In *22nd AIAA Atmospheric Flight Mechanics Conference*, New Orleans, Louisiana, USA, August 1997. doi:10.2514/6.1997-3712.

Silva, W. A. Simultaneous excitation of multiple-input/multiple-output CFD-based unsteady aerodynamic systems. *Journal of Aircraft*, 45(4):1267–1274, 2008. doi:10.2514/1.34328.

Silva, W. A. and R. E. Bartels. Development of reduced-order models for aeroelastic analysis and flutter prediction using the CFL3Dv6.0 code. *Journal of Fluids and Structures*, 19(6):729–745, 2004. doi:10.1016/j.jfluidstructs.2004.03.004.

Silva W. A. and S. Dunn. Higher-order spectral analysis of F-18 flight flutter data. In *45th AIAA/ASME/ASCE/AHS/ASC Structures, Structural Dynamics, and Materials Conference*, Austin, Texas, USA, 18–21 April 2005. doi:10.2514/6.2005-2014.

Silva, W. A., P. S. Beran, C. E. S. Cesnik, R. E. Guendel, A. Kurdila, and R. J. Prazenica. Reduced-order modeling: Cooperative research and development at NASA Langley Research Center. In *International Forum on Aeroelasticity and Structural Dynamics*, Madrid, Spain, 5–7 June 2001.

Simiriotis N., and R. Palacios. A numerical investigation on direct and data-driven flutter prediction methods. *Journal of Fluids and Structures*, 117:103835, 2023 doi:10.1016/j.jfluidstructs.2023.103835.

Simó, J. C. and L. Vu-Quoc. On the dynamics of flexible beams under large overall motions – The plane case: Part II. *ASME Journal of Applied Mechanics*, 53:855–863, 1986. doi:10.1115/1.3171871.

Simó, J. C. and L. Vu-Quoc. On the dynamics in space of rods undergoing large motions – A geometrically exact approach. *Computer Methods in Applied Mechanics and Engineering*, 66(2):125–161, 1988. doi:10.1016/0045-7825(88)90073-4.

Simpson, R. J. S., R. Palacios, and S. Maraniello. State-space realizations of potential-flow unsteady aerodynamics with arbitrary kinematics. In *58th AIAA/ASME/ASCE/AHS/ASC Structures, Structural Dynamics, and Materials Conference*, Kissimmee, Florida, USA, 9–13 January 2017. doi:10.2514/6.2017-1595.

Simpson, R. J. S., R. Palacios, and J. Murua. Induced drag calculations in the unsteady vortex-lattice method. *AIAA Journal*, 51(7):1775–1779, 2013. doi:10.2514/1.J052136.

Simpson, R. J. S., R. Palacios, H. Hesse, and P. J. Goulart. Predictive control for alleviation of gust loads on very flexible aircraft. In *55th AIAA/ASME/ASCE/AHS/ASC Structures, Structural Dynamics, and Materials Conference*, National Harbor, Maryland, USA, 13–17 April 2014. doi:10.2514/6.2014-0843.

Skogestad, S. and I. Postlethwaite. *Multivariate Feedback Control, Analysis & Design*. John Wiley & Sons, New York, New York, USA, 2nd edition, 2005.

Skujins, T. and C. E. S. Cesnik. Reduced-order modeling of unsteady aerodynamics across multiple Mach regimes. *Journal of Aircraft*, 51(6):1681–1704, 2014. doi:10.2514/1.C032222.

Sleeper, R. K. Spanwise measurements of vertical components of atmospheric turbulence. Technical report 2963, NASA, April 1990.

Smilg, B. and L. S. Wasserman. Application of three-dimensional flutter theory to aircraft structures. Air Corps Technical report 4798, 1942.

Smith, M. J., D. H. Hodges, and C. E. S. Cesnik. An evaluation of computational algorithms to interface between CFD and CSD methodologies. WL-TR-96-3055, Flight Dynamics Directorate, Wright Laboratory, Wright-Patterson Air Force Base, Ohio, USA, 1995.

Smith, M. J., C. E. S. Cesnik, and D. H. Hodges. Evaluation of computational algorithms suitable for fluid-structure interactions. *Journal of Aircraft*, 37(2):282–294, 2000. doi:10.2514/2.2592.

Spalart, P. R. Airplane trailing vortices. *Annual Review of Fluid Mechanics*, 30:107, 1998.

Spielberg, I. N. The two-dimensional incompressible aerodynamic coefficients for oscillatory changes in airfoil camber. *Journal of the Aeronautical Sciences*, 20:432–434, June 1953.

Stanford, B. K. Optimal aircraft control surface layouts for maneuver and gust load alleviation. In *AIAA Science and Technology Forum and Exposition*, Orlando, Florida, USA, 6–10 January 2020. doi:10.2514/6.2020-0448.

Stengel, R. F. *Flight Dynamics*. Princeton University Press, Princeton, New Jersey, USA, 2004.

Stevens, B. L., F. L. Lewis, and Johnson E. N. *Aircraft Control and Simulation*. John Wiley & Sons, Hoboken, New Jersey, USA, 3rd edition, 2016.

Strganac, T. W. and D. T. Mook. Numerical model of unsteady subsonic aeroelastic behavior. *AIAA Journal*, 28(5):903–909, 1990. doi:10.2514/3.25137.

Strganac, T. W., P. G. Cizmas, C. Nichkawde, J. Gargoloff, and P. S. Beran. Aeroelastic analysis for future air vehicle concepts using a fully nonlinear methodology. In *46th AIAA/ASME/ASCE/AHS/ASC Structures, Structural Dynamics, and Materials Conference*, Austin, Texas, USA, 18–21 April 2005. doi:10.2514/6.2005-2171.

Strong, D. D., R. M. Kolonay, L. J. Huttsell, and P. M. Flick. Flutter analysis of wing configurations using pre-stressed frequencies and mode shapes. In *46th AIAA/ASME/ASCE/AHS/ASC Structures, Structural Dynamics, and Materials Conference*, Austin, Texas, USA, 18–21 April 2005. doi:10.2514/6.2005-2173.

Su, W. Development of an aeroelastic formulation for deformable airfoils using orthogonal polynomials. *AIAA Journal*, 55(8):2793–2807, 2017. doi:10.2514/1.J055665.

Su, W. and C. E. S. Cesnik. Nonlinear aeroelasticity of a very flexible blended-wing-body aircraft. *Journal of Aircraft*, 47(5):1539–1553, 2010. doi:10.2514/1.47317.

Su, W. and C. E. S. Cesnik. Dynamic response of highly flexible flying wings. *AIAA Journal*, 49(2):324–339, 2011a. doi:10.2514/1.J050496.

Su, W. and C. E. S. Cesnik. Strain-based geometrically nonlinear beam formulation for modeling very flexible aircraft. *International Journal of Solids and Structures*, 48:2349–2360, 2011b. doi:10.1016/j.ijsolstr.2011.04.012.

Suryakumar, V. S., Y. Babbar, T. W. Strganac, and A. S. Mangalam. Unsteady aerodynamic model based on the leading-edge stagnation point. *Journal of Aircraft*, 53(6):1626–1637, 2016. doi:10.2514/1.C033602.

Tang, D. M. and E. H. Dowell. Experimental and theoretical study on aeroelastic response of high-aspect-ratio wings. *AIAA Journal*, 39(8):1430–1441, 2001. doi:10.2514/2.1484.

Tang, D. M. and E. H. Dowell. Limit cycle oscillations of two-dimensional panels in low subsonic flow. *International Journal of Non-Linear Mechanics*, 37(7):1199–1209, October 2002.

Taylor, G. The spectrum of turbulence. *Proceedings of the Royal Society of London*, 164:476–490, 1938.

Taylor, P. F., R. Moreno, N. Banavara, R. K. Narisetti, and L. Morgan. Flutter flight testing at Gulfstream Aerospace using advanced signal processing techniques. In *58th AIAA/ASME/ASCE/AHS/ASC Structures, Structural Dynamics, and Materials Conference*, Grapevine, Texas, USA, 9–13 January 2017. doi:10.2514/6.2017-1823.

Teixeira, P. C. and C. E. S. Cesnik. Propeller effects on the response of high-altitude long-endurance aircraft. *AIAA Journal*, 57(10):4328–4342, 2019. doi:10.2514/1.J057575.

Teixeira, P. C. and C. E. S. Cesnik. Propeller influence on the aeroelastic stability of high altitude long endurance aircraft. *Aeronautical Journal*, 124(1275):703–730, 2020. doi:10.1017/aer.2019.165.

Tessler, A. and S. B. Dong. On a hierarchy of conforming Timoshenko beam elements. *Computers & Structures*, 14(3–4):335–344, 1981. doi:10.1016/0045-7949(81)90017-1.

Teufel, P., M. Hanel, and K. H. Well. Integrated flight mechanic and aeroelastic modelling and control of a flexible aircraft considering multidimensional gust input. Technical report, RTO-MP-36, In *RTO Meeting Proceedings on Structural Aspects of Flexible Aircraft Control*, DTIC Document, 2000.

Theodorsen, T. General theory of aerodynamic instability and the mechanism of flutter. Report 496, N.A.C.A. 1935.

Theodorsen, T. and I. E. Garrick. General potential theory of arbitrary wing sections. Report 452, N.A.C.A. 1934.

Theodorsen, T. and I. E. Garrick. Mechanism of flutter. A theoretical and experimental investigation of the flutter problem. Report 685, N.A.C.A. 1938.

Thomas, J. P., E. H. Dowell, and K. C. Hall. Nonlinear inviscid aerodynamic effects on transonic divergence, flutter and limit cycle oscillations. *AIAA Journal*, 40(4):638–646, 2002.

Thormann, R. and M. Widhalm. Linear-frequency-domain predictions of dynamic-response data for viscous transonic flows. *AIAA Journal*, 51(11):2540–2557, 2013. doi:10.2514/1.J051896.

Tiffany, S. H. and W. M. Adams. Nonlinear programming extensions to rational function approximation methods for unsteady aerodynamic forces. Technical Publication 2776, NASA, 1988.

Tong, Y. L. The bivariate normal distribution. In *The Multivariate Normal Distribution*, page 6–22. Springer, New York, New York, USA, 1990. doi:10.1007/978-1-4613-9655-0_2.

Torenbeek, E. *Synthesis of Subsonic Aircraft Design*. Kluwer Academic Publishers, Dordrecht, The Netherlands, 1982.

Tuzcu, I., P. Marzocca, E. Cestino, G. Romeo, and G. Frulla. Stability and control of a high-altitude, long-endurance UAV. *Journal of Guidance, Control, and Dynamics*, 30(3):713–721, 2007. doi:10.2514/1.25814.

van der Schaft, A. and D. Jeltsema. Port-Hamiltonian systems theory: An introductory overview. *Foundations and Trends in Systems and Control*, 1(2–3):173–378, June 2014. doi:10.1561/2600000002.

van Schoor, M. C. and A. H. von Flotow. Aeroelastic characteristics of a highly flexible aircraft. *Journal of Aircraft*, 27(10):901–908, 1990. doi:10.2514/3.45955.

van Zyl, L. H. and E. H. Matthews. Aeroelastic analysis of T-tails using an enhanced doublet lattice method. *Journal of Aircraft*, 48(3):823–831, 2011. doi:10.2514/1.C001000.

Vartio, E. J., A. Shimko, C. P. Tilmann, and P. M. Flick. Structural modal control and gust load alleviation for a SensorCraft concept. In *46th AIAA/ASME/ASCE/AHS/ASC Structures, Structural Dynamics, and Materials Conference*, Austin, Texas, USA, 18–21 April 2005. AIAA Paper 2005-1946.

Veers, P. S. Three-dimensional wind simulation. Sandia Report SAND88-0152, Sandia National Laboratories, 1988.

Verwaal, N. W., G. J. van der Veen, and J. W. van Wingerden. Predictive control of an experimental wind turbine using preview wind speed measurements. *Wind Energy*, 18(3):385–398, 2015. doi:10.1002/we.1702.

Vizzaccaro, A. *Numerical Methods for Nonlinear Vibration in Aircraft Engines: Dynamics of Blades in Large Amplitude Vibration and Blade-Casing Interaction*. PhD thesis, Imperial College London, London, UK, April 2021.

von Baumhauer, A. F. and C. Koning. On the stability of oscillations of an airplane wing. In *International Air Congress*, Marseille, France, 1922. Royal Aeronautical Society.

von Kármán, T. *Aerodynamics: Selected Topics in the Light of Their Historical Development*. Cornell University Press, 1954.

von Kármán, T. and J. M. Burgers. *General Aerodynamic Theory – Perfect Fluids: Aerodynamic Theory. Vol. 2, edited by W. Durand*. Dover, New York, 1968.

Wagner, H. Ueber die entstehung des dynamischen auftriebes von tragflugeln. *Zietschrift für Angewandte Mathematik und Mechanick*, 5(1):17–35, 1925.

Waite, J. M., J. Grauer, R. E. Bartels, and B. K. Stanford. Aeroservoelastic control law development for the integrated adaptive wing technology maturation wind-tunnel test. In *AIAA Science and Technology Forum and Exposition*, Nashville, Tennessee, USA, 4–8 January 2021. doi:10.2514/6.2021-0609.

Walker, W. P. and M. J. Patil. Unsteady aerodynamics of deformable thin airfoils. *Journal of Aircraft*, 51(6):1673–1680, November–December 2014. doi:10.2514/1.C031434.

Wang, Q., M. A. Sprague, J. Jonkman, N. Johnson, and B. Jonkman. BeamDyn: A high-fidelity wind turbine blade solver in the FAST modular framework. *Wind Energy*, 20(8):1439–1462, 2017. doi:10.1002/we.2101.

Wang, S. T. and W. Frost. Atmospheric turbulence simulation techniques with application to flight analysis. Contractor Report 3309, NASA, September 1980.

Wang, Y., R. Palacios, and A. Wynn. A method for normal-mode-based model reduction in nonlinear dynamics of slender structures. *Computers & Structures*, 159:26–40, 2015. doi:10.1016/j.compstruc.2015.07.001.

Wang, Y., A. Wynn, and R. Palacios. Nonlinear aeroelastic control of very flexible aircraft using model updating. *Journal of Aircraft*, 55(4), 2018. doi:10.2514/1.C034684.

Wang, Z., P. C. Chen, D. D. Liu, and D. T. Mook. Nonlinear-aerodynamics/nonlinear-structure interaction methodology for a high-altitude long-endurance wing. *Journal of Aircraft*, 47(2): 556–566, 2010. doi:10.2514/1.45694.

Waszak, M. R. and J. Fung. Parameter estimation and analysis of actuators for the BACT wind-tunnel model. *21st AIAA Atmospheric Flight Mechanics Conference*, 29–31 July 1996. doi:10.2514/6.1996-3362.

Waszak, M. R. and D. K. Schmidt. Flight dynamics of aeroelastic vehicles. *Journal of Aircraft*, 25(6):563–571, 1988.

Weisshaar, T. A. and D. H. Lee. Aeroelastic tailoring of joined-wing configurations. In *43rd AIAA/ASME/ASCE/AHS/ASC Structures, Structural, Dynamics, and Materials Conference*, Denver, Colorado, USA, 22–25 April 2002. doi:10.2514/6.2002-1207.

Welch, G. and G. Bishop. An introduction to the Kalman filter. *Proc. Siggraph Course* 8, Association for Computer Machinery, 2001.

Williams, P. D. and M. M. Joshi. Intensification of winter transatlantic aviation turbulence in response to climate change. *Nature Climate Change*, 3, 2013. doi:10.1038/nclimate1866.

Willis, D. J., J. Peraire, and J. K. White. A combined pFFT-multipole tree code, unsteady panel method with vortex particle wakes. *International Journal for Numerical Methods in Fluids*, 53:1399–1422, 2007. doi:10.1002/fld.1240.

Wilson, E. B. Report on behavior of aeroplanes in gusts. Part II: Theory of an aeroplane encoutering gusts. N.A.C.A. Report No. 1, 1916.

Wirsching, P. H., T. L. Paez, and K. Ortiz. *Random Vibrations: Theory and Practice*. John Wiley & Sons, New York, New York, USA, 1995.

Wright, J. R. and J. E. Cooper. *Introduction to Aircraft Aeroelasticity and Loads*. John Wiley & Sons, Chichester, England, 2007.

Wright, W. *The Papers of Wilbur and Orville Wright, Including the Chanute-Wright letters, Volume 1: 1899–1905*. Edited by M. W. McFarland, McGraw-Hill, New York, New York, USA, 2001.

Wu, T. Y. Hydromechanics of swimming propulsion. Part 1. Swimming of a two-dimensional flexible plate at variable speeds in an inviscid fluid. *Journal of Fluid Mechanics*, 46(2):337–355, 1971. doi:10.1017/S0022112071000570.

Wyngaard, J. C. *Turbulence in the Atmosphere*. Cambridge University Press, Cambridge, UK, 2010.

Wynn, A., Y. Wang, R. Palacios, and P. J. Goulart. An energy-preserving description of non-linear beam vibrations in modal coordinates. *Journal of Sound and Vibration*, 332(21): 5543–5558, 2013. doi:10.1016/j.jsv.2013.05.021.

Yue, C. and Y. Zhao. An improved aeroservoelastic modeling approach for state-space gust analysis. *Journal of Fluids and Structures*, 99:103148, November 2020. doi:10.1016/j.jfluidstructs.2020.103148.

Zeiler, T. A. Results of Theodorsen and Garrick revisited. *Journal of Aircraft*, 37(5):918–920, 2000.

Zhou, K., J. C. Doyle, and K. Glover. *Robust and Optimal Control*. Prentice-Hall, Upper Saddle River, New Jersey, USA, 1st edition, 1995.

Zou, F., V. A. Riziotis, S. G. Voutsinas, and J. Wang. Analysis of vortex-induced and stall-induced vibrations at standstill conditions using a free wake aerodynamic code. *Wind Energy*, 18(12):2145–2169, 2015. doi:10.1002/we.1811.

Zwölfer, A. and J. Gerstmayr. Preconditioning strategies for linear dependent generalized component modes in 3D flexible multibody dynamics. *Multibody System Dynamics*, 47(1): 65–93, 2019. doi:10.1007/s11044-019-09680-6.

Index

Printed in the United States
by Baker & Taylor Publisher Services